Nutrition: Health and Metabolism

Nutrition: Health and Metabolism

Edited by **Dave Stewart**

Callisto Reference
New York

Published by Callisto Reference,
106 Park Avenue, Suite 200,
New York, NY 10016, USA
www.callistoreference.com

Nutrition: Health and Metabolism
Edited by Dave Stewart

International Standard Book Number: 978-1-63239-711-9 (Hardback)

Printed in China.

Contents

Preface

The world is advancing at a fast pace like never before. Therefore, the need is to keep up with the latest developments. This book was an idea that came to fruition when the specialists in the area realized the need to coordinate together and document essential themes in the subject. That's when I was requested to be the editor. Editing this book has been an honour as it brings together diverse authors researching on different streams of the field. The book collates essential materials contributed by veterans in the area which can be utilized by students and researchers alike.

Nutrition is the branch of science that studies the effect of nutrients present in food on the growth and health of living organisms. Nutrition improves and maintains health and metabolism. Health is the ability of human beings to adapt and manage themselves when subjected to a change in their surroundings whether physical, mental or social whereas metabolism is the sum of all the reactions taking place inside a human body. The objective of this book is to give a general view of the different areas of nutrition and metabolism. It covers in detail some existent theories and innovative concepts revolving around this field. A number of latest researches have been included in this book to keep the readers up-to-date with the global concepts in this area of study. This book, with its detailed analysis and data, will prove immensely beneficial to professionals and students involved in this area at various levels. Those in search of information to further their knowledge will be greatly assisted by this book.

Each chapter is a sole-standing publication that reflects each author´s interpretation. Thus, the book displays a multi-facetted picture of our current understanding of applications and diverse aspects of the field. I would like to thank the contributors of this book and my family for their endless support.

Editor

1

Validity of Nutritional Screening Tools for Hospitalized Children

Nathania Wonoputri, Julistio T. B. Djais, and Ina Rosalina

Department of Child Health, Universitas Padjadjaran, Dr. Hasan Sadikin General Hospital, Jalan Pasteur No. 38, Bandung 40161, Indonesia

Correspondence should be addressed to Nathania Wonoputri; nathania.wonop@gmail.com

Academic Editor: Johannes B. van Goudoever

Background. Malnutrition in hospitalized children can be prevented if children with risk of malnutrition are identified. Every hospital is recommended to have a standard nutritional screening tool. Numerous simple screening tools have been developed, namely Paediatric Yorkhill Malnutrition Score (PYMS), Screening Tool for the Assessment of Malnutrition in Paediatrics (STAMP), and Screening Tool for Risk on Nutritional Status and Growth (STRONG-kids). None has been accepted as a universal tool. Our study aims to determine the best screening tools compared to Subjective Global Nutrition Assessment (SGNA), an assessment tool which is more complex as our gold standard. *Methods.* This diagnostic study involved 116 patients aged 1–15 years. Three screening tools and SGNA were examined to each subject. Statistical analysis was used to determine sensitivity, specificity, and likelihood ratio (LR) by results from screening tools divided into low and moderate-high risk of malnutrition compared to results from SGNA divided into no and moderate-severe malnutrition. *Results.* PYMS showed superior agreement to SGNA resulting in sensitivity 95.32%, specificity 76.92%, positive LR 4.13, and negative LR 0.061. STAMP resulted in sensitivity, specificity, positive LR, and negative LR, respectively, as 100%, 11.54%, 1.13, and 0 and STRONG-kids resulted in 100%, 7.7%, 1.083, and 0. *Conclusion.* PYMS was the most reliable screening tool.

1. Introduction

Substrates such as food and vitamins are required by children as the materials in daily metabolism. Increased cost of growth leads to risks of malnutrition especially in hospitalization [1]. Malnutrition has been associated with increased mortality, morbidity, length of stay, and, eventually, cost itself [1–6]. Previous studies reported prevalence of malnutrition in pediatric hospitals ranging from 15 to 50% [3, 5–9]. The European Society for Clinical Nutrition and Metabolism (ESPEN) [10], American Society for Parenteral and Enteral Nutrition (ASPEN) [1, 8], and European Society for Paediatric Gastroenterology, Hepatology and Nutrition (ESPGHAN) [6] recommend nutritional screening, which is rapid and simple, to be used to determine patients who are nutritionally at risk. Nutritional assessment using an assessment tool is then indicated for children nutritionally at risk, facilitating early detection of subsequent nutrition deterioration and finally adequate therapy. An assessment tool is a complete clinical assessment of nutritional status including several steps with detailed medical and dietary history and physical examination including anthropometric, body composition measurement, and laboratory data. The process is lengthy and more time-consuming making it impossible to evaluate every child admitted to the hospital [1, 11]. An assessment tool that has been adapted for children, Subjective Global Nutritional Assessment in Children (SGNA), has shown its ability to identify malnourished children, identifying those at higher risk of nutrition-associated complications, and prolonged hospitalizations [12, 13]. The usage of SGNA has also been validated in our hospital [14]. However, assessment tools are more complex and time-consuming [15]. Therefore, a nutritional screening tool is needed to evaluate every child admitted to the hospital [10].

Numerous screening tools have been developed in previous years, but no universally accepted screening tools are available [9–11]. Paediatric Yorkhill Malnutrition Score (PYMS) was developed and validated in United Kingdom for hospitalized children aged 1–16 years old [16]. Screening Tool for the Assessment of Malnutrition in Paediatrics (STAMP) was also developed and validated in the United Kingdom for hospitalized children [17]. Screening Tool for Risk on Nutritional Status and Growth (STRONG-kids) was developed and validated in Dutch hospitals screening children between 1 month and 16 years [18]. Validation of these three screening tools has never been conducted in our hospital. Therefore, our study aims to determine concurrent validity between the three nutritional screening tools to SGNA, an nutritional assessment tool in Dr. Hasan Sadikin General Hospital, Bandung, Indonesia, a tertiary general hospital.

2. Materials and Methods

This diagnostic study was conducted from January to February 2014 in Department of Child Health, Dr. Hasan Sadikin General Hospital, Bandung, Indonesia. This study has been approved by the Health Research Ethics Committee, Faculty of Medicine, Padjadjaran University, Bandung, Indonesia. The inclusion criteria were patients between 1 and 17 years old, admitted to the pediatric ward for at least 24 hours and whose parents had agreed to participate in the study.

Sampling technique used was consecutive sampling. Patient's characteristic data were recorded for gender, age, underlying disease, weight, height, and body mass index. Anthropometry measurements were taken in a standardized way and then plotted on the WHO Growth Chart or WHO Growth Reference to determine malnutrition. Our study used the definition of acute and chronic malnutrition according to WHO [19]. Acute malnutrition was defined as a weight for height (WFH) SD score lower than −2. Body mass index for age was used if information was not available. Chronic malnutrition was defined as height for age (HFA) SD score lower than −2.

Three screening tools (STAMP, PYMS, and STRONG-kids) and one assessment tool (SGNA) were evaluated in each participant by a single operator. The STAMP [20] and PYMS [2] tools consist of 5 steps with questions and points for every answer yielding a sum total which is divided into 3 categories: high, medium, and low risk of malnutrition. The STRONG-kids tool consists of 4 questions with an allocated score resulting in the same 3 categories [18]. The SGNA assessment tool consisted of nutrition-focused medical history and physical examination yielding an overall ranking as normal, moderate, and severe malnutrition [21]. Statistical analyses were used to determine sensitivity, specificity, and likelihood ratio by defining 2 × 2 tables, which compares the results from low and moderate-high risk of malnutrition categories to the results from no malnutrition and moderate-severe malnutrition categories. Kappa value was used to determine the ability of each of the screening tools to pick up acute and chronic malnutrition according to WHO. Results from low and medium-high risk are compared to normal and acute/chronic malnutrition for determining kappa value.

3. Results

Subject characteristics of 116 participants are shown in Table 1. A total 116 patients participated in the study ranging from 1 to 15 years old. The malnutrition prevalence in this study is found to be 28.44%. A total of 33 children which are considered to be malnourished based on WHO criteria were grouped based on their underlying disease as shown in Table 2, most of who suffered from oncology disorder.

Table 3 showed the kappa value used to determine the ability of each screening tool and SGNA to pick up acute/chronic malnutrition. PYMS was the screening tool that showed best agreement with the anthropometric measurement of acute and chronic malnutrition.

The sensitivity, specificity, and likelihood ratio between nutritional screening tools and SGNA are shown at Table 4. PYMS showed superior results than other nutritional screening tools evaluated.

4. Discussion

As shown in Table 1, malnutrition prevalence found for our study is 28.44%. This value is still in the range from previous reports, which ranges from 15 to 50% [3, 5–9, 22]. Most of the children that are found to be malnourished are the children with oncology disorder, followed by infectious disease, and neurological disorder with infectious disease (Table 2). Malnutrition in patients with oncology disorder is common [23], while children with infectious disease were found to be one of the largest proportions of malnourished patients as in previous study [22]. Poor growth and nutritional status are often seen in children with neurological disorder [23], especially with combined diagnosis [22].

Anthropometric measurements are used for assessing nutritional status worldwide [9]. By applying the definition of acute and chronic malnutrition from WHO based on anthropometric measurements, our study tried to determine the ability of each nutritional screening tool to pick up the types of malnutrition. In Table 3, PYMS showed superior results compared with other nutritional screening tools based on its kappa value.

Similar result is also shown by sensitivity, specificity, and LR parameters. When being compared with SGNA as the gold standard, PYMS is the best screening tool that is used in our hospital. The PYMS showed sensitivity 95.31% implying false negative 4.69% and specificity 76.92% implying false positive 23.08%. The STAMP showed sensitivity 100% implying no false negative, however specificity 11.54% implying false positive 88.46%. The STRONG-kids showed sensitivity 100% implying no false negative, however specificity 7.7% implying false positive 92.3%. PYMS showed the smallest false positive. Estimation of the probability of the disease in individual patients is described by the predictive values [24]. The PYMS showed both high positive (83.56%) and negative predictive values (93.02%). Even though STAMP and STRONG-kids showed very high negative predictive value (100%), they showed positive predictive value with lower value than that of PYMS. Sensitivity, specificity, positive predictive value, and negative predictive value vary according to prevalence;

TABLE 1: Subjects characteristics of the participants.

Characteristics	Total (n = 116)	Percentage (%)
Age		
1–≤5 years	42	36.21
5–≤10 years	45	38.79
>10 years	29	25
Gender		
Male	66	56.90
Female	50	43.10
Underlying disease		
Oncology disorder	50	43.10
Renal disease	12	10.34
Infection disease	17	14.65
Bleeding diathesis	6	5.17
Neurological disorder	3	2.59
Neurological disorder with infection disease	6	5.17
Neurological disorder with respiratory condition	3	2.59
Cardiac disease	3	2.59
Immunological disorder	2	1.72
Endocrinology disease	1	0.86
Gastrointestinal disease	5	4.31
Respiratory disorder	8	6.90
Acute malnutrition (WFH < −2 SD)		
Malnutrition prevalence	33	28.44
Moderate malnutrition	17	14.66
Severe malnutrition	16	13.79
Chronic malnutrition (HFA < −2 SD)		
Stunted	24	20.69
Severely stunted	24	20.69

TABLE 2: Malnutrition prevalence based on the underlying disease.

Underlying disease	Severe malnutrition	Moderate malnutrition	Total (n = 33)
Oncology disorder	6	5	11
Infection disease	3	6	9
Neurological disorder	1	1	2
Neurological disorder with infection disease	2	1	3
Neurological disorder with respiratory condition	2	0	2
Gastrointestinal disease	1	1	2
Renal disease	1	0	1
Cardiology disease	0	1	1
Respiratory disorder	0	2	2

TABLE 3: Kappa value between each of the screening tools and SGNA compared to acute/chronic malnutrition.

Tools	Kappa value (95% CI) for acute malnutrition (wasting)	Kappa value (95% CI) for chronic malnutrition (stunted)
PYMS	0.348 (0.191–0.506)	0.125 (0–0.299)
STAMP	0.018 (0–0.140)	0 (0–0.140)
STRONG-kids	0.028 (0–0.149)	0 (0–0.144)

hence a stable characteristic of a diagnostic test is needed, namely, likelihood ratio that gives strength of the test [25]. Likelihood ratio of PYMS concluded fair strength from the positive LR (4.13) and excellent strength from negative LR (0.061). Positive LR of STAMP concluded useless strength (1.13) and excellent strength from negative LR (0). Positive LR of STRONG-kids concluded useless strength (1.083) and excellent strength from negative LR (0). Eventually, usage of STAMP and STRONG-kids might lead to overdiagnosis;

Table 4: Sensitivity, specificity, and likelihood ratio between nutritional screening tools and SGNA.

	Sensitivity (95% CI)	Specificity (95% CI)	Positive predictive value (95% CI)	Negative predictive value (95% CI)	Positive LR (95% CI)	Negative LR (95% CI)
PYMS	95.31 (0.87–0.98)	76.92 (0.63–0.86)	83.56 (0.73–0.9)	93.02 (0.81–0.97)	4.13 (2.507–6.804)	0.061 (0.02–0.186)
STAMP	100 (0.94–1)	11.54 (0.05–0.23)	58.2 (0.48–0.67)	100 (0.61–1)	1.13 (~)	0 (0–~)
STRONG-kids	100 (0.94–1)	7.7 (0.03–0.18)	57.14 (0.479–0.659)	100 (0.51–1)	1.083 (1.036–1.222)	0 (0–0.757)

LR: likelihood ratio.

therefore, PYMS was the best screening tool without too much overdiagnosis according to our study.

Until now, there is no accepted gold standard for the assessment of the nutritional status of a child [9]. Based on our knowledge, no previous study has used SGNA as a gold standard to compare screening tools in children. There is a lot of debate among professionals on how to validate screening tools, especially in validating a nutritional screening tool if it can predict current nutritional status. Many have validated their screening tools with full dietetic assessment. However, it is questionable if this is the gold standard, especially as not all countries have dieticians and their role may vary depending on the country [9], as in Indonesia. In this study, SGNA is chosen as the gold standard. SGNA was initially designed as a screening tool. The SGNA consists of both subjective and objective components in a detailed questionnaire and complete physical examination; then each child is classified as well-nourished, moderately malnourished, or severely malnourished. Completion of SGNA is lengthy and time-consuming [11]. More in-depth than nutrition screening tool, SGNA is now used to assess nutritional status of children who may be at risk of malnutrition such as those who are hospitalized [21]. SGNA has been evaluated in 175 children admitted for a major surgical procedure. Malnourished children had higher rates of nutrition-associated complications such as infectious complications and postoperative length than well-nourished children [12]. Our hospital has also evaluated SGNA in 2010. SGNA was evaluated in 320 children at our hospital showing that malnourished children had longer length of hospital stay than well-nourished children [14]. Since then, SGNA has been the assessment tool in our hospital. Therefore, our study chose SGNA as our gold standard.

All previous studies showed comparisons of nutritional screening tools with full nutritional assessment as their gold standard. Gerasimidis et al. [2] compared STAMP, PYMS, and SGNA to full nutritional assessment. The study included 2,174 children from pediatric and surgery ward concluding that PYMS with sensitivity 85% and specificity 87% is an acceptable screening tool for identifying children at risk of malnutrition without producing unmanageable numbers of false-positive cases. Children from cardiology, renal, orthopaedic specialties, and critical care were not included. A 2 × 2 table was made from the results of STAMP, PYMS, and SGNA to determine sensitivity and specificity, which divided the result into low-medium risk and high risk, compared to full nutritional assessment [2]. Our study included all patients in the pediatric ward. However, no patients were listed for surgery in our study. Our study divided the 2 × 2 table differently into low and medium-high risk to determine sensitivity and specificity. Another study conducted by Moeeni et al. [13] comparing PYMS, STAMP, and STRONG-kids to full nutritional assessment which included 119 children in Iran suggested that STRONG-kids was the most useful and reliable tool. Weight for height z-score correlated with all three tools ($P < 0.001$), but only STRONG-kids correlated with height for age z-score ($P = 0.04$). The STRONG-kids detected more children with moderate undernutrition compared to PYMS ($P = 0.0001$) and STAMP ($P = 0.05$). High risk determined by STRONG-kids was also associated with length of stay ($P = 0.004$). PYMS was superior in detection of the severely undernourished children compared to STRONG-kids ($P = 0.003$) [13]. Another study conducted by Moeeni et al. [26] that evaluated 162 children with PYMS, STAMP, and STRONG-kids to full nutritional assessment in New Zealand recommended that STRONG-kids was the most reliable tool. Only high and moderate risk children using PYMS ($P = 0.01$ and $P = 0.001$, resp.) and STRONG-kids ($P = 0.003$ and $P = 0.002$, resp.) had significantly longer admissions than low risk group [26]. Both studies conducted by Moeeni et al. [13, 26] were cross-sectional studies comparing hospitalized and nonhospitalized children. The study recorded the length of stay and excluded oncology disorder in their population. However, in this diagnostic study the oncology patients were included.

According to ESPEN, screening tools are used to detect protein and energy undernutrition and/or to predict whether undernutrition is likely to develop/worsen under the present and future conditions of the patient/client. Therefore, screening tools should consist of four main principles.

(1) The current condition can be described in the height and weight allows the calculation of body mass index.

(2) Is the condition stable? This concerns any recent weight loss that can be obtained from patient's history or previous measurements in medical records.

TABLE 5: Comparison of screening tools based on 4 main principles of a screening tool according to ESPEN [9, 10].

Tools	Main principles according to ESPEN: Current nutritional status	Weight loss	Reduced intake	Disease severity	Aim of each screening tool: Identify nutritional status	Identify need for nutritional intervention	Predict clinical outcome without intervention
PYMS	√	√	√	√*	√	√	√
STAMP	√		√	√	√	√	
STRONG-kids	√#	√	√	√		√	√

#Current nutritional status by STRONG-kids is not evaluated by BMI but by subjective clinical assessments.
*Disease severity in PYMS is not evaluated based on underlying disease that has nutritional implications as in STAMP or STRONG-kids, but on whether the child's nutrition will be affected by recent condition for at least the next week.

(3) Will the condition worsen? This question might be answered by asking whether food intake has been decreased up to the time of screening and if so by approximately how much and how long.

(4) Will the disease process accelerate nutritional deterioration? This item concerns the underlying disease process which may increase nutritional requirement due to the stress metabolism associated with the severity of the underlying disease, causing nutritional status to worsen more rapidly or to develop a poor nutritional status.

Variables 1–3 should be included in all screening tools, while variable 4 is relevant mainly in hospital settings. Each variable must be given a score in every screening tool, thereby quantifying the degree of risk and a direct course of action [10]. Each nutritional screening tool used in our study is compared according to the main principles based on ESPEN of a screening tool and their aims showed in Table 5. STRONG-kids also used subjective clinical assessment and is the only screening tool that does not include height and weight which some may consider it as its disadvantages [3, 27]. Pediatricians may believe that they can recognise a malnourished child but the fact does not always agree with this. Reproducibility in clinical assessment of nutritional status is poor in a study conducted by Cross et al. [28], especially in assessing the more severely malnourished children. Clinical evaluation of nutritional status alone is inadequate for accurate assessment and anthropometry is prominent [27]. The screening tool PYMS does not include the impact of underlying disease. The other two screening tools included lists of underlying disease in their components [13], which has been considered as a disadvantage [11, 13]. However, it must be noted that PYMS still asked about condition effect on nutrition whether due to decreased intake caused by orofacial trauma or severe nausea, increased gut loss, increased energy requirements, or intention to perform major abdominal surgery causing minimal intake [16]. This was emphasized by a review that concluded that PYMS also evaluates disease severity [9].

In conclusion, we recommend PYMS as the most reliable screening tool in Dr. Hasan Sadikin General Hospital in Bandung.

Conflict of Interests

The authors declare that there is no conflict of interests regarding the publication of this paper.

Authors' Contribution

Nathania Wonoputri and Julistio T. B. Djais proposed the conception of the study and participated in its design. Nathania Wonoputri also carried out the studies, drafted the paper, and analyzed the data. Ina Rosalina helped to draft the paper. All authors read and approved the final paper.

Acknowledgments

The authors would like to thank all participating children and their parents for their cooperation and all the dieticians that supported the study. No source of funding participated in the study.

References

[1] M. R. Corkins, K. C. Griggs, S. Groh-Wargo et al., "Standards for nutrition support: pediatric hospitalized patients," *Nutrition in Clinical Practice*, vol. 28, no. 2, pp. 263–276, 2013.

[2] K. Gerasimidis, O. Keane, I. MacLeod, D. M. Flynn, and C. M. Wright, "A four-stage evaluation of the Paediatric Yorkhill Malnutrition Score in a tertiary paediatric hospital and a district general hospital," *British Journal of Nutrition*, vol. 104, no. 5, pp. 751–756, 2010.

[3] C. Hartman, R. Shamir, C. Hecht, and B. Koletzko, "Malnutrition screening tools for hospitalized children," *Current Opinion in Clinical Nutrition and Metabolic Care*, vol. 15, no. 3, pp. 303–309, 2012.

[4] I. Sermet-Gaudelus, A.-S. Poisson-Salomon, V. Colomb et al., "Simple pediatric nutritional risk score to identify children at risk of malnutrition," *The American Journal of Clinical Nutrition*, vol. 72, no. 1, pp. 64–70, 2000.

[5] G. A. Rocha, E. J. M. Rocha, and C. V. Martins, "The effects of hospitalization on the nutritional status of children," *Jornal de Pediatria*, vol. 82, no. 1, pp. 70–74, 2006.

[6] C. Agostoni, I. Axelson, V. Colomb et al., "The need for nutrition support teams in pediatric units: a commentary by the ESPGHAN committee on nutrition," *Journal of Pediatric Gastroenterology and Nutrition*, vol. 41, no. 1, pp. 8–11, 2005.

[7] N. N. M. Nesa, G. L. Sidiartha, E. P. Prawirohartono, and K. G. Suandi, "Accuracy of modified simple pediatric nutritional risk score to detect in-hospital malnutrition," *Paediatrica Indonesiana*, vol. 50, no. 5, pp. 305–309, 2010.

[8] N. M. Mehta, M. R. Corkins, B. Lyman et al., "Defining pediatric malnutrition: a paradigm shift toward etiology-related definitions," *Journal of Parenteral and Enteral Nutrition*, vol. 37, no. 4, pp. 460–481, 2013.

[9] K. F. M. Joosten and J. M. Hulst, "Nutritional screening tools for hospitalized children: methodological considerations," *Clinical Nutrition*, vol. 33, no. 1, pp. 1–5, 2014.

[10] J. Kondrup, S. P. Allison, M. Elia, B. Vellas, and M. Plauth, "ESPEN guidelines for nutrition screening 2002," *Clinical Nutrition*, vol. 22, no. 4, pp. 415–421, 2003.

[11] V. Moeeni and A. S. Day, "Nutritional risk screening tools in hospitalized children," *International Journal of Child Health and Nutrition*, vol. 1, pp. 39–43, 2012.

[12] D. J. Secker and K. N. Jeejeebhoy, "Subjective global nutritional assessment for children," *The American Journal of Clinical Nutrition*, vol. 85, no. 4, pp. 1083–1089, 2007.

[13] V. Moeeni, T. Walls, and A. S. Day, "Assessment of nutritional status and nutritional risk in hospitalized Iranian children," *Acta Paediatrica*, vol. 101, no. 10, pp. e446–e451, 2012.

[14] F. M. Andriyani, *Status gizi berdasarkan Subjective Global Assessment sebagai faktor yang mempengaruhi lama perawatan pasien rawat inap anak [Thesis]*, Universitas Padjadjaran, Bandung, Indonesia, 2010, (Indonesian).

[15] H. Lochs, S. P. Allison, R. Meier et al., "Introductory to the ESEN guidelines on enteral nutrition: terminology, definitions and general topics," *Clinical Nutrition*, vol. 25, no. 2, pp. 180–186, 2006.

[16] K. Gerasimidis, I. Macleod, A. Maclean et al., "Performance of the novel Paediatric Yorkhill Malnutrition Score (PYMS) in hospital practice," *Clinical Nutrition*, vol. 30, no. 4, pp. 430–435, 2011.

[17] H. McCarthy, M. Dixon, I. Crabtree, M. J. Eaton-Evans, and H. McNulty, "The development and evaluation of the Screening Tool for the Assessment of Malnutrition in Paediatrics (STAMP ©) for use by healthcare staff," *Journal of Human Nutrition and Dietetics*, vol. 25, no. 4, pp. 311–318, 2012.

[18] J. M. Hulst, H. Zwart, W. C. Hop, and K. F. M. Joosten, "Dutch national survey to test the STRONGkids nutritional risk screening tool in hospitalized children," *Clinical Nutrition*, vol. 29, no. 1, pp. 106–111, 2010.

[19] World Health Organization, *Management of Severe Malnutrition: A Manual for Physician and Other Health Workers*, World Health Organization, Geneva, Switzerland, 1999.

[20] Central Manchester University Hospitals, "STAMP instructions," 2010, http://www.stampscreeningtool.org/data/pdfs/stamp_tool.pdf.

[21] D. J. Secker and K. N. Jeejeebhoy, "How to perform subjective global nutritional assessment in children," *Journal of the Academy of Nutrition and Dietetics*, vol. 112, no. 3, pp. 424–431, 2012.

[22] I. Pawellek, K. Dokoupil, and B. Koletzko, "Prevalence of malnutrition in paediatric hospital patients," *Clinical Nutrition*, vol. 27, no. 1, pp. 72–76, 2008.

[23] K. F. M. Joosten and J. M. Hulst, "Prevalence of malnutrition in pediatric hospital patients," *Current Opinion in Pediatrics*, vol. 20, no. 5, pp. 590–596, 2008.

[24] A. K. Akobeng, "Understanding diagnostic tests 1: Sensitivity, specificity and predictive values," *Acta Paediatrica*, vol. 96, no. 3, pp. 338–341, 2007.

[25] D. Mayer, *Essential Evidence-Based Medicine*, Cambridge University Press, New York, NY, USA, 2nd edition, 2010.

[26] V. Moeeni, T. Walls, and A. S. Day, "Nutritional status and nutrition risk screening in hospitalized children in New Zealand," *Acta Paediatrica*, vol. 102, no. 9, pp. e419–e423, 2013.

[27] P. B. Sullivan, "Malnutrition in hospitalised children," *Archives of Disease in Childhood*, vol. 95, no. 7, pp. 489–490, 2010.

[28] J. H. Cross, C. Holden, A. MacDonald, G. Pearmain, M. C. G. Stevens, and I. W. Booth, "Clinical examination compared with anthropometry in evaluating nutritional status," *Archives of Disease in Childhood*, vol. 72, no. 1, pp. 60–61, 1995.

2

Effects of *Astrocaryum aculeatum* Meyer (Tucumã) on Diet-Induced Dyslipidemic Rats

Geórgia Craveiro Holanda Malveira Maia,[1] Mozer da Silva Campos,[2] Janice Barros-Monteiro,[3] Juan Eduardo Lucas Castillo,[1] Murilo Soares Faleiros,[1] Rejane Souza de Aquino Sales,[1] Denise Moraes Lopes Galeno,[1] Edson Lira,[4] Francisca das Chagas do Amaral Souza,[5] Carmen Ortiz,[6] Luisa Morales,[7] and Rosany Piccolotto Carvalho[1]

[1] *Programa Multi-Institucional de Pós-graduação em Biotecnologia, Universidade Federal do Amazonas, Avenue General Rodrigo Octávio Jordão Ramos, 3000 Coroado, Manaus, AM, Brazil*
[2] *Departamento de Ciências Fisiológicas, Instituto de Ciências Biológicas, Universidade Federal do Amazonas, Avenue General Rodrigo Octávio Jordão Ramos, 3000 Coroado, Manaus, AM, Brazil*
[3] *Biochemistry Department, Ponce School of Medicine and Health Sciences, P.O. Box 7004, Ponce, PR 00732-7004, USA*
[4] *Fundação de Hematologia e Hemoterapia do Estado do Amazonas (FHEMOAM), Nucleo de Estatística, Avenue Constantino Nery 3240, 69050-001 Manaus, Brazil*
[5] *Instituto Nacional de Pesquisas da Amazônia, Coordenação Sociedade Saúde Ambiente-Laboratório de Alimentos, Avenue André Araújo 2.936, Petrópolis, 69067-375 Manaus, AM, Brazil*
[6] *Physiology, Pharmacology, and Toxicology Department, Ponce School of Medicine and Health Sciences, P.O. Box 7004, Ponce, PR 00732-7004, USA*
[7] *Public Health Department, Ponce School of Medicine and Health Sciences, P.O. Box 7004, Ponce, PR 00732-7004, USA*

Correspondence should be addressed to Carmen Ortiz; caortiz@stu.psm.edu
and Rosany Piccolotto Carvalho; prosany@ufam.edu.br

Academic Editor: C. S. Johnston

An *in vivo* study was conducted to assess the effects of the consumption of *Astrocaryum aculeatum Amazon Meyer* (*tucumã*) in the treatment of diet-induced dyslipidemia in sedentary and exercised Wistar rats. With an average weight of 350 grams, 40 male rats were divided into 4 subgroups of 10. The sedentary control group (SCG) was fed with commercial feed, while the sedentary treatment group (STG) was fed with a ration of tucumã. In addition to the sedentary groups, two exercise groups were formed. The Exercised control group (ECG) was fed with commercial food and the exercised treatment group (ETG) was fed with a ration of tucumã. Body weight gain and food intake were monitored during the experiment. Plasma was analyzed for cholesterol, triglycerides, HDL-C, LDL-C, VLDL, total protein, glucose, insulin, and leptin concentrations. Our results show that the ECG group tended to consume more food, while the groups that were fed with tucumã pulp (STG and ETG) presented a greater tendency to gain body mass. ECG group showed a tendency towards a higher concentration of cholesterol in plasma, while STG and ETG presented higher absolute values for triglycerides and VLDL. No hypolipiemic effect was observed related to tucuma ingestion.

1. Introduction

Dyslipidemias are characterized by alterations in lipid metabolism, with repercussions on the level of lipoproteins present in the bloodstream. They are directly related to the development of cardiovascular alterations, therefore, promoting the development of chronic cardiovascular and cerebrovascular diseases, such as atherosclerosis, acute myocardial infarction, ischemic disease, and encephalic vascular accident [1–3].

Nontransmissible chronic diseases (NTCDs) are one of the main causes of death worldwide. In Brazil, NTCDs are responsible for approximately 72% of deaths, mostly caused by diseases of the circulatory system (30%), therefore, increasing public health expenses. These diseases cause several alterations in the organism, leading to premature death or even disability, eventually, limiting the individual's working opportunities and consequently increasing social care expenditure [4, 5]. Risk factors triggering NTCDs include tobacco and alcohol consumption, sedentary lifestyle, and unhealthy food habits. It is crucial for the Brazilian public health system to develop strategies aiming to decrease the risk of these diseases in the population and to incorporate healthcare services targeting these diseases [6, 7].

Limited exercise and inadequate food habits such as a high-fat and low-fiber diet can lead to the accumulation of adipose tissue and the consequent development of obesity. This triggers alterations in the lipid metabolism and accelerates the onset of hypercholesterolemia and triglyceridemia among other metabolic alterations leading to the development of coronary diseases [8, 9]. Most carriers of hypercholesterolemia do not show signs or symptoms that can be directly attributed to a lipid alteration, and the diagnosis is based almost solely on a serological analysis of lipids in plasma. However, some patients with more severe hyperlipidemias, usually of hereditary nature, may present clinical manifestations such as fat accumulation in several tissues (i.e., eyes, skin, tendons, and nervous system) that are externally visible, which can lead to suspect a lipid alteration even before serological determination [10].

According to Sposito et al., previous studies have demonstrated that a reduction of cholesterol levels can decrease the risk of cardiovascular events, the incidence of cardiac ischemic disease (by 25%–60%), and the risk of death by 30%. Cholesterol reduction can also decrease other morbid effects such as angina and cerebrovascular accidents among other conditions [3]. Soares and coworkers found that a 1% reduction in serum cholesterol levels results in a 2% reduction in the risk of coronary diseases [11]. Given the crucial role of cholesterol in these morbidities, it is vital to find alternatives to allow its control and treatment.

Evidence shows that pharmacological treatment can reduce the risk factors related to dyslipidemia. However, they can be prevented with lifestyle changes, including physical exercise in combination with nutritional therapy [3]. Therefore, preventive measures that decrease fat accumulation like changes in diet and physical activity can contribute to balance the blood lipoprotein concentrations. Moreover, the beneficial effects of these measures have been well established for normolipidemic and dyslipidemic patients alike [12].

It is known that the intake of fiber, especially insoluble ones, as well as the moderate consumption of unsaturated fatty acids is beneficial in the reduction of lipids in plasma. Fibers reduce the absorption of intestinal cholesterol in food, regularizing the enteric food flow and promoting glycemic and cholesterolemic control [13]. In addition to these physiological effects, the ingestion of polyunsaturated or monounsaturated fatty acids reduces total blood cholesterol, as well as low density lipoprotein C (LDL-C) and triglycerides that can help to avoid lipid oxidation [3].

Foods with favorable characteristics for the treatment of diseases are described as functional foods. The tucumã (*Astrocaryum aculeatum* Meyer) palm can be found in degraded environments among secondary vegetation in the Central and Western Amazonian ecosystem. Its fruit is widely consumed by the Amazonian population *in natura*, in sandwiches and tapioca, desserts, and ice cream. Indians apply the oil extracted from the pulp of the fruit and the seeds over their body and hair [14].

Tucumã is a smooth drupe with an epicarp and mesocarp that can vary greatly in color (yellowish to dark orange and red, or whitish), shape (oval, spherical), mesocarp thickness (2.8 to 10 mm), endocarp, and seed type. The consistency of the fruit is mucilaginous, oily, and sweet to the taste with a characteristic flavor. Its weight varies from 20 to 100 g, of which 22.4% is edible [15–17]. As to nutritional composition, the high crude fiber content stands out. Furthermore, in some fruits of the same genus, there are unsaturated fatty acids which are beneficial if taken moderately [18].

According to Ferreira et al., the Palm tucumã (*Astrocaryum vulgare Mart*) is composed of fatty acids that may induce a reduction of cholesterol in plasma, triglycerides, and LDL-cholesterol, as well as an increase of the HDL-c concentration [18]. This could be due to the fact that, among all the fatty acids that constitute its crude oil, a concentration of 68% monounsaturated fatty acids and only 29% saturated fatty acids is found. These data were also evident in the studies by Guedes [19].

From the centesimal composition analysis of the Palm tucumã (*Astrocaryum vulgare* Mart), and according to studies by Aguiar et al., and Leitão [17, 20], it can be assumed that the key components are present in similar concentrations, as shown in Table 1.

Apart from the beneficial fiber concentration, Guedes demonstrated that another type of tucumã (*Astrocaryum vulgare* Mart) of the same genus as the one studied is composed of 29.6 g of palmitic acid, 3.0 g of stearic acid, 58.5 g of oleic acid, 3.8 g of linoleic acid, and 5.5 g of linolenic acid [19]. This suggests that the intake of the tucumã (*Astrocaryum aculeatum Meyer*) may be beneficial, given that unsaturated fatty acids, when taken moderately, may be favorable to patients with dyslipidemias.

Since there is no previous research on the use of the Amazonas' tucumã as a functional food for the control or reduction of dyslipidemia, the relevance of our project becomes apparent. Hence, the objective of this study is to assess the effects of tucumã fruit (*Astrocaryum aculeatum* Meyer) intake in the treatment of diet-induced dyslipidemia in sedentary and exercised Wistar rats.

2. Material and Methods

2.1. Animals. Forty male rats of the Wistar strain, obtained from the Animal House of the Amazonas Federal University (UFAM), weighting between 250 and 350 grams, were used to perform the study. They were kept under a 12-hour light-dark cycle, with free (*ad libitum*) access to food and water under

TABLE 1: Proximate composition of the tucumã fruit *in natura*.

Determination	Aguiar et al. (1980) [17] $n =$	Leitão (2008) [20] $n =$	Ferreira et al. (2008) [18] $n =$
Humidity (%)	38.5	50.25 ± 11.49*	44.90 ± 0.30*
Protein (g)	5.5	3.39 ± 1.08*	3.54 ± 0.07*
Fat (g)	47.2	25.19 ± 10.91*	40.49 ± 0.54*
Crude fiber (g)	19.2	28.71 ± 1.86*	10.93 ± 0.10*
Ash (g)	2.0	1.23 ± 0.27*	2.53 ± 0.05*
Total carbohydrates (g)	6.8	19.25 ± 1.69*	8.54 ± 0.61*
Energy (Kcal/100 g)	474.0	320.0 ± 98.75*	412.73 ± 2.12*

*Mean ± SD.

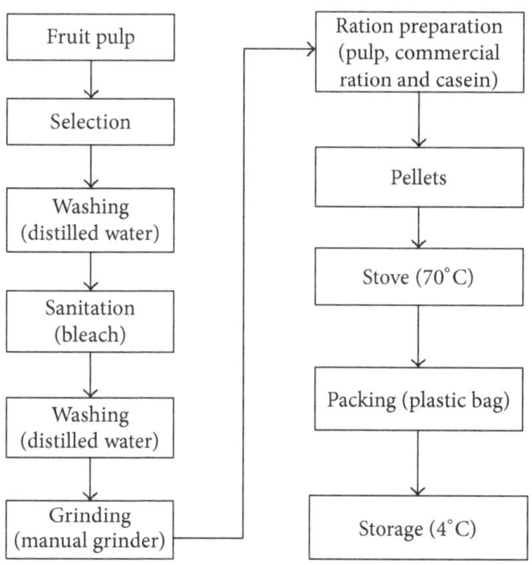

FIGURE 1: Fruit processing and ration preparation.

adequate light and temperature conditions. The experimental protocol was approved by the Animal Experimentation Ethics Committee of the same university, protocol number 020/2011.

2.2. Experimental Diets. All diets contained Labina commercial feed for laboratory animals, produced by Agribrands do Brasil Ltda of Paulínia, SP. The experimental protocol was approved by the Animal Experimentation Ethcontrol groups. In the first part of the experiment, a hyperlipidic food portion was prepared using pork lard to induce dyslipidemia. It was composed of 36% of Sadia lard, 28% of Synth casein, and 36% of commercial feed [21].

The treatment portion was composed of 12% of casein, 28% of commercial feed, and 60% of tucumã pulp from a farm located in the Municipality of Rio Preto da Eva (AM), and posteriorly processed at UFAM according to the methodology proposed by Yuyama et al. with some modifications (Figure 1) [14].

2.3. Determination of Proximate Composition of Tucumã Based Diets. To determine the fruit pulp concentration necessary for the treatment, rations with different concentrations of tucumã pulp (20%, 30%, and 60%) were prepared, followed by physicochemical analysis at the Fish Technology Laboratory at UFAM. Humidity, protein, fat, ash, crude fiber, and carbohydrate content were analyzed according to procedures of the Fish Technology Laboratory. In addition, the proportions of soluble and insoluble fibers of the rations with different fruit concentrations were also investigated at the National Research Institute of Amazonia, according to the methodology adopted by the institute.

2.4. Experimental Groups. Four groups of 10 animals were used in the study. Each group was designated according to the treatment used and the amount of physical activity as sedentary control group (SCG), which were fed commercial feed and did not perform physical activity, exercised control group (ECG), which were fed commercial feed and exercised, sedentary tucumã group (STG), which were fed a portion of tucumã pulp and did not perform physical activity, and exercised tucumã group (ETG), which were fed a portion of tucumã pulp and exercised. Food intake was monitored over time and body mass gain was determined.

2.5. Physical Activity Protocol. Animals were submitted to an aerobic exercise program. The protocol consisted of a daily session of swimming for sixty minutes, three days a week, during eight weeks, in a plastic tank with a height of 120 cm and a diameter of 80 cm filled with water at 30°C. During the sessions, the animals had a resistance equivalent to 3 to 5% of their body weight attached to the tail as described by Rodrigues and Zambon et al., with minor modifications [21, 22].

2.6. Determination of Selected Plasma Biomolecules. Plasma samples were collected at the beginning, after 30 days, and on the last day of the experiment. Blood was collected by punction of the animals' tails, stocked in 1.5 mL heparinized microtubes, and centrifuged at 3000 rpm for 5 minutes for plasma separation and immediate analysis of the concentrations of cholesterol, triglycerides, HDL-C, LDL-C, VLDL, total protein, glucose, insulin, and leptin. These analyses were performed using a colorimetric enzymatic method using an In Vitro Diagnno Dia S/A (Itabira, MG) commercial kit [23].

TABLE 2: Physicochemical characterization of the dry food portion based on tucumã pulp concentration.

Tucumã pulp (%)	Humidity (%)	Ash (%)	Protein (%)	Lipids (%)	Carbohydrates (%)	Fibers (%)
20%	7.97 ± 0.15*	7.41 ± 0.04*	19.39 ± 1.14*	4.71 ± 0.85*	60.52 ± 0.92*	2.23 ± 0.09*
30%	7.53 ± 0.58*	7.08 ± 0.11*	21.67 ± 0.19*	6.31 ± 0.51*	58.41 ± 1.14*	2.33 ± 0.08*
60%	8.47 ± 0.07*	5.27 ± 0.07*	27.10 ± 0.65*	14.71 ± 0.94*	44.45 ± 0.61*	2.49 ± 0.19*

Asterisks (*) denote the difference with the control group.
Values are presented as average values ± S.E., where $P \leq 0.05$. Note: S.E.: standard error.

2.7. Statistical Analysis. To assess the influence of tucumã and physical exercise in our study, variance analysis (ANOVA) and Tukey's test were used to establish statistical differences among the groups. The difference was considered significant when $P < 0.05$. Results were expressed as average ± standard deviation.

3. Results

After conducting this study for 90 consecutive days, plasma and proximate composition analyses were performed, and food intake and body mass gain were assessed for the different experimental groups.

3.1. Proximate Composition of Food Portion with Increasing Concentration of Tucumã. The three samples with different concentrations of tucumã showed similar physicochemical properties. However, a higher concentration of fat was reported for the 60% sample, which is twice of the concentration in the 30% sample, and three times higher than that of the 20% sample. In addition, a higher proportion of carbohydrates was found in the 20% sample and a much lower proportion in the 60% sample. As to fibers, the concentration was similar among the three samples (Table 2).

3.2. Food Intake and Body Mass Gain. Figure 2 illustrates food intake and shows no statistically significant difference among the studied groups ($P = 0.495$). In addition, the results show a tendency of higher food intake in the ECG when compared to the other groups.

Figure 3 shows the average body mass gain in the groups under study. In the SCG, there was a statistically significant difference between body mass gain only between day 0 and day 90. In the ECG, there was a statistically significant difference among the three time intervals of the experiment. STG and ETG did not show significant body mass differences between day 30 and day 90. However, these two time points were significantly different from the body mass at day 0. The descriptive analysis demonstrated that the groups that consumed the diet with tucumã fruit showed a higher tendency to gain body mass, whereas animals belonging to the exercise groups (ECG and ETG) showed a reduced tendency to gain body mass.

3.3. Biochemical Analysis

3.3.1. Lipid Profile. The lipids profile of the groups under study was analyzed separately at days 0, 30, and 90 (Table 3).

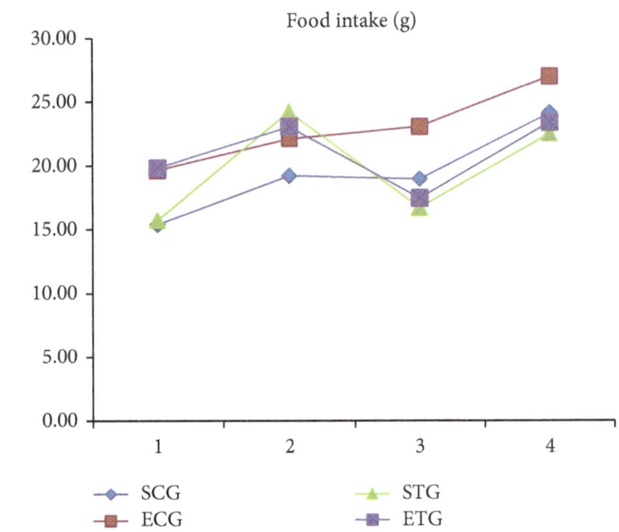

FIGURE 2: Average values of food intake (g) among the studied groups throughout the experimental period. Note: SCG: sedentary control group; ECG: exercised control group; STG: sedentary tucumã group; ETG: exercised tucumã group. P value < 0.001 for all groups.

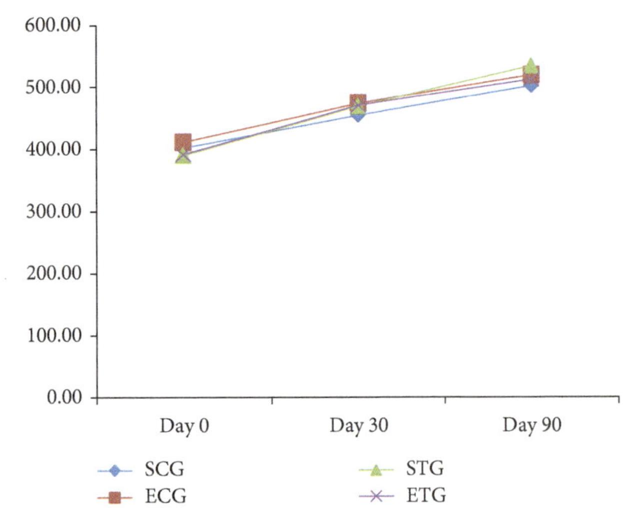

FIGURE 3: Average body mass gain values (g) of the groups evaluated at days 0, 30, and 90. Different letters in superscript denote significant difference for $P < 0.05$. Note: SCG: sedentary control group; ECG: exercised control group; STG: sedentary tucumã group; ETG: exercised tucumã group.

TABLE 3: Lipid profile of the groups under study at different time points (days 0, 30, and 90).

GROUPS	Day 0 Chol.	Day 0 TG	Day 0 VLDL	Day 30 Chol.	Day 30 TG	Day 30 VLDL	Day 90 Chol.	Day 90 TG	Day 90 VLDL
SCG	56.72 ± 4.47^{a} $n = 9$	52.40 ± 3.96^{a} $n = 8$	10.08 ± 0.81^{a} $n = 9$	82.09 ± 12.70^{a} $n = 8$	72.02 ± 13.94^{a} $n = 8$	18.62 ± 4.67^{a} $n = 9$	186.61 ± 13.17^{a} $n = 8$	174.90 ± 17.31^{a} $n = 6$	24.31 ± 2.70^{a} $n = 9$
ECG	70.87 ± 4.60^{a} $n = 9$	$83.35 \pm 13.98^{a,b}$ $n = 10$	$16.67 \pm 2.80^{a,b}$ $n = 10$	122.26 ± 12.55^{a} $n = 9$	96.85 ± 15.74^{a} $n = 10$	19.37 ± 3.15^{a} $n = 10$	219.44 ± 16.91^{a} $n = 7$	80.18 ± 14.31^{b} $n = 10$	$16.04 \pm 2.86^{a,b}$ $n = 10$
STG	71.77 ± 2.95^{a} $n = 10$	$75.78 \pm 12.17^{a,b}$ $n = 9$	$16.74 \pm 2.69^{a,b}$ $n = 10$	101.84 ± 8.50^{a} $n = 10$	114.14 ± 24.25^{a} $n = 9$	$26.77 \pm 5.86^{a,b}$ $n = 10$	260.79 ± 22.93^{a} $n = 8$	106.74 ± 18.82^{b} $n = 9$	$24.04 \pm 4.31^{a,b}$ $n = 10$
ETG	56.72 ± 4.47^{a} $n = 9$	52.40 ± 3.96^{a} $n = 8$	10.08 ± 0.81^{a} $n = 9$	82.09 ± 12.70^{a} $n = 8$	72.02 ± 13.94^{a} $n = 8$	18.62 ± 4.67^{a} $n = 9$	186.61 ± 13.17^{a} $n = 8$	174.90 ± 17.31^{a} $n = 6$	24.31 ± 2.70^{a} $n = 9$
P value among groups	70.87 ± 4.60^{a} $n = 9$	$83.35 \pm 13.98^{a,b}$ $n = 10$	$16.67 \pm 2.80^{a,b}$ $n = 10$	122.26 ± 12.55^{a} $n = 9$	96.85 ± 15.74^{a} $n = 10$	19.37 ± 3.15^{a} $n = 10$	219.44 ± 16.91^{a} $n = 7$	80.18 ± 14.31^{b} $n = 10$	$16.04 \pm 2.86^{a,b}$ $n = 10$

Values shown represent the average ± S.E. expressed in mg/dL. ^{a,b}Values followed by different letters in superscript present significant differences among the different time points for $P < 0.05$. Note: SCG: sedentary control group; ECG: exercise control group; STG: sedentary tucumã group; ETG: exercise tucumã group; S.E.: standard error; n: number of animals evaluated per group.

Table 4: Total protein, glucose, insulin, and leptin concentrations of the different groups on day 90.

Groups	SCG	ECG	STG	ETG	P value
Protein (g/dL)	6.35 ± 0.45^{a} $n = 9$	6.83 ± 0.37^{a} $n = 10$	6.15 ± 0.36^{a} $n = 10$	5.98 ± 0.46^{a} $n = 9$	<0.500
Insulin (ng/dL)	3.17 ± 0.39^{a} $n = 9$	2.30 ± 0.26^{a} $n = 7$	2.43 ± 0.53^{a} $n = 9$	2.47 ± 0.55^{a} $n = 9$	<0.516
Glucose (mg/dL)	171.91 ± 1.91^{a} $n = 9$	162.36 ± 3.27^{b} $n = 10$	$170.46 \pm 1.39^{a,b}$ $n = 10$	$165.58 \pm 3.51^{a,b}$ $n = 9$	<0.040
Leptin (ng/dL)	$7.70 \pm 0.94^{a,b}$ $n = 9$	5.15 ± 0.44^{a} $n = 9$	9.86 ± 1.11^{b} $n = 9$	$8.43 \pm 1.19^{a,b}$ $n = 9$	<0.007

^{a,b}Values (average ± S.E.) followed by different letters in superscript present significant differences for $P < 0.05$. Note: SCG: sedentary control group; ECG: exercise control group; STG: sedentary tucumã group; ETG: exercise tucumã group; S.E.: standard error; n: number of animals evaluated per group.

At the beginning of the experiments (day 0), there was no statistically significant difference among the groups in relation to cholesterol levels, which demonstrates the homogeneity of this parameter among the experimental groups. At this time point, triglycerides and VLDL of ETG and SCG were statistically different, ETG presenting higher concentration of these lipids.

After the induction period on day 30, it can be seen that there was no statistically significant difference among groups for any of the lipids evaluated. Although ECG showed a tendency for higher concentration of cholesterol in plasma, STG and ETG presented higher absolute values for triglycerides and VLDL.

At the end of the experiment (day 90), we found that the variance analysis of cholesterol levels did not reveal a statistically significant difference among the groups. However, it was observed that the groups that consumed tucumã had a higher concentration of this lipid. On the other hand, the plasma triglyceride analysis revealed that the SCG had much higher values in comparison with ECG, STG, and ETG. Furthermore, descriptive statistics analysis showed that groups that exercised presented lower triglyceride levels in comparison to sedentary groups. VLDL was shown to be significantly different between SCG and ETG. Exercised groups also presented lower values in comparison to the sedentary ones.

Among the groups that exercised (ECG and ETG), the group consuming tucumã presented a reduction of the lipid concentration in plasma.

3.4. Protein, Glucose, Insulin, and Leptin. Table 4 shows that the values related to protein and insulin concentration were not significantly different among the experimental groups. However, concentration of glucose in plasma was statistically different among the control groups (SCG and ECG). Leptin concentration in plasma was found to be different between ECG and STG.

4. Discussion

This study aimed to assess the effect of tucumã (*Astrocaryum aculeatum* Meyer) intake in lipid metabolism. Since it is a regional fruit from Amazonia and complies with the characteristics established by the Dyslipidemia Brazilian Society, tucumã should be considered a functional food due to its medicinal properties that can promote benefits to health [3]. Functional foods possess beneficial properties in addition to its basic nutritional properties, presented in the form of common foods. They are part of conventional diets but they show the capacity of regulating body functions and aiding in the protection against diseases such as hypertension, diabetes, cancer, osteoporosis, and coronaropathy [24, 25].

It can be seen that food intake (Figure 2) among the groups was probably regulated by the ingestion of calories, because although they are present *ad libitum* there was not a higher intake by any experimental group. Our data agree with the study of Rocha et al., who also recorded a proportional food intake among all studied groups [26]. An increase in body mass was observed after the beginning of the treatment in all animals (Figure 3). Hence, the significant reduction among exercised groups in relation to sedentary groups and among groups that ingested tucumã was not apparent in this study. These same results were found in studies by Bell et al. and Pellizzon et al., who recorded weight gain by animals at the end of the experiment even in those that exercised and those that had a diet with polyunsaturated fatty acids [27, 28].

Although there was no significant difference in food intake and body mass among all groups, STG showed a greater tendency to body mass gain (Figure 3). This could be due to the addition of tucumã to their diet. Taking into consideration the high energy value of tucumã and the absence of physical activity in this group, we could provide a possible explanation for the accumulation of fat and body mass gain in these animals.

Using an animal model, Franco et al. demonstrated that the consumption of large quantities of saturated fatty acids promotes a greater accumulation of fat and consequently a greater body mass gain, when compared to the ingestion of a diet rich in polyunsaturated and monounsaturated fatty acids [29]. In contrast, Lombardo and Chicco relate a greater weight gain and obesity in rats fed on a diet with polyunsaturated fatty acids when compared to saturated fatty acids [30]. However, since the analysis of the type of fatty acids that compose the tucumã (*Astrocaryum aculeatum* Meyer) was not performed, it cannot be asserted exactly which one was responsible for the body mass increase in the STG.

Therefore, the lesser tendency to body mass gain of ECG and ETG (Figure 3) is probably related to the performance of light physical exercise, such as swimming, and not to the intake of tucumã, since it is already established in the literature that aerobic exercises lead to increased mitochondrial function, favoring fatty acid oxidation. Furthermore, these exercises facilitate fat mobilization, stimulating the action of the hormone-sensitive lipase [29].

In this study, as in those by Curi et al. and Morais et al., hyperglycemia was noticed (Table 4) in all groups, even in those subjected to physical activity [31, 32]. Since all animals showed an increase in body mass, a tendency to develop obesity and probably a metabolic alteration (i.e., resistance to insulin) could be established.

It is known that when physical exercise is performed, the available fat is used preferentially for energy production (oxidation), decreasing glycogen degradation and glucose oxidation [29]. When the amount of fat increases, the organism can develop resistance to the action of insulin, thus, favoring an increase in the concentration of glucose in plasma. Previous research has shown that obesity, with the accumulation of visceral fat, is one of the main drivers for the development of this resistance. These alterations were detected in our study, allowing us to establish an association between these factors and the development of hyperglycemia in the experimental groups [33–35].

However, insulin and leptin (a hormone that participates in energy balance regulation by reducing food ingestion and increasing energy expenditure) both remained unaltered in relation to plasma concentration (Table 4). This could be due to the short time interval used for this experiment. We believe it was not enough to promote that alteration. When obesity develops, a decreased sensitivity to leptin occurs in parallel leading to a lesser activity, which may justify the body mass increase and reduced activity [35, 36].

The results obtained for protein in plasma (Table 4) did not express alteration of basal values when compared to studies performed to evaluate biochemical parameters of rats of the same strain [37]. As to the lipidogram, it can be seen that STG and ETG, both fed with tucumã, showed a relatively higher increase in total cholesterol when compared with SCG and ECG (Table 2).

The absence of hypocholesterolemic effect was also found in the study by Rocha et al. with the use of fruta-de-lobo (*Solanum lycocarpum* A. St.-Hit) that, even after 42 days of experimentation, did not reduce total cholesterol levels [26]. However, other studies demonstrate the inhibitory effect of the HMG CoA reductase enzyme derived from the consumption of vegetables like eggplant (*Solanum melongena*), (*Solanum gilo*), onion (*Allium cepa*), and carrot (*Daucus carota*) [38]. Other fruits have been found to promote the reduction of triglycerides and total cholesterol through other metabolic pathways, due to their physicochemical characteristics, such as the camucamu (*Myrciaria dubia*) [39]. Therefore, we can state that, due to its physicochemical characteristics, the ingestion of tucumã in high concentrations promotes metabolic alterations favorable to the development of dyslipidemia. So, we expected that given the fatty acid composition of the Pará tucumã, the Amazonas tucumã, which is a different species of the same genus, could also reduce lipids in plasma. However, this assumption was not confirmed, which allows us to suggest that the Amazonas tucumã has a different fatty acid composition that can lead to the development of obesity, resistance to insulin, and dyslipidemia. Therefore, nutrient unbalance in the diet is associated with metabolic alterations, such as overweight, hyperlipidemia with high levels of triglycerides in plasma, low levels of HDL, resistance to insulin, and retardation in glucose clearance, probably due to an increase of free circulating fatty acids, related to a reduction in life expectancy [40–42].

Some of these alterations, like body weight increase (Figure 3), leading to the development of obesity, as well as hyperlipidemia (Table 3) and hyperglycemia (Table 4), were found in this experiment, mainly in sedentary groups (SCG and STG), but markedly in the sedentary group fed with tucumã (STG).

5. Conclusion

An adequate diet in combination with changes in lifestyle can promote positive alterations in lipid metabolism, as well as the balance of organic functions and consequent reduction of risk factors for the development of hypercholesterolemia and cardiovascular diseases. This study showed that the tucumã fruit (*Astrocaryum aculeatum* Meyer) does not have a hypolipidemic effect, since animals presented high lipid concentrations. Furthermore, it is apparent that low intensity aerobic physical activity (i.e., swimming) exerts a positive effect in the lipid profile of Wistar rats, since it was observed that exercised animals' showed reduction in some biochemical parameters. However, we conclude that the duration of the exercise was not sufficient, since it promoted slightly significant alterations.

Disclosure

All authors have read and approved the final paper.

Conflict of Interests

The authors declare that they do not have conflict of interests regarding the publication of this paper.

Acknowledgments

Research was supported by the RCMI Program G12MD007579 (JBM) and the MBRS-RISE Program funded by RISE Grant R25GM082406 from the National Institutes of Health (CO) at Ponce School of Medicine and Health Sciences and Universidade Federal do Amazonas.

References

[1] R. D. Almeida, E. S. Prado, L. A. Melo, and A. C. C. Oliveira, "Lipoproteína(A) e massa corporal de ratos submetidos à hipercolesteronemia e treinamentos físicos de força e aeróbico," *Fitness & Performance Journal*, vol. 7, no. 3, pp. 137–144, 2008.

[2] H. M. Magalhaes Junior, "Aprova o Protocolo Clínico e Diretrizes Terapêuticas da Dislipidemia para a prevenção de eventos cardiovasculares e pancreatite," in *Ministério da Saúde*, 2013.

[3] A. C. Sposito, B. Caramelli, F. A. H. Fonseca et al., "IV Brazilian guideline to dyslipidemias and prevention of atherosclerosis: department of atherosclerosis of the Brazilian society of cardiology," *Arquivos Brasileiros de Cardiologia*, vol. 88, supplement 1, pp. 2–19, 2007.

[4] D. C. Malta, O. L. Morais Neto, and J. B. Silva Junior, "Apresentação do plano de ações estratégicas para o enfrentamento das doenças crônicas não transmissíveis no Brasil, 2011 a 2022," *Epidemiologia e Servicios de Saúde*, vol. 20, no. 4, pp. 425–438, 2011.

[5] M. I. Schmidt, B. B. Duncan, G. A. E Silva et al., "Chronic noncommunicable diseases in Brazil: burden and current challenges," *The Lancet*, vol. 377, no. 9781, pp. 1949–1961, 2011.

[6] A. D. Gepner, M. E. Piper, H. M. Johnson, M. C. Fiore, T. B. Baker, and J. H. Stein, "Effects of smoking and smoking cessation on lipids and lipoproteins: outcomes from a randomized clinical trial," *The American Heart Journal*, vol. 161, no. 1, pp. 145–151, 2011.

[7] D. C. Malta, A. C. Cezário, L. Moura, O. L. Morais Neto, and J. B. Silva Jr., "A construção da vigilância e prevenção das doenças crônicas não transmissíveis no contexto do Sistema Único de Saúde," *Epidemiologia e Servicios de Saúde*, vol. 15, no. 1, pp. 47–65, 2006.

[8] A. N. M. Neto, M. O. S. Simões, A. C. D. Medeiros, A. S. Portela, and A. M. Souza, "Obesidade, envelhecimento e risco cardiovascular no Brasil: possíveis soluções para problemas atuais," *Revista Saúde.com*, vol. 4, no. 1, pp. 57–63, 2008.

[9] M. H. Teixeira, G. V. da Veiga, and R. Sichieri, "Fat intake and hypercholesterolemia in a probabilistic sample of adolescents from Niterói, Rio de Janeiro," *Arquivos Brasileiros de Endocrinologia e Metabologia*, vol. 51, no. 1, pp. 65–71, 2007.

[10] L. Tuneu, M. A. Gastelurrutia, and F. Fernández-Llimós, *Guía de Seguimiento Farmacoterapéutico sobre Dislipemias*, Universidad de Granada, 2003.

[11] D. Soares, G. Marthendal, M. C. Zimmerman, and A. L. B. Zeni, "Estudo dos níveis lipídicos em ratos após tratamento com infusão de algumas plantas medicinais de uso popular," *Revista Brasileira de Farmácia*, vol. 86, pp. 71–74, 2005.

[12] J. L. Durstine, "Effect of aerobic exercise on high-density lipoprotein cholesterol: a meta-analysis," *Clinical Journal of Sport Medicine*, vol. 18, no. 1, pp. 107–108, 2008.

[13] M. O. Weickert and A. F. H. Pfeiffer, "Metabolic effects of dietary fiber consumption and prevention of diabetes," *Journal of Nutrition*, vol. 138, no. 3, pp. 439–442, 2008.

[14] L. K. O. Yuyama, R. N. Maeda, L. Pantoja, J. P. L. Aguilar, and H. A. Marinho, "Processamento e avaliação da vida-de-prateleira do tucumã (Astrocaryum aculeatum Meyer) desidratado e pulverizado," *Ciência e Tecnologia de Alimentos*, vol. 28, no. 2, 2008.

[15] R. R. Lima, L. C. Trassato, and V. Coelho, *O Tucumã (Astrocaryum vulgare Mart.): Principais Características e Potencialidade Agroindustrial*, vol. 75, EMBRAPA-CPATU, 1983.

[16] J. E. M. Ferrão, *Fruticultura Tropical: Espécies Com Frutos Comestíveis*, Lisboa, Portugal, 1999.

[17] J. P. L. Aguiar, H. A. Marinho, Y. S. Rebelo, and R. Shrimpton, "Aspectos nutritivos de alguns frutos da Amazonia," *Acta Amazonica*, vol. 10, no. 4, pp. 755–758, 1980.

[18] E. S. Ferreira, V. G. Lucien, A. S. Amaral, and C. S. Silveira, "Caracterização físico-química do fruto e do óleo extraído de tucumã (Astrocaryum vulgare Mart)," *Alimentos e Nutrição Araraquara*, vol. 19, no. 4, pp. 427–433, 2008.

[19] A. M. M. Guedes, *Estudo da Extração de Óleo da Polpa de Tucumã por CO_2 Supercrítico*, Universidade Federal do Pará, Pará, Brazil, 2006.

[20] A. M. Leitão, *Caracterização Morfológica e Físico-Química de Frutos e Sementes de Astrocaryum aculeatum Meyer (Arecaceae), de uma Floresta Secundária*, Universidade Federal do Amazonas, Amazonas, Brazil, 2008.

[21] M. J. M. Rodrigues, *Exercícios de natação previnem alterações cardiovasculares, bioquímicas e histológicas induzidas por dieta em ratos Wistar*, Universidade Estadual de Campinas, São Paulo, Brazil, 2009.

[22] L. Zambon, F. O. Duarte, L. F. Freitas et al., "Efeitos de dois tipos de treinamento de natação sobre a adiposidade e o perfil lipídico de ratos obesos exógenos," *Revista de Nutrição*, vol. 22, no. 5, pp. 707–715, 2009.

[23] M. R. Eufrásio, M. F. P. Barcelos, R. V. Sousa, W. C. Abreu, M. A. C. Lima, and M. C. A. Pereira, "Efeito de diferentes tipos de fibras sobre frações lipídicas do sangue e fígado de ratos wistar," *Ciência e Agrotecnologia*, vol. 33, no. 6, 2009.

[24] P. H. M. Souza, M. H. Souza Neto, and G. A. Maia, "Componentes funcionais nos alimentos," in *Boletim da SBCTA*, vol. 37, pp. 127–135, 2003.

[25] C. V. B. Pimentel, V. M. Francki, and A. P. B. Gollucke, *Alimentos Funcionais—Introdução Às Principais Substâncias Bioativa em Alimentos*, Varela, São Paulo, Brazil, 2005.

[26] D. A. Rocha, C. M. P. Abreu, R. V. Sousa, A. D. Correa, and E. W. N. Fonseca, "Avaliação preliminar *in vivo* do efeito hipocolesterolêmico do polvilho da fruta-de-lobo (*Solanum lycocarpum* A. St.-Hil.)," *Revista Brasileira de Plantas Medicinais*, vol. 14, no. 2, pp. 362–369, 2012.

[27] R. R. Bell, M. J. Spencer, and J. L. Sherriff, "Voluntary exercise and monounsaturated canola oil reduce fat gain in mice fed diets high in fat," *Journal of Nutrition*, vol. 127, no. 10, pp. 2006–2010, 1997.

[28] M. Pellizzon, A. Buison, F. Ordiz Jr., L. Santa Ana, and K. L. Jen, "Effects of dietary fatty acids and exercise on body-weight regulation and metabolism in rats," *Obesity Research*, vol. 10, no. 9, pp. 947–955, 2002.

[29] L. D. P. Franco, J. A. D. B. Campos, and A. Demonte, "Teor lipídico da dieta, lipídios séricos e peso corporal em ratos exercitados," *Revista de Nutrição*, vol. 22, no. 3, pp. 359–366, 2009.

[30] Y. B. Lombardo and A. G. Chicco, "Effects of dietary polyunsaturated n-3 fatty acids on dyslipidemia and insulin resistance in rodents and humans. A review," *Journal of Nutritional Biochemistry*, vol. 17, no. 1, pp. 1–13, 2006.

[31] R. Curi, C. J. Lagranha, J. R. J. G. Rodriguez et al., "Ciclo de Krebs como fator limitante na utilização de ácidos graxos durante o exercício aeróbico," *Arquivos Brasileiros de Endocrinologia & Metabologia*, vol. 47, no. 2, 2003.

[32] C. S. N. Morais, M. F. P. Barcelos, R. V. Sousa, H. M. Lima, and A. L. Lima, "Efeitos das fontes e níveis de lipídios nas dietas de ratos machos da linhagem wistar (rattus norvegicus) sobre frações lipídicas do sangue," *Ciência e Agrotecnologia*, vol. 27, no. 5, pp. 1082–1088, 2003.

[33] E. P. Weiss, J. Brandauer, O. Kulaputana et al., "FABP2 Ala54Thr genotype is associated with glucoregulatory function and lipid

oxidation after a high-fat meal in sedentary nondiabetic men and women," *The American Journal of Clinical Nutrition*, vol. 85, no. 1, pp. 102–108, 2007.

[34] L. O. P. Lancha, *Obesidade e resistência à ação da insulina: alterações moleculares, bioquímicas e estruturais*, Universidade de São Paulo, São Paulo, Brazil, 2009.

[35] A. J. Borba, *Efeito da Dieta Hiperlipídico-Protéica No Metabolismo de Ratos Wistar Adultos*, Universidade Federal do Triângulo Mineiro, Uberaba, Brazil, 2008.

[36] R. Eguchi, N. C. Cheik, L. M. Oyama et al., "Efeitos do Exercício Crônico Sobre a Concentração Circulante da Leptina e Grelina em Ratos com Obesidade Induzida por Dieta," *Revista Brasileira de Medicina do Esporte*, vol. 14, no. 3, pp. 182–187, 2008.

[37] G. D. Carvalho, A. P. B. Masseno, S. F. Zanini, L. C. Porfirio, J. P. Machado, and H. Mauad, "Clinical evaluation of laboratory rats (Rattus novergicus Wistar Strain): sanitary, biological and physiological parameters," *Revista Ceres*, vol. 56, no. 1, pp. 51–57, 2009.

[38] M. C. R. Gonçalves, M. F. F. M. Diniz, J. D. C. Borba, X. P. Nunes, and J. M. Barbosa-Filho, "Eggplant (*Solanum melongena* L.)—myth or reality in the fight against the dislipidemy?" *Revista Brasileira de Farmacognosia*, vol. 16, no. 2, pp. 252–257, 2006.

[39] M. C. Schwertz, J. R. P. Maia, R. F. S. Sousa, J. P. L. Aguiar, L. K. O. Yuyama, and E. S. Lima, "Efeito hipolipidêmico do suco de camu-camu em ratos," *Revista de Nutrição*, vol. 25, no. 1, 2012.

[40] J. S. Flier, "Obesity wars: molecular progress confronts an expanding epidemic," *Cell*, vol. 116, no. 2, pp. 337–350, 2004.

[41] E. Oda, "The metabolic syndrome as a concept of adipose tissue disease," *Hypertension Research*, vol. 31, no. 7, pp. 1283–1291, 2008.

[42] A. J. Borba, M. G. M. Rocha, M. F. Sivla et al., "Dieta hiperlipídico-proteica utilizada para emagrecimento induz obesidade em ratos," *Revista de Nutrição*, vol. 24, no. 4, 2011.

3

Evaluation of Beneficial Metabolic Effects of Berries in High-Fat Fed C57BL/6J Mice

Lovisa Heyman,[1] Ulrika Axling,[1] Narda Blanco,[2] Olov Sterner,[2] Cecilia Holm,[1] and Karin Berger[1]

[1] *Department of Experimental Medical Science, Lund University, Lund, Sweden*
[2] *Centre for Analysis and Synthesis Organic Chemistry, Department of Chemistry, Lund University, P.O. Box 124, 22100 Lund, Sweden*

Correspondence should be addressed to Lovisa Heyman; lovisa.heyman@med.lu.se

Academic Editor: Duo Li

Objective. The aim of the study was to screen eight species of berries for their ability to prevent obesity and metabolic abnormalities associated with type 2 diabetes. *Methods*. C57BL/6J mice were assigned the following diets for 13 weeks: low-fat diet, high-fat diet or high-fat diet supplemented (20%) with lingonberry, blackcurrant, bilberry, raspberry, açai, crowberry, prune or blackberry. *Results*. The groups receiving a high-fat diet supplemented with lingonberries, blackcurrants, raspberries or bilberries gained less weight and had lower fasting insulin levels than the control group receiving high-fat diet without berries. Lingonberries, and also blackcurrants and bilberries, significantly decreased body fat content, hepatic lipid accumulation, and plasma levels of the inflammatory marker PAI-1, as well as mediated positive effects on glucose homeostasis. The group receiving açai displayed increased weight gain and developed large, steatotic livers. Quercetin glycosides were detected in the lingonberry and the blackcurrant diets. *Conclusion*. Lingonberries were shown to fully or partially prevent the detrimental metabolic effects induced by high-fat diet. Blackcurrants and bilberries had similar properties, but to a lower degree. We propose that the beneficial metabolic effects of lingonberries could be useful in preventing obesity and related disorders.

1. Introduction

During the last decades, the prevalence of obesity and type 2 diabetes mellitus has increased dramatically. This epidemic of lifestyle-related disorders is affecting all parts of the world, and 439 million people are estimated to suffer from diabetes mellitus in 2030 [1]. Obesity is a strong risk factor for type 2 diabetes with 90% of affected patients being overweight or obese. Obesity is also associated with increased risk of various metabolic disorders including insulin resistance, chronic low-grade inflammation, dyslipidemia, nonalcoholic fatty liver disease (NAFLD), and cardiovascular disease. Oxidative stress, inflammatory response, and altered gut microbiota can play a significant role in the development of obesity-related disorders [2–4]. Type 2 diabetes is a multifactorial disease; however, it appears clear that prevention is possible by avoiding overeating and a sedentary lifestyle to maintain a healthy body weight [5]. The difficulty for many individuals to comply with dietary and lifestyle changes makes it of great interest to identify new foods with well-established effects on preventing the development of obesity and thereby type 2 diabetes and its associated metabolic complications.

Dietary patterns with high consumption of polyphenol rich foods (fruits, vegetables, and berries) have been associated with reduced risk of type 2 diabetes. In general, berries are rich in polyphenols which are suggested to play a role in health benefits of plant-based diets [4, 6–8]. Plant phenolics are a large group of secondary metabolites which provide color and taste in fruits and berries and include flavonoids (anthocyanins, flavonols, and flavanols), tannins, stilbenoids, and phenolic acids [9]. The antioxidant effect of berry anthocyanins has been studied extensively, but still little is known about the biological activities linking berries and polyphenols to the prevention of type 2 diabetes [4, 10, 11].

The aim of the present study was to perform a screening to investigate and compare metabolic effects of different berries in the C57BL/6J mouse. The C57BL/6J is a mouse strain that develops obesity and prediabetes when fed a diet rich in fat, thus mimicking detrimental effects of an energy dense western diet [12]. By supplementing high-fat diets with potentially health-promoting berries, we sought to identify berries capable of ameliorating risk factors from excessive energy intake. Here we report that different berries, possibly due to their polyphenol composition, have varying ability to prevent weight gain and metabolic disorders ultimately leading to diabetes.

2. Methods

2.1. Animals and Study Design. The study was approved by the Animal Ethics Committee in Lund, Sweden, (Permit Number: M185-11) and is in accordance with the Council of Europe Convention (ETS 123). Male C57BL/6JBomTac mice, 6 weeks old, 21.2 ± 1.1 g were obtained from Taconic (Skensved, Denmark). The animals were housed in a controlled environment (12 h light/dark cycle, light on 7 a.m.). After 9 days of acclimatization the mice were divided into 10 groups of 12 mice each housed in groups of 6 mice per cage. The mice were fed high-fat diets (45 kcal% fat) (Research diets, New Brunswick, NJ, USA) supplemented (20% w/w) with eight different freeze dried berries *ad libitum* for 13 weeks. A control group was fed a macronutrient-matched, isocaloric diet without supplements. One group received a low-fat diet (10 kcal% fat) as an internal control to the high-fat diet induced obesity. Body weight and food intake were monitored weekly throughout the study period. The energy intake was calculated based on registered food consumption. Mice were housed with minimal bedding material and feces was quantitatively collected for two consecutive days at the end of the study and stored at −20°C prior to lyophilization, weighing and powdering with a mortar. At the end of the study, 4 h-fasted animals were anesthetized with an intraperitoneal injection of midazolam (Midazolam Panpharma 5 mg/mL, Panpharma S.A., Luitré, France) and a mixture of fluanisone 10 mg/mL and fentanyl citrate 0.315 mg/mL (Hypnorm, VetaPharma, Leeds, UK). Body composition was determined with dual-energy X-ray absorptiometry (DEXA) technique using a Lunar PIXImus (GE Lunar, Madison, WI, USA). Blood samples were taken by intraorbital puncture. The animals were sacrificed by cervical dislocation and liver, cecum, spleen, and epididymal fat pads were excised, weighed, and snap frozen in liquid nitrogen.

2.2. Preparation and Analysis of Diets. The experimental diets were high-fat diets supplemented with one of eight freeze dried berries; Lingonberry (*Vaccinium vitis-idaea*), blackcurrant (*Ribes nigrum*), raspberry (*Rubus idaeus*), bilberry (*Vaccinium myrtillus*), and blackberry (*Rubus fruticosus*) were obtained from MOLDA AG (Dahlenburg, Germany). Crowberries (*Empetrum nigrum*) were from Olle Svenssons Partiaffär AB (Olofström, Sweden) and prunes (*Prunus domestica*) from Semper AB (Sundbyberg, Sweden). Freeze dried açai berry powder (*Euterpe oleracea*) was purchased from Superfruit Scandinavia AB (Sweden). Information about origin and processing of the berries can be found in Table S2 (see Supplementary Material available online at http://dx.doi.org/10.1155/2014/403041). Based on data of macronutrient composition of the berries (obtained from supplier and/or Covance, Madison, WI, USA) the diets were designed to have identical macronutrient composition (Tables 1 and S1, Supporting information). After manufacturing, all diets were analyzed for soluble and insoluble dietary fiber content by Eurofins (Lidköping, Sweden) (Table 1).

2.3. Analysis of Plasma Samples and Assessment of Insulin Resistance. Plasma was prepared by immediate centrifugation of blood samples. Glucose, total cholesterol, triacylglycerol, high-density lipoprotein (HDL) cholesterol, alanine aminotransferase (ALT) (Infinity, Thermo Fisher Scientific, Waltham, MA, USA), and nonesterified fatty acid (NEFA) (NEFA-HR, Wako Chemicals, Neuss, Germany) concentrations in plasma were measured using kits. Low-density lipoprotein (LDL) cholesterol was estimated by the Friedewald formula [13]. Insulin was measured using an enzyme-linked immunosorbent assay kit (Mercodia, Uppsala, Sweden). Plasma levels of tumor necrosis factor-alpha (TNF-α) and plasminogen activator inhibitor-1 (PAI-1) were analyzed using Luminex technology (LX200, Luminex Corporation, Austin, TX, USA). Insulin resistance was assessed by the homeostasis model assessment (HOMA), a mathematical model describing the degree of insulin resistance from fasting plasma glucose and insulin [14, 15]. Homeostasis model assessment-estimated insulin resistance (HOMA-IR) was calculated by multiplying fasting plasma insulin (mU/L) with fasting plasma glucose (mmol/L) divided by 22.5.

2.4. Fecal Analyses. Lipids were extracted from feces using a modified version of the protocol by Hara and Radin [16]. In short, around 100 mg lyophilized, grounded feces from each sample were extracted in hexane-isopropanol (3 : 2 v/v) with 0,005% 2,6-di-tert-butyl-4-methylphenol. Five mL of the extract was dried under N_2 after which the remaining residue was redissolved in 100 μL isopropanol containing 1% Triton X-100. The solution was analyzed in triplicates using the triacylglycerol and cholesterol kits used for plasma samples.

2.5. Extraction and Quantification of Liver Lipids. The frozen livers were grounded to a powder in a mortar under liquid nitrogen. Samples were prepared in duplicates from ten randomly selected livers from each group and subjected to lipid extraction as previously described. For extraction of hepatic lipids, 20 mg of frozen liver powder was used. The solution containing redissolved lipids was first analyzed in triplicates for cholesterol. After adding another 90 μL isopropanol 1% Triton X-100 the solution was analyzed for triacylglycerol.

2.6. Extraction of Polyphenols from Diets. Five g of each diet was grounded, weighed, and extracted using 25 mL of heptane, ethyl acetate, and methanol, respectively. Samples were stirred during 24 h at room temperature and the extracts were filtered, concentrated, and weighed. Finally, 25 mL of methanol : water : acetic acid (85 : 15 : 0.5; v/v) was used for

Table 1: Composition of diets[1].

	Lingonberry	Blackcurrant	Bilberry	Raspberry	Açai	Crowberry	Prune	Blackberry	Control	LFD
Calculated energy (kcal)										
Protein	812.0	812.0	812.0	812.0	812.0	812.0	812.0	812.0	812.0	812.0
Carbohydrate	1422.4	1422.4	1422.4	1422.4	1422.4	1422.4	1422.4	1422.4	1422.4	2840.4
Starch	731.2	731.2	731.2	731.2	731.2	731.2	731.2	731.2	731.2	2149.2
Sucrose	347.2	347.2	347.2	347.2	347.2	347.2	347.2	347.2	347.2	347.2
Fructose	172.0	172.0	172.0	172.0	172.0	172.0	172.0	172.0	172.0	172.0
Glucose	172.0	172.0	172.0	172.0	172.0	172.0	172.0	172.0	172.0	172.0
Fat	1822.5	1822.5	1822.5	1822.5	1822.5	1822.5	1822.5	1822.5	1822.5	405.0
Fiber	0.0	0.0	0.0	0.0	0.0	0.0	0.0	0.0	0.0	0.0
Other	0.0	0.0	0.0	0.0	0.0	0.0	0.0	0.0	0.0	0.0
Total kcals	4057	4057	4057	4057	4057	4057	4057	4057	4057	4057
Calculated energy per gram diet (kcal/g)										
kcal/g	4.2	4.1	4.3	4.3	4.2	4.2	4.6	4.5	4.5	3.7
Calculated energy (kcal%)										
Protein	20	20	20	20	20	20	20	20	20	20
Carbohydrate	35	35	35	35	35	35	35	35	35	70
Fat	45	45	45	45	45	45	45	45	45	10
Fiber	0	0	0	0	0	0	0	0	0	0
Analyzed fiber[2] (g/100 g diet)										
Insoluble fiber	7.4	8.2	9.3	10.8	8.8	12.5	6.3	11.9	10.1	8.5
Soluble fiber	1.8	2.5	<1	1.2	1.6	<1	1.1	<1	<1	<1
Total fiber	9.2	10.7	10.0	12.0	10.4	13.0	7.4	12.6	10.5	9.1

[1]All diets were designed to have an equal caloric content of fat, protein, and carbohydrates (including glucose, fructose, and sucrose). LFD: low-fat diet.
[2]Fiber analyzed by Eurofins, Sweden.

polyphenol analysis. The extraction was carried out using 15 min of sonication. Ethyl acetate extracts were dissolved in 2 mL methanol and the soluble fraction was filtered through 0.2 μm GH polypro membrane to remove insoluble particles and kept at −18°C until analysis.

2.7. LC-MS Analysis of Polar Polyphenols (Anthocyanins). The liquid chromatography-tandem mass spectrometry (LC-MS) analyses were performed on a LC-MS Agilent technologies 1260 infinity equipped with an quaternary pump, autosampler, thermostatted column compartment, diode array detector tandem quadrupole LC/MS. A 250 × 4.6 mm i.d. Zorbax SB-C_{18} column, 3.5 μm particle size, was used at 40°C. The method used is described by Wu et al. [17] using mobile phase A (5% formic acid in water) and B (methanol). The flow rate was 1 mL/min, the temperature used was 40°C, and UV detection was at 520 nm. The gradient used was 5% B, 0–2 min; 5–20% B, 2–10 min; 20% B, 10–15 min; 20–30% B, 15–30 min; 30% B, 30–35 min; 30–45% B, 35–50 min; 45% B, 50–55 min, 45–5% B 55–65 min, 5% B. The injection volume was 20 μL in all samples. Enhanced product ion mass spectrometry (EPI-MS) analysis was performed in positive mode using a capillary voltage of 3000 V, nebulizer pressure 40 psig, drying gas flow 12 L/min, and drying gas temperature 300°C. The UV-vis, reference times, and mass spectra were used for identification of the peaks and compared to anthocyanin data in the literature [17–19].

2.8. LC-MS Analysis of Medium Polar Polyphenols. The LC-MS system together with a 2.1 × 100 mm i.d. 3 μm Atlantis C18 column was used for the analysis of the mobile phase A (0.1% formic acid in water) and mobile phase B (methanol). The flow rate was 0.3 mL/min, the temperature used was 30°C, and the UV detector was fixed at 360 nm. The injection volume was 20 μL. The initial gradient elution was used at 6–12% B, 20 min; 12–55% B, 20–50 min. The conditions for MS were set in both positive and negative mode (method described in [20]). Qualitative standards (resveratrol, apigenin, and quercetin-3-O-glucoside) as well as comparison of MS data in the literature were used to enable identification of compounds.

2.9. Statistical Analysis. Data are presented as mean ± standard error of the mean (SEM). Unless stated otherwise, results were analyzed by one-way analysis of variance (ANOVA) in conjunction with Dunnett's multiple comparisons test. In cases where Gaussian distribution could not be assumed, groups were compared using Kruskal-Wallis and Dunn's post test. All results are compared to the high-fat control group. Differences with a P value < 0.05 were considered significant. $^{*}P < 0.05$, $^{**}P < 0.01$ and $^{***}P < 0.001$. Statistical analyses were performed using GraphPad Prism versions 5.0 and 6.0 (GraphPad Software, San Diego, CA, USA).

3. Results

3.1. Energy Intake, Body Weight, and Body Fat Content. The high-fat control diet induced obesity in mice was compared to the low-fat diet group (Figure 1(a)). In week 12 of the experiment, the mean body weight in the low-fat group was 32±0.9 g, whereas the mean weight of the high-fat fed control group was 42 ± 1.2 g. Body weight was significantly lower in groups receiving lingonberry (33 ± 0.9 g), blackcurrant (36 ± 0.7 g), raspberry (37 ± 1.5 g), and bilberry (38 ± 1.1 g) compared to the high-fat control. Consumption of the açai diet resulted in significantly increased body weight (48 ± 0.6 g). The DEXA scan showed significantly decreased body fat content in the groups receiving lingonberry, blackcurrant, and bilberry compared to the high-fat control. In fact, mice fed the lingonberry-supplemented diet had a body fat content as low as the low-fat diet group. The lean body mass was similar in all groups (Figure 1(c)), except in the group receiving açai where lean mass was increased (+15%) compared to the control. The mass of the epididymal fat pads was expressed per gram lean tissue to take into account differences in body size. The relative size of the fat pad was lower in groups receiving lingonberry and açai, compared to the high-fat control (Figure 1(d)). The accumulated mean energy intake per body weight was similar for all groups, except for mice receiving blackcurrant and bilberry supplementation where a higher food and energy intake was registered (Figure S1).

3.2. Plasma Parameters and Insulin Resistance Index. Four-hour fasting plasma glucose levels were significantly reduced in groups receiving lingonberry- and blackcurrant-supplemented diets compared to the high-fat control diet (Figure 2(a)). These groups together with the bilberry and raspberry groups had reduced fasting insulin levels (Figure 2(b)). In addition, the lingonberry, blackcurrant, and bilberry supplementation resulted in a lower HOMA index of insulin resistance (Figure 2(c)). The lingonberry and blackcurrant groups had glucose, insulin, and insulin resistance levels very similar to the group receiving a low-fat diet. A tendency of increased HOMA-IR, glucose, and insulin was observed in mice consuming açai-supplemented high-fat diet compared to control; however, the increase was not significant (P value; 0.07, 0.25 and 0.55, resp.).

The plasma lipid profiles are shown in Table 2. Compared to control, the total plasma cholesterol was significantly lower in groups fed lingonberry, blackcurrant, and low-fat diet whereas it tended to be higher in the acai group ($P = 0.05$). The lingonberry, blackcurrant, and low-fat groups displayed decreased levels of LDL and HDL cholesterol whereas açai had increased HDL cholesterol. However, there were no significant changes regarding the calculated LDL/HDL ratio. Plasma triacylglycerol levels were significantly increased in the group receiving blackcurrant and in the low-fat control. There were no significant differences in circulating nonesterified fatty acids.

3.3. Liver Lipid Accumulation and Liver Function. The liver masses (relative to lean body mass) were significantly lower in mice fed a diet supplemented with lingonberries compared to control ($P = 0.009$), whereas açai supplementation led to significantly increased liver mass ($P < 0.0001$) (Figure 3(a)). The plasma levels of the enzyme ALT, a marker of liver

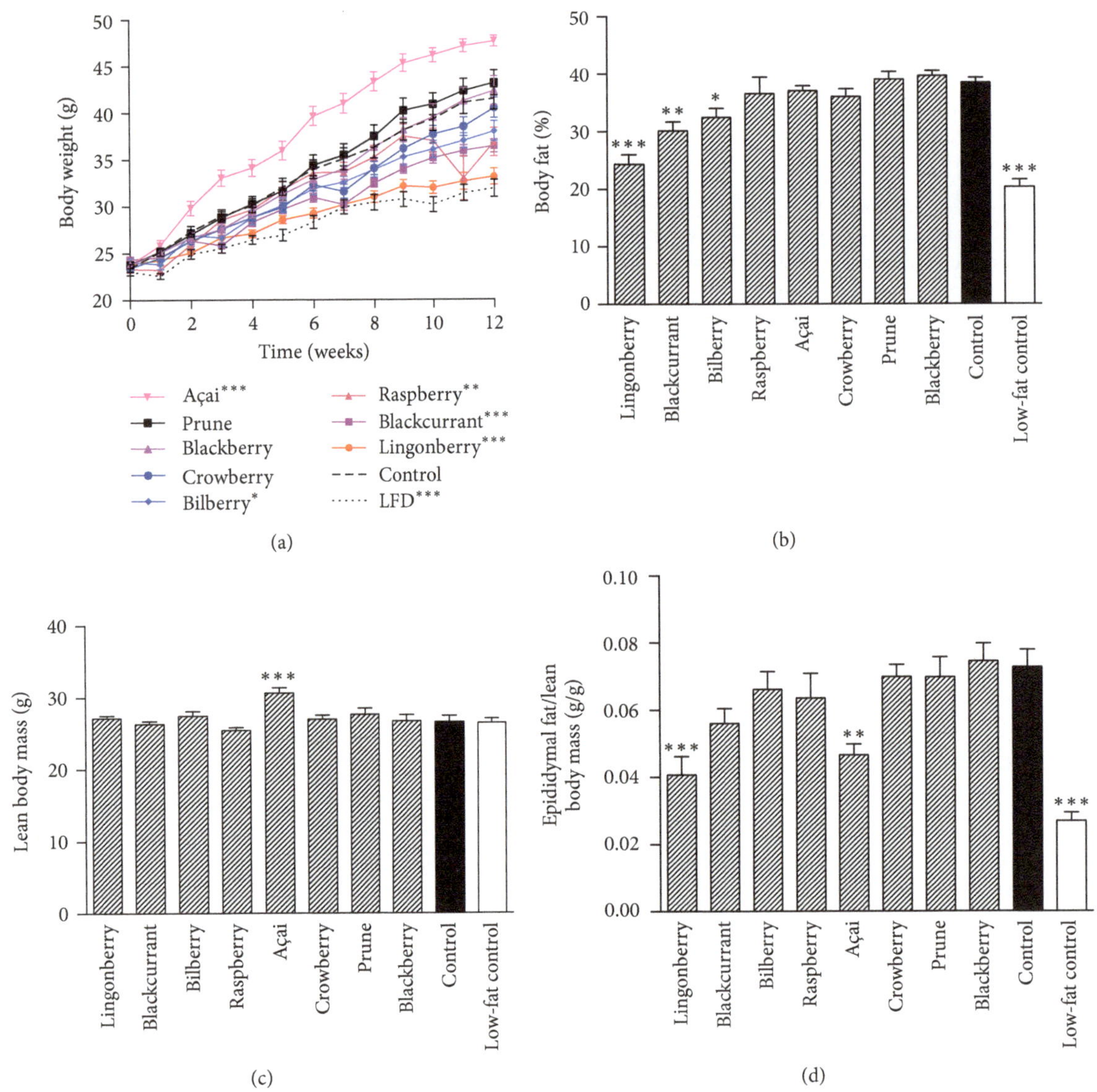

FIGURE 1: Body weight and composition after 13 weeks of high-fat diet and berry supplementation. (a) Weekly body weight registration. Statistical comparisons of body weight compared to control were made using a two-way ANOVA with Bonferroni post test. The stars represent significant differences in body weight at the last time point of weight registration before ending the study. Body fat (b) and lean body mass (c) were recorded using DEXAscan technique at week 13 of the study. (d) Epididymal white adipose tissue weight related to lean body mass after 13 weeks on the different diets ($n = 11$-12). Values represent mean ± SEM, $n = 12$ mice/group. Mean values significantly different from the control are denoted with $^{*}P < 0.05$, $^{**}P < 0.01$ or $^{***}P < 0.001$.

dysfunction, were significantly elevated in the açai group compared to all groups except the blackberry and control groups (Figure 3(b)). Compared to the high-fat control, ALT levels were significantly reduced in groups receiving lingonberry and blackcurrant as well as bilberry. The liver contents of triacylglycerol (Figure 3(c)) were markedly decreased in the mice receiving supplement of lingonberries, blackcurrant, and, to a lower degree, bilberries whereas açai induced an increase in liver triacylglycerol. Lingonberries, bilberries, and crowberries diets reduced liver cholesterol content compared to the high-fat control (Figure 3(d)). All the studied liver parameters were decreased in the low-fat diet compared to the high-fat control.

3.4. Effect of Berry Supplementation on Fecal Excretion and Cecal Weight. The total amount of feces collected over 24 h as well as fecal excretion of triacylglycerol and cholesterol is presented in Table 3. There were no significant differences in total amount of excreted feces, whereas fecal content of cholesterol was elevated in the bilberry, açai, crowberry, blackberry, and low-fat control group. Compared to control, the amount of excreted triacylglycerol was significantly higher in all groups except the low-fat control and the group receiving prune. The first part of the large intestine (cecum) is a site for bacterial fermentation. The mass of the cecum, including content, was increased in all groups compared to control except in the raspberry group (Table 3). In this study, bilberry

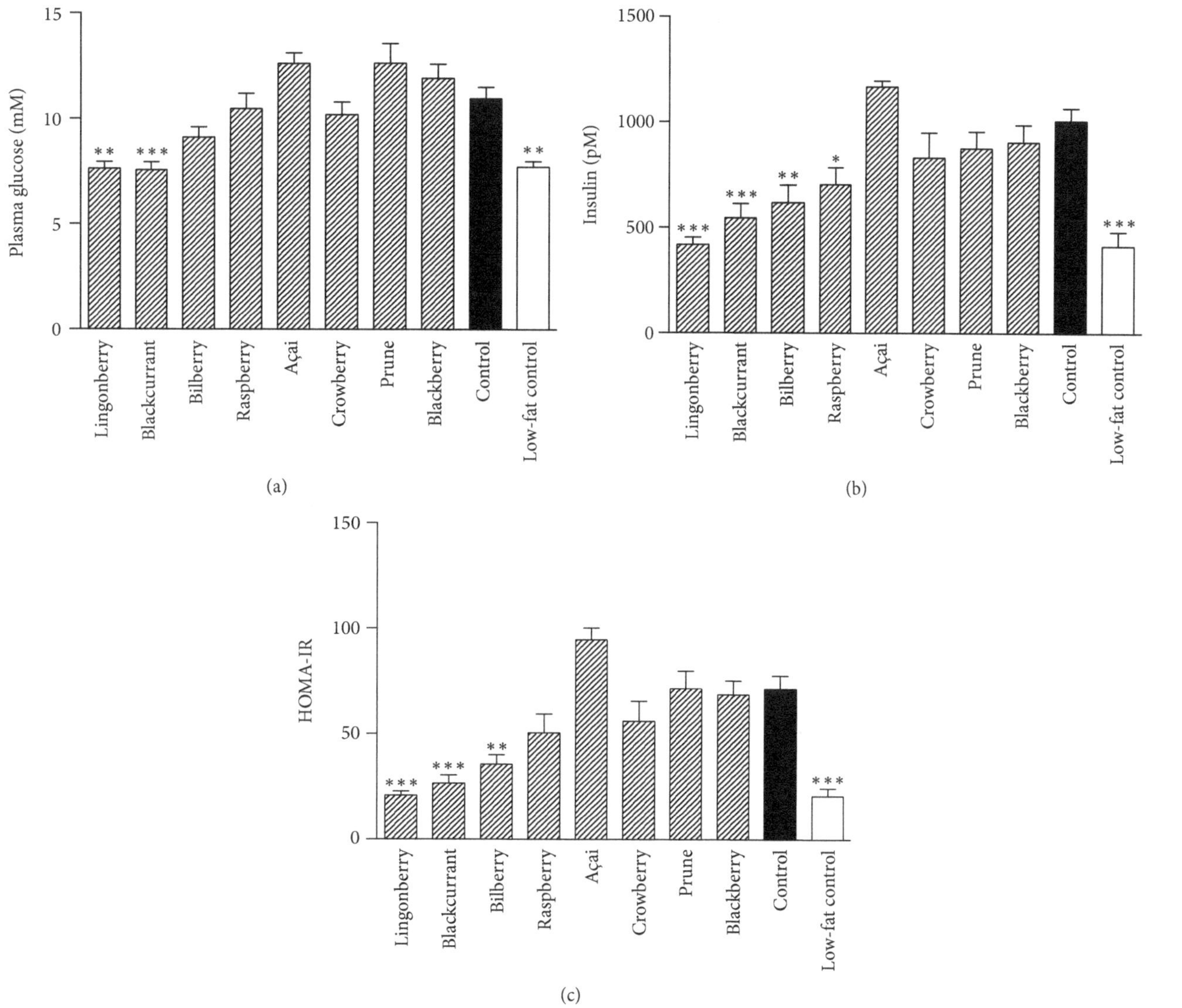

FIGURE 2: Glycemic control in mice after 13 weeks of high-fat diet and berry supplementation. Plasma glucose (a) and insulin (b) concentrations after 4-hour fasting were used to calculate HOMA-IR index (c). Results represent mean ± SEM, $n = 11$-12. Values significantly different from the control are denoted: $^{*}P < 0.05$, $^{**}P < 0.01$, $^{***}P < 0.001$.

TABLE 2: Plasma lipid profiles (mM) in mice after 13 weeks on high-fat diets supplemented with berries[1].

	Triacylglycerol		Total cholesterol		HDL cholesterol		LDL cholesterol		LDL/HDL ratio		NEFA	
	Mean	SEM	Mean	SEM	Mean	SEM	Mean	SEM	Mean	SEM	Mean	SEM
Lingonberry	0.65	0.06	2.6^{***}	0.11	1.2^{***}	0.04	1.0^{**}	0.06	0.84	0.04	0.56	0.04
Blackcurrant	0.83^{*}	0.04	2.9^{**}	0.14	1.4^{**}	0.06	1.1^{*}	0.09	0.76	0.06	0.66	0.03
Bilberry	0.59	0.06	3.4	0.15	1.7	0.06	1.5	0.08	0.87	0.03	0.57	0.06
Raspberry	0.66	0.03	3.6	0.24	1.8	0.13	1.5	0.14	0.87	0.05	0.54	0.04
Açai	0.59	0.03	4.3	0.13	2.1^{*}	0.08	1.9	0.07	0.90	0.04	0.49	0.03
Crowberry	0.56	0.04	4.1	0.11	2.0	0.05	1.8	0.08	0.91	0.04	0.54	0.05
Prune	0.57	0.03	4.1	0.15	2.0	0.06	1.8	0.15	0.94	0.08	0.45	0.02
Blackberry	0.67	0.02	4.0	0.17	1.9	0.08	1.8	0.13	0.92	0.06	0.60	0.04
Control	0.64	0.03	3.7	0.16	1.8	0.07	1.6	0.11	0.88	0.05	0.56	0.05
Low-fat control	0.84^{*}	0.09	2.4^{***}	0.15	1.1^{***}	0.06	0.85^{***}	0.13	0.76	0.12	0.67	0.05

[1]Mean values significantly different from control are depicted with $^{*}P < 0.05$, $^{**}P < 0.01$, respectively, $^{***}P < 0.001$. SEM: standard error of the mean. $n = 12$ mice from all groups except $n = 11$ in the lingonberry group in low density-lipoprotein LDL and low-/high-density lipoprotein LDL/HDL cholesterol. NEFA: nonesterified fatty acids.

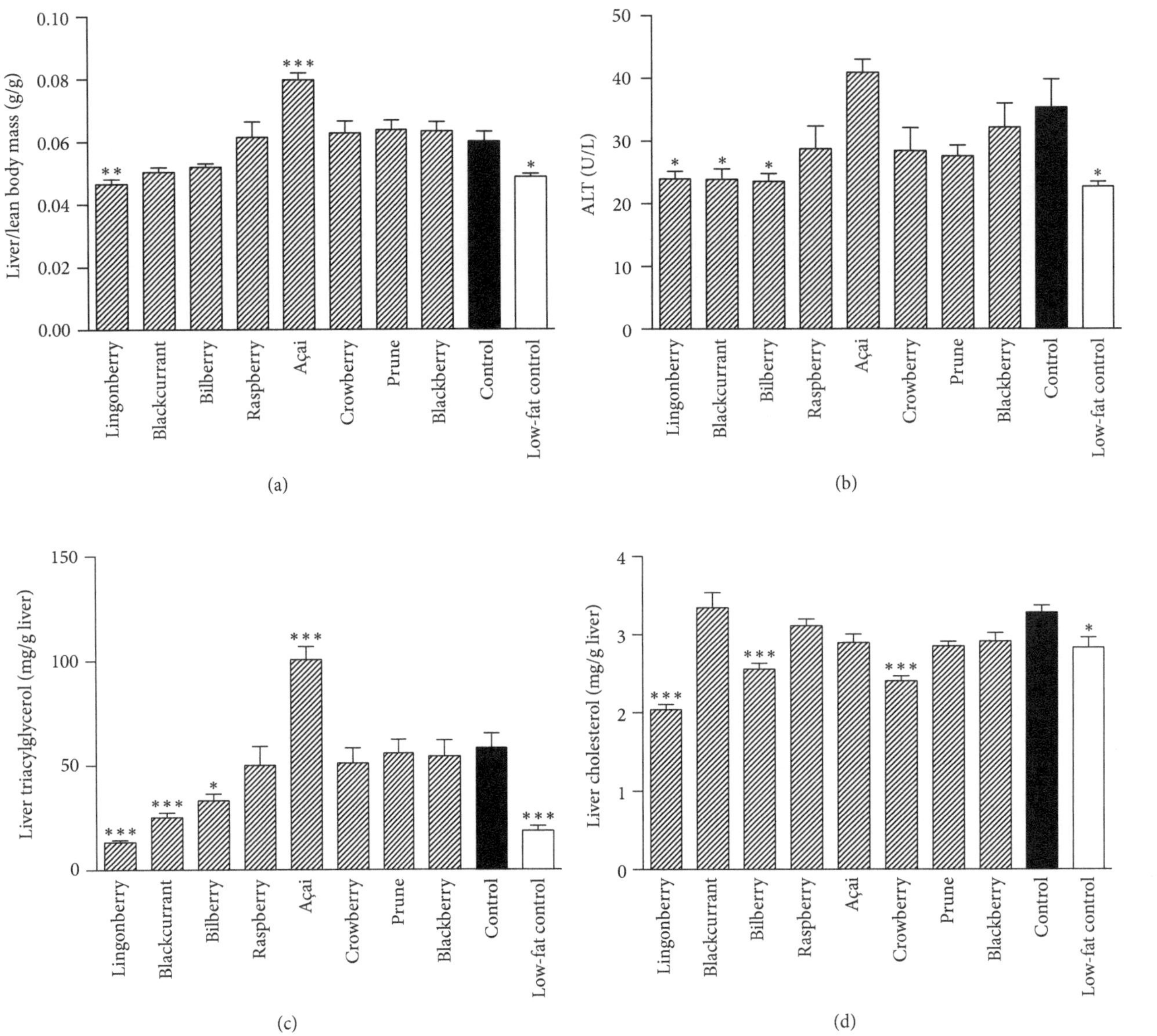

FIGURE 3: Effects of berry supplementation on liver size, function and lipid accumulation. (a) Liver weights, expressed as gram per gram lean body mass, after 13 weeks of high-fat diet supplemented with different berries. (b) Plasma alanine aminotransferase (ALT) concentration (units per liter). Liver triacylglycerol (c) and cholesterol (d) content, expressed as mg per g liver weight. Data are means ± SEM, $n = 10$–12. $^{*}P < 0.05$, $^{**}P < 0.01$, $^{***}P < 0.001$.

and lingonberry-supplemented diets gave rise to the largest cecum masses.

3.5. Effects of Berry Supplementation on Inflammation. In order to assess the effects of berry supplementation on low-grade inflammation, plasma levels of the inflammation markers PAI-1 (Figure 4(a)) and TNF-α were measured. The PAI-1 concentration was lower in plasma from mice receiving lingonberries, blackcurrants, bilberries, and low-fat diet compared to control. The PAI-1 levels in mice receiving açai were elevated compared to other groups, although not significantly compared to control. Plasma TNF-α was below the detection limit in all samples. Spleen size is sometimes analyzed as a reflection of inflammatory activity [21]. In this study, the spleen mass relative to lean tissue mass was significantly lower in the lingonberry group compared to high-fat control (Figure 4(b)).

3.6. Polyphenolic Compounds in Berry Diets. The polyphenols detected in the different berry diets are displayed in Table 4. Anthocyanins were present in extracts from all experimental diets, except the prune diet. However, medium polar polyphenols were only detected in the lingonberry and blackcurrant diets. Quercetin-3-O-glucoside and quercetin-3-O-galactoside were identified in both lingonberry and blackcurrant diets and in agreement with the earlier

TABLE 3: Effect of berry supplementation on cecum weight, total fecal excretion, and lipid excretion[1].

	Cecum weight (g)		Dry feces (g/c/24 h)		Fecal triacylglycerol (g/c/24h)		Fecal cholesterol (g/c/24 h)	
	Mean (n = 12)	SEM	Mean (n = 4†)	SD	Mean (n = 4†)	SD	Mean (n = 4†)	SD
Lingonberry	0.64***	0.028	2.44	0.15	6.22***	1.22	3.74	0.93
Blackcurrant	0.35**	0.022	2.12	0.45	10.68***	3.45	2.80	0.39
Bilberry	0.78***	0.035	2.86	0.25	15.18***	1.61	5.88***	1.21
Raspberry	0.29	0.013	2.85	0.40	7.89***	1.36	3.74	0.34
Açai	0.50***	0.031	3.18	0.55	5.57**	1.64	5.75***	1.03
Crowberry	0.42***	0.030	3.41	0.21	9.50***	1.83	7.52***	0.93
Prune	0.38***	0.023	2.25	0.12	0.96	0.26	3.91	0.53
Blackberry	0.35**	0.019	3.15	0.34	4.73**	1.16	2.33*	0.41
Control	0.21	0.012	2.85	0.24	1.40	0.55	3.60	0.21
Low-fat control	0.32*	0.036	2.68	0.30	0.77	0.14	2.26*	0.15

[1]Statistical comparisons of cecum weight were made using ANOVA and Dunnett's posttest, all groups compared to control, $^{*}P < 0.05$, $^{**}P < 0.01$, $^{***}P < 0.001$. Remaining statistical comparisons to control were made using repeated-measures two-way ANOVA and Sidak's posttest. SEM: standard error of the mean; SD: standard deviation. †The number refers to the number of observations (two cages (c) per group analyzed over 24 hours, two days in a row). Six mice were housed in each cage.

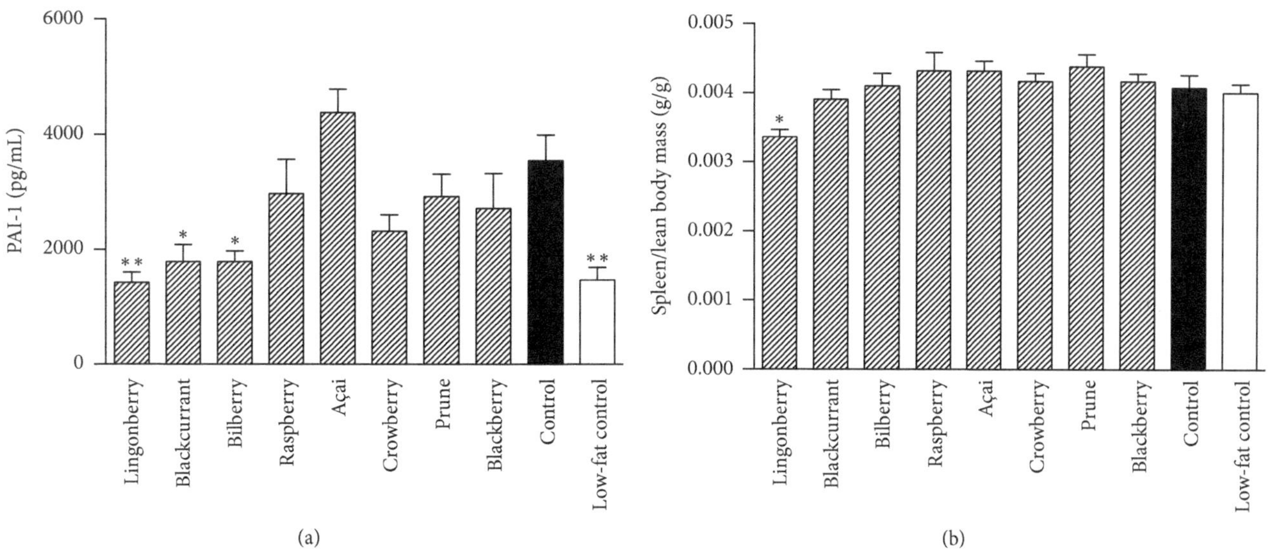

FIGURE 4: Effect of berry supplementation on inflammatory markers. (a) The plasma concentration of PAI-1 (plasminogen activator inhibitor 1), mirroring low-grade inflammation, was reduced in mice receiving lingonberries, blackcurrants, and bilberries compared to control. Spleen weight (b) related to lean body mass was reduced in the lingonberry group compared to the control. Data are means ± SEM, n = 10–12. $^{*}P < 0.05$, $^{**}P < 0.01$.

literature [20] quercetin-3-O-α-rhamnoside (quercitrin), kaempferol-deoxyhexoside, and quercetin-3-O-(4$''$-HMG)-α-rhamnoside were identified in the lingonberry diet. Resveratrol and apigenin were not detected in our analysis. No polyphenol signal was detected in the control diets.

4. Discussion

In this study, we show that intake of certain species of berries can prevent weight gain and counteract the metabolic derangements induced by a high-fat diet. Dietary intake of lingonberries prevented adiposity, hepatic lipid accumulation, alleviated hyperglycemia, hyperinsulinemia, and dyslipidemia and decreased plasma PAI-1 and ALT levels in mice fed a high-fat diet. Blackcurrants and, to a lower extent, bilberries and raspberries had similar effects, whereas neither crowberries, prunes, nor blackberries caused any significant improvements of metabolic parameters in this study.

The almost complete prevention of body weight gain observed in the group receiving lingonberry-supplemented diet is an effect of reduced adiposity. Our finding is in agreement with a study in Wistar rats, in which it was shown that lingonberry extracts favorably affected antioxidant defense enzymes, but it was also apparent that the lingonberry extracts reduced weight gain compared to the control [22]. However, to the best of our knowledge, there

Table 4: Polyphenolic composition of berry diets.

Anthocyanins	Lingonberry	Blackcurrant	Bilberry	Raspberry	Açai	Crowberry	Prune	Blackberry
Cyanidin-3-glucoside	x	x	x	x	x	x		x
Cyanidin-3-galactoside	x		x			x		
Cyanidin-3-arabinoside	x		x			x		
Cyanidin-3-sophoroside				x				
Cyanidin-3-rutinoside				x	x			
Cyanidin-3- (6″-malonyl)-glucoside						x		x
Delphinidin-3-rutinoside		x		x				
Delphinidin-3-glucoside			x			x		
Delphinidin-3-galactoside			x			x		
Delphinidin-3-arabinoside			x			x		
Petunidin-3-rutinoside				x				
Petunidin-3-galactoside			x			x		
Petunidin-3-glucoside			x					
Petunidin-3-arabinoside			x			x		
Peonidin-3-rutinoside				x	x			
Peonidin-3-galactoside			x			x		
Peonidin-3-glucoside			x					
Peonidin-3-arabinoside						x		
Malvidin-3-galactoside			x			x		
Malvidin-3-glucoside			x					
Malvidin-3-arabinoside			x			x		
Medium polar polyphenols								
Quercetin-3-O-glucoside	x	x						
Quercetin-3-O-galactoside	x	x						
Quercetin-3-O-α-rhamnoside	x							
Kaempferol-deoxyhexoside	x							
Quercetin-3-O-(4″-HMG)-α-rhamnoside	x							

are no other publications addressing the antiobesity effect of lingonberries. Increased fat excretion is unlikely to entirely explain the protection against adiposity since fecal excretion of triacylglycerol was not elevated in the lingonberry group compared to groups receiving some of the other berries. Also, there was no correlation between body weight and dietary intake of soluble, insoluble, or total fiber (data not shown). Interestingly, blackcurrant supplementation efficiently prevented weight gain and body fat accumulation despite an increased energy intake. However, the monitoring of caloric intake was based on registered food intake and due to texture, the blackcurrant diet gave rise to a higher spillage than other diets which could be incorrectly interpreted as a higher food intake. The bilberry group also had a higher food intake compared to control. The increased excretion of triacylglycerol in bilberries, blackcurrants, and raspberries could be caused by reduced energy absorption and explain some of the beneficial effects on adiposity. This potential mechanism should be further investigated since increased fat excretion also was observed by supplementation with crowberry, acai, and blackberries, without preventing weight gain. Polyphenol-rich extracts have been shown to inhibit pancreatic lipase [23, 24] although it remains to be established if this mechanism operates also in vivo.

Approximately 70–80% of patients with type 2 diabetes suffer from nonalcoholic fatty liver disease (NAFLD), which is linked to increased risk of cardiovascular disease [25, 26]. Fat accumulation in the liver is associated with impaired hepatic insulin sensitivity [27–29] and production of inflammatory markers [30, 31]. Interestingly, lingonberry supplementation significantly reduced liver mass and liver lipid accumulation. Elevated ALT levels in plasma correlate with reduced liver function and steatosis and predicts type 2 diabetes [32–34]. In this study, the significant reduction in liver triacylglycerol and ALT levels in the lingonberry, blackcurrant, and bilberry groups implies protection against liver steatosis and subsequent improved liver function. Our findings are supported by a 20-week dietary intervention study where a diet rich in berries (bilberries, sea buckthorns, blackcurrants, and lingonberries) reduced ALT in overweight women [35]. However, in a follow-up study using only bilberries and sea buckthorn no effects on ALT-levels were observed; thus, the authors concluded that either the lingonberries and/or blackcurrants were responsible for modulating hepatic lipid metabolism in a positive direction with lower ALT as a result [36]. In contrast to the beneficial effects observed following intake of lingonberry, blackcurrant, or bilberry, livers dissected from mice receiving supplementation with açai were large and visibly more whitish than other groups, indicating liver steatosis. Analysis revealed that the livers weighed more and had higher triacylglycerol content, and mice receiving açai had significantly higher plasma ALT levels than the rest of the berry groups, but not compared to the control group. Taken together, this suggests that açai promotes rather than prevents fatty liver. However, lipid accumulation per se may not be causal for the insulin resistance associated with fatty liver [37].

Mice fed lingonberries, blackcurrants, and bilberries had lower fasting glucose and/or insulin, resulting in a lower HOMA-insulin resistance index (Figure 2). Thus our data indicate that these three berries protected against high-fat induced insulin resistance. The improved glucose control may be an effect of the reduced adiposity. However, in humans, ingestion of lingonberries has been shown to abolish and decrease, respectively, the postprandial hyperglycemic response of ingesting glucose or sucrose in normal-weight healthy subjects [38, 39]. A cell-assay based bioactivity screening of plants traditionally used to treat symptoms related to diabetes in a native Canadian population identified an extract of lingonberries as having the highest antidiabetic potential [40]. A follow-up study by Eid et al. revealed that a lingonberry extract stimulated glucose uptake into muscle cells by activating the AMP-activated protein kinase (AMPK) pathway [41]. Also, a bilberry extract fed to diabetic KK-A^y mice ameliorated hyperglycemia and enhanced insulin sensitivity accompanied by AMPK activation [10].

In the polyphenolic characterization of the diets, quercetin-3-O-glucoside and quercetin-3-O-galactoside were found in the lingonberry and the blackcurrant diets. Lingonberries and bilberries belong to the *Vaccinium* genus and Eid et al. proposed that the traditional use of this genus to treat diabetes is due to the ability of quercetin and some of its glucosides to transiently inhibit ATP synthase and activate AMPK [41]. In a study by Kobori et al., supplementation with 0.05% quercetin alleviated hepatic fat accumulation and decreased visceral fat deposition, hyperglycemia, hyperinsulinemia, dyslipidemia, and inflammation in C57BL/6J mice fed a high-fat Western diet [42]. Since lingonberries and blackcurrants had the most beneficial effects in our study, the finding that quercetin glycosides were detected only in the lingonberry and blackcurrant diets is interesting and will be further evaluated. Moreover, Erlund et al. have shown that daily intake of lingonberries, blackcurrants, and bilberries significantly increase, serum quercetin levels in humans [43].

Berries are in general rich in anthocyanins and several anthocyanins were found in the experimental diets. Many of the detected anthocyanins and quercetin glycosides are associated with relevant bioactivities [11, 42, 44, 45], but our study design does not permit elaboration of causal relationships between specific compounds and health outcomes. The fact that resveratrol was not detected in the diets is somewhat unexpected since its presence in some of the studied berries has been demonstrated [46]. However, contents of polyphenols are known to vary considerably due to different varieties, cultivars, and growing conditions [47].

Obesity and type 2 diabetes are associated with a systemic low-grade inflammation, and it has been suggested that an altered gut microbiota could be the underlying cause of this inflammation [48–50]. Berries are high in fiber and phenolic compounds which could promote or inhibit growth of certain species of bacteria and thereby alter the microbiota. The finding that all berry diets, except raspberries, increased cecum mass suggests a change in microbial fermentative activity. Dietary administration of lingonberries, blackcurrants, and bilberries reduced plasma levels of PAI-1, which indicate an anti-inflammatory effect. In addition, the significantly lower spleen mass observed in the lingonberry-group may reflect a reduced systemic inflammation. Many previous studies on berries have focused on the capacity of antioxidative polyphenols to reduce oxidative stress and prevent cardiovascular disease. Increased PAI-1 concentrations result in reduced fibrinolytic activity and play a key role in atherothrombosis [51]. Elevated PAI-1 is also associated with NAFLD [52]. The reduction of PAI-1 in combination with lower plasma cholesterol levels by lingonberry and blackcurrant supplementation suggests that these berries may be useful in preventing cardiovascular events. However, there was no significant effect on LDL/HDL cholesterol ratio and in the blackcurrant group an increase in plasma triacylglycerol was observed. In contrast, mice receiving açai had increased total cholesterol and PAI-1 in plasma, further questioning the health aspects of this berry.

In conclusion, this study demonstrated that daily supplementation with lingonberries and also blackcurrants and bilberries had pronounced antiobesity and beneficial metabolic effects in high-fat fed C57BL/6J mice. The mechanisms behind the effects should be further evaluated, taking into account lower doses and reproducibility in humans. The capacity of lingonberries to counteract the negative outcomes of an unhealthy diet could be useful in designing dietary intervention strategies aimed at preventing development of obesity and type 2 diabetes.

Conflict of Interests

The authors declare that there is no conflict of interests regarding the publication of this paper.

Acknowledgments

The authors thank Birgitta Danielsson and Sara Larsson for excellent technical assistance. Thanks are due to you Kristina Andersson for support with the Luminex measurements. They would also like to thank Anders Högberg at Orkla and Håkan Almén at Semper for supply of berries. This study was supported by the Antidiabetic Food Centre; a VINNOVA VINN Excellence Centre at Lund University, the Swedish Research Council (project no. 112849), the Swedish Diabetes Association, A. Påhlsson Foundation, the Royal Physiographic Society in Lund, and the Crafoord Foundation.

References

[1] J. E. Shaw, R. A. Sicree, and P. Z. Zimmet, "Global estimates of the prevalence of diabetes for 2010 and 2030," *Diabetes Research and Clinical Practice*, vol. 87, no. 1, pp. 4–14, 2010.

[2] K. E. Wellen and G. S. Hotamisligil, "Inflammation, stress, and diabetes," *The Journal of Clinical Investigation*, vol. 115, no. 5, pp. 1111–1119, 2005.

[3] P. D. Cani and N. M. Delzenne, "Interplay between obesity and associated metabolic disorders: new insights into the gut microbiota," *Current Opinion in Pharmacology*, vol. 9, no. 6, pp. 737–743, 2009.

[4] A. Dembinska-Kiec, O. Mykkänen, B. Kiec-Wilk, and H. Mykkänen, "Antioxidant phytochemicals against type 2 diabetes," *British Journal of Nutrition*, vol. 99, supplement 1, pp. ES109–ES117, 2008.

[5] International Diabetes Federation, http://www.idf.org/prevention.

[6] N. M. Wedick, A. Pan, A. Cassidy et al., "Dietary flavonoid intakes and risk of type 2 diabetes in US men and women," *The American Journal of Clinical Nutrition*, vol. 95, no. 4, pp. 925–933, 2012.

[7] P. Knekt, J. Kumpulainen, R. Järvinen et al., "Flavonoid intake and risk of chronic diseases," *The American Journal of Clinical Nutrition*, vol. 76, no. 3, pp. 560–568, 2002.

[8] K. Esposito, C.-M. Kastorini, D. B. Panagiotakos, and D. Giugliano, "Prevention of type 2 diabetes by dietary patterns: a systematic review of prospective studies and meta-analysis," *Metabolic Syndrome and Related Disorders*, vol. 8, no. 6, pp. 471–476, 2010.

[9] N. P. Seeram, "Berry fruits: compositional elements, biochemical activities, and the impact of their intake on human health, performance, and disease," *Journal of Agricultural and Food Chemistry*, vol. 56, no. 3, pp. 627–629, 2008.

[10] M. Takikawa, S. Inoue, F. Horio, and T. Tsuda, "Dietary anthocyanin-rich bilberry extract ameliorates hyperglycemia and insulin sensitivity via activation of amp-activated protein kinase in diabetic mice," *Journal of Nutrition*, vol. 140, no. 3, pp. 527–533, 2010.

[11] B. Jayaprakasam, S. K. Vareed, L. K. Olson, and M. G. Nair, "Insulin secretion by bioactive anthocyanins and anthocyanidins present in fruits," *Journal of Agricultural and Food Chemistry*, vol. 53, no. 1, pp. 28–31, 2005.

[12] R. S. Surwit, C. M. Kuhn, C. Cochrane, J. A. McCubbin, and M. N. Feinglos, "Diet-induced type II diabetes in C57BL/6J mice," *Diabetes*, vol. 37, no. 9, pp. 1163–1167, 1988.

[13] W. T. Friedewald, R. I. Levy, and D. S. Fredrickson, "Estimation of the concentration of low-density lipoprotein cholesterol in plasma, without use of the preparative ultracentrifuge," *Clinical Chemistry*, vol. 18, no. 6, pp. 499–502, 1972.

[14] D. R. Matthews, J. P. Hosker, A. S. Rudenski, B. A. Naylor, D. F. Treacher, and R. C. Turner, "Homeostasis model assessment: insulin resistance and β-cell function from fasting plasma glucose and insulin concentrations in man," *Diabetologia*, vol. 28, no. 7, pp. 412–419, 1985.

[15] T. M. Wallace, J. C. Levy, and D. R. Matthews, "Use and abuse of HOMA modeling," *Diabetes Care*, vol. 27, no. 6, pp. 1487–1495, 2004.

[16] A. Hara and N. S. Radin, "Lipid extraction of tissues with a low toxicity solvent," *Analytical Biochemistry*, vol. 90, no. 1, pp. 420–426, 1978.

[17] X. Wu, L. Gu, R. L. Prior, and S. McKay, "Characterization of anthocyanins and proanthocyanidins in some cultivars of Ribes, Aronia, and Sambucus and their antioxidant capacity," *Journal of Agricultural and Food Chemistry*, vol. 52, no. 26, pp. 7846–7856, 2004.

[18] A. G. Schauss, X. Wu, R. L. Prior et al., "Phytochemical and nutrient composition of the freeze-dried amazonian palm berry, Euterpe oleraceae Mart. (Acai)," *Journal of Agricultural and Food Chemistry*, vol. 54, no. 22, pp. 8598–8603, 2006.

[19] K. Ogawa, H. Sakakibara, R. Iwata et al., "Anthocyanin composition and antioxidant activity of the crowberry (Empetrum nigrum) and other berries," *Journal of Agricultural and Food Chemistry*, vol. 56, no. 12, pp. 4457–4462, 2008.

[20] S. Ek, H. Kartimo, S. Mattila, and A. Tolonen, "Characterization of phenolic compounds from lingonberry (Vaccinium vitis-idaea)," *Journal of Agricultural and Food Chemistry*, vol. 54, no. 26, pp. 9834–9842, 2006.

[21] B. Siegmund, F. Rieder, S. Albrich et al., "Adenosine kinase inhibitor GP515 improves experimental colitis in mice," *Journal of Pharmacology and Experimental Therapeutics*, vol. 296, no. 1, pp. 99–105, 2001.

[22] C. Mane, M. Loonis, C. Juhel, C. Dufour, and C. Malien-Aubert, "Food grade lingonberry extract: polyphenolic composition and in vivo protective effect against oxidative stress," *Journal of Agricultural and Food Chemistry*, vol. 59, no. 7, pp. 3330–3339, 2011.

[23] C. Juhel, M. Armand, Y. Pafumi, C. Rosier, J. Vandermander, and D. Lairon, "Green tea extract (AR25) inhibits lipolysis of triglycerides in gastric and duodenal medium in vitro," *Journal of Nutritional Biochemistry*, vol. 11, no. 1, pp. 45–51, 2000.

[24] G. J. McDougall, N. N. Kulkarni, and D. Stewart, "Current developments on the inhibitory effects of berry polyphenols on digestive enzymes," *BioFactors*, vol. 34, no. 1, pp. 73–80, 2008.

[25] P. Angulo, "Treatment of nonalcoholic fatty liver disease," *Annals of Hepatology*, vol. 1, no. 1, pp. 12–19, 2002.

[26] G. Targher, L. Bertolini, R. Padovani et al., "Prevalence of nonalcoholic fatty liver disease and its association with cardiovascular disease among type 2 diabetic patients," *Diabetes Care*, vol. 30, no. 5, pp. 1212–1218, 2007.

[27] L. Ryysy, A.-M. Häkkinen, T. Goto et al., "Hepatic fat content and insulin action on free fatty acids and glucose metabolism rather than insulin absorption are associated with insulin requirements during insulin therapy in type 2 diabetic patients," *Diabetes*, vol. 49, no. 5, pp. 749–758, 2000.

[28] G. Marchesini, M. Brizi, G. Blanchi et al., "Nonalcoholic fatty liver disease: a feature of the metabolic syndrome," *Diabetes*, vol. 50, no. 8, pp. 1844–1850, 2001.

[29] A. Seppälä-Lindroos, S. Vehkavaara, A.-M. Häkkinen et al., "Fat accumulation in the liver is associated with defects in insulin suppression of glucose production and serum free fatty acids independent of obesity in normal men," *Journal of Clinical Endocrinology and Metabolism*, vol. 87, no. 7, pp. 3023–3028, 2002.

[30] D. Greco, A. Kotronen, J. Westerbacka et al., "Gene expression in human NAFLD," *American Journal of Physiology—Gastrointestinal and Liver Physiology*, vol. 294, no. 5, pp. G1281–G1287, 2008.

[31] A. Kotronen and H. Yki-Järvinen, "Fatty liver: a novel component of the metabolic syndrome," *Arteriosclerosis, Thrombosis, and Vascular Biology*, vol. 28, no. 1, pp. 27–38, 2008.

[32] L.-O. Ohlson, B. Larsson, P. Bjorntorp et al., "Risk factors for type 2 (non-insulin-dependent) diabetes mellitus. Thirteen and one-half years of follow-up of the participants in a study of Swedish men born in 1913," *Diabetologia*, vol. 31, no. 11, pp. 798–805, 1988.

[33] N. Sattar, O. Scherbakova, I. Ford et al., "Elevated alanine aminotransferase predicts new-onset type 2 diabetes independently of classical risk factors, metabolic syndrome, and C-reactive protein in the West of Scotland Coronary Prevention Study," *Diabetes*, vol. 53, no. 11, pp. 2855–2860, 2004.

[34] A. J. G. Hanley, K. Williams, A. Festa et al., "Elevations in markers of liver injury and risk of type 2 diabetes: the insulin resistance atherosclerosis study," *Diabetes*, vol. 53, no. 10, pp. 2623–2632, 2004.

[35] H.-M. Lehtonen, J.-P. Suomela, R. Tahvonen et al., "Berry meals and risk factors associated with metabolic syndrome," *European Journal of Clinical Nutrition*, vol. 64, no. 6, pp. 614–621, 2010.

[36] H.-M. Lehtonen, J.-P. Suomela, R. Tahvonen et al., "Different berries and berry fractions have various but slightly positive effects on the associated variables of metabolic diseases on overweight and obese women," *European Journal of Clinical Nutrition*, vol. 65, no. 3, pp. 394–401, 2011.

[37] B. A. Neuschwander-Tetri, "Hepatic lipotoxicity and the pathogenesis of nonalcoholic steatohepatitis: the central role of nontriglyceride fatty acid metabolites," *Hepatology*, vol. 52, no. 2, pp. 774–788, 2010.

[38] K. M. Linderborg, R. Jarvinen, H. M. Lehtonen, M. Viitanen, and H. P. Kallio, "The fiber and/or polyphenols present in lingonberries null the glycemic effect of the sugars present in the berries when consumed together with added glucose in healthy human volunteers," *Nutrition Research*, vol. 32, pp. 471–478, 2012.

[39] R. Torronen, M. Kolehmainen, E. Sarkkinen, H. Mykkanen, and L. Niskanen, "Postprandial glucose, insulin, and free fatty acid responses to sucrose consumed with blackcurrants and lingonberries in healthy women," *The American Journal of Clinical Nutrition*, vol. 96, pp. 527–533, 2012.

[40] D. Harbilas, L. C. Martineau, C. S. Harris et al., "Evaluation of the antidiabetic potential of selected medicinal plant extracts from the Canadian boreal forest used to treat symptoms of diabetes: part II," *Canadian Journal of Physiology and Pharmacology*, vol. 87, no. 6, pp. 479–492, 2009.

[41] H. M. Eid, L. C. Martineau, A. Saleem et al., "Stimulation of AMP-activated protein kinase and enhancement of basal glucose uptake in muscle cells by quercetin and quercetin glycosides, active principles of the antidiabetic medicinal plant vaccinium vitis-idaea," *Molecular Nutrition and Food Research*, vol. 54, no. 7, pp. 991–1003, 2010.

[42] M. Kobori, S. Masumoto, Y. Akimoto, and H. Oike, "Chronic dietary intake of quercetin alleviates hepatic fat accumulation associated with consumption of a Western-style diet in C57/BL6J mice," *Molecular Nutrition and Food Research*, vol. 55, no. 4, pp. 530–540, 2011.

[43] I. Erlund, J. Marniemi, P. Hakala, G. Alfthan, E. Meririnne, and A. Aro, "Consumption of black currants, lingonberries and bilberries increases serum quercetin concentrations," *European Journal of Clinical Nutrition*, vol. 57, no. 1, pp. 37–42, 2003.

[44] J. Joven, E. Espinel, A. Rull et al., "Plant-derived polyphenols regulate expression of miRNA paralogs miR-103/107 and miR-122 and prevent diet-induced fatty liver disease in hyperlipidemic mice," *Biochimica et Biophysica Acta*, vol. 1820, pp. 894–899, 2012.

[45] C.-C. Chuang, K. Martinez, G. Xie et al., "Quercetin is equally or more effective than resveratrol in attenuating tumor necrosis factor-α-mediated inflammation and insulin resistance in primary human adipocytes," *The American Journal of Clinical Nutrition*, vol. 92, no. 6, pp. 1511–1521, 2010.

[46] A. M. Rimando, W. Kalt, J. B. Magee, J. Dewey, and J. R. Ballington, "Resveratrol, pterostilbene, and piceatannol in Vaccinium berries," *Journal of Agricultural and Food Chemistry*, vol. 52, no. 15, pp. 4713–4719, 2004.

[47] S. Ehala, M. Vaher, and M. Kaljurand, "Characterization of phenolic profiles of Northern European berries by capillary electrophoresis and determination of their antioxidant activity," *Journal of Agricultural and Food Chemistry*, vol. 53, no. 16, pp. 6484–6490, 2005.

[48] F. Bäckhed, H. Ding, T. Wang et al., "The gut microbiota as an environmental factor that regulates fat storage," *Proceedings of the National Academy of Sciences of the United States of America*, vol. 101, no. 44, pp. 15718–15723, 2004.

[49] P. J. Turnbaugh, R. E. Ley, M. A. Mahowald, V. Magrini, E. R. Mardis, and J. I. Gordon, "An obesity-associated gut microbiome with increased capacity for energy harvest," *Nature*, vol. 444, no. 7122, pp. 1027–1031, 2006.

[50] R. E. Ley, F. Bäckhed, P. Turnbaugh, C. A. Lozupone, R. D. Knight, and J. I. Gordon, "Obesity alters gut microbial ecology," *Proceedings of the National Academy of Sciences of the United States of America*, vol. 102, no. 31, pp. 11070–11075, 2005.

[51] D. E. Vaughan, "PAI-1 and atherothrombosis," *Journal of Thrombosis and Haemostasis*, vol. 3, no. 8, pp. 1879–1883, 2005.

[52] G. Targher, L. Bertolini, L. Scala et al., "Plasma PAI-1 levels are increased in patients with nonalcoholic steatohepatitis," *Diabetes Care*, vol. 30, no. 5, pp. e31–e32, 2007.

4

Understanding Barriers and Facilitators to Healthy Eating and Active Living in Rural Communities

Rebecca Seguin,[1] Leah Connor,[1] Miriam Nelson,[2] Andrea LaCroix,[3] and Galen Eldridge[4]

[1] *Division of Nutritional Sciences, Cornell University, Ithaca, NY 14853, USA*
[2] *Friedman School of Nutrition Science and Policy, Tufts University, Boston, MA 02111, USA*
[3] *University of California, San Diego, CA 92093, USA*
[4] *Montana State University, Bozeman, MT 59717, USA*

Correspondence should be addressed to Rebecca Seguin; rebeccaseguin@hotmail.com

Academic Editor: H. K. Biesalski

Objective. Studies demonstrate that people's food and physical activity (PA) environments influence behavior, yet research examining this in rural communities is limited. *Methods.* Focus groups of 8–15 women were conducted in rural communities in seven US states. Questions were designed to identify factors within residents' food and PA environments they felt helped or hindered them from eating healthfully and being physically active. *Results.* Participants were aged 30–84 years; mean (SD) = 61 (14) ($N = 95$). On average, communities had fewer than 5,000 residents. Limited time, social norms, and distances from or lack of exercise facilities were common PA barriers. Facilitators for PA included social support, dog walking, and availability of affordable facilities. Healthy eating barriers included the perception that healthy foods were too expensive; calorically dense large portion sizes served at family meals; and frequency of eating foods away from home, which were perceived as generally unhealthy. Healthy eating supports included culture/value around local food gathering (e.g., hunting and gardening) and preservation (e.g., canning and smoking). Friends and family were frequently identified as key influencers of eating and PA behavior. *Conclusions.* Targeting both social and built environment factors, particularly those unique to rural locales, may enhance support for healthy eating and PA behavior change interventions.

1. Introduction

The consequences of obesity are well known. They include risk of developing chronic diseases such as cardiovascular disease, hypertension, diabetes, and several types of cancer [1]. The prevalence of obesity is higher in adults who live in rural areas compared to those who reside in urban areas [2, 3], and rural adults are less likely to meet physical activity guidelines than urban residents [1].

The influence of the built environment on physical activity and healthy eating behavior is an important issue, with the Centers for Disease Control, Institute of Medicine, and World Health Organization all prioritizing changes in built environment as one of the top recommendations for increasing physical activity [4–6]. Many aspects of built environment barriers to physical activity and healthy eating are common to both rural and urban areas, including cost of accessible food and recreation, access to healthy foods, and the "walkability" and "bikeability" of communities [7–10]. However, rural residents may face additional obstacles, such as travel distances from recreational facilities and lack of facilities themselves [11, 12]. Research also suggests that people in rural areas who aim to exercise outside (as a remedy for the lack of or distance from exercise facilities) may face barriers such as weather extremes and safety issues such as busy roads, lack of sidewalks and lighting on streets, loose dogs, and the presence of hunters during hunting season [13–22]. Likewise, access to healthy, affordable foods may be a problem in rural areas: in small grocery, convenience, or village stores there may be a lack of high-quality, healthy options or high costs. Stores and restaurants with more nutritious selections may be a long distance away [16, 23–25].

Despite these unique considerations, research examining barriers and facilitators of physical activity and healthy eating in rural communities is limited. Much of the emphasis has been on youth [26–39] and areas with large population groups [9, 12]. Additionally, research in rural areas has often included participants from only one or two communities and/or limited geographic scope [14–16, 21, 40]. The present study aimed to better identify and understand common rural barriers to healthy eating and active living in a geographically diverse, rural sample.

2. Methods

The research team traveled to seven rural communities in seven states in the USA as part of a larger project, called the StrongWomen Change Club project (manuscript currently in review). StrongWomen Change Clubs (SWCC) were developed through an academic-community research partnership guided by community-based participatory research principles. Local leaders, all cooperative extension educators with whom our research team has been working for the past decade, each recruited 10–15 residents to undertake a project to improve some aspect of the nutrition or physical activity environment. Most SWCC participants had limited (or no) experience in civic engagement and/or policy work. At 6- and 12-month after implementation, the research team conducted key informant interviews with each SWCC leader to capture their perceptions of program process, benchmark achievement, and self-efficacy. The first step in implementing the project was to better understand barriers and facilitators to healthy eating and active living among residents in those rural townships, defined as a population of fewer than 15,000 people. Focus groups were conducted with community members in each of the following states: Alaska, Arkansas, Kansas, Missouri, Montana, Pennsylvania, and Wisconsin (Figure 1).

Seven focus groups (one in each location, with SWCC participants) were conducted (by RAS or MEN), lasting approximately 45–60 minutes per focus group. Researchers asked open-ended questions about factors that influence physical activity and healthy eating among residents. They asked residents to think about both the barriers and facilitators that contribute to their exercise and eating habits in their community. In addition to participating in the focus groups, participants were asked to complete a brief, self-report questionnaire that contained demographic and health behavior questions derived from standard inventories (e.g., Behavioral Risk Factor Surveillance System (BRFSS)). All materials and procedures were approved by the Tufts University Institutional Review Board. All study activities were conducted with the understanding and consent of the study subjects.

FIGURE 1: Geographic representation of study sample.

2.1. Analysis. Qualitative data analysis was conducted by LMC and RAS between May and October 2013. Transcript files were entered and coded using NVivo (QSR International, version 9.0, 2011), led by LMC with a subset of double coding and agreement checks between RAS and LMC. Manifest content analysis was used to analyze the data, employing the following process: verbatim transcription; transcription review by multiple coauthors; development of and connection between codes and themes (based upon the main focus group questions and the addition of emergent subthemes); and data interpretation and review by the research team in an iterative process until agreement was reached. Quantitative (survey) data were analyzed by RAS using SPSS version 21.

3. Results

The main focus group questions informed the initial development of the coding structure with the original questions serving as the main themes: (1) barriers to physical activity; (2) facilitators to physical activity; (3) barriers to healthy eating; and (4) facilitators to healthy eating. Reoccurring subthemes were identified through an iterative process and confirmed for agreement. Themes and subthemes are displayed in Table 1.

Demographic characteristics of participants are displayed in Table 2. Focus groups contained 8–18 female participants ($N = 95$) per focus group. The average (SD) age of the 95 women was 59.3 years (14.0). The majority of the women were white (94.7%), married (68.1%), and educated beyond high school (92.6%). Focus group participants (82.4%) described the area they lived in as very rural or mostly rural. The average population size of the seven towns was 4,747 residents. Demographic characteristics of participants were compared to the overall population using census data for each town (American Community Survey 2010).

Survey responses related to physical activity and nutrition are displayed in Table 3. Survey responses indicated that more than half of the women were overweight or obese (57.9%) based on self-report height and weight (shown as body mass index (BMI)). Women self-reported their health status as excellent (20.0%), very good (49.5%), or good (28.4%), based upon a 5-point Likert scale. The majority of participants (63.2%) considered themselves active, defined as "generally active daily and exercise two or more times per week." Exercising was usually performed in a combination of settings including inside at home, outdoors, and/or inside at a facility (71.0%). Most participants (51.6%) engaged in physical activity in a combination of ways including alone, with one other person, and/or in a group or class. Thirty-eight percent of participants reported excellent access to physical activity facilities and 22.3% reported excellent access to healthy foods within their neighborhood or community.

Table 1: Themes and subthemes identified through focus groups.

Barriers to physical activity
Lack of time and competing priorities
Competition with activities that promote sedentary behavior
Social norms and stigma
Built environment barriers and geographic isolation
Facilitators to physical activity
Building physical activity into daily routines
Social support
Affordable and accessible fitness venues
Barriers to healthy eating
Lack of time and competing priorities
Cost of healthy food
Adjusting habits to favor a healthier diet
Food environment barriers and geographic isolation
Difficulty avoiding unhealthy food at community venues or gatherings
Facilitators to healthy eating
Ability to grow and produce food
Access to farmers' markets and farm shares

Table 2: Demographic characteristics of focus group participants ($N = 95$).

Age (mean, SD)	59.3	14.0
	N	%
Married	64	68.1
Race		
White	89	94.7
African American	1	1.1
American Indian/Alaskan Native	4	4.3
Education		
High school	7	7.4
Some college/technical school	19	20.0
Associate's degree	8	8.4
Bachelor's degree	26	27.4
Master's degree or higher	30	31.6
Graduate or professional	5	5.3
Rural classification[a]		
Very rural	24	26.4
Mostly rural	51	56.0
Suburban	4	4.4
Somewhat urban	12	13.2

[a]$N = 91$.

Table 3: Survey responses of focus group participants ($N = 95$).

BMI		
Underweight, ≤18.5	1	1.1
Normal, 18.5–24.9	34	35.8
Overweight, 25–29.9	34	35.8
Obese, ≥30	21	22.1
Missing	5	5.3
Health status		
Excellent	19	20.0
Very good	47	49.5
Good	27	28.4
Fair	1	1.1
Poor	1	1.1
Diet		
Very healthy	10	10.5
Mostly healthy	55	57.9
Sometimes healthy, sometimes unhealthy	30	31.6
Somewhat unhealthy	0	0.0
Unhealthy	0	0.0
Physical activity		
Very active	17	17.9
Active	60	63.2
Sometimes active	17	17.9
Mostly inactive	1	1.1
Inactive	0	0.0
Physical activity habits[a]		
Mostly alone	17	18.3
Mostly with one other person	7	7.5
Mostly in group or class	21	22.6
Combination	48	51.6
Physical activity venues[a]		
Mostly inside at home	5	5.4
Mostly outdoors	12	12.9
Mostly inside at facility	10	10.8
Combination	66	71.0
Excellent access to physical activity facilities[b]	36	38.3
Excellent access to healthy foods[b]	21	22.3

[a]$N = 93$.
[b]$N = 94$.

3.1. Perceived Barriers to Physical Activity. Barriers to physical activity included lack of time and competing priorities; activities that promote sedentary behavior; social norms and stigma; and built environment barriers.

Lack of Time due to Busy Work Schedules and Competing Priorities. Time emerged as a major barrier to engaging in physical activity. Very busy schedules often made exercising at the end of a work day seem impossible.

> *. . .people work constantly to make ends meet.*

> *Whatever I do has to be at the end of the day. And some days, at the end of the day, all I want to do is go home. . .ya know, because my day has been busy.*

Several women describe a tradeoff between engaging in physical activity and forfeiting some of their already limited sleep time when exercising in the morning is their "best option."

> *Because if I do not (exercise in the morning) I can think of a million excuses.*
>
> *Yeah, cause being active for me means less sleep, which means getting up in the morning and doing it and that is hard to give up.*
>
> *...we are not talking giving up from like eight hours of sleep or ten hours of sleep, we are getting like five hours of sleep, so you are giving up from that, so now you are getting four hours of sleep so that you can exercise.*

Competition with Activities That Promote Sedentary Time. Women discuss how sedentary activities that they enjoy like watching television and reading consume their time and make it difficult to engage in activity outside of the household.

> *...the television is on all of the time, there is always this stimulation we do not necessarily need with the television in our bedroom and umm, just kind of being addicted to...having this tube speak at you or speak around you all of the time, it keeps us kind of in the house.*
>
> *...I think that screen time, computer time, any kind of sedentary time has become a big part of our lives.*

Social Norms and Stigma. Using active transport, such as bicycling to work, had a negative stigma associated with it. One woman suggests that cycling to work must be caused from losing a driver's license.

> *So this is my question, if I get a bicycle and start riding it to work, are you all going to think that I lost my driver's license? I go by this house and this man evidentially, bless his heart, lost his driver's license for several years, because I have driven that route before and I have seen him riding (his bicycle) for over a year.*

Built Environment Barriers and Geographic Isolation. Furthermore, cycling and walking in some communities were considered unsafe due to fast traffic, busy roads, and poor street safety features. Rural residents often have to travel long distances to access exercise facilities due to the separation between commercial areas and residential housing.

> *...it has been several months now, but we did have a high school kid (driving a car) hit a woman on a bicycle.*
>
> *...that is one of the problems too, getting a route from my house to work that is safe.*
>
> *I do not want her (my daughter) to walk by herself because she has to cross a four lane highway and I am not ready for that to happen yet.*
>
> *On the other hand, if you live a long ways out, are you gonna get up, come in town, exercise, get all hot and sweaty, drive back home, shower, and get yourself back to work, I am not going to do that.*
>
> *When we moved here forty years ago, the post office was in the middle of town, the shopping A&P was, you could walk. Those days I walked, pulled a little wagon with the kids in it, cannot do that anymore.*

3.2. Perceived Facilitators to Physical Activity. Facilitators to physical activity included building physical activity into daily routines; social support; and affordable and accessible fitness venues.

Building Physical Activity into Daily Routines. Building physical activity into daily routines was reported by many of the women. Women reported that they received a lot of their physical activity through yard work, gardening, and household chores. Some women consciously modified their normal habits to actively incorporate more physical activity into their daily lives.

> *...I try to get up every commercial and go do something...you would be amazed at how much you can get done if you get up and do it during those three to five minute segments. You can almost clean your entire house in a few hours.*
>
> *I volunteer at the hospital and I have tried to walk up and down the stairs as much as I can.*
>
> *...park a couple of blocks down the street when you are on Main Street from where you want to go and walk the whole Main Street.*
>
> *In the summer, I swim on my lunch hour. And when I started doing that, I told my bosses that that was what I was going to do, and I would come back with wet hair and it wasn't an issue, it wasn't an issue for them at all, which was good.*

Social Support. Many people report that their pets keep them active. In addition, living in a small community allows interaction among people who have common interests which further promotes physical activity.

> *I walked before I had a dog, but the dog made me walk even more. And then I met more people when I was walking the dog and then we kind of all walked together.*
>
> *And I had some old guy who is retired now who stopped once; he used to ride his bicycle around. He stopped once...(A friend) and I walked every morning, and he stopped to tell us that he was going to be gone on vacation, so we wouldn't worry if we did not go by him.*

...you have to have that encouragement from somebody else.

Affordable and Accessible Fitness Venues. Access to affordable physical activity venues contributes to the promotion of physical activity within communities.

I did think about our community as a whole and how much healthier our community has gotten since they put the fitness club in, since they have offered these physical activities.

And they open it (fitness center) up...you do not have to be a member; you can just come once and pay three dollars.

...they are making it affordable and a lot of people are taking...their spinning classes were all full from what I understand.

(A friend's) insurance pays for her to go to fitness.

3.3. Perceived Barriers to Healthy Eating. Barriers to healthy eating included lack of time and competing priorities; cost of healthy food; adjusting habits to favor a healthier diet; and geographic isolation.

Lack of Time and Competing Priorities. The time required to shop and prepare food was identified as a major barrier for people who already struggled with busy family and work schedules. Commute time for people who reside in rural communities compounded the issue. Eating out or relying on quick microwave meals often took the place of eating a home-cooked meal.

...all of the handy dandy little things that you can buy and have in the cupboard and then your kid can just pop in the microwave and they got their own dinner in just a minute, ya know? I think that that is a barrier for people...the convenience.

I think that our restaurants are a big challenge, ya know, because of my work schedule, we do eat out once or twice a week sometimes.

I have worked a lot with people with kids that are very active with school activities, tons of after school activities. These parents run constantly. Family time for these families is in the car on the way to a ball game eating something from a fast food place.

And part of it in the rural areas is if you have kids active in sports, you are not driving to facilities in the same town, you are driving thirty miles, an hour away for these ball games. You get off work, you get in the car, and when you get home that night, you are not putting something in the crockpot for the next night. And so ya know, all of the driving and distance we have.

Cost of Healthy Food. The cost of fresh food was identified as a barrier to eating healthy, especially among low-income members of the community. Purchasing produce at the local farmers' market was costly and the Supplemental Nutrition Assistance Program (SNAP) benefits were not accepted at most markets. Furthermore, offering a healthier substitute is a great way to promote healthy eating; however, when the healthier option is more costly, it can be seen as a barrier.

And what is adequate access for some of us does not mean that it is for everyone. Some of us have more resources than a lot of the community.

But like our farmers' market does not do EBT (electronic benefit transfer) cards and things like that.

The farmers' market here is more expensive then supermarket food sometimes.

...the soft drink comes free (with the meal), if I want to get them (kids) milk, that costs me extra.

Difficulty Adjusting Habits to Favor a Healthier Diet. Adjusting long-term habits of cooking and eating was difficult for many women. Women reported cooking large, high calorie meals for their husbands that farmed or engaged in physically demanding labor. As men got older and duties on the farm subsided, adjusting for the reduction in calorie requirements was difficult.

...some of us here were raised on a farm and when four men have been out working all day and so even though my husband does not farm fulltime now, I think I have to cook...and so that mindset is really hard to get through, and I can blame it on my husband, but it is really my mindset.

That was the way with me, we lived on a farm for years and we had mashed potatoes, chicken, gravy, sometimes twice a day, and then ya know, as you get older, you do not feel like you need all that, or you should not.

Food Environment Barriers and Geographic Isolation. Access to healthy food options is often limited by the long distances that residents from rural communities must travel to get to these sources.

I wouldn't say that we have excellent access, I would say we have adequate access. There are a lot of things that you cannot get in (small city) that you can get in (larger city) which is forty minutes away.

But again, they (farmers' markets) are not everywhere, you have to drive to them. That's just our ruralness, I understand that too, but...

Difficulty Avoiding Unhealthy Food at Community Venues and Gatherings. Making the decision to select a healthy option is not always easy. Furthermore, it can be difficult to avoid unhealthy food at community events or through organizations that offer food as rewards or incentives.

> *A lot of the churches cook and people go and eat at the church. And so you eat what they offer because you do not have to cook it. But it is lasagna and things of that type, chicken pot pies.*

> *The bank is a barrier for me. Cause there are fresh cookies on the inside. . .and I have three small children who know that.*

3.4. Perceived Facilitators to Healthy Eating. Facilitators to healthy eating common to focus group participants included ability to grow and produce food and access to farmers' markets and farm shares.

Ability to Grow and Produce Food. Several women reported growing and producing their own food. They not only had access to fresh food, but were able to control the production process as well.

> *I raised one hundred chickens this summer, but I pumped only what I wanted into them.*

> *I have a 105 (square) foot asparagus patch and blueberries.*

Access to Farmers' Markets and Farm Shares. Participants felt that having access to local farmers' markets and having the opportunity to join a community farm share have supported healthy eating.

> *You sign up (for the CSA-Community Supported Agriculture) in the beginning of the year and you pay for kind of a share of the farm. . .and they have a wonderful newsletter, which for me is useful for recipes on how to use this stuff.*

> *I think we have accessibility too to fresh meats, meats that are locally grown.*

> *We also have some farmers doing options now where you get combined things in bulk. They were doing those two or three times a week during the summer.*

4. Discussion

Much of the research examining barriers and facilitators to active living and healthy eating among residents in rural communities has only focused on one geographic location [15, 16, 22, 40]. This study aimed to gather perspectives from a broad geographic range—from Alaska to Pennsylvania and in between.

Interestingly, there were findings that are similar to what has been reported in studies not focused on rural locales—particularly lack of time and competing priorities to exercising and eating healthfully.

Our study also supports and expands upon existing research related to barriers to physical activity and healthy eating specific to rural communities such as busy roads and fast traffic, poor access to fitness facilities, expensive costs of healthy food, difficulty adjusting eating habits like consuming large portions, lack of variety and quality of food at local stores, and difficulty avoiding unhealthy food at community events and gatherings [14–16, 22, 24, 40]. Primary facilitators to active living and healthy eating in this study also confirm prior findings, highlighting factors such as accessible and affordable fitness facilities, social support, and access to fresh and local food [15, 24, 40].

Participant and community characteristics from prior studies differed from our sample of participants with respect to gender [24, 40], education [24, 40], geographic location [14–16, 22, 24, 40], and race [15, 40]. This suggests that similar barriers and facilitators exist around healthy eating and active living in rural communities and across a range of demographic characteristics.

A major theme that was discussed in four of the seven focus groups was the attempt to build physical activity into daily routines. In some instances, this meant that participants proactively sought ways to increase their activity such as waking up early to work out, parking further away from destinations, taking the stairs when an elevator is an option, and engaging in activity on lunch breaks. In other cases, physical activity was naturally a part of daily routines like housework, laundry, gardening, and walking the dog.

Several themes identified may be specific to rural communities. A unique finding that emerged from the focus groups is the stigma associated with being physically active in some of these small rural communities. If a person was routinely seen riding a bicycle or walking to work it might be assumed that he/she had lost his/her driver's license. This finding may be specific to rural communities where active transport is not common and residents in the community can easily identify one another.

Changing dietary habits is difficult and the reduction of perceived barriers, increased overall health concern, social modeling, and self-efficacy may be necessary to take action and maintain change [41]. Our findings suggest that making dietary changes can be difficult, especially when women are accustomed to preparing large portions of high-caloric food for men who have physically demanding jobs such as farming and field labor. As these men age and transition away from these physically demanding occupations, adjusting cooking practices can be a difficult challenge unique to rural and farming households.

As documented in other studies, rural residents may have improved access to healthy foods due to the availability of fresh and local produce offered through farmers' markets and roadside stands [24, 40]. Our participants discussed in detail how they produced their own food. Joining farm shares and acquiring local meats and produce were commonly reported as well. Although access to healthy foods may be limited in traditional food stores, other alternative options to acquiring fresh produce and meats are available and may be more common among rural communities.

Many of the subthemes identified in our study were interconnected across other subthemes and main themes illustrated in this project. For example, living in a rural community may require long distance travel to access food or physical activity venues which may put a strain on budgets and time. Our participants identified the cost of healthy food and lack of time as barriers to engaging in physical activity and healthy eating. The interconnection among these themes may amplify the barriers that rural residents face. In prior research, Wilcox et al. found that many of the facilitators identified in focus groups were opposite to the barriers and vice versa [15]. Some of our results indicated a similar pattern. For example, access to farmers' markets and farm shares was identified as a facilitator to eating healthfully, while long distances from food outlets was discussed as a barrier. Social support was viewed as a positive contributor to engaging in physical activity while social norms and stigma acted as a barrier to engaging in active transport such as cycling.

The main limitation to this study is the inclusion of primarily white, educated women, which hinders the generalizability of our findings. However, the women were generally representative of the communities visited. Also, we identified several common barriers and facilitators in this work compared to past research that suggests that key themes are stable across demographic groups. Another limitation is the lack of urban comparison groups. Because of this, we are not able to conclude that the themes that emerged are specifically associated with rural communities. Lastly, social desirability bias may have occurred if focus group participants felt that they could not express personal barriers or if they overreported or exaggerated positive physical activity or healthy eating behaviors.

Overall, this study fills an important gap in knowledge related to barriers and facilitators to physical activity and healthy eating in rural communities. Future intervention strategies should target the identified factors and place specific emphasis on the overlapping themes that emerged from our study and past studies. In addition, addressing stigma related active transport, adjusting cooking practices and habits, and increasing awareness of local food options could be of particular interest when working with rural communities.

Conflict of Interests

The authors declare that there is no conflict of interests regarding the publication of this paper.

Acknowledgments

Dr. Rebecca Seguin worked on this project while employed at Fred Hutchinson Cancer Research Center and Cornell University. Dr. Rebecca Seguin was supported by a grant from the National Heart, Lung, and Blood Institute (K01 HL108807). The project was also supported, in part, by the Omidyar Foundation, Brookdale Foundation, and Tisch College of Citizenship and Public Service, Tufts University.

References

[1] National Center for Health Statistics, *Special Feature on Socioeconomic Status and Health*, National Center for Health, Hyattsville, Md, USA, 2012.

[2] C. A. Befort, N. Nazir, and M. G. Perri, "Prevalence of obesity among adults from rural and urban areas of the United States: findings from NHANES (2005–2008)," *The Journal of Rural Health*, vol. 28, no. 4, pp. 392–397, 2012.

[3] J. E. Jackson, M. P. Doescher, A. F. Jerant, and L. G. Hart, "A national study of obesity prevalence and trends by type of rural county," *Journal of Rural Health*, vol. 21, no. 2, pp. 140–148, 2005.

[4] World Health Organization Regional Office for Europe Copenhagen: Health 21: The health for all policy framework for the WHO European Region European Health for All Series no. 6, 1999.

[5] D. Keener, K. Goodman, A. Lowry, S. Zaro, and L. Kettel Khan, *Recommended Community Strategies and Measurements to Prevent Obesity in the United States: Implementation and Measurement Guide*, Department of Health and Human Services, Centers for Disease Control and Prevention, Atlanta, Ga, USA, 2009.

[6] Institute of Medicine, *Accelerating Progress in Obesity Prevention: Solving the Weight of the Nation*, The National Academies Press, Washington, DC, USA, 2012.

[7] R. E. Walker, C. R. Keane, and J. G. Burke, "Disparities and access to healthy food in the United States: a review of food deserts literature," *Health & Place*, vol. 16, no. 5, pp. 876–884, 2010.

[8] M. ver Ploeg, V. Breneman, P. Dutko et al., "Access to affordable and nutritious food: updated estimates of distance to supermarkets using 2010 data," Economic Research Report no. ERR-143, United States Department of Agriculture, ERS, 2010.

[9] E. D. Mansfield, N. Ducharme, and K. G. Koski, "Individual, social and environmental factors influencing physical activity levels and behaviours of multiethnic socio-economically disadvantaged urban mothers in Canada: a mixed methods approach," *International Journal of Behavioral Nutrition and Physical Activity*, vol. 9, article 42, 2012.

[10] J. F. Sallis and K. Glanz, "Physical activity and food environments: solutions to the obesity epidemic," *Milbank Quarterly*, vol. 87, no. 1, pp. 123–154, 2009.

[11] S. S. Frost, R. T. Goins, R. H. Hunter et al., "Effects of the built environment on physical activity of adults living in rural settings," *The American Journal of Health Promotion: AJHP*, vol. 24, no. 4, pp. 267–283, 2010.

[12] S. Wilcox, C. Castro, A. C. King, R. Housemann, and R. C. Brownson, "Determinants of leisure time physical activity in rural compared with urban older and ethnically diverse women in the United States," *Journal of Epidemiology & Community Health*, vol. 54, no. 9, pp. 667–672, 2000.

[13] K. R. Evenson, O. L. Sarmiento, M. L. Macon, K. W. Tawney, and A. S. Ammerman, "Environmental, policy, and cultural factors related to physical activity among Latina immigrants," *Women and Health*, vol. 36, no. 2, pp. 43–57, 2002.

[14] R. E. Aronson and R. F. Oman, "Views on exercise and physical activity among rural-dwelling senior citizens," *The Journal of Rural Health*, vol. 20, no. 1, pp. 76–79, 2004.

[15] S. Wilcox, L. Oberrecht, M. Bopp, S. K. Kammermann, and C. T. McElmurray, "A qualitative study of exercise in older African American and white women in Rural South Carolina:

perceptions, barriers, and motivations," *Journal of Women and Aging*, vol. 17, no. 1-2, pp. 37–53, 2005.

[16] M. Maley, B. S. Warren, and C. M. Devine, "Perceptions of the environment for eating and exercise in a rural community," *Journal of Nutrition Education and Behavior*, vol. 42, no. 3, pp. 185–191, 2010.

[17] A. A. Eyler and J. R. Vest, "Environmental and policy factors related to physical activity in rural white women," *Women and Health*, vol. 36, no. 2, pp. 111–121, 2002.

[18] D. E. Whaley and P. P. Haley, "Creating community, assessing need: preparing for a community physical activity intervention," *Research Quarterly for Exercise and Sport*, vol. 79, no. 2, pp. 245–255, 2008.

[19] T. M. Kruger, M. Swanson, R. E. Davis, S. Wright, K. Dollarhide, and N. E. Schoenberg, "Formative research conducted in rural Appalachia to inform a community physical activity intervention," *The American Journal of Health Promotion*, vol. 26, no. 3, pp. 143–151, 2012.

[20] B. Sanderson, M. A. Littleton, and L. Pulley, "Environmental, policy, and cultural factors related to physical activity among rural, African American women," *Women and Health*, vol. 36, no. 2, pp. 75–90, 2002.

[21] M. Bopp, S. Wilcox, L. Oberrecht, S. Kammermann, and C. T. McElmurray, "Correlates of strength training in older rural African American and Caucasian women," *Women and Health*, vol. 40, no. 1, pp. 1–20, 2004.

[22] E. C. Paluck, M. Allerdings, K. Kealy, and H. Dorgan, "Health promotion needs of women living in rural areas: an exploratory study," *Canadian Journal of Rural Medicine*, vol. 11, no. 2, pp. 111–116, 2006.

[23] USDA Economic Research Service, *Access to Affordable and Nutritious Food: Measuring and Understanding Food Deserts and Their Consequences*, Department of Agriculture, Washington, DC, USA, 2009.

[24] C. Smith and L. W. Morton, "Rural food deserts: low-income perspectives on food access in Minnesota and Iowa," *Journal of Nutrition Education and Behavior*, vol. 41, no. 3, pp. 176–187, 2009.

[25] C. Neill, B. D. Leipert, A. C. Garcia, and M. Kloseck, "Using photovoice methodology to investigate facilitators and barriers to food acquisition and preparation by rural older women," *Journal of Nutrition in Gerontology and Geriatrics*, vol. 30, no. 3, pp. 225–247, 2011.

[26] H. Baslington, "School travel plans: overcoming barriers to implementation," *Transport Reviews*, vol. 28, no. 2, pp. 239–258, 2008.

[27] S. M. Burke and G. E. Mandich, "Physical activity among Latino children: a qualitative examination of barriers and facilitators," *Journal of Sport & Exercise Psychology*, vol. 34, p. S209, 2012.

[28] N. E. Findholt, Y. L. Michael, L. J. Jerofke, and V. W. Brogoitti, "Environmental influences on children's physical activity and eating habits in a rural Oregon county," *American Journal of Health Promotion*, vol. 26, no. 2, pp. e74–e85, 2011.

[29] L. B. Kaye, C. M. Tucker, M. A. Bragg, and A. C. Estampador, "Low-income children's reported motivators of and barriers to healthy eating behaviors: a focus group study," *Journal of the National Medical Association*, vol. 103, no. 9-10, pp. 941–951, 2011.

[30] M. R. U. Meyer, J. R. Sharkey, M. S. Patterson, and W. R. Dean, "Understanding contextual barriers, supports, and opportunities for physical activity among Mexican-origin children in Texas border colonias: a descriptive study," *BMC Public Health*, vol. 13, article 14, 2013.

[31] J. B. Moore, S. B. Jilcott, K. A. Shores, K. R. Evenson, R. C. Brownson, and L. F. Novick, "A qualitative examination of perceived barriers and facilitators of physical activity for urban and rural youth," *Health Education Research*, vol. 25, no. 2, pp. 355–367, 2010.

[32] J. A. O'Dea, "Why do kids eat healthful food? Perceived benefits of and barriers to healthful eating and physical activity among children and adolescents," *Journal of the American Dietetic Association*, vol. 103, no. 4, pp. 497–501, 2003.

[33] C. Peeters, H. Marchand, H. Tulloch et al., "Perceived facilitators, barriers, and changes in a randomized exercise trial for obese youth: a qualitative inquiry," *Journal of Physical Activity and Health*, vol. 9, no. 5, pp. 650–660, 2012.

[34] N. Pierre, O. Receveur, A. C. Macaulay, and L. Montour, "Identification of barriers and facilitators of healthy food choices among children aged 6 to 12 years: from the kahnawake schools diabetes prevention project," *Ecology of Food and Nutrition*, vol. 46, no. 2, pp. 101–123, 2007.

[35] E. Rawlins, G. Baker, M. Maynard, and S. Harding, "Perceptions of healthy eating and physical activity in an ethnically diverse sample of young children and their parents: the DEAL prevention of obesity study," *Journal of Human Nutrition and Dietetics*, vol. 26, no. 2, pp. 132–144, 2013.

[36] J. Shepherd, A. Harden, R. Rees et al., "Young people and healthy eating: a systematic review of research on barriers and facilitators," *Health Education Research*, vol. 21, no. 2, pp. 239–257, 2006.

[37] D. Simons, P. Clarys, I. de Bourdeaudhuij, B. de Geus, C. Vandelanotte, and B. Deforche, "Factors influencing mode of transport in older adolescents: a qualitative study," *BMC Public Health*, vol. 13, no. 1, article 323, 2013.

[38] K. Skinner, R. M. Hanning, and L. J. S. Tsuji, "Barriers and supports for healthy eating and physical activity for first nation youths in northern Canada," *International Journal of Circumpolar Health*, vol. 65, no. 2, pp. 148–161, 2006.

[39] O. Stewarta, A. V. Moudon, and C. Claybrooke, "Common ground: eight factors that influence walking and biking to school," *Transport Policy*, vol. 24, pp. 240–248, 2012.

[40] M. C. Kegler, C. Escoffery, I. Alcantara, D. Ballard, and K. Glanz, "A qualitative examination of home and neighborhood environments for obesity prevention in rural adults," *International Journal of Behavioral Nutrition and Physical Activity*, vol. 5, article 65, 2008.

[41] L. A. Sporny and I. R. Contento, "Stages of change in dietary fat reduction (Social psychological correlates)," *Journal of Nutrition Education*, vol. 27, no. 4, pp. 191–199, 1995.

5

A Randomized Clinical Trial of Nutrition Education for Improvement of Diet Quality and Inflammation in Iranian Obese Women

Majid Mohammadshahi,[1] Fatemeh Haidari,[2] Majid Karandish,[2] Sara Ebrahimi,[3] and Mohammad-Hosein Haghighizadeh[4]

[1] *Hyperlipidemia Research Center, Ahvaz Jundishapur University of Medical Sciences, Ahvaz 61357-15794, Iran*
[2] *Nutrition and Metabolic Diseases Research Center, Ahvaz Jundishapur University of Medical Sciences, Ahvaz 61357-15794, Iran*
[3] *Department of Nutritional Science, Arvand International Division of Ahvaz Jundishapur University of Medical Sciences, Ahvaz 61357-15794, Iran*
[4] *School of Health, Ahvaz Jundishapur University of Medical Sciences, Ahvaz 61357-15794, Iran*

Correspondence should be addressed to Sara Ebrahimi; saraebrahimi123@gmail.com

Academic Editor: H. K. Biesalski

Background. Obesity is considered as a low grade inflammation condition. The aim of this study was to investigate the effect of nutritional education on diet quality and biomarkers of inflammation in Iranian obese women. *Method.* Sixty obese women voluntarily participated in this randomized clinical trial and were randomly assigned to intervention or control group ($n = 30$). Intervention group was instructed to attend nutrition education sessions (1 hr/wk, for 3 months) in small groups. Diet quality scores were measured by Healthy Eating Index (HEI). Anthropometric indices and serum concentration of hs-CRP, TNF-α, and adiponectin were measured at the baseline and end of the intervention. *Results.* There were no significant differences in anthropometric indices of participants between the two groups at the end of intervention ($P > 0.05$). However, the total HEI score was significantly higher in the educated group compared to the control group after intervention ($P < 0.05$). The educated group also showed significant lower concentration of TNF-α and hs-CRP and higher levels of adiponectin than the control group at the end of study ($P < 0.05$). *Conclusions.* Our results provide limited evidence that higher dietary quality contributes to reduced inflammation in obese women. This effect could be independent of the weight loss.

1. Background

The link between unhealthy eating habits and obesity is well established [1–4]. Since human diets contain many components that may work synergistically to prevent or promote disease, assessing that diet quality may be an informative strategy when studying the relation between nutrition and obesity [5–8].

Indexes of dietary quality have been developed to address this need in nutrition research. Two of these, the Healthy Eating Index (HEI) and the Diet Quality Index (DQI), were developed to measure adherence to dietary guidelines and were shown to adequately measure overall diet quality [9, 10]. The HEI measures adherence to the Food Guide Pyramid developed in the mid-1990s, whereas the DQI reflects a person's adherence to the Diet and Health recommendations of the National Academy of Sciences [11].

Obesity is a low-grade inflammatory condition; however, limited studies to date have considered the overall quality of the diet and its relation to obesity-induced inflammation; these have been mainly limited to the cross-sectional studies [1, 12, 13]. According to findings from the National Health and Nutrition Examination Survey III, the HEI score was inversely associated with C-reactive protein (CRP) concentration, which was largely attributed to total grain consumption [13]. Fung et al., however, found no association between HEI score and various markers of inflammation and endothelial dysfunction [14]. Fargnoli et al. showed that adherence to a healthier diet, as reflected by a higher HEI score, is associated with higher plasma adiponectin and lower plasma resistin, CRP, and E-selectin levels independent of obesity and lifestyle factors [2]. Thus, dietary quality may be inversely associated with inflammatory conditions independent of BMI [12].

Although concentrations of inflammatory markers and adipokines were previously associated with modifiable lifestyle factors (or dietary patterns) in cross-sectional studies, diet quality has not been previously studied in relation to proinflammatory factors and adiponectin during nutritional intervention. According to Tehran Lipid and Glucose Study (TLGS), the diet quality of most people in Tehran, Iran, (74%) needs improvement and thus nutritional education to improve diet quality is recommended [15].

The purpose of the study reported here was to evaluate if nutrition education could improve diet quality, as measured by HEI score, and thereby it would be independent of weight loss associated with improving of inflammatory markers (hs-CRP and TNF-α) and adiponectin concentration.

2. Methods

2.1. Study Setting and Design. This study was designed as a randomized clinical trial to investigate the effects of nutritional education for 3 months on the improvement of HEI scores and serum levels of inflammatory markers in obese women. At first, a call for participating in the study was provided and the aim of the study was described completely to who attended to the Health and Nutrition Clinic of Ahvaz, Iran. Then regarding the inclusion and exclusion criteria, subjects were selected. Inclusion criteria were women with body mass index (BMI) ≥30, having an initial HEI score <80, under conditions of weight stability for at least 3 months, and subject satisfaction. Exclusion criteria included chronic disease (diabetes, cardiovascular disease, cancer, arthritis, lung disease, asthma, or serious allergies), being active in exercise or weight reduction programs, taking weight loss medications, taking drugs that are known to affect the immune system, including nonsteroidal anti-inflammatory medications and corticosteroids, consumption of nutritional supplements at least 6 months before sampling, alcohol consumption, vegan diet, smoking, and lack of subject follow-up (being absent in 2 or more education sessions). We also excluded under and over reporters of energy intakes [15].

2.2. Study Population, Sample Size, and Sampling. Finally, sixty obese women, aged 20–45 y, who had volunteered, were randomly divided into two groups (education group: $n = 30$ and noneducation group: $n = 30$) using a table of random numbers with block size not revealed to the investigators. Those randomized to the intervention (education) group were instructed to attend nutrition education sessions (1 hr/wk for 3 months, 12 sessions) in small groups (6 persons per session) conducted by a registered dietitian. All subjects in this group received the same information over the course of 3 months of participation. Each 1 hr session included an oral presentation on nutrition information relevant to healthy diet and HEI recommendations. In this study, the nutrition education program was based on the modified My Plate (regarding to Iranian food culture) and food pyramid guideline. All participants in the education group were also provided with a designated booklet related to nutrition education. The control group in this study did not receive the nutrition education program and they were educated for food recording only.

2.3. Data Collection and Procedure. In this study, all participants were requested not to change their physical activity habits during the study. At baseline and the end of the study, anthropometric measurements (weight, height, waist circumference (WC), and hip circumference (HC)) were measured and then waist-to-hip ratio (WHR) was determined. Body mass index (BMI) was calculated from measured height and weight. Participants' body fat percent was also measured by body state set (Quad Scan 4000).

Dietary intakes were assessed by the seven consecutive days' food record. On the basis of 7 days' food record, each HEI component score was calculated for each individual. The HEI total score was also calculated as the sum of these scores over the 12 components. Then, for each component score and for the total score, the mean score was taken. Briefly, the HEI-2005 comprises 12 components, which are scored on a scale from 0 to maximum (M), where M is 5, 10, or 20 according to the component. Thus, the composite HEI score can potentially range from a minimum of zero to a maximum score of 100, with 100 points referring to perfect diet quality and lower results indicating larger deviations from the recommended intakes [7]. For each individual and each component, the ratio of the reported intake of food group (relevant to the HEI component considered) to the reported energy intake was also calculated. Then, the mean of these ratios over the individuals was taken. All of the dietary and anthropometric data were collected by an expert registered dietitian. In this study, a modified Nutritionist IV program was used to estimate dietary intake of participants.

2.4. Biochemical Analysis. At baseline and the end of the study, fasting blood samples were also collected from all the participants, and sera were separated for biochemical analyses. Serum concentrations of hs-CRP, TNF-α, and adiponectin were measured using a commercially available enzyme-linked immunosorbent assays (ELISA) method (Labor Diagnostika Nord for hs-CRP and Orgenium laboratories-Finland for TNF-α and adiponectin). All assays were performed according to the manufacturer's instructions.

2.5. Statistical Analysis. All statistical analyses were performed with Statistical Package for Social Sciences (SPSS Inc., Chicago, IL, USA) Program version 18 for windows. At first, normal distribution of all variables was checked with the Kolmogorov-Smirnov test. Group variable means were compared with each other using both independent sample t-test and ANCOVA in the adjusted models, which were adjusted for confounders (age, weight, and energy intake). The end values of each variable were also compared with the

TABLE 1: Sociodemographic and anthropometric characteristics of the study groups at the baseline and end of study.

Variables	Educated group (n = 30)	Noneducated group (n = 30)	$P1$	$P2$
Age (y)	33.13 ± 5.60	35.13 ± 4.97	0.211	—
Income [n (%)]			0.516	0.306
Low	4 (13.33)	6 (19.98)		
Medium	22 (73.26)	19 (63.27)		
High	4 (13.33)	5 (16.65)		
Educational level [n (%)]			0.911	0.783
<6 years	1 (3.33)	0		
6–12 years	5 (16.65)	7 (23.31)		
>12 years	24 (79.92)	23 (76.59)		
Physical activity [n (%)]			0.852	0.771
Light	22 (73.26)	21 (69.93)		
Moderate	6 (19.98)	8 (26.64)		
Heavy	2 (6.66)	1 (3.33)		
Weight (kg)				
Baseline	91.1 ± 10.6	92.5 ± 14.6	0.672	0.581
End	89.8 ± 12.8	91.8 ± 13.5	0.533	0.671
$P3$	0.145	0.348		
BMI				
Baseline	34.9 ± 3.9	34.7 ± 5.07	0.877	0.676
End	34.3 ± 4.3	34.5 ± 5.11	0.852	0.845
$P3$	0.090	0.466		
WC (cm)				
Baseline	107.3 ± 10.8	107.6 ± 13.2	0.910	0.721
End	106.2 ± 8.3	108.2 ± 13.5	0.510	0.511
$P3$	0.161	0.416		
HC (cm)				
Baseline	110.2 ± 10.4	108.8 ± 10.1	0.593	0.804
End	109.5 ± 7.6	109 ± 9.9	0.812	0.973
$P3$	0.744	0.832		
WHR				
Baseline	0.97 ± 0.10	0.98 ± 0.06	0.655	0.690
End	0.97 ± 0.09	0.99 ± 0.07	0.417	0.553
$P3$	0.802	0.424		
Fat percent (%)				
Baseline	34.6 ± 6.3	33.5 ± 7.2	0.211	0.189
End	33.8 ± 6.6	33.4 ± 6.2	0.425	0.385
$P3$	0.253	0.837		

BMI: body mass index; WC: waist circumference; HC: hip circumference; and WHR: waist to hip ratio.
Data were expressed as mean ± SD or percentages.
$P1$ resulted from independent sample t-test or chi-squared test as appropriate; $P2$ resulted from ANCONA test (adjustment for age and energy intake); and $P3$ resulted from paired sample t-test.

baseline values using paired sample t-test. The differences with P values <0.05 were considered as significant.

2.6. Ethnic. The study was approved by and performed under the guidelines of the Research Ethics Committee of Ahvaz Jundishapur University of Medical Sciences, Iran (ETH-9115). A written consent was also obtained from all the participants.

3. Results

In this study, 60 participants with 34.15 ± 5.34 years old were included in the study. Sociodemographic and anthropometrics characteristics of the study participants at baseline and the end of the study are described in Table 1. There were no significant differences in baseline and end of study measures

TABLE 2: Dietary intakes of the study groups at the baseline and end of study.

Variables	Educated group (n = 30)	Noneducated group (n = 30)	$P1$	$P2$
Energy (Kcal)				
Baseline	2232 ± 593	2086 ± 320	0.086	0.105
End	2041 ± 315	1931 ± 249	0.140	0.246
$P3$	0.066	0.073		
Protein (gr)				
Baseline	86.8 ± 25.8	84.1 ± 16.4	0.632	0.483
End	81.4 ± 12.5	79.4 ± 14.2	0.620	0.909
$P3$	0.234	0.061		
Protein (%)				
Baseline	14.8 ± 2.3	16.0 ± 1.7	0.091	0.063
End	16.0 ± 1.4	16.5 ± 2.1	0.307	0.242
$P3$	0.324	0.651		
Carbohydrate (gr)				
Baseline	260.2 ± 78.5	219 ± 44.2	0.116	0.522
End	206.7 ± 40.7	196 ± 37.1	0.297	0.103
$P3$	**0.001**	**0.006**		
Carbohydrate (%)				
Baseline	44.5 ± 6.3	42.0 ± 5.5	0.071	0.132
End	40.4 ± 4.0	40.6 ± 5.8	0.202	0.731
$P3$	0.001	0.082		
Fat (gr)				
Baseline	109.4 ± 30.2	101.7 ± 18.4	0.240	0.407
End	73.3 ± 18.7	97 ± 15.7	**0.000**	**0.000**
$P3$	**0.000**	**0.045**		
Fat (%)				
Baseline	34.6 ± 6.3	33.5 ± 7.2	0.201	0.083
End	33.8 ± 6.6	33.4 ± 6.2	0.251	0.206
$P3$	0.132	0.282		

Data were expressed as mean ± SD.
$P1$ resulted from independent sample t-test; $P2$ resulted from ANCONA test in the adjusted models; and $P3$ resulted from paired sample t-test.

of participants (age, income, education, physical activity, weight, BMI, WC, HC, WHR, and fat percent) between the two groups. The anthropometric indices did not also decrease within groups at the end of study when compared to the baseline values.

Dietary intakes of the study participants are presented in Table 2. Total carbohydrate and total fat intake of participants decreased at the end of the study compared to the educated and the noneducated groups at baseline ($P < 0.05$). At the end of study, dietary intakes of total fat also decreased significantly in the educated group when compared to the control group ($P = 0.000$). This result remained significant after adjustment for confounding variables (age and energy intake).

As shown in Table 3, the mean total HEI score was 60.58 ± 6.31 in educated group and 62.05 ± 5.7 in noneducated group at the baseline. Therefore, the group differences in baseline total HEI scores were not significant in this respect. Mean component scores were not significantly different between the two groups at the baseline. At the end of study, the total fruits, total vegetables, dark green and orange vegetables and legumes, whole grains, milk, meat and beans, saturated fats, sodium, and SoFAAS (calories from solid fats, alcoholic beverages, and added Sugars) component scores and consequently the total HEI score were significantly higher in the educated group compared to the control ($P < 0.05$). These results remained significant after adjustment for confounders. After 3 months of nutrition education, the mean total HEI score was also significantly increased in the educated group, compared to baseline values (83.34 ± 5.12 versus 60.58 ± 6.31, $P = 0.000$).

The relative intake of each HEI component to the energy intake is presented in Table 4. The average intake of total fruits, whole fruits, total vegetables, dark green and orange vegetables and legumes, total grains, whole grains, and milk per 1,000 kcal significantly increased after intervention in the educated group compared to baseline intakes ($P < 0.05$). Furthermore, the relative intake of oils, saturated fat, and sodium components significantly decreased in this group ($P < 0.05$). After intervention, the comparison between two groups also showed that the relative intake of total fruits, whole fruits, total vegetables, dark green and orange

TABLE 3: Component score of HEI in the study groups at the baseline and end of study.

Components of HEI	Educated group (n = 30)	Noneducated group (n = 30)	$P1$	$P2$
Total fruit				
Baseline	3.10 ± 1.4	3.14 ± 1.09	0.898	0.322
End	4.91 ± 0.33	3.78 ± 1.17	**0.000**	**0.000**
$P3$	**0.000**	0.110		
Whole fruit				
Baseline	4.27 ± 0.85	4.34 ± 1.25	0.797	0.872
End	4.77 ± 0.69	4.77 ± 0.69	0.734	0.500
$P3$	**0.003**	0.057		
Total vegetables				
Baseline	3.18 ± 1.14	3.44 ± 1.02	0.354	0.119
End	4.68 ± 0.52	3.28 ± 0.98	**0.000**	**0.000**
$P3$	**0.000**	0.400		
Dark green and orange vegetables and legumes				
Baseline	1.53 ± 1.41	1.64 ± 1.27	0.750	0.960
End	4.36 ± 0.86	1.64 ± 1.13	**0.000**	**0.000**
$P3$	**0.000**	0.598		
Total grain				
Baseline	4.92 ± 0.22	4.80 ± 0.35	0.802	0.792
End	4.94 ± 0.47	4.93 ± 0.33	0.858	0.702
$P3$	0.873	0.532		
Whole grain				
Baseline	0.4 ± 0.1	0.13 ± 0.07	0.036	0.021
End	3.1 ± 1.3	0.13 ± 0.49	**0.000**	**0.000**
$P3$	**0.000**	0.256		
Milk				
Baseline	0.89 ± 0.62	0.77 ± 0.34	0.344	0.099
End	5.52 ± 2.23	0.64 ± 0.53	**0.000**	**0.000**
$P3$	**0.000**	0.242		
Meat and beans				
Baseline	6.44 ± 1.95	6.36 ± 2.31	0.881	0.329
End	7.86 ± 2.03	5.9 ± 1.99	**0.000**	**0.000**
$P3$	**0.005**	0.237		
Oils				
Baseline	7.25 ± 1.81	7.60 ± 1.12	0.377	0.429
End	8.41 ± 2.25	8.32 ± 1.15	0.843	0.754
$P3$	**0.025**	0.032		
Saturated fat				
Baseline	7.38 ± 1.54	7.31 ± 1.66	0.849	0.544
End	8.65 ± 1	7.56 ± 1.41	**0.001**	**0.000**
$P3$	**0.001**	0.520		
Sodium				
Baseline	4.12 ± 2.76	5.80 ± 3.44	0.040	0.028
End	6.97 ± 2.54	4.61 ± 3.15	**0.002**	**0.024**
$P3$	**0.000**	0.031		
SoFAAS				
Baseline	17.20 ± 2.12	16.66 ± 1.97	0.211	0.487
End	19.10 ± 1.2 5	17.76 ± 1.83	**0.002**	**0.002**
$P3$	**0.000**	0.051		

Table 3: Continued.

Components of HEI	Educated group (n = 30)	Noneducated group (n = 30)	$P1$	$P2$
Total score				
Baseline	60.58 ± 6.31	62.05 ± 5.7	0.350	0.283
End	83.34 ± 5.12	63.44 ± 7.49	**0.000**	**0.000**
$P3$	**0.000**	0.203		

SoFAAS.
Calories from solid fats, alcoholic beverages, and added sugars.
Data were expressed as mean ± SD.
$P1$ resulted from independent sample t-test; $P2$ resulted from ANCONA test (adjustment for age and energy intake); and $P3$ resulted from paired sample t-test.

vegetables and legumes, total grain, whole grains, milk, and meet and beans per 1,000 kcal was significantly higher in the educated group. In contrast, a significantly lower relative intake of oils, saturated fat, sodium, and SoFAAS per 1,000 kcal was observed in the educated group, compared to the control group, at the end of study ($P < 0.05$). These differences even remained significant after adjustment for confounders.

As presented in Table 5, significant differences were observed between the two groups regarding serum levels of inflammatory biomarkers at the end of intervention. Serum levels of TNF-α and hs-CRP were significantly lower in the educated group compared to the control group (P = 0.044 and P = 0.021, resp.). This significance remained even after adjustment for age, weight, and energy intake (P = 0.011 and P = 0.039, resp.). The educated group also showed significant higher adiponectin concentrations compared to the control group at the end of study (P = 0.035, in the adjusted models). Statistical analyses (within groups) also showed that the serum levels of TNF-α and hs-CRP significantly decreased and the mean concentration of adiponectin significantly increased after intervention in the educated group ($P < 0.05$).

4. Discussion

The results of this study showed that 3-month nutrition education improved diet quality, as measured by the HEI. Consequently, the improved diet quality was associated with the lower levels of hs-CRP and TNF-α and the higher levels of adiponectin in obese women ($P < 0.05$). This effect was independent of weight changes.

The relationship of diet quality to obesity has been inconsistent, and lack of association between diet quality and anthropometric indicators of obesity has also been reported in other studies [1, 7, 16–19]. In a cross-sectional study conducted by Boynton et al., the HEI scores were modestly correlated with BMI, yet not with percent body fat in postmenopausal women. However, individuals with lower BMI or lower percent body fat had more healthful diets, as measured by the DQI [1]. One possible reason for these discrepant results is that while the DQI generally emphasizes moderation, half of the HEI score concerns meeting or exceeding the recommended amount of grains, fruits, vegetables, meat, and milk. Thus, an individual who eats more in general receives a higher HEI score, which may explain why no clear associations between HEI and body composition were observed [1, 17, 18].

Although recent studies have shown an important role for inflammation in obesity, limited studies have evaluated the association between diet quality and circulating level of inflammatory markers with respect to obesity [20]. Since human diets contain many components that may work synergistically to prevent or promote disease, assessing diet quality may be informative [6]. The results from a crossover study that was conducted in healthy women have shown that, after adjustment for age and energy intake, women with the highest adherence to the HEI had 24% higher plasma adiponectin and 41% lower plasma CRP than did women with the lowest adherence to the HEI. Inverse association between the HEI and TNF-α was also reported, but it was not significant after adjustment for body mass index [2].

Some studies have examined the contribution of major dietary patterns to markers of systemic inflammation. Lopez-Garcia et al., in a crossover study, showed that the prudent pattern that is characterized by higher intakes of fruit, vegetables, legumes, fish, poultry, and whole grains is inversely associated with plasma concentrations of CRP. Contrary to this finding, the Western pattern that is characterized by higher intakes of red and processed meats, sweets, desserts, French fries, and refined grains showed a positive relation with CRP and interleukin 6 after adjustment for confounders [21]. The prudent pattern and the Western pattern in Lopez-Garcia et al. study were relatively comparable to perfect diet quality and imperfect diet quality in the present study, respectively. In another study, median plasma adiponectin concentrations were 23% higher in women who most closely followed a Mediterranean-type diet than in low adherers [22]. Similar results were also reported by others [14, 23]. Similarly, Esposito et al., in a 2-y randomized controlled trial, reported that a Mediterranean-type diet accompanied by increased physical activity significantly increased adiponectin concentrations in obese postmenopausal women, even after accounting for the decreased body weight associated with the intervention [24]. The results of this study are consistent with those of the above studies and extend them by showing that adherence to a healthier diet, which is achieved by nutrition education in the educated group, was associated with higher adiponectin concentrations and lower hs-CRP and TNF-α levels. Several recent studies have shown that fiber, antioxidant, flavonoids,

TABLE 4: Component relative intake of HEI in the study groups at the baseline and end of study.

Components of HEI	Educated group (n = 30)	Noneducated group (n = 30)	$P1$	$P2$
Total fruit (cup per 1000 kcal)				
Baseline	0.53 ± 0.300	0.51 ± 0.19	0.246	0.689
End	1.53 ± 0.62	0.64 ± 0.24	**0.000**	**0.000**
$P3$	**0.000**	0.084		
Whole fruit (cup per 1000 kcal)				
Baseline	0.25 ± 0.12	0.18 ± 0.09	0.235	0.315
End	0.43 ± 0.27	0.20 ± 0.29	**0.000**	**0.000**
$P3$	**0.012**	0.205		
Total vegetables (cup per 1000 kcal)				
Baseline	0.73 ± 0.33	0.77 ± 0.25	0.565	0.339
End	1.31 ± 0.32	0.76 ± 0.36	**0.000**	**0.000**
$P3$	**0.000**	0.775		
Dark green and orange vegetables and legumes (cup per 1000 kcal)				
Baseline	0.12 ± 0.12	0.13 ± 0.10	0.706	0.6 53
End	0.59 ± 0.30	0.13 ± 0.09	**0.000**	**0.000**
$P3$	**0.000**	0.998		
Total grain (oz per 1000 kcal)				
Baseline	5.92 ± 0.91	6.87 ± 0.89	0.551	0.004
End	7.30 ± 0.72	6.61 ± 1.26	**0.000**	**0.000**
$P3$	**0.049**	0.063		
Whole grain (oz per 1000 kcal)				
Baseline	0.32 ± 0.07	0.62 ± 0.16	0.000	0.000
End	0.85 ± 0.42	0.28 ± 1.04	**0.000**	**0.000**
$P3$	**0.000**	0.256		
Milk (cup per 1000 kcal)				
Baseline	0.11 ± 0.08	0.10 ± 0.04	0.138	0.103
End	0.61 ± 0.244	0.08 ± 0.06	**0.000**	**0.000**
$P3$	**0.000**	0.242		
Meet and beans (oz per 1000 kcal)				
Baseline	1.62 ± 0.52	1.66 ± 0.73	0.613	0.591
End	1.76 ± 0.56	1.48 ± 0.50	**0.030**	**0.025**
$P3$	0.318	0.130		
Oils (gr per 1000 kcal)				
Baseline	9.81 ± 4.74	11.12 ± 2.42	0.553	0.127
End	7.13 ± 2.32	13.80 ± 3.75	**0.000**	**0.000**
$P3$	**0.009**	0.000		
Saturated fat (% of energy)				
Baseline	12.85 ± 2.13	13.47 ± 2.12	0.268	0.419
End	8.57 ± 2.02	13.31 ± 1.80	**0.000**	**0.000**
$P3$	**0.000**	0.614		
Sodium (gr per 1000 kcal)				
Baseline	1.82 ± 1.20	1.64 ± 1.09	0.554	0.356
End	1.02 ± 0.71	1.72 ± 0.97	**0.011**	**0.025**
$P3$	**0.000**	0.341		
SoFAAS (% of energy)				
Baseline	4.84 ± 1.50	6.72 ± 2.32	0.001	0.013
End	5.17 ± 2.07	7.88 ± 1.83	**0.000**	**0.000**
$P3$	0.432	0.044		

Data were expressed as mean ± SD.
$P1$ resulted from independent sample t-test; $P2$ resulted from ANCONA test (adjustment for age and energy intake); and $P3$ resulted from paired sample t-test.

TABLE 5: Serum levels of inflammatory markers in the study groups at the baseline and end of study.

Variables	Educated group (n = 30)	Noneducated group (n = 30)	$P1$	$P2$
Adiponectin				
Baseline	8.52 ± 2.28	8.10 ± 3.74	0.609	0.995
End	11.72 ± 4.24	7.26 ± 3.87	**0.013**	**0.035**
$P3$	**0.027**	0.330		
TNF-α				
Baseline	10.05 ± 4.60	9.86 ± 3.42	0.858	0.809
End	7.91 ± 2.46	9.65 ± 3.85	**0.044**	**0.011**
$P3$	**0.031**	0.605		
Hs-CRP				
Baseline	7.50 ± 3.01	7.00 ± 3.43	0.561	0.535
End	5.12 ± 2.1	7.41 ± 2.96	**0.021**	**0.039**
$P3$	**0.019**	0.271		

Data were expressed as mean ± SD.
$P1$ resulted from independent sample t-test; $P2$ resulted from ANCONA test (adjustment for age, weight and energy intake); and $P3$ resulted from paired sample t-test.

TABLE 6: Healthy eating index components and standards for scoring.

Component	Maximum points	Standard for maximum score	Standard for minimum score of zero
Total fruit (includes 100% juice)	5	≥0.8 cup/1000 kcal	No fruit
Whole fruit (not juice)	5	≥0.4 cup/1000 kcal	No whole fruit
Total vegetables	5	≥1.1 cups/1000 kcal	No vegetables
Dark green and orange vegetables and legumes	5	≥0.4 cup/1000 kcal	No dark green or orange vegetables or legumes
Total grains	5	≥3.0 cups/1000 kcal	No grains
Whole grains	5	≥1.5 oz/1000 kcal	No whole grains
Milk	10	≥1.3 cups/1000 kcal	No milk
Meat and beans	10	≥2.5 oz/1000 kcal	No meat or beans
Oils	10	≥12 grams/1000 kcal	No oil
Saturated fat	10	≤7% of energy	≥15% of energy
Sodium	10	≤0.7 gram/1000 kcal	≥2.0 grams/1000 kcal
Calories from solid fat, alcohol, and added sugar (SoFAAS)	20	≤20% of energy	≥50% of energy

folate, vitamin C, beta carotene, selenium, magnesium, and cupper, which was found mainly in the HEI components (fruits, vegetables, legumes, and whole grains), are positively associated with improved metabolic responses and have beneficial effects on markers of inflammation [25, 26]. However, since foods are not eaten in isolation, recent studies affirmed the importance of the overall dietary pattern or dietary quality, rather than that of specific food groups or nutrients, to inflammatory markers [22]. Although, in the present study, the relation between total HEI components scores and inflammatory markers was not assessed, individuals with higher total HEI scores had significantly higher adiponectin concentration and lower hs-CRP and TNF-α level (Table 6). In this study, it was attempted to account for confounders that were likely associated with concentration of inflammatory markers or consumption of a diet. However, the potential for remaining confounders by uncontrolled covariates was possible, and the present study was also limited by the small sample size, short duration of the intervention, and the self-report of food intake. Finally, the sample used in this randomized clinical trial does not represent a random sample of the Iran population; thus future studies with larger sample size are needed to identify these determinants.

5. Implications for Research and Practice

The results of this study support the hypothesis that beneficial effects of improved dietary quality with respect to obesity may be partially mediated by improvements in plasma concentrations of adiponectin and other biomarkers of systemic inflammation. This effect could be independent of weight loss and the other anthropometric variations. Future studies should extend these findings by investigating potential mechanisms underlying these relations and by examining whether prevention of the diabetes, atherosclerosis, and metabolic syndrome by diet quality improvements is mediated through changes in clinically important biomarkers, including adiponectin, hs-CRP, and TNF-α concentration.

Conflict of Interests

The authors declare that there is no conflict of interests regarding the publication of this paper.

Acknowledgments

This study is issued from the Master of Science thesis for Sara Ebrahimi. Special thanks are due to Arvand International Division of Ahvaz Jundishapur University of Medical Sciences for the financial support (Grant no. B-9115).

References

[1] A. Boynton, M. L. Neuhouser, B. Sorensen, A. McTiernan, and C. M. Ulrich, "Predictors of diet quality among overweight and obese postmenopausal women," *Journal of the American Dietetic Association*, vol. 108, no. 1, pp. 125–130, 2008.

[2] J. L. Fargnoli, T. T. Fung, D. M. Olenczuk, J. P. Chamberland, F. B. Hu, and C. S. Mantzoros, "Adherence to healthy eating patterns is associated with higher circulating total and high-molecular-weight adiponectin and lower resistin concentrations in women from the Nurses' Health Study," *The American Journal of Clinical Nutrition*, vol. 88, no. 5, pp. 1213–1224, 2008.

[3] A. Drewnowski, E. C. Fiddler, L. Dauchet, P. Galan, and S. Hercberg, "Diet quality measures and cardiovascular risk factors in France: applying the healthy eating index to the SU.VI.MAX study," *Journal of the American College of Nutrition*, vol. 28, no. 1, pp. 22–29, 2009.

[4] S. M. George, M. L. Neuhouser, S. T. Mayne et al., "Postdiagnosis diet quality is inversely related to a biomarker of inflammation among breast cancer survivors," *Cancer Epidemiology Biomarkers and Prevention*, vol. 19, no. 9, pp. 2220–2228, 2010.

[5] S. Y. Bu, "Transitional changes in energy intake, skeletal muscle content and nutritional behavior in college students during course-work based nutrition education," *Clinical Nutrition Research*, vol. 2, no. 2, pp. 125–134, 2013.

[6] A. P. Tardivo, J. Nahas-Neto, E. A. P. Nahas, N. Maesta, M. A. H. Rodrigues, and F. L. Orsatti, "Associations between healthy eating patterns and indicators of metabolic risk in postmenopausal women," *Nutrition Journal*, vol. 9, no. 1, article 64, 2010.

[7] N. A. Tek, H. Yildiran, G. Akbulut et al., "Evaluation of dietary quality of adolescents using Healthy Eating Index," *Nutrition Research and Practice*, vol. 5, no. 4, pp. 322–328, 2011.

[8] A. Boynton, M. L. Neuhouser, M. H. Wener et al., "Associations between healthy eating patterns and immune function or inflammation in overweight or obese postmenopausal women," *American Journal of Clinical Nutrition*, vol. 86, no. 5, pp. 1445–1455, 2007.

[9] A. K. Kant and B. I. Graubard, "A comparison of three dietary pattern indexes for predicting biomarkers of diet and disease," *Journal of the American College of Nutrition*, vol. 24, no. 4, pp. 294–303, 2005.

[10] M. L. Neuhouser, R. E. Patterson, I. B. King, N. K. Horner, and J. W. Lampe, "Selected nutritional biomarkers predict diet quality," *Public Health Nutrition*, vol. 6, no. 7, pp. 703–709, 2003.

[11] E. Kennedy, "Putting the pyramid into action: the healthy eating index and food quality score," *Asia Pacific Journal of Clinical Nutrition*, vol. 17, supplement 1, pp. 70–74, 2008.

[12] S. Taechangam, U. Pinitchun, and C. Pachotikarn, "Development of nutrition education tool: healthy eating index in Thailand," *Asia Pacific Journal of Clinical Nutrition*, vol. 17, supplement 1, pp. 365–367, 2008.

[13] E. S. Ford, A. H. Mokdad, and S. Liu, "Healthy Eating Index and C- reactive protein concentration: findings from the National Health and Nutrition Examination Survey III, 1988–1994," *European Journal of Clinical Nutrition*, vol. 59, no. 2, pp. 278–283, 2005.

[14] T. T. Fung, M. L. McCullough, P. K. Newby et al., "Diet-quality scores and plasma concentrations of markers of inflammation and endothelial dysfunction," *The American Journal of Clinical Nutrition*, vol. 82, no. 1, pp. 163–173, 2005.

[15] L. Azadbakht, P. Mirmiran, F. Hosseini, and F. Azizi, "Diet quality status of most Tehranian adults needs improvement," *Asia Pacific Journal of Clinical Nutrition*, vol. 14, no. 2, pp. 163–168, 2005.

[16] S. J. Woodruff, R. M. Hanning, K. McGoldrick, and K. S. Brown, "Healthy eating index-C is positively associated with family dinner frequency among students in grades 6–8 from Southern Ontario, Canada," *European Journal of Clinical Nutrition*, vol. 64, no. 5, pp. 454–460, 2010.

[17] K. M. Hurley, S. E. Oberlander, B. C. Merry, M. M. Wrobleski, A. C. Klassen, and M. M. Black, "The healthy eating index and youth healthy eating index are unique, nonredundant measures of diet quality among low-income, African American adolescents," *Journal of Nutrition*, vol. 139, no. 2, pp. 359–364, 2009.

[18] S. C. de Andrade, M. B. de Azevedo Barros, L. Carandina, M. Goldbaum, C. L. G. Cesar, and R. M. Fisberg, "Dietary quality index and associated factors among adolescents of the state of Sao Paulo, Brazil," *Journal of Pediatrics*, vol. 156, no. 3, pp. 456–460, 2010.

[19] P. Togo, M. Osler, T. I. A. Sørensen, and B. L. Heitmann, "Food intake patterns and body mass index in observational studies," *International Journal of Obesity*, vol. 25, no. 12, pp. 1741–1751, 2001.

[20] A. Esmaillzadeh and L. Azadbakht, "Dairy consumption and circulating levels of inflammatory markers among Iranian women," *Public Health Nutrition*, vol. 13, no. 9, pp. 1395–1402, 2010.

[21] E. Lopez-Garcia, M. B. Schulze, T. T. Fung et al., "Major dietary patterns are related to plasma concentrations of markers of inflammation and endothelial dysfunction," *The American Journal of Clinical Nutrition*, vol. 80, no. 4, pp. 1029–1035, 2004.

[22] C. S. Mantzoros, C. J. Williams, J. E. Manson, J. B. Meigs, and F. B. Hu, "Adherence to the Mediterranean dietary pattern is positively associated with plasma adiponectin concentrations in diabetic women," *The American Journal of Clinical Nutrition*, vol. 84, no. 2, pp. 328–335, 2006.

[23] C. Chrysohoou, D. B. Panagiotakos, C. Pitsavos, U. N. Das, and C. Stefanadis, "Adherence to the Mediterranean diet attenuates inflammation and coagulation process in healthy adults: the ATTICA study," *Journal of the American College of Cardiology*, vol. 44, no. 1, pp. 152–158, 2004.

[24] K. Esposito, A. Pontillo, C. Di Palo et al., "Effect of weight loss and lifestyle changes on vascular inflammatory markers in obese women: a randomized trial," *The Journal of the American Medical Association*, vol. 289, no. 14, pp. 1799–1804, 2003.

[25] N. M. McKeown, J. B. Meigs, S. Liu, E. Saltzman, P. W. F. Wilson, and P. F. Jacques, "Carbohydrate nutrition, insulin resistance, and the prevalence of the metabolic syndrome in the framingham offspring cohort," *Diabetes Care*, vol. 27, no. 2, pp. 538–546, 2004.

[26] N. M. McKeown, J. B. Meigs, S. Liu, P. W. F. Wilson, and P. F. Jacques, "Whole-grain intake is favorably associated with metabolic risk factors for type 2 diabetes and cardiovascular disease in the Framingham Offspring Study," *The American Journal of Clinical Nutrition*, vol. 76, no. 2, pp. 390–398, 2002.

6

Body Composition and Basal Metabolic Rate in Women with Type 2 Diabetes Mellitus

Marina de Figueiredo Ferreira,[1] Filipe Detrano,[1] Gabriela Morgado de Oliveira Coelho,[1,2,3,4] Maria Elisa Barros,[1] Regina Serrão Lanzillotti,[5] José Firmino Nogueira Neto,[6] Emilson Souza Portella,[1] Haydée Serrão Lanzillotti,[1] and Eliane de Abreu Soares[1]

[1] *Nutrition and Health of Nutrition Institute, State University of Rio de Janeiro, St. São Francisco Xavier, No. 524, Block D, 12° Floor, Maracanã, 20559-900 Rio de Janeiro, RJ, Brazil*
[2] *Nutrition Institute, Federal University of Rio de Janeiro, Aloízio Avenue, No. 50, Granja dos Cavaleiros, 27930-560 Macaé, RJ, Brazil*
[3] *Nutrition Department, Estácio de Sá University, St. Bispo, No. 83, 20261-063 Rio de Janeiro, RJ, Brazil*
[4] *Nutrition Department, Arthur Sá Earp Neto College, Barão do Rio Branco Avenue, No. 1003, 25680-120 Petrópolis, RJ, Brazil*
[5] *Mathematics and Statistics Institute, State University of Rio de Janeiro, St. São Francisco Xavier, No. 524, Block B, 6° Floor, Maracanã, 20559-900 Rio de Janeiro, Brazil*
[6] *Lipid Laboratory, Faculty of Medical Science, State University of Rio de Janeiro, Avenida Marechal Rondon, No. 381, 20950-003 Rio de Janeiro, RJ, Brazil*

Correspondence should be addressed to Gabriela Morgado de Oliveira Coelho; gabimorgado@yahoo.com.br

Academic Editor: Michael J. Pagliassotti

Objective. The aim of this study was to determine which of the seven selected equations used to predict basal metabolic rate most accurately estimated the measured basal metabolic rate. *Methods.* Twenty-eight adult women with type 2 diabetes mellitus participated in this cross-sectional study. Anthropometric and biochemical variables were measured as well as body composition (by absorptiometry dual X-ray emission) and basal metabolic rate (by indirect calorimetry); basal metabolic rate was also estimated by prediction equations. *Results.* There was a significant difference between the measured and the estimated basal metabolic rate determined by the FAO/WHO/UNU ($P_{\text{value}} < 0.021$) and Huang et al. ($P_{\text{value}} \leq 0.005$) equations. *Conclusion.* The calculations using Owen et al's. equation were the closest to the measured basal metabolic rate.

1. Introduction

Type 2 diabetes mellitus (T2DM) is the most common type of diabetes mellitus, characterized mostly by obesity and/or a high abdominal fat percentage [1]. The primary strategy for treatment of obese individuals with type 2 diabetes is the loss of body mass (BM), which is associated with better glycemic control [2].

A precondition for an appropriate dietary prescription with the goal of reducing BM is knowing the daily energy needs of individuals with T2DM, which are determined by the total energy expenditure (TEE) of these individuals. Calculating the TEE of individuals or populations requires knowledge of the basal metabolic rate (BMR), which is the major component of TEE [3].

The BMR is influenced by different factors such as hormonal and body composition changes [4], which are features found in obesity and T2DM. In sedentary individuals, the BMR is about 60 to 70% of TEE [5], and a small change can lead to an energy imbalance and changes in BM [6].

The use of prediction equations is the fastest, simplest, and cheapest way to estimate BMR, taking into consideration factors such as gender, age, mass, height, and LBM. However, several authors have shown that these equations can generate errors that overestimate or underestimate the result, without, however, clarifying the magnitude of these errors [3, 7–9].

These deviations can happen because the characteristics of the population, which we want to evaluate, often differ from the characteristics of individuals who participated in the study which originated these equations [8].

Research conducted in different ethnic groups found that the equations from Harris & Benedict [10], FAO/WHO/UNU [11], Schofield [12], and Henry & Rees [13] overestimated the BMR values particularly in individuals living in tropical countries [13–16]. Wahrlich & Anjos [16] justified these differences by the fact that the equations are derived mostly from samples of American and European populations which show differences in body composition, besides living in different environmental conditions.

In Brazilian women, aged 19–27 years of age, living in Niterói (RJ), the equations from Harris & Benedict [10], FAO/WHO/UNU [11], and Henry & Rees [13] overestimated BMR in 18.9%, 12.5%, and 7.2%, respectively [15]. Similarly, women living in tropical regions, ranging in age from three to 60 years, the equation from Schofield [12] overestimated BMR by 5.4% [13] and by 3.8% in Australian who aged 18 to 30 years [14]. Furthermore, the study from Wahrlich et al. [17], performed with Brazilian residents in the southwestern United States, aged 20–60 years, showed that the equations from Harris & Benedict [10], Schofield [12], and Henry & Rees [13] overestimated the BMR by 8.5% to 15%.

Considering the data presented, it can be inferred that the choice of prediction equations for calculating BMR should be careful. The researchers recommend the use of indirect calorimetry or the development of specific equations for the population of interest or even to validate the prediction equation for the population studied [8, 18].

Several studies have evaluated the BMR of adult women without T2DM using prediction equations [7, 17–20]. However, only a few studies have compared the basal metabolism in adult women with T2DM with the prediction equations [6, 21–23]. Therefore, the aim of the present study is to indicate which of the selected equations most accurately reflects basal metabolism.

2. Methods

This cross-sectional study included 92 adult women with T2DM, aged 30–60 years, attended by the Public Health System of Rio de Janeiro (Brazil), from March to December of 2011. Among these, 12 (13.0%) declined to participate. Of the 80 women who accepted the invitation, only 28 (35.0%) were evaluated; 52 were unable to participate because of the following exclusion criteria: 18.8% (n = 15) were using insulin; 18.8% (n = 15) had thyroid dysfunction; 7.5% (n = 6) had cardiomyopathy; 1.2% (n = 1) had liver disease; 1.2% (n = 1) had nephropathy; and 17.5% (n = 14) were smokers.

Informed consent was obtained from all participants in the study. This project was approved by the Ethics Committee in Research from State University of Rio de Janeiro (Brazil) under the number 020.3.2010.

The participants answered a questionnaire to assess personal characteristics, lifestyle habits, physical activity, menstrual cycle, and use of supplements and medication. Sedentary is defined as physical inactivity at the minimum necessary to promote and maintain health (moderate-intensity aerobic physical activity for a minimum of 30 min on five days a week or high-intensity aerobic physical activity for a minimum of 20 min on three days a week).

Anthropometric and body composition evaluations, BMR measurements, and blood samplings were performed at the Interdisciplinary Laboratory of Nutritional Assessment (ILNA) of the Nutrition Institute, State University of Rio de Janeiro (Brazil). The BM (in kilograms) was measured on an electronic platform scale (Filizola; Indústrias Filizola S.A., São Paulo, Brazil) with a maximum capacity of 180 kg and precision of 100 g. Height (in centimeters) was measured using a mobile stadiometer (Alturexata; Alturexata Ltda., MG, Brazil) with an extension of 2 m. Waist circumference (in centimeters) was measured using an inelastic and flexible tape of 0.7 cm width and 0.1 cm accuracy. Measurements were performed by a trained technician with barefoot volunteers wearing minimal clothing and free from accessories, according to the standards of Lohman et al. [24]. The classifications of nutritional status based on body mass index (BMI = kg/m^2) followed the recommendations of WHO [25].

The body composition assessment to estimate lean body mass (LBM) and fat mass (FM) was performed by the method of dual absorptiometry emission X-ray (DXA; Lunar IDXA, GE, USA; Encore software version 12.2). The BMR was measured by indirect calorimetry using the Vmax Encore 29 Calorimeter System (Viasys Healthcare, Inc., Yorba Linda, CA) calibrated every day before collecting data. The measurement was performed in the morning in a relaxed, temperature-controlled, low-light, and noiseless environment. Following the protocol for measuring BMR, the participants were instructed to arrive at the ILNA by car or public transport upon waking up after sleeping for six to eight hours and fasting for at least 12 hours and avoiding heavy physical exercise and alcoholic intake the day before. Adherence to this protocol was checked before starting the measurement of BMR. After the participants remained initially at rest (supine position) for 20 minutes, the gas exchange was measured for 30 minutes by a canopy. Throughout the test, the participants could not sleep, get up, and/or talk. The volumes of CO_2 and O_2 obtained in the first ten minutes were discarded, and the gases obtained during the subsequent 20 minutes of the test were used to determine the BMR in kilocalorie per minute [26]. The means of these values were multiplied by 1440 to obtain the 24-hour BMR. In volunteers who had menstruated, BMR was measured in the follicular phase of the cycle. Two measures of BMR were made in two consecutive days in nine volunteers to assess the quality of the data measured; no variability occurred between the first and second measure ($P_{value} < 0.05$).

The BMR was estimated by seven commonly used prediction equations for women: Harris and Benedict [10], FAO/WHO/UNU [11], Owen et al. [19], Mifflin et al. [20], Gougeon et al. [23], Huang et al. [6], and Rodrigues et al. [18] (Table 1).

Biochemical tests were performed to determine the level of metabolic control; blood samples were collected after an overnight fast of 12 hours on the same day as the BMR

Table 1: Selected prediction equations for estimating basal metabolic rate in women with type 2 diabetes mellitus.

References	Equation for estimation BMR (kcal/day)
Harris and Benedict [10]	655.0955 + (9.5634 × BM) + (1.8496 × Ht) − (4.6756 × Age)
FAO/WHO/UNU [11]	(8.7 × BM) + 829
Owen et al. [19]	795 + (7.18 × BM)
Mifflin et al. [20]	(10 × BM) + (6.25 × Ht) − (5 × Age) − 161
Gougeon et al. [23]	375 + (85 × BM) − (48 × FM) + (63 × FPG)
Huang et al. [6]	71.767 − (2.337 × Age) + (257.293 × 0) + (9.996 × BM) + (4.132 × Ht) + (145.959 × 1)
Rodrigues et al. [18]	BMI > 35 kg/m^2: 172.19 + (10.93 × BM) + (3.10 × Ht) − (2.55 × Age) BMI < 35 kg/m^2: 407.57 + (9.58 × BM) + (2.05 × Ht) − (1.74 × Age)

BM: body mass (kg); Ht: height (cm); FPG: fasting plasma glucose (mM)*; BMI: body mass index (kg/m^2).
*Unit of measure described in the original article of the equation.

measurement. Volunteers were requested to suspend intake of oral hypoglycemic medications on the morning of the biochemical exam day until after blood collection and intake of offered snacks (fruit drink with no added sugar and a whole wheat bread sandwich with no added sugar and white cheese). The following biochemical analyses were performed in the Laboratory of Lipids, Faculty of Medical Sciences, State University of Rio de Janeiro (Brazil): fasting plasma glucose (glucose oxidase/peroxidase method), glycated hemoglobin (immunoturbidimetric assay), total cholesterol (cholesterol oxidase/peroxidase method), HDL-c (direct detergent method), and triglycerides (glycerol phosphate oxidase/peroxidase method). The LDL-c values were calculated by the Friedewald equation, once the plasma triglyceride levels were less than 400 mg/dL. Normal values of biochemical analyses were as follows: fasting plasma glucose of 70–99 mg/dL; glycated hemoglobin < 7.0%; total cholesterol < 200 mg/dL; HDL-c > 50 mg/dL; LDL-c < 100 mg/dL; triglycerides < 150 mg/dL.

The difference between the BMR estimated by equations and BMR measured by indirect calorimetry was calculated (estimated BMR − measured BMR). The percentage of deviation between estimated BMR values for each prediction equation and the measured BMR were calculated as follows: [(estimated BMR − measured BMR)/measured BMR] × 100. The *Kolmogorov-Smirnov* normality test was used to determine the distribution of the variables. Paired Student's t-test was used to estimate the statistical significance of the mean difference between the measured and estimated BMR for each prediction equation. The Bland and Altman method [27] was used to evaluate the agreement between the results of the measured and estimated BMR, and Pearson's correlation coefficient was used to assess the correlation between them. The participants were divided into three categories: normal weight, overweight, and obese, according to the classification of nutritional status [25]. The differences between the means of the BM, LBM, FM, fasting plasma glucose, and glycated hemoglobin and the means of BMR (measured, adjusted by BM, and adjusted by LBM) were evaluated by one-way ANOVA in the three categories of nutritional status, followed by Tukey's test. For inferences, a confidence level of 95% was adopted. We used Pearson's correlation between the dependent variable (BMR) and the independent variables of age, height, BM, BMI, waist circumference, FM, LBM, fasting plasma glucose, and glycated hemoglobin.

Table 2: Characteristics of women with type 2 diabetes mellitus.

Variable	Mean	95% CI
Age (year)	51	(49; 54)
Body mass (kg)	78.4	(72.9; 84.0)
Height (cm)	156.0	(153.8; 158.3)
BMI (kg/m^2)	32.1	(30.0; 34.2)
WC (cm)	95.7	(91.6; 99.8)
LBM (kg)	45.0	(42.5; 47.5)
FM (kg)	33.0	(29.6; 36.3)
FPG (mg/dL)	141.6	(117.4; 165.8)
A1C (%)	7.7	(6.7; 8.6)
Total cholesterol (mg/dL)	195.3	(181.0; 209.6)
HDL-c (mg/dL)	48.7	(45.5; 51.9)
LDL-c (mg/dL)	119.0	(106.1; 132.0)
Triglycerides (mg/dL)	137.6	(111.3; 163.9)

95% CI: 95% confidence interval; BMI: body mass index; WC: waist circumference; LBM: lean body mass; FM: fat mass; FPG: fasting plasma glucose; A1C: glycated hemoglobin.

3. Results

The women in the study were all sedentary (n = 28); 39.3% (n = 11) were of childbearing age; of those on hypoglycemic medication, 71.4% (n = 20) used metformin, 7.1% (n = 2) used glibenclamide, and 21.4% (n = 6) used both. The ages of the volunteers ranged from 37 to 59 years, and BMI results showed that 17.9% (n = 5) were normal weight, 17.9% (n = 5) were preobese, 32.1% (n = 9) were obese class I, 25.0% (n = 7) were obese class II, and 7.1% (n = 2) were obese class III (Table 2). Total cholesterol and plasma triglyceride levels were within established limits. However, blood glucose, glycated hemoglobin, and LDL-c were elevated, and the HDL-c value was below the normal range, which confirms poor metabolic control. Other biochemical tests were in the normal range (Table 2).

TABLE 3: Comparison between the estimated and measured BMR in women with type 2 diabetes mellitus.

Variable	Mean	(95% CI)	P_{value}
Measured BMR (kcal in 24 h)	1411.9	(1300.2; 1523.2)	
Estimated BMR (kcal in 24 h)			
Harris and Benedict [10]	1453.9	(1397.8; 1511.1)	
Difference (1) (kcal in 24 h)	42.3	(−36.1; 120.6)	0.278
Deviation % (2)	5.9	(−0.7; 12.6)	
FAO/WHO/UNU [11]	1511.5	(1463.5; 1559.4)	
Difference (1) (kcal in 24 h)	99.8	(16.5; 183.1)	0.021*
Deviation % (2)	10.6	(2.9; 18.2)	
Owen et al. [19]	1358.2	(1318.6; 1397.8)	
Difference (1) (kcal in 24 h)	−53.5	(−140.5; 33.5)	0.218
Deviation % (2)	−0.5	(−7.5; 6.5)	
Mifflin et al. [20]	1342.1	(1276.7; 1407.4)	
Difference (1) (kcal in 24 h)	−69.6	(−146.6; 7.4)	0.075
Deviation % (2)	−2.6	(−8.4; 3.2)	
Gougeon et al. [23]	1419.1	(1339.1; 1499.0)	
Difference (1) (kcal in 24 h)	7.4	(−69.2; 83.9)	0.845
Deviation % (2)	2.8	(−3.7; 9.3)	
Huang et al. [6]	1526.7	(1466.0; 1587.1)	
Difference (1) (kcal in 24 h)	115.0	(36.9; 193.1)	0.005*
Deviation % (2)	11.3	(4.2; 18.4)	
Rodrigues et al. [18]	1394.0	(1336.3; 1451.8)	
Difference (1) (kcal in 24 h)	−17.6	(−96.4; 61.1)	0.649
Deviation % (2)	1.5	(−4.9; 8.0)	

Paired Student's t-test: $^*P_{value} < 0.05$. BMR: basal metabolic rate; 95% CI: 95% confidence interval.
(1) (Estimated − measured) (kcal in 24 h).
(2) (Difference/measured) × 100 (%).

The variables assumed a normal distribution according to the *Kolmogorov-Smirnov test* (data not shown).

Paired Student's t-test showed a significant difference between the estimated and measured BMR for FAO/WHO/UNU [11] and Huang et al. [6] prediction equations (Table 3).

According to the deviation percentage, the prediction equation that overestimated the measured BMR the most was that of Huang et al. [6] (11.26%; 4 to 18), followed by FAO/WHO/UNU [11] (10.58%; 3 to 18). Similarly, the equation that underestimated the measured BMR the most was that of Mifflin et al. [20] (−2.58%; −8 to 3), and the equation that estimated most closely the measured BMR was that of Owen et al. [19] The coefficient of variation was 20.62% for measured BMR and 7.51 to 12.56% for estimated BMR (Table 3).

The graphs demonstrating the agreement between the values of measured and estimated BMR suggest a poor correlation between the two methods, with wide limits of agreement. However, strong negative correlations were observed ($P_{value} < 0.01$) between methods (Figure 1).

In women with diabetes classified as obese, mean BM, LBM, FM, and measured and estimated BMR were significantly higher ($P_{value} < 0.05$) than in women who were nonobese. However, when the BMR was adjusted for BM and LBM, there was no significant difference between groups (Table 4). There was also no significant difference between groups, when we compared the BMR differences and the percentage deviations.

The correlation shows the association between dependent and independent variables, and BMR was significantly correlated ($P_{value} < 0.01$) with BM ($r = 0.729$), BMI ($r = 0.640$), waist circumference ($r = 0.705$), FM ($r = 0.705$), and LBM ($r = 0.642$). There were no significant correlations between BMR and fasting plasma glucose or between BMR and glycated hemoglobin.

4. Discussion

There is little research that compares the BMR measured by indirect calorimetry with that estimated by prediction equations in adult women with T2DM. Therefore, we selected five equations developed for healthy adult women with different BM [10, 11, 18–20] but only two equations from populations of obese adults with T2DM [6, 23].

Of the two specific equations for populations with T2DM, the estimations determined by Huang et al. [6] equation were significantly different from the BMR measured by indirect calorimetry in the investigated sample. As for Gougeon et al. [23] equation, there was no significant difference, with overestimation of only 2.80%.

When comparing the prediction equations for the assessment of BMR in adult healthy women [10, 11, 18–20] with the measured BMR in women with T2DM investigated in this study, the results were controversial. Although BMR values were overestimated when determined by Harris and Benedict [10] and Rodrigues et al. [18] equations and underestimated when determined by Mifflin et al. [20] equation, these calculated values were not significantly different from the measured BMR. Only the results determined by the FAO/WHO/UNU [11] equation were significantly different from the measured BMR values. In Ryan et al. [22] study, the FAO/WHO/UNU [11] equation also overestimated the BMR measured in French individuals with T2DM in both genders. However, contradicting the results of this study, the Harris and Benedict [10] equation underestimated the BMR measured in individuals with T2DM in other studies [6, 21]. The Harris and Benedict [10] and FAO/WHO/UNU [11] equations, when used for adult women without diabetes mellitus, tend to overestimate the BMR measured by indirect calorimetry by 5 to 15% [7, 17–20]. The authors of these studies justify this variability by noting that these equations were applied to populations of different racial groups with different body composition and life style. Among the equations selected in this study, that of Owen et al. [19] resulted in values that are the closest to those measured by indirect calorimetry, according to the deviation percentage, with the population of his study being 44 healthy American women, who aged 18 to 65 years, classified as lean to obese.

This study revealed that most of the estimations from the selected equations differed from those measured by indirect calorimetry because the equations cannot estimate values with the same consistency and magnitude as the results

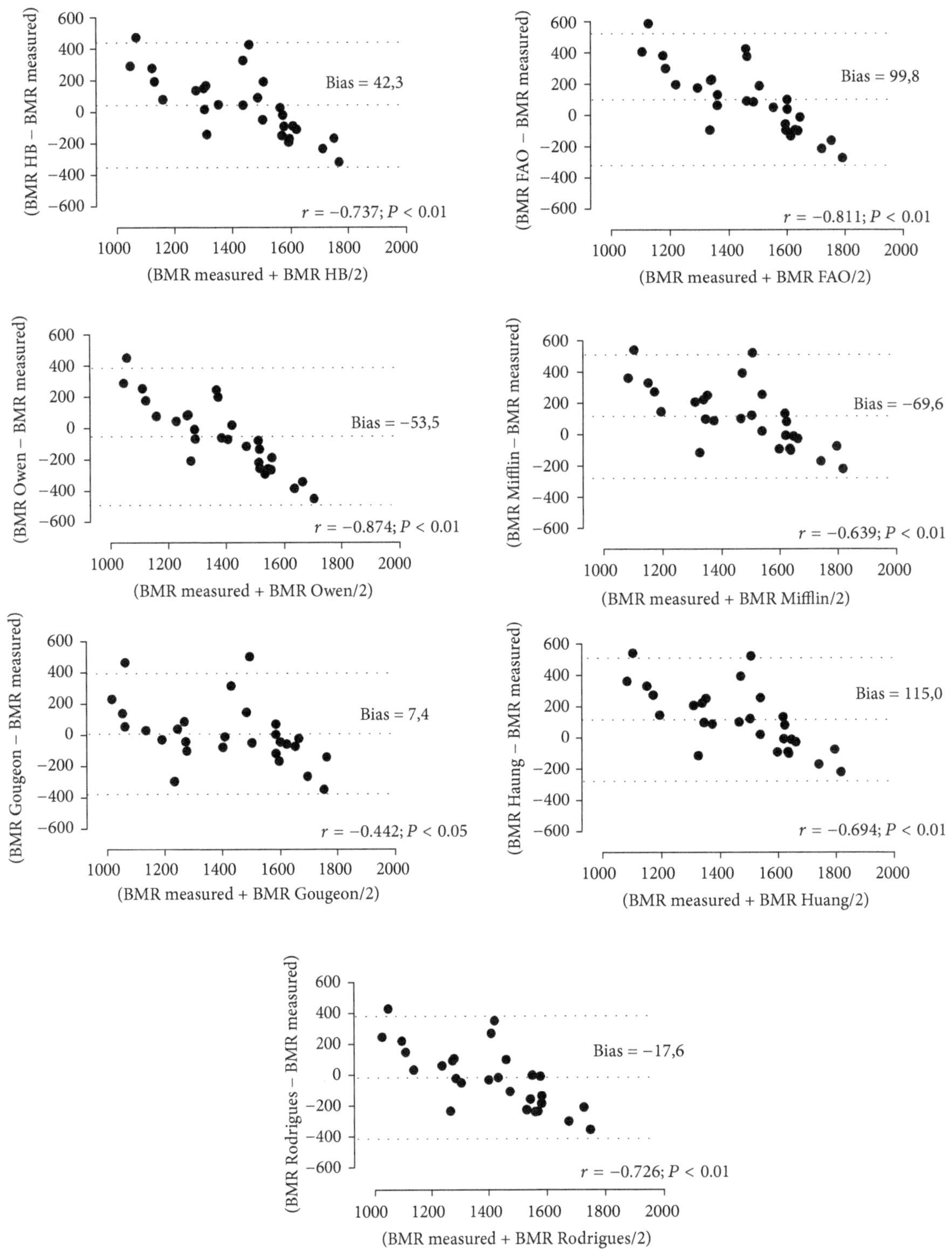

FIGURE 1: Analysis of Bland and Altman association and the difference between the estimated and measured BMR and the difference between the two methods in women with type 2 diabetes mellitus. BMR: basal metabolism rate; HB: Harris e Benedict; FAO, FAO/WHO/UNU; *r*: Pearson's correlation coefficients.

determined by gas exchange. Such discrepancies were also observed by Wahrlich et al. [17] among Brazilian women living in the United States. Bland and Altman [27] warned that discrepancies such as these should be interpreted carefully since it may be clinically relevant. Although the current study has revealed poor agreement between the two methods, it is important to emphasize the high negative correlation found between them. However, the correlation indicates only

TABLE 4: Difference between the means of anthropometric, biochemical, and body composition of women with diabetes mellitus type 2 classified according to nutritional status [18].

	Normal weight (n = 5)		Pre-obese (n = 5)		Obese (n = 18)	
	Mean	(95% CI)	Mean	(95% CI)	Mean	(95% CI)
Age (years)	52.6	(43.2; 62.0)	52.2	(43.0; 61.4)	50.7	(47.6; 53.9)
BM (kg)	57.0	(52.0; 62.0)^{ab}	67.5	(62.2; 72.9)^{ac}	87.4	(83.7; 91.2)^{bc}
LBM (kg)	36.8	(34.9; 38.8)^{b}	39.5	(35.9; 43.1)^{c}	48.8	(46.7; 50.9)^{bc}
FM (kg)	20.0	(14.2; 25.8)^{ab}	27.8	(23.4; 32.2)^{ac}	38.0	(35.6; 40.5)^{bc}
FPG (mg/dL)	129.0	(67.2; 190.8)	102.8	(95.7; 109.9)	155.9	(121.1; 190.7)
A1C (%)	7.4	(5.6; 9.2)	6.2	(5.5; 7.0)	8.1	(6.7; 9.5)
BMR (kcal)	1129.2	(905.0; 1352.4)^{b}	1211.8	(1044.9; 1378.7)^{c}	1545.7	(1418.7; 1672.7)^{bc}
BMR/BM ($kcal \cdot kg^{-1}$)	19.9	(15.2; 24.6)	18.0	(15.7; 20.2)	17.7	(16.4; 18.9)
BMR/LBM ($kcal \cdot kg^{-1}$)	30.7	(24.3; 37.1)	30.6	(28.4; 32.8)	31.8	(29.1; 34.4)
Estimated BMR (kcal)						
Harris and Benedict [10]	1239.9	(1165.5; 1314.2)^{b}	1342.5	(1274.3; 1410.7)^{c}	1544.3	(1502.3; 1586.4)^{bc}
Difference $^{(1)}$	110.7	(−90.2; 311.5)	130.7	(2.3; 259.1)	−1.3	(−133.0; 110.4)
Deviation % $^{(2)}$	11.7	(−7.7; 31.1)	11.7	(−1.4; 24.8)	2.7	(−6.7; 12.1)
FAO/WHO/UNU [11]	1324.6	(1281.1; 1368.0)^{ab}	1416.6	(1369.9; 1463.2)^{ac}	1589.7	(1557.2; 1622.2)^{bc}
Difference $^{(1)}$	195.4	(−36.6; 427.3)	204.8	(55.9; 353.7)	44.1	(−69.9; 158.0)
Deviation % $^{(2)}$	19.7	(−4.0; 43.5)	18.0	(2.0; 34.1)	6.0	(−4.3; 16.3)
Owen et al. [19]	1203.9	(1168.1; 1239.9)^{ab}	1279.9	(1241.5; 1318.4)^{ac}	1422.8	(1396.0; 1449.6)^{bc}
Difference $^{(1)}$	74.8	(−155.3; 304.8)	68.1	(−83.0; 219.3)	−122.9	(−238.5; −7.2)
Deviation % $^{(2)}$	8.8	(−12.7; 30.3)	6.7	(−8.0; 21.3)	−5.1	(−14.4; 4.2)
Mifflin et al. [20]	1111.9	(1010.3; 1213.6)^{b}	1218.4	(1120.7; 1316.1)^{c}	1440.3	(1386.4; 1494.3)^{bc}
Difference $^{(1)}$	−17.2	(−228.9; 194.5)	6.6	(−115.6; 128.9)	−105.3	(−216.2; 5.5)
Deviation % $^{(2)}$	0.1	(−17.6; 17.8)	1.2	(−9.3; 11.7)	−4.4	(−12.8; 4.0)
Gougeon et al. [23]	1126.5	(1078.5; 1174.5)^{b}	1235.2	(1149.3; 1321.2)^{c}	1551.4	(1495.9; 1606.8)^{bc}
Difference $^{(1)}$	−2.7	(−240.2; 234.9)	23.4	(−97.5; 144.4)	5.7	(−106.5; 117.9)
Deviation % $^{(2)}$	1.9	(−19.1; 22.9)	2.6	(−8.0; 13.1)	3.1	(−6.3; 12.5)
Huang et al. [6]	1303.1	(1224.2; 1381.9)^{b}	1408.8	(1333.1; 1484.6)^{c}	1621.5	(1575.1; 1667.8)^{bc}
Difference $^{(1)}$	173.9	(−50.5; 398.2)	197.1	(67.8; 326.3)	75.8	(−34.9; 186.5)
Deviation % $^{(2)}$	17.6	(−4.6; 39.8)	17.2	(3.5; 30.9)	7.8	(−2.01; 17.7)
Rodrigues et al. [18]	1178.8	(1116.3; 1241.3)^{b}	1280.4	(1219.4; 1341.4)^{c}	1485.4	(1441.3; 1529.4)^{bc}
Difference $^{(1)}$	49.6	(−173.4; 272.6)	68.6	(−66.7; 203.9)	−60.3	(−171.0; 50.5)
Deviation % $^{(2)}$	6.4	(−13.7; 26.6)	6.6	(−6.3; 19.4)	−1.2	(−10.2; 7.8)

BM: body mass; LBM: lean body mass; FM: fat mass; FPG: fasting plasma glucose; A1C: glycated hemoglobin; BMR: basal metabolic rate; BMR/BM: basal metabolic rate adjusted for BM; BMR/LBM: basal metabolic rate adjusted for LBM; 95% CI: 95% confidence interval.

One-way ANOVA-Tukey: $^{a\ b\ c}$same letters express significant difference between groups.

$^{(1)}$ (Estimated − measured) (kcal in 24 h).

$^{(2)}$ (Difference/measured) × 100 (%).

how the two methods are linearly interacted not expressing properly the agreement between them.

There is no scientific evidence indicating how the presence of diabetes mellitus may influence basal metabolism. However, some authors have confirmed higher BMR values in subjects with T2DM compared with controls without the disease [6, 21, 23, 28]. The reason for this increase in BMR is not yet well established, and several mechanisms have been proposed to explain it, such as increased protein turnover [29] and elevated plasma concentrations of free fatty acids in fasting [30] and increased gluconeogenesis in patients with T2DM, which is known to be an energy-consuming metabolic pathway [31]. Consoli et al. [32] observed that the increased gluconeogenesis increases BMR by more than 50% in subjects with T2DM. Another factor to consider is the association between T2DM and excessive body weight, as some authors have shown that obese people have both increased FM and LBM, which contributes to the increased BMR [33, 34].

The LBM, which is the most active metabolic tissue of the body, composed of intra- and extracellular water, proteins, carbohydrates, mineral tissues, and essential lipids [35], is the main determinant of BMR [36]. In the present study, when women with T2DM were classified into three groups according to nutritional status (normal weight, overweight, and obese), measured BMR was significantly lower in normal weight and preobese women than in obese women. However, these differences disappeared when the BMR was adjusted for BM and LBM, confirming the evidence found in the literature that LBM is the main determinant of BMR and indicating also that, for women with T2DM, the BM seems to be a determinant of BMR. However, when we tried to understand if the differences in body composition influenced the estimation error of the selected equations in the study, we found no difference between the BMI groups. Thus, we could not elucidate which of the equations had a lower error between normal weight, overweight, and obese groups.

In women with T2DM evaluated in this study, the best correlation found with BMR was BM ($r = 0.729$). This result is in agreement with the study done with severely obese Australian adults with and without T2DM [6] that found a better correlation ($r = 0.694$) of BM with BMR. These findings corroborate the importance of using the BM as an independent variable in the prediction equations to correctly estimate the BMR of women with T2DM since the equations selected in this study included BM as independent variable [6, 10, 11, 18–20, 23]. It is vital to estimate more accurately the BMR of women with T2DM and preobesity and/or obesity to provide an individualized program for food planning aimed at glycemic and BM control in these patients.

Gougeon et al. [23] evaluated the BMR of women with T2DM, proposing an equation to predict BMR that tested plasma glucose and glycated hemoglobin levels as some of its independent variables, justifying a better adjustment in the model equation. Huang et al. [6] indicated that both the plasma glucose levels and the glycated hemoglobin should be included in the model. However, in this study, although these variables were also considered, there were no significant correlations ($P_{\text{value}} = 0.283$ and 0.251) for fasting plasma glucose and glycated hemoglobin, respectively, with BMR. This suggests that other metabolic factors, not controlled in this study, could influence the BMR of women with T2DM.

To obtain a more homogeneous study population and, therefore, observe the characteristics displayed by the evaluated group without the influence of other factors that could affect the basal metabolism, strict inclusion criteria were adopted in this study in the selection of volunteers, which was not always observed in other studies. The rigidity in the selection criteria resulted in a reduction of the sample size, which is one of the limitations of this study. However, it is emphasized that some authors [6, 21, 23], evaluating the BMR of patients with diabetes mellitus, did not take into account the difference between genders, the type of diabetes, or the presence of other diseases.

It is necessary in future research to compare the BMR of individuals with and without T2DM to elucidate the association of T2DM with obesity and other intercurrent factors. Likewise, it is necessary to validate the BMR prediction equations, including women with T2DM with their inherent characteristics, in study populations normally found in public or private clinics.

The findings showed that among the selected prediction equations, the BMR estimated by Owen et al. [19] equation was the closest to the measured BMR as assessed by the percentage deviation.

Abbreviations

T2DM:	Type 2 diabetes mellitus
BM:	Body mass
TEE:	Total energy expenditure
BMR:	Basal metabolic rate
ILNA:	Interdisciplinary Laboratory of Nutritional Assessment
BMI:	Body mass index
LBM:	Lean body mass
FM:	Fat mass
DXA:	Dual absorptiometry emission X-ray
β_{value}:	Regression coefficients
R^2:	Coefficient of determination
Ht:	Height
FPG:	Fasting plasma glucose
95% CI:	95% confidence interval
WC:	Waist circumference
A1C:	Glycated hemoglobin
BMR/BM:	Basal metabolic rate adjusted for BM
BMR/LBM:	Basal metabolic rate adjusted for LBM
HB:	Harris-Benedict equation
FAO/WHO/UNU:	Food and Agriculture Organization/World Health Organization/United Nations University equation
r:	Pearson's correlation coefficients.

Conflict of Interests

The authors declare that there is no conflict of interests regarding the publication of this paper.

Authors' Contribution

The authors responsibilities were as follows: Marina de Figueiredo Ferreira, Filipe Detrano, Maria Elisa Barros, Regina Serrão Lanzillotti, José Firmino Nogueira Neto, Emilson Souza Portella, Haydée Serrão Lanzillotti, Gabriela Morgado de Oliveira Coelho, and Eliane de Abreu Soares designed the research; Marina de Figueiredo Ferreira performed literature review; Marina de Figueiredo Ferreira, Filipe Detrano, and José Firmino Nogueira Neto participated in the development; Marina de Figueiredo Ferreira, Filipe Detrano, Regina Serrão Lanzillotti, and Haydée Serrão Lanzillotti performed statistical analysis; Maria Elisa Barros, José Firmino Nogueira Neto, Emilson Souza Portella, Haydée Serrão Lanzillotti, Gabriela Morgado de Oliveira Coelho, and Eliane de Abreu Soares performed critical revision of intellectual content. All authors read and approved the final paper.

Acknowledgments

The teams of Laboratory of Lipids, Faculty of Medical Sciences, State University of Rio de Janeiro (Brazil) and the Interdisciplinary Laboratory of Nutritional Assessment (ILNA), Nutrition Institute, State University of Rio de Janeiro (Brazil) and the State of Rio de Janeiro Carlos Chagas Filho Research Foundation (Fundação Carlos Chagas Filho de Amparo à Pesquisa do Estado do Rio de Janeiro (FAPERJ)), no. E-26/111.621/2010, are acknowledged.

References

[1] World Health Organization, *Definition, Diagnosis and Classification of Diabetes Mellitus and Its Complications*, Report of a WHO Consultation, WHO, Geneva, Switzerland, 1999.

[2] C. A. Maggio and F. X. Pi-Sunyer, "The prevention and treatment of obesity: application to type 2 diabetes," *Diabetes Care*, vol. 20, no. 11, pp. 1744–1766, 1997.

[3] V. Wahrlich and L. A. Anjos, "Aspectos históricos e metodológicos da medição e estimativa da taxa metabólica basal: uma revisão da literatur," *Cadernos de Saúde Pública*, vol. 17, no. 4, pp. 801–817, 2001.

[4] H. K. M. Antunes, R. F. Santos, R. A. Boscolo, O. F. A. Buenol, and M. T. de Mello, "Análise da taxa metabólica basal e composição corporal de idosos do sexo masculino antes e seis meses após exercício de resistência," *Revista Brasileira de Medicina do Esporte*, vol. 11, no. 1, pp. 71–75, 2005.

[5] E. Ravussin and B. A. Swinburn, "Pathophysiology of obesity," *The Lancet*, vol. 340, no. 8816, pp. 404–408, 1992.

[6] K.-C. Huang, N. Kormas, K. Steinbeck, G. Loughnan, and I. D. Caterson, "Resting metabolic rate in severely obese diabetic and nondiabetic subjects," *Obesity Research*, vol. 12, no. 5, pp. 840–845, 2004.

[7] M. Siervo, V. Boschi, and C. Falconi, "Which REE prediction equation should we use in normal-weight, overweight and obese women?" *Clinical Nutrition*, vol. 22, no. 2, pp. 193–204, 2003.

[8] C. L. Kien and F. Ugrasbul, "Prediction of daily energy expenditure during a feeding trial using measurements of resting energy expenditure, fat-free mass, or Harris-Benedict equations," *American Journal of Clinical Nutrition*, vol. 80, no. 4, pp. 876–880, 2004.

[9] D. Frankenfield, L. Roth-Yousey, and C. Compher, "Comparison of predictive equations for resting metabolic rate in healthy nonobese and obese adults: a systematic review," *Journal of the American Dietetic Association*, vol. 105, no. 5, pp. 775–789, 2005.

[10] J. A. Harris and F. G. Benedict, *A Biometric Study of Basal Metabolism in Man*, Carnigie Institution of Washington, Boston, Mass, USA, 1919.

[11] Food and Agriculture Organization, World Health Organization, and United Nations University, "Energy and protein requirements," WHO Technical Report Series 724, WHO, Geneve, Switzerland, 1985.

[12] W. N. Schofield, "Predicting basal metabolic rate, new standards and review of previous work," *Human Nutrition: Clinical Nutrition*, vol. 39, supplement 1, pp. 5–41, 1985.

[13] C. J. K. Henry and D. G. Rees, "New predictive equations for the estimation of basal metabolic rate in tropical peoples," *European Journal of Clinical Nutrition*, vol. 45, no. 4, pp. 177–185, 1991.

[14] L. S. Piers, B. Diffey, M. J. Soares et al., "The validity of predicting the basal metabolic rate of young Australian men and women," *European Journal of Clinical Nutrition*, vol. 51, no. 5, pp. 333–337, 1997.

[15] C. M. Cruz, A. F. Da Silva, and L. A. Dos Anjos, "A taxa metabólica basal é superestimada pelas equações preditivas em universitárias do Rio de Janeiro, Brasil," *Archivos Latinoamericanos de Nutricion*, vol. 49, no. 3, pp. 232–237, 1999.

[16] V. Wahrlich and L. A. Anjos, "Validação de equações de predição da taxa metabólica basal em mulheres residentes em Porto Alegre, RS, Brasil," *Revista de Saúde Pública*, vol. 35, no. 1, pp. 39–45, 2001.

[17] V. Wahrlich, L. A. Anjos, S. B. Going, and T. G. Lohman, "Basal metabolic rate of Brazilians living in the Southwestern United States," *European Journal of Clinical Nutrition*, vol. 61, no. 2, pp. 289–293, 2007.

[18] A. E. Rodrigues, M. C. Mancini, L. Dalcanale et al., "Padronização do gasto metabólico de repouso e proposta de nova equação para uma população feminina brasileira," *Arquivos Brasileiros de Endocrinologia & Metabologia*, vol. 54, pp. 470–476, 2010.

[19] O. E. Owen, E. Kavle, R. S. Owen et al., "A reappraisal of caloric requirements in healthy women," *The American Journal of Clinical Nutrition*, vol. 44, no. 1, pp. 1–19, 1986.

[20] M. D. Mifflin, S. T. St Jeor, L. A. Hill, B. J. Scott, S. A. Daugherty, and Y. O. Koh, "A new predictive equation for resting energy expenditure in healthy individuals," *The American Journal of Clinical Nutrition*, vol. 51, no. 2, pp. 241–247, 1990.

[21] V. Rigalleau, C. Lasseur, S. Pécheur et al., "Resting energy expenditure in uremic, diabetic, and uremic diabetic subjects," *Journal of Diabetes and its Complications*, vol. 18, no. 4, pp. 237–241, 2004.

[22] M. Ryan, A. Sallé, G. Guilloteau, M. Genaitay, M. B. E. Livingston, and P. Ritz, "Resting energy expenditure is not increased in mildly hyperglycaemic obese diabetic patients," *British Journal of Nutrition*, vol. 96, no. 5, pp. 945–948, 2006.

[23] R. Gougeon, M. Lamarche, J.-F. Yale, and T. Venuta, "The prediction of resting energy expenditure in type 2 diabetes mellitus is improved by factoring for glycemia," *International Journal of Obesity*, vol. 26, no. 12, pp. 1547–1552, 2002.

[24] T. G. Lohman, A. F. Roche, and R. Martorel, *Anthropometric Standardization Reference Manual*, Human Kinetics Books, Champaign, Ill, USA, 1988.

[25] World Health Organization, *Obesity: Preventing and Managing the Global Epidemic*, Report of a WHO Consultation on Obesity, World Health Organization, Geneva, Switzerland, 1998.

[26] J. B. Weir, "New methods for calculating metabolic rate with special reference to protein metabolism," *The Journal of Physiology*, vol. 109, no. 1-2, pp. 1–9, 1949.

[27] J. M. Bland and D. G. Altman, "Statistical methods for assessing agreement between two methods of clinical measurement," *The Lancet*, vol. 1, no. 8476, pp. 307–310, 1986.

[28] C. Weyer, C. Bogardus, and R. E. Pratley, "Metabolic factors contributing to increased resting metabolic rate and decreased insulin-induced thermogenesis during the development of type 2 diabetes," *Diabetes*, vol. 48, no. 8, pp. 1607–1614, 1999.

[29] P. R. Payne and J. C. Waterlow, "Relative energy requirements for maintenance, growth, and physical activity," *The Lancet*, vol. 2, no. 7717, pp. 210–211, 1971.

[30] E. E. Tredget, "The metabolic effects of thermal injury," *World Journal of Surgery*, vol. 16, no. 1, pp. 68–79, 1992.

[31] P. Felig, J. Wahren, and R. Hendler, "Influence of maturity-onset diabetes on splanchnic glucose balance after oral glucose ingestion," *Diabetes*, vol. 27, no. 2, pp. 121–126, 1978.

[32] A. Consoli, N. Nurjhan, F. Capani, and J. Gerich, "Predominant role of gluconeogenesis in increased hepatic glucose production in NIDDM," *Diabetes*, vol. 38, no. 5, pp. 550–557, 1989.

[33] E. Ferrannini, "Physiological and metabolic consequences of obesity," *Metabolism*, vol. 44, no. 3, pp. 15–17, 1995.

[34] J.-P. Flatt, "Differences in basal energy expenditure and obesity," *Obesity*, vol. 15, no. 11, pp. 2546–2548, 2007.

[35] V. H. Heyward and L. M. Stolarczyk, "Body composition basics," in *Applied Body Composition Assessment*, V. H. Heyward and L. M. Stolarczyk, Eds., Human Kinetics, Champaign, Ill, USA, 1996.

[36] E. Ravussin, S. Lillioja, T. E. Anderson, L. Christin, and C. Bogardus, "Determinants of 24-hour energy expenditure in man. Methods and results using a respiratory chamber," *Journal of Clinical Investigation*, vol. 78, no. 6, pp. 1568–1578, 1986.

7

Sex-Related Differences in the Effects of the Mediterranean Diet on Glucose and Insulin Homeostasis

Alexandra Bédard,[1,2] Louise Corneau,[1] Benoît Lamarche,[1,2] Sylvie Dodin,[1,3] and Simone Lemieux[1,2]

[1] *Institute of Nutrition and Functional Foods (INAF), Laval University, 2440 Hochelaga Boulevard, Québec, QC, Canada G1V 0A6*
[2] *Department of Food Science and Nutrition, Pavillon Paul-Comtois, Laval University, 2425 rue de l'Agriculture, Québec, QC, Canada G1V 0A6*
[3] *Department of Obstetrics and Gynaecology, Pavillon Ferdinand-Vandry, Laval University, 1050 Medicine Avenue, Québec, QC, Canada G1V 0A6*

Correspondence should be addressed to Simone Lemieux; simone.lemieux@fsaa.ulaval.ca

Academic Editor: Duo Li

Objective. To document sex differences in the impact of the Mediterranean diet (MedDiet) on glucose/insulin homeostasis and to verify whether these sex-related effects were associated with changes in nonesterified fatty acids (NEFA). *Methods*. All foods were provided to 38 men and 32 premenopausal women (24–53 y) during 4 weeks. Variables were measured during a 180 min OGTT before and after the MedDiet. *Results*. A sex-by-time interaction for plasma insulin iAUC was found (men: −17.8%, $P = 0.02$; women: +9.4%, $P = 0.63$; P for sex-by-time interaction = 0.005). A sex-by-time interaction was also observed for insulin sensitivity (Cederholm index, $P = 0.03$), for which only men experienced improvements (men: +8.1%, $P = 0.047$; women: −5.9%, $P = 0.94$). No sex difference was observed for glucose and C-peptide responses. Trends toward a decrease in NEFA AUC ($P = 0.06$) and an increase in NEFA suppression rate ($P = 0.06$) were noted, with no sex difference. Changes in NEFA were not associated with change in insulin sensitivity. *Conclusions*. Results suggest that the more favorable changes in glucose/insulin homeostasis observed in men compared to women in response to the MedDiet are not explained by sex differences in NEFA response. This clinical trial is registered with clinicaltrials.gov NCT01293344.

1. Introduction

Diabetes has reached epidemic proportion with 382 million people worldwide affected by this disease [1]. An important sex difference has been highlighted in the health consequences of type 2 diabetes mellitus with a 50% higher risk of coronary heart disease mortality in diabetic women compared to diabetic men [2]. In addition to this higher mortality risk found in diabetic women, there is growing evidence that the response to early lifestyle interventions aiming at counteracting insulin resistance and thereby reducing the risk of developing diabetes and its cardiovascular complications may be sex-specific, with women having generally fewer benefits than men [3]. In fact, greater improvements in insulin sensitivity have been observed in men compared to women in response to exercise training interventions [4, 5]. Moreover, it has been suggested that body weight loss improves insulin sensitivity to a greater extent in men [6]. Very few studies have yet documented potential sex-related differences in response to the adoption of healthy dietary habits. Moreover, the few studies that have examined this issue were generally achieved in uncontrolled nutritional contexts and gender differences in the adherence to prescribed dietary habits in these studies may have confounded the interpretation of results obtained.

The adoption of the traditional Mediterranean diet (MedDiet) is now widely recommended for its various health benefits [7]. Among others, the adoption of this dietary pattern has been identified as a useful approach to improve glycemic control and prevent the development of insulin resistance and type 2 diabetes mellitus [8, 9]. However, little is known about how the MedDiet leads to improvements in glucose and insulin homeostasis. One possible mechanism is a decrease in nonesterified fatty acid (NEFA) concentration in response to the MedDiet. In fact, from a metabolic standpoint, elevated NEFA concentration is established as one of the main contributing factors for the development of insulin resistance through the impairment of both hepatic glucose production and insulin action in peripheral tissues [10].

Using a fully controlled nutritional context, the aim of this study was therefore to examine how sex modifies the short-term impact of a MedDiet on variables related to glucose and insulin homeostasis. We also verified if changes in glucose and insulin homeostasis were related to changes in NEFA concentration in men and women.

2. Subjects and Methods

2.1. Subjects. Thirty-eight men and 32 premenopausal women aged from 24 to 53 years were recruited. In addition to having slightly elevated plasma low-density lipoprotein- (LDL-) cholesterol concentration (between 3.4 and 4.9 mmol/L) or a total cholesterol to high-density lipoprotein- (HDL-) cholesterol ratio ≥5.0, participants had at least one of the four following cardiovascular risk factors: (1) waist circumference > 94 cm in men and >80 cm in women [12]; (2) triacylglycerol (TAG) > 1.7 mmol/L; (3) fasting glycemia between 6.1 and 6.9 mmol/L; and (4) blood pressure concentrations ≥ 130/85 mm Hg. Exclusion criteria included a significant weight change (>2.5 kg) in the three months before the study, cardiovascular events, prior diagnosis of type 2 diabetes mellitus, use of medication that could affect dependent variables under study (namely, lipid-lowering, hypoglycemic, insulin sensitizers and antihypertensive medications), smoking, pregnancy, and use of systemic hormonal contraceptives. Premenopausal status was determined by a regular menstrual cycle for the last three months and a FSH measurement <20 IU/L during the early follicular phase. Power analyses for repeated measures and within-between interactions showed that a total sample size of $n = 68$ is sufficient to detect significant differences in the outcomes with a small effect-size estimate (Cohen's d of 0.15 [13]) and with an $\alpha = 0.05$ and a power ($1 - \beta$ error probability) of 0.95 (G*Power Version 3.0.10, Franz Faul, Universität Kiel, Germany). The present study was conducted according to the guidelines laid down in the Declaration of Helsinki (1964). All procedures involving human subjects were approved by the Laval University Research Ethics Committee on human experimentation. Written informed consent was obtained from all subjects. This clinical trial was registered at clinicaltrials.gov as NCT01293344.

2.2. Study Design. The design and methods of this study were previously described in detail [11]. Firstly, during a 4-week run-in period, participants had to comply with the recommendations of Canada's Food Guide [14] as prescribed by a registered dietician. Canada's Food Guide is an educational tool which promotes a healthy eating pattern in order to reduce the risk of many chronic diseases and to achieve overall health and vitality. Participants were also instructed to maintain constant their body weight and physical activity participation. The purpose of this run-in period was to ensure similar dietary habits between men and women prior to the controlled MedDiet phase, a goal which has been achieved as reported in a previous publication [11].

The 4-week controlled MedDiet followed immediately the run-in period. All foods and drinks were provided for participants according to a 7-day cyclic menu. It has been shown that 4 weeks of nutritional intervention under controlled conditions is sufficient to obtain significant changes in glucose/insulin homeostasis [15–18]. Details about the composition of the MedDiet have been previously reported [11]. The percentages of energy derived from lipids, carbohydrates, proteins, and alcohol were, respectively, of 32%, 46%, 17%, and 5%. Participants were instructed to consume only foods and drinks provided and their entire meals. Compliance of participants was evaluated with a checklist throughout the controlled MedDiet phase. Participants were also instructed to maintain their usual physical activity.

The controlled MedDiet phase aimed at being isocaloric in order to control for the potential confounding effect of body weight change. On each weekday, body weight was measured immediately before lunch and foods and energy provided were revised if necessary to minimize body weight fluctuations. In women, all tests were carried out in the early follicular phase of their menstrual cycle (from the third to ninth day of the menstrual cycle; mean duration of the feeding period in women is 28.8 ± 4.3 days) since fluctuations in female hormones may influence glucose and insulin homeostasis [19].

2.3. Laboratory Measurements. After a 12-hour overnight fast, basic lipid profile was measured after the run-in period (i.e., just before the controlled MedDiet phase). Serum cholesterol, HDL-cholesterol, and TAG concentrations were measured using commercial reagents on a Modular P chemistry analyzer (Roche Diagnostics, Mannheim, Germany). LDL-cholesterol was obtained by the equation of Friedewald et al. [20]. Moreover, a 180 min oral glucose tolerance test (OGTT) with 75 g of glucose in solution was performed just before and after the controlled MedDiet. Blood samples were collected into vacutainer tubes containing ethylene diamine tetra-acetic acid (EDTA) at −15, 0, 15, 30, 45, 60, 90, 120, 150, and 180 min. At each time, plasma glucose concentration was determined using the hexokinase-glucose-6-phosphate dehydrogenase method [21], plasma insulin concentration was measured by radioimmunoassay [22], plasma C-peptide concentration was measured by a modification of the method of Heding [23] with polyclonal antibody A-4741 from Ventrex (Portland, ME, USA) and polyethylene glycol precipitation [22], and plasma NEFA concentration was assessed with an enzymatic detection kit (ZenBio, Research Triangle Park, NC, USA). Fasting values were computed as the mean of −15 and 0 min values. Incremental areas under the curve (iAUC) for glucose, insulin, and C-peptide and the area under the curve (AUC) for NEFA were calculated for the 180 min interval using the trapezoidal method.

The incremental peak was calculated as the difference between the concentration of glucose, insulin, and C-peptide at peak and the concentration in the fasting state during the OGTT. The percentage of suppression of NEFA was calculated in each participant using this formula: [(fasting NEFA − the lowest NEFA concentration during the OGTT)/fasting NEFA] × 100. Hepatic insulin sensitivity was assessed by the homeostasis model assessment (HOMA-IS) approach index (1/[fasting glucose (mmol/L) × fasting insulin (mIU/L)/22.5]). Peripheral insulin sensitivity was assessed with the Cederholm index according to the following formula: [75 000 + (fasting glucose − glucose 120 min postload) × 1.15 × 180 × 0.19 × body weight]/[120 × log (mean insulin) × mean glucose] in which glucose concentrations are expressed in mmol/L, insulin concentrations in mIU/L, and body weight in kg [24].

As proposed by the Canadian Diabetes Association [25], normal glucose tolerance was defined as a fasting plasma glucose <6.1 mmol/L and plasma glucose 120 min postload <7.8 mmol/L, isolated impaired fasting glucose as fasting glucose between 6.1 and 6.9 and glucose 120 min postload <7.8 mmol/L, isolated impaired glucose tolerance as fasting glucose <6.1 mmol/L and glucose 120 min postload between 7.8 and 11.0 mmol/L, and both impaired fasting glucose and impaired glucose tolerance as fasting glucose between 6.1 and 6.9 and glucose 120 min postload between 7.8 and 11.0 mmol/L and type 2 diabetes mellitus was defined as fasting glucose ≥7.0 mmol/L or glucose 120 min postload ≥11.1 mmol/L.

2.4. Anthropometric and Blood Pressure Measurements. After the run-in period (i.e., just before the controlled MedDiet phase), body weight, height, and waist circumference measurements were performed using standardized methods [26]. Systolic and diastolic blood pressure levels were measured on the right arm using an automated blood pressure monitor (BPM 300-BpTRU: Vital Signs Monitor) as previously described [11].

2.5. Statistical Analyses. Data were collected after the run-in period (i.e., just before the controlled MedDiet phase) and immediately after the controlled MedDiet. For variables not normally distributed, a transformation was performed. To determine differences between men and women for characteristics before the controlled MedDiet phase, Student's t-test for unpaired data was used. Time and sex-by-time interaction effects on dependent variables were assessed by using MIXED procedures for repeated measurements followed by Tukey-Kramer tests. Statistical analyses were adjusted for values before the MedDiet phase when a difference between men and women was observed. Associations between variables were assessed by Pearson's correlation analyses. Although the controlled MedDiet phase aimed at being isocaloric, both men and women had a small but significant weight loss (−1.2 kg or 1.3% of initial body weight in men and −0.5 kg or 0.7% in women). However, no significant change was observed in both men and women for waist circumference (−0.3 cm or 0.3% of initial waist circumference in men and −0.8 cm or 0.8% in women). All results are adjusted for this small change in body weight during the controlled MedDiet phase. We excluded one man from our analyses due to illness, which led to a significant reduction in food intake during several days at the end of the controlled MedDiet phase. Therefore, 37 men and 32 women were included in the analyses. A $P \leq 0.05$ (two-sided) was considered as statistically significant. All analyses were performed with the SAS statistical package version 9.2 (SAS Institute Inc., Cary, NC, USA).

3. Results

3.1. Participant Characteristics after the Run-In Period. As previously reported in another publication [11], men and women had similar age and BMI (Table 1). Waist circumference measurements were higher in men than in women. On average, both men and women were characterized by a slightly deteriorated lipid profile and normal blood pressure levels, but some sex differences were noted for these metabolic variables. Specifically, men displayed higher values for TAG and systolic and diastolic blood pressure and a lower value for HDL-cholesterol concentration than women. For variables related to glucose and insulin homeostasis, men and women had comparable values before the controlled MedDiet, except for fasting glucose, for which men had a higher value than women. There was no sex difference for C-peptide and NEFA concentrations in the fasting state.

Forty-one participants had a normal glucose tolerance (20 men and 21 women), 10 had an isolated impaired fasting glucose (nine men and one woman), nine were characterized by an isolated impaired glucose tolerance (five men and four women), and four participants had both an impaired fasting glucose and an impaired glucose tolerance (two men and two women). Even if none of the participants had received a diagnosis of type 2 diabetes mellitus prior to their inclusion in this study, one man and four women met diagnosis criteria for type 2 diabetes mellitus.

3.2. Glucose, Insulin, and C-Peptide Responses to the MedDiet. There was no sex-by-time interaction or time effect for fasting glucose, insulin, and C-peptide concentrations (Figure 1). During the 75 g OGTT, a sex-by-time interaction for plasma insulin iAUC was found, with men having a significant decrease whereas women experienced a nonsignificant increase (men −17.8%, $P = 0.02$, women +9.4%, $P = 0.63$, P for sex-by-time interaction = 0.005) (Figure 1). More precisely, when statistical analyses were performed at each time point of the OGTT, sex-by-time interaction effects were found at 60 and 120 min (resp., $P = 0.02$ and $P = 0.03$). Specifically, a significant increase in insulin concentrations at 60 min was observed in women (+21.1%, $P = 0.04$) whereas men showed a nonsignificant decrease (−3.4%, $P = 0.95$). For insulin concentration at 120 min, men showed a greater reduction after MedDiet than women (−25.4% in men, $P =$ 0.03 and −10.3% in women, $P = 0.99$). There was no sex-by-time interaction or time effect for glucose and C-peptide

Table 1: Characteristics of men and women after the run-in period.

	Men (n = 37)		Women (n = 32)		Sex difference
	Mean	SD	Mean	SD	P value
Age (years)[a]	42.6	7.3	41.2	7.3	0.43
Body weight (kg)[a,b]	91.8	14.0	78.0	14.7	**<0.0001**
BMI (kg/m^2)[a,b]	29.2	3.2	29.6	5.4	0.97
Waist circumference (cm)[a,b]	102.6	10.7	96.4	10.5	**0.01**
TAG (mmol/L)[a,b]	1.86	1.17	1.36	0.63	**0.03**
Total cholesterol (mmol/L)[a]	5.56	0.91	5.40	0.60	0.38
LDL cholesterol (mmol/L)[a]	3.61	0.72	3.47	0.52	0.36
HDL cholesterol (mmol/L)[a,b]	1.09	0.31	1.30	0.26	**0.002**
Systolic blood pressure (mm Hg)[a]	117.1	12.6	108.6	10.4	**0.004**
Diastolic blood pressure (mm Hg)[a]	80.3	9.0	73.5	9.0	**0.003**
Fasting glucose (mmol/L)[a,b]	5.89	0.37	5.68	0.63	**0.04**
Glucose 120 min postload (mmol/L)[a]	6.36	1.81	7.02	2.48	0.21
Fasting insulin (pmol/L)[a,b]	96.0	57.6	86.2	79.3	0.19
Insulin 120 min postload (pmol/L)[a,b]	493.8	443.0	520.9	588.7	0.84
Fasting C-peptide (nmol/L)[b]	1.04	0.37	0.94	0.53	0.21
Fasting NEFA (umol/L)[b]	497.8	180.8	563.8	199.7	0.15

SD, standard deviation; BMI, body mass index; TAG, triacylglycerol; LDL, low-density lipoprotein; HDL, high-density lipoprotein; NEFA, nonesterified fatty acids.
These characteristics were measured after the run-in period that is immediately before the controlled MedDiet phase. Differences between men and women were assessed by Student's t-test for unpaired data.
[a]These characteristics have been reported in a previous publication [11].
[b]Analysis was performed on transformed values.

Table 2: Glucose and insulin homeostasis response to the 4-week controlled Mediterranean diet.

	Men (n = 37)				Women (n = 32)				Time	Sex-by-time interaction
	Before	SEM	After	SEM	Before	SEM	After	SEM		P value
Incremental peak										
Glucose (mmol/L)[a,b]	4.66	0.25	4.66	0.26	3.35	0.30	3.53	0.28	0.36	0.32
Insulin (pmol/L)[a]	864	95	787	69	715	99	688	77	0.52	0.57
C-peptide (nmol/L)[b,c]	4.33	0.31	4.14	0.24	3.47	0.24	3.35	0.17	0.37	0.87
HOMA-IS index[a,d]	0.408	0.105	0.356	0.041	0.606	0.172	0.496	0.071	0.55	0.68
Cederholm index[a]	20.9	1.1	22.6	0.9[e]	24.6	1.5	23.1	1.2	0.18	**0.03**
NEFA suppression rate (%)	73.3	2.0	74.1	2.3	70.5	2.1	75.1	2.0	0.06	0.18

SEM, standard error of the mean; HOMA-IS index, homeostasis model assessment for insulin sensitivity index; NEFA, nonesterified fatty acids.
Time and sex-by-time interaction effects on dependent variables were assessed by using MIXED procedures for repeated measurements followed by Tukey-Kramer tests.
[a]Analysis was performed on transformed values.
[b]Differences between men and women were observed with Student's t-test before the controlled MedDiet phase; $P = 0.0004$ for the incremental peak of glucose and $P = 0.04$ for the incremental peak of C-peptide. For these variables, statistical analyses were adjusted for values before the controlled MedDiet phase.
[c]For women, $n = 31$ for the incremental peak of C-peptide due to a missing value.
[d]Results about HOMA-IS have been reported in a previous publication [11].
[e]A significant decrease was observed for the Cederholm index in men, $P = 0.047$.

iAUCs (Figure 1) as well as for incremental peaks of glucose, insulin, and C-peptide in response to the MedDiet (Table 2).

3.3. Insulin Sensitivity Response to the MedDiet. No time or sex-by-time interaction effect was observed for hepatic insulin sensitivity as calculated with the HOMA-IS index (Table 2). However, a sex-by-time interaction was found for peripheral insulin sensitivity as calculated by the Cederholm index ($P = 0.03$), with men showing an increase (+8.1%, $P = 0.047$) whereas a nonsignificant decrease was observed in women (−5.9%, $P = 0.94$) (Table 2).

3.4. NEFA Response to the MedDiet. No time or sex-by-time interaction was observed for fasting NEFA concentration. A trend toward a decrease in NEFA AUC was noted in men and women combined (−6.4%; time effect: $P = 0.06$) but no sex-by-time interaction was observed for this variable (Figure 1). Moreover, a tendency for an increase was found for the NEFA

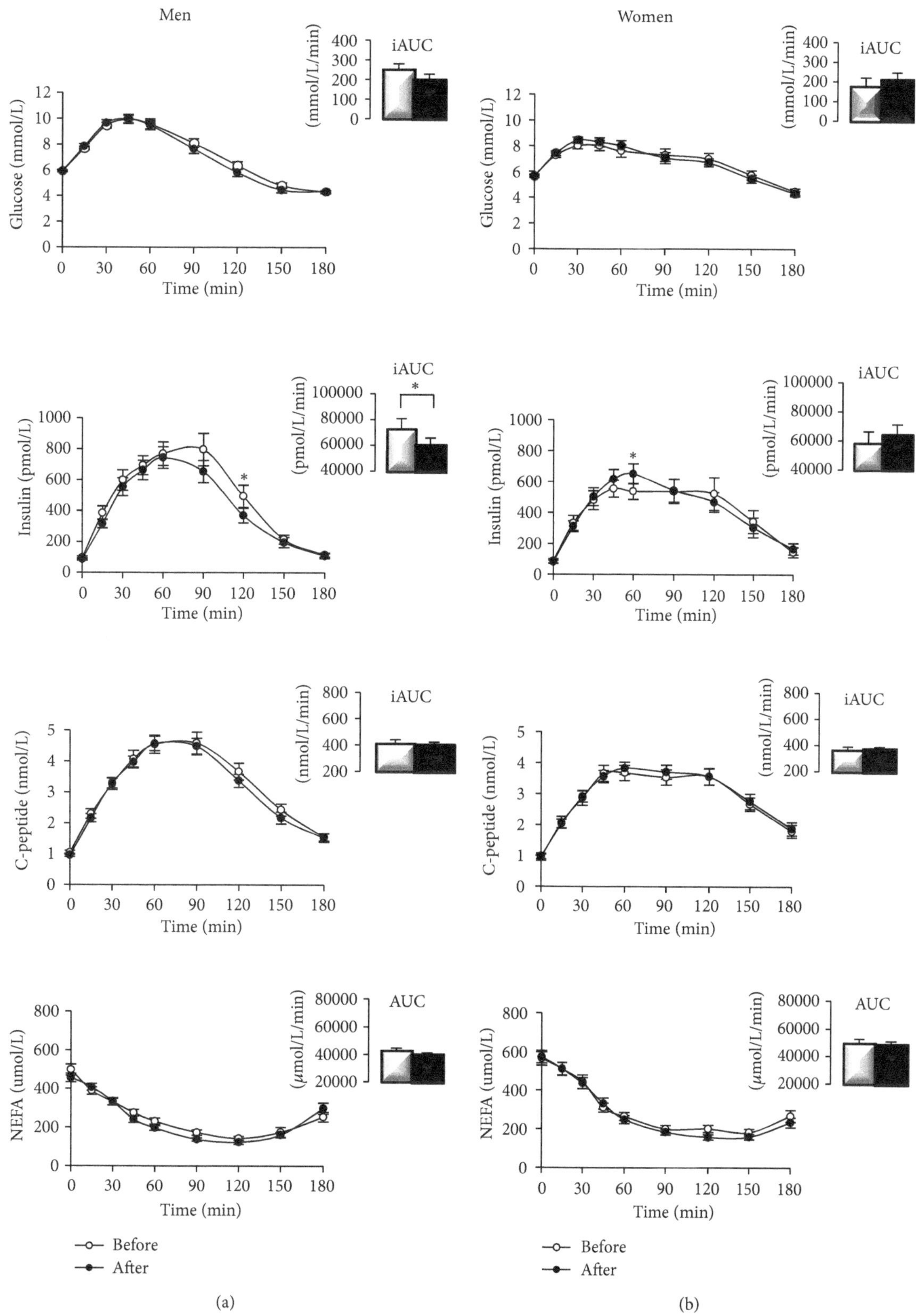

FIGURE 1: Glucose, insulin, C-peptide, and nonesterified fatty acid (NEFA) concentrations observed in men (left, $n = 37$) and women (right, $n = 32$) before (-o-) and after (-●-) the 4-week controlled Mediterranean diet. Analyses were adjusted for values before the controlled phase and body weight change during the controlled phase. MIXED procedures for repeated measurements followed by Tukey-Kramer tests were used. Sex-by-time interaction effects were found for insulin iAUC ($P = 0.005$) as well as for insulin 60 min postload ($P = 0.02$) and insulin 120 min postload ($P = 0.03$). *Decreases in insulin iAUC and insulin 120 min postload were found only in men (resp., $P = 0.02$ and $P = 0.03$) whereas only women experienced an increase in insulin 60 min postload ($P = 0.04$).

TABLE 3: Pearson's correlation coefficients for the associations between changes in nonesterified fatty acids (NEFA) concentrations and changes in variables related to the glucose and insulin homeostasis in men and women in response to the Mediterranean diet.

	Men			Women		
	Fasting NEFA	NEFA AUC	NEFA suppression rate	Fasting NEFA	NEFA AUC	NEFA suppression rate
	r	*r*	*r*	*r*	*r*	*r*
Glucose						
Fasting	0.11	−0.15	0.25	−0.32	−0.19	**−0.36***
Peak	**0.46****	0.24	**0.36***	−0.11	0.07	−0.01
120 min	0.23	0.07	0.08	0.02	**0.42***	−0.15
iAUC	**0.42***	0.23	0.26	0.31	**0.44***	0.21
Insulin						
Fasting	0.12	0.07	0.12	−0.32	−0.37	−0.06
Peak	0.14	0.07	0.16	0.12	−0.09	0.08
120 min	0.11	0.02	0.05	0.08	−0.07	0.22
iAUC	0.23	0.18	0.27	0.22	0.08	0.24
C-peptide						
Fasting	0.03	−0.14	0.18	**−0.45***	**−0.41***	−0.30
Peak	0.05	0.14	0.06	0.01	0.16	−0.09
120 min	−0.02	0.10	0.01	0.09	0.08	0.06
iAUC	0.08	0.15	0.09	0.16	0.28	−0.02
HOMA-IS index	−0.07	0.10	−0.11	0.10	0.15	−0.02
Cederholm index	**−0.39***	−0.18	−0.29	0.15	−0.12	−0.07

*$P < 0.05$, **$P < 0.01$.

NEFA, nonesterified fatty acids; AUC, area under the curve; iAUC, incremental area under the curve; HOMA-IS index, homeostasis model assessment index for insulin sensitivity.

$n = 37$ for men, except for associations including incremental peaks of glucose, insulin and C-peptide, iAUC for glucose, insulin, C-peptide and NEFA, fasting C-peptide, C-peptide 120 min postload, and the Cederholm index for which $n = 35$ due to missing values.

$n = 32$ for women, except for associations including peaks of glucose ($n = 31$), insulin ($n = 31$) and C-peptide ($n = 30$), iAUC for glucose ($n = 29$), insulin ($n = 29$), C-peptide ($n = 28$) and NEFA ($n = 28$, except for associations with variables related to C-peptide, $n = 27$), fasting C-peptide ($n = 31$), C-peptide 120 min postload ($n = 31$), and the Cederholm index ($n = 29$) due to missing values.

suppression rate when the whole sample was considered (+3.5%; time effect: $P = 0.06$) (Table 2), but no sex-by-time interaction effect was noted.

3.5. Associations between Changes in NEFA and Changes in Glucose and Insulin Homeostasis in Response to the MedDiet. In men, changes in fasting NEFA concentration in response to the MedDiet were positively associated with concurrent variations in the incremental peak of glucose and glucose iAUC and negatively associated with changes in the peripheral insulin sensitivity as measured by the Cederholm index (Table 3). In women, changes in fasting NEFA concentration and NEFA AUC with MedDiet were negatively associated with concurrent variations in fasting C-peptide concentration. Moreover, changes in NEFA AUC were positively associated with variations in glucose concentration at 120 min and glucose iAUC in women. Finally, changes in NEFA suppression rate were positively associated with variations in the incremental peak of glucose in men and negatively associated with variations in fasting glucose concentration in women.

4. Discussion

The present study reports for the first time, in the context of a fully controlled nutritional feeding protocol, detailed sex-related differences in glucose and insulin homeostasis in response to consumption of a traditional MedDiet, a healthy food pattern recognized to prevent the development of insulin resistance and type 2 diabetes mellitus [8, 9]. Specifically, our results showed that, in men, MedDiet improves peripheral insulin sensitivity as measured by the Cederholm index and reduces insulin iAUC during an oral glucose challenge. In contrast, women had no beneficial change for these variables in response to the MedDiet. Sex differences in insulin homeostasis found in this study are not explained by sex differences in NEFA response to the MedDiet. However, sex-specific pattern of associations between changes in variables related to glucose and insulin homeostasis and changes in NEFA concentrations were found.

The beneficial effects of the MedDiet on peripheral insulin sensitivity in men are in concordance with the previous literature. In fact, many characteristics of the MedDiet, such as its high content in monounsaturated fatty acids, fibers, and polyphenolic compounds and the moderate alcohol intake habitually found with this dietary pattern, have been previously identified as improving insulin sensitivity through several mechanisms and independently of weight loss, as previously reviewed [27]. However, our results add to the previous literature since we showed that, when considered as a whole, the MedDiet leads to beneficial effects on insulin homeostasis, but these benefits seem to be sex-specific, with men having more favorable effects than women.

Improvements in peripheral insulin sensitivity observed in men after the MedDiet suggest a higher effectiveness of insulin-stimulated glucose uptake in peripheral tissues, leading to a decrease in insulin needs. This decrease in insulin needs can in turn explain the decrease in insulin concentrations observed in men during the OGTT. Two mechanisms can explain this decrease in insulin concentrations with MedDiet: reduced insulin secretion or enhanced insulin clearance from the circulation. Since C-peptide concentration has been identified as an indirect indicator of in vivo pancreatic beta-cell insulin secretion [28] and no significant change in C-peptide concentration occurred in this study, our results suggest that reduced insulin concentrations in men in response to the MedDiet are likely to be primarily due to enhanced insulin clearance. Insulin clearance was not measured in the present study and further work is warranted to confirm this hypothesis.

Elevated circulating NEFA concentrations have been identified as one of the main contributing metabolic factors for the development of insulin resistance. More precisely, NEFA induce insulin resistance through activation of a serine-kinase cascade, leading to the inhibition of the insulin signaling pathway in adipose tissue, skeletal muscle, and liver [29]. Results from the present study suggest that sex differences in insulin homeostasis are not due to sex differences in NEFA response to the MedDiet. Other potential factors related to insulin resistance may explain these sex differences in the response to MedDiet. Among others, it has been demonstrated that estrogen favors insulin sensitivity, via its numerous beneficial effects on insulin and glucose homeostasis, adipose tissue distribution, oxidative stress, and inflammatory status [30]. Since a decrease in estrogen levels has been previously highlighted in response to the MedDiet in women [31], we can speculate that a potential decrease in estrogen levels may have counteracted the beneficial effects of the MedDiet in premenopausal women from our study.

The lack of a significant reduction in glucose response to the MedDiet in men, despite improvements in insulin homeostasis, may be explained by the fact that the majority of our subjects had a normal glucose tolerance before the controlled MedDiet phase. In fact, previous studies have highlighted that the MedDiet leads to beneficial effects on glucose concentrations in individuals with type 2 diabetes mellitus [8]; however, the impact of the MedDiet in individuals without type 2 diabetes is still unknown [32–37]. In a weight loss context, Shai and collaborators have observed decreases in both glucose and insulin concentrations in individuals characterized by type 2 diabetes mellitus with the adoption of the MedDiet whereas only reduced insulin concentrations were found in nondiabetic individuals [32]. Given the limited number of our participants who met diagnosis criteria for type 2 diabetes mellitus, we cannot verify this issue in the present study.

Sex differences in the pattern of associations between variables related to glucose and insulin homeostasis and NEFA concentrations have been reported in this study in response to the MedDiet. Our results showed that changes in fasting NEFA concentrations after the MedDiet are negatively associated with parallel changes in peripheral insulin sensitivity in men, meaning that men who decreased the most their fasting NEFA in response to the MedDiet had greater improvements in peripheral insulin sensitivity. This association was not observed in women.

Since women characterized by type 2 diabetes mellitus are at higher risk of coronary heart disease mortality than diabetic men [2], sex differences highlighted in this study have therefore important clinical implications. Indeed, women have been underrepresented in the majority of clinical trials; therefore we know little about efficient strategies aiming at improving glucose and insulin homeostasis in women. Thus our results, along with those from other studies, highlight the importance of including both men and women in future studies in order to identify effective interventions for women in prevention of diabetes and its cardiovascular-related risk. However, despite these results, it is important to stress out the fact that the adoption of the MedDiet brings beneficial effects on the cardiovascular health with regard to the lipid profile and blood pressure levels in both men and women, with no sex difference for these cardiovascular risk factors [11], suggesting beneficial cardiovascular effects of the MedDiet, regardless of the sex.

A major strength of this study is its strictly controlled design during the MedDiet phase, ensuring optimal control over energy intake and diet quality. One limitation is that peripheral insulin sensitivity was assessed using the Cederholm index, a surrogate marker, rather than the gold-standard euglycemic-hyperinsulinemic clamp. However, the Cederholm index has been demonstrated to correlate closely with specific measures of insulin sensitivity derived from the euglycemic-hyperinsulinemic clamp [38, 39]. Another limitation is the heterogeneity of the study group, consisting of subjects with various degrees of glucose tolerance and insulin sensitivity. However, for markers related to glucose and insulin homeostasis, except for fasting glucose, men and women had comparable values before the controlled MedDiet, which facilitates the comparison between sexes.

5. Conclusions

In summary, short-term consumption of a traditional MedDiet leads to beneficial effects on insulin homeostasis in men but not in premenopausal women. Sex differences in insulin homeostasis are not due to sex differences in NEFA response to the MedDiet. Since hyperinsulinemia increases the risk of cardiovascular disease and type 2 diabetes mellitus [40], our results suggest that the MedDiet may have particular benefits for cardiovascular health in men through its favorable effects on postprandial insulin concentrations. Further studies are needed to document underlying mechanisms behind this sex difference in response to the MedDiet.

Conflict of Interests

The authors declare that there is no conflict of interests regarding the publication of this paper.

Acknowledgments

The authors would like to thank Johanne Marin for her contribution in the laboratory. This research was supported by the Canadian Institutes of Health Research (Grant no. MOP 84568) and the Heart and Stroke Foundation of Quebec (Grant no. 2007-180). Alexandra Bédard is a recipient of a studentship from Canadian Institutes of Health Research (CIHR).

References

[1] L. Guariguata, D. R. Whiting, I. Hambleton, J. Beagley, U. Linnenkamp, and J. E. Shaw, "Global estimates of diabetes prevalence for 2013 and projections for 2035," *Diabetes Research and Clinical Practice*, vol. 103, no. 2, pp. 137–149, 2014.

[2] R. Huxley, F. Barzi, and M. Woodward, "Excess risk of fatal coronary heart disease associated with diabetes in men and women: meta-analysis of 37 prospective cohort studies," *British Medical Journal*, vol. 332, no. 7533, pp. 73–76, 2006.

[3] B. Mittendorfer, "Insulin resistance: sex matters," *Current Opinion in Clinical Nutrition and Metabolic Care*, vol. 8, no. 4, pp. 367–372, 2005.

[4] D. E. Hurlbut, M. E. Lott, A. S. Ryan et al., "Does age, sex, or ACE genotype affect glucose and insulin responses to strength training?" *Journal of Applied Physiology*, vol. 92, no. 2, pp. 643–650, 2002.

[5] N. G. Boulé, S. J. Weisnagel, T. A. Lakka et al., "Effects of exercise training on glucose homeostasis: the HERITAGE family study," *Diabetes Care*, vol. 28, no. 1, pp. 108–114, 2005.

[6] L. Perreault, M. Yong, S. Dagogo-Jack et al., "Sex differences in diabetes risk and the effect of intensive lifestyle modification in the diabetes prevention program," *Diabetes Care*, vol. 31, no. 7, pp. 1416–1421, 2008.

[7] F. Sofi, R. Abbate, G. F. Gensini, and A. Casini, "Accruing evidence on benefits of adherence to the Mediterranean diet on health: an updated systematic review and meta-analysis," *The American Journal of Clinical Nutrition*, vol. 92, no. 5, pp. 1189–1196, 2010.

[8] K. Esposito, M. I. Maiorino, A. Ceriello, and D. Giugliano, "Prevention and control of type 2 diabetes by Mediterranean diet: a systematic review," *Diabetes Research and Clinical Practice*, vol. 89, no. 2, pp. 97–102, 2010.

[9] O. Ajala, P. English, and J. Pinkney, "Systematic review and meta-analysis of different dietary approaches to the management of type 2 diabetesl-3," *The American Journal of Clinical Nutrition*, vol. 97, no. 3, pp. 505–516, 2013.

[10] J. Delarue and C. Magnan, "Free fatty acids and insulin resistance," *Current Opinion in Clinical Nutrition and Metabolic Care*, vol. 10, no. 2, pp. 142–148, 2007.

[11] A. Bédard, M. Riverin, S. Dodin, L. Corneau, and S. Lemieux, "Sex differences in the impact of the Mediterranean diet on cardiovascular risk profile," *The British Journal of Nutrition*, vol. 108, no. 8, pp. 1428–1434, 2012.

[12] International Diabetes Federation, "IDF worldwide definition of the metabolic syndrome," 2006, http://www.idf.org/.

[13] K. D. Bird, "Confidence intervals for effect sizes in analysis of variance," *Educational and Psychological Measurement*, vol. 62, no. 2, pp. 197–226, 2002.

[14] Minister of Health Canada, *Eating Well with Canada's Food Guide*, 2007, http://www.hc-sc.gc.ca/.

[15] T. J. Hartman, P. S. Albert, Z. Zhang et al., "Consumption of a legume-enriched, low-glycemic index diet is associated with biomarkers of insulin resistance and inflammation among men at risk for colorectal cancer," *Journal of Nutrition*, vol. 140, no. 1, pp. 60–67, 2010.

[16] J. C. Lovejoy, M. M. Windhauser, J. C. Rood, and J. A. de La Bretonne, "Effect of a controlled high-fat versus low-fat diet on insulin sensitivity and leptin levels in African-American and Caucasian women," *Metabolism: Clinical and Experimental*, vol. 47, no. 12, pp. 1520–1524, 1998.

[17] C. P. Marinangeli and P. J. Jones, "Whole and fractionated yellow pea flours reduce fasting insulin and insulin resistance in hypercholesterolaemic and overweight human subjects," *British Journal of Nutrition*, vol. 105, no. 1, pp. 110–117, 2011.

[18] A. M. Louheranta, U. S. Schwab, E. S. Sarkkinen et al., "Insulin sensitivity after a reduced-fat diet and a monoene-enriched diet in subjects with elevated serum cholesterol and triglyceride concentrations," *Nutrition, Metabolism and Cardiovascular Diseases*, vol. 10, no. 4, pp. 177–187, 2000.

[19] J. M. E. Pulido and M. A. Salazar, "Changes in insulin sensitivity, secretion and glucose effectiveness during menstrual cycle," *Archives of Medical Research*, vol. 30, no. 1, pp. 19–22, 1999.

[20] W. T. Friedewald, R. I. Levy, and D. S. Fredrickson, "Estimation of the concentration of low-density lipoprotein cholesterol in plasma, without use of the preparative ultracentrifuge.," *Clinical Chemistry*, vol. 18, no. 6, pp. 499–502, 1972.

[21] R. Richterich and H. Dauwalder, "Determination of plasma glucose by hexokinase-glucose-6-phosphate dehydrogenase method," *Schweizerische Medizinische Wochenschrift*, vol. 101, no. 17, pp. 615–618, 1971.

[22] B. Desbuquois and G. D. Aurbach, "Use of polyethylene glycol to separate free and antibody-bound peptide hormones in radioimmunoassays.," *Journal of Clinical Endocrinology and Metabolism*, vol. 33, no. 5, pp. 732–738, 1971.

[23] L. G. Heding, "Radioimmunological determination of human C peptide in serum," *Diabetologia*, vol. 11, no. 6, pp. 541–548, 1975.

[24] J. Cederholm and L. Wibell, "Insulin release and peripheral sensitivity at the oral glucose tolerance test," *Diabetes Research and Clinical Practice*, vol. 10, no. 2, pp. 167–175, 1990.

[25] Canadian Diabetes Association, *2008 Clinical Practice Guidelines for the Prevention and Management of Diabetes in Canada*, 2008, http://www.diabetes.ca/.

[26] T. Lohman, A. Roche, and R. Martorel, *The Airlie (VA) Consensus Conference: Standardization of Anthropometric Measurements*, Human Kinetics, Champaign, Ill, USA, 1988.

[27] T. Thomas and A. F. Pfeiffer, "Foods for the prevention of diabetes: How do they work?" *Diabetes-Metabolism Research and Reviews*, vol. 28, no. 1, pp. 25–49, 2012.

[28] R. Hovorka and R. H. Jones, "How to measure insulin secretion," *Diabetes/Metabolism Reviews*, vol. 10, no. 2, pp. 91–117, 1994.

[29] C. Capurso and A. Capurso, "From excess adiposity to insulin resistance: the role of free fatty acids," *Vascular Pharmacology*, vol. 57, no. 2–4, pp. 91–97, 2012.

[30] E. B. Geer and W. Shen, "Gender differences in insulin resistance, body composition, and energy balance," *Gender Medicine*, vol. 6, no. 1, pp. 60–75, 2009.

[31] G. Carruba, O. M. Granata, V. Pala et al., "A traditional Mediterranean diet decreases endogenous estrogens in healthy postmenopausal women," *Nutrition and Cancer*, vol. 56, no. 2, pp. 253–259, 2006.

[32] I. Shai, D. Schwarzfuchs, Y. Henkin et al., "Weight loss with a low-carbohydrate, Mediterranean, or low-fat diet," *The New England Journal of Medicine*, vol. 359, no. 3, pp. 229–241, 2008.

[33] F. Pérez-Jiménez, J. López-Miranda, M. D. Pinillos et al., "A mediterranean and a high-carbohydrate diet improve glucose metabolism in healthy young persons," *Diabetologia*, vol. 44, no. 11, pp. 2038–2043, 2001.

[34] A. Ambring, P. Friberg, M. Axelsen et al., "Effects of a Mediterranean-inspired diet on blood lipids, vascular function and oxidative stress in healthy subjects," *Clinical Science*, vol. 106, no. 5, pp. 519–525, 2004.

[35] N. di Daniele, L. Petramala, L. di Renzo et al., "Body composition changes and cardiometabolic benefits of a balanced Italian Mediterranean Diet in obese patients with metabolic syndrome," *Acta Diabetologica*, vol. 50, no. 3, pp. 409–416, 2013.

[36] Z. Djuric, J. Ren, J. Blythe, G. VanLoon, and A. Sen, "A Mediterranean dietary intervention in healthy American women changes plasma carotenoids and fatty acids in distinct clusters," *Nutrition Research*, vol. 29, no. 3, pp. 156–163, 2009.

[37] K. Esposito, A. Pontillo, C. Di Palo et al., "Effect of weight loss and lifestyle changes on vascular inflammatory markers in obese women—a randomized trial," *Journal of the American Medical Association*, vol. 289, no. 14, pp. 1799–1804, 2003.

[38] M. È. Piché, S. Lemieux, L. Corneau, A. Nadeau, J. Bergeron, and S. J. Weisnagel, "Measuring insulin sensitivity in postmenopausal women covering a range of glucose tolerance: comparison of indices derived from the oral glucose tolerance test with the euglycemic-hyperinsulinemic clamp," *Metabolism*, vol. 56, no. 9, pp. 1159–1166, 2007.

[39] E. Pigeon, M. E. Riou, J. St-Onge et al., "Validation of a simple index (SIisOGTT) of insulin sensitivity in a population of sedentary men," *Diabetes and Metabolism*, vol. 35, no. 5, pp. 398–403, 2009.

[40] G. I. Uwaifo and R. E. Ratner, "The roles of insulin resistance, hyperinsulinemia, and thiazolidinediones in cardiovascular disease," *The American Journal of Medicine*, vol. 115, no. 8, pp. 12–19, 2003.

8

Measuring Outcomes in Adult Weight Loss Studies That Include Diet and Physical Activity: A Systematic Review

Rachel A. Millstein

SDSU/UCSD Joint Doctoral Program in Clinical Psychology, 6363 Alvarado Court, San Diego, CA 92120, USA

Correspondence should be addressed to Rachel A. Millstein; rmillste@ucsd.edu

Academic Editor: Michael B. Zemel

Background. Measuring success of obesity interventions is critical. Several methods measure weight loss outcomes but there is no consensus on best practices. This systematic review evaluates relevant outcomes (weight loss, BMI, % body fat, and fat mass) to determine which might be the best indicator(s) of success. *Methods.* Eligible articles described adult weight loss interventions that included diet and physical activity and a measure of weight or BMI change and body composition change. *Results.* 28 full-text articles met inclusion criteria. Subjects, settings, intervention lengths, and intensities varied. All studies measured body weight (−2.9 to −17.3 kg), 9 studies measured BMI (−1.1 to −5.1 kg/m^2), 20 studies measured % body fat (−0.7 to −10.2%), and 22 studies measured fat mass (−0.9 to −14.9 kg). All studies found agreement between weight or BMI and body fat mass or body fat % decreases, though there were discrepancies in degree of significance between measures. *Conclusions.* Nearly all weight or BMI and body composition measures agreed. Since body fat is the most metabolically harmful tissue type, it may be a more meaningful measure of health change. Future studies should consider primarily measuring % body fat, rather than or in addition to weight or BMI.

1. Introduction

The obesity treatment literature includes many sophisticated analyses, methods, and conclusions, yet the problem persists [1–3]. Given all of the information now known, to move forward in intervention development and evaluation, more accurate measures of success are needed to monitor changes. The field of obesity treatment often has redundancy of interventions and measures but heterogeneity of outcome measures, making it difficult to combine results and move toward the ultimate goal of achieving healthy weights [3–5]. Most weight loss studies measure weight and/or BMI to assess intervention-related changes, given their ease of measurement and interpretation [4, 5]. Though well correlated with body composition, weight and BMI only inform total loss or change, which could include lean body mass in addition to fat loss [6, 7]. BMI may also inaccurately reflect intervention-related change, in that it does not account for bone and muscle density, frame size, and fat distribution [8, 9].

Loss of fat is the major outcome of interest for a variety of health reasons. Body fat can have varying degrees of benefit and harm, depending on location, amount, and time of fat disposition [10–12]. While babies and young children depend on body fat to promote brain and tissue growth, as children age, the percent of fat that is beneficial decreases [13]. Among older adults, body fat is also metabolically harmful, with an exception that it may offer some protection against bone loss and fractures [14]. However, for adults, excess body fat is widely acknowledged to be associated with such diseases as type 2 diabetes, stroke, hypertension, cardiovascular disease, and arthritis [15, 16]. In particular, abdominal fat has been associated with increased morbidity and mortality among adults [11, 12]. Body fat is best captured by measures other than weight and BMI, such as bioelectrical impedance (BIA), dual-energy X-ray absorptiometry (DEXA), underwater weighing, air displacement, and skinfold thickness [17]. Measuring body fat as part of weight loss interventions is a common but not universal practice and it is usually viewed as a secondary outcome, with weight or BMI being primary [5, 18].

Compounding the measurement issues described above, adult weight loss intervention studies primarily target dietary interventions, rather than (or more than) physical activity (PA) interventions [5]. Although diet can lead to reduced body fat, it can also lead to overall weight loss that can include reduced fat-free mass (bone and muscle) [19]. Overall weight loss does not distinguish between types of tissue mass lost. Much research shows that PA is a key driver of fat loss and maintenance or increase of fat-free mass [19, 20]. Ideally, interventions should include components of both diet and PA, to reduce body fat and maintain or increase fat-free mass [18, 20]. Therefore, studies that include PA as a component of interventions (in addition to diet) should theoretically include a measure other than weight or BMI to potentially best capture changes.

The objectives of this review were to address the following questions:

(1) What are the best or most consistent measures of success in adult weight loss interventions that include diet and PA: weight or BMI, or body composition? (2) Are weight loss or BMI changes or body composition changes adequate measures of intervention success? (3) Do the studies that include a body composition measure in addition to weight or BMI reach the same conclusions? A systematic review was conducted to examine and evaluate all of the existing findings on these topics. This review does not include a meta-analysis because intervention studies contain many heterogeneous components (duration, intensity, length, and setting of interventions), so that compiling results may lead to incomplete or inconclusive findings.

2. Methods

2.1. Search Strategy. The Cochrane library was searched for existing reviews on this topic and information from related systematic reviews. As there were no available or registered review papers on these specific questions, the following search methods were employed in July and August 2012. This search was primarily conducted using online scientific literature databases. PubMed (1953–present) and PsychInfo (1806–present) were used to do an exhaustive search combining all terms. Additional search resources were used to acquire remaining papers meeting review criteria: the online database at Google Scholar (1992–present) and searching through references from eligible papers found (ancestry search). The overarching goal of the search was to identify studies of adult weight loss interventions that included diet and PA and a measure of weight or BMI and body composition. The following title and abstract keyword search terms were used in all databases, with limitations to humans, English language, randomized controlled trials (RCTs) or clinical trials, and adults over age 18: "weight loss" AND "overweight" OR "obese" OR "obesity" AND "diet" OR "dietary" OR "calorie restriction" AND "exercise" OR "physical activity" OR "fitness" AND "BMI" OR "body mass index" AND "fat loss" OR "body composition" OR "skinfold" OR "skinfold thickness" or "tricep skinfold" OR "DEXA" OR "DXA" OR "underwater weigh*" OR "bioelectrical impedance" OR "BIA."

2.2. Study Selection. Eligibility criteria for inclusion into this review were as follows. Study types were limited to randomized controlled trials published in English. There was no lower limit on year of publication. The types of participants intended for this review were adults ages 18–65 who were overweight or obese (BMI ≥ 25.0). Studies of adults with comorbid health conditions (i.e., prediabetes and prehypertension) were included because many interventions target such populations. All interventions were eligible if they targeted weight loss and included a diet and aerobic PA component. No restrictions were placed on duration, intensity, or setting of intervention (i.e., inpatient and outpatient). The eligible outcome measures were at least one measure of overall weight change (taken both before and after intervention), pounds or kilograms of weight or BMI reduction, and at least one body fat measure (taken both before and after intervention): skinfold thickness, DEXA, underwater weighing, BIA, or air displacement. While waist or hip circumference is sometimes considered to be a body composition measure, it can be too gross to reliably identify changes in short-term studies, so it was excluded from this search.

Exclusion criteria were studies not published in English, inclusion of children or older adults, interventions that included only diet or only PA, interventions in which the only PA was strength training, studies that included only one outcome measure (weight or BMI or body composition), nonintervention studies or designs other than RCTs (reviews, position papers, and cross-sectional or noncontrolled studies), studies that included dietary supplements or drugs to assist in weight loss, secondary or redundant data analyses (in that case, only the primary results were included in this review), and the case when the full-text article was unavailable from interlibrary loan, online sources, or correspondence with the author.

The study selection process, all completed by the author, began with general keyword (title and abstract) searches in the databases and reference lists of appropriate papers. Of those titles that appeared relevant, a more thorough abstract review was conducted. Of abstracts that appeared relevant, a full-text review was conducted, when available, and all eligible full-text articles were included in this paper. Throughout the study selection process, duplicates were removed. If one trial published multiple papers, only the primary outcome paper or the most recent (whichever was most relevant) was included.

2.3. Data Extraction and Assessment of Studies. The author extracted and compiled the detailed data items and study characteristics from all articles. The data extraction tables included, as available, study type, sample size including gender breakdown, baseline ages, SES or race, comorbidities, intervention setting and length, body composition measure, baseline weight and/or BMI (and standard deviations (SD) or range), baseline body composition (and SD or range), follow-up weight and/or BMI (and SD), and follow-up body composition (and SD or range). For the principle summary measures, the variables were extracted from the studies when the data were published, but when the average weight and/or BMI change, average body composition change, and summary of agreement were not explicitly listed, they were calculated from the available data (e.g., calculating

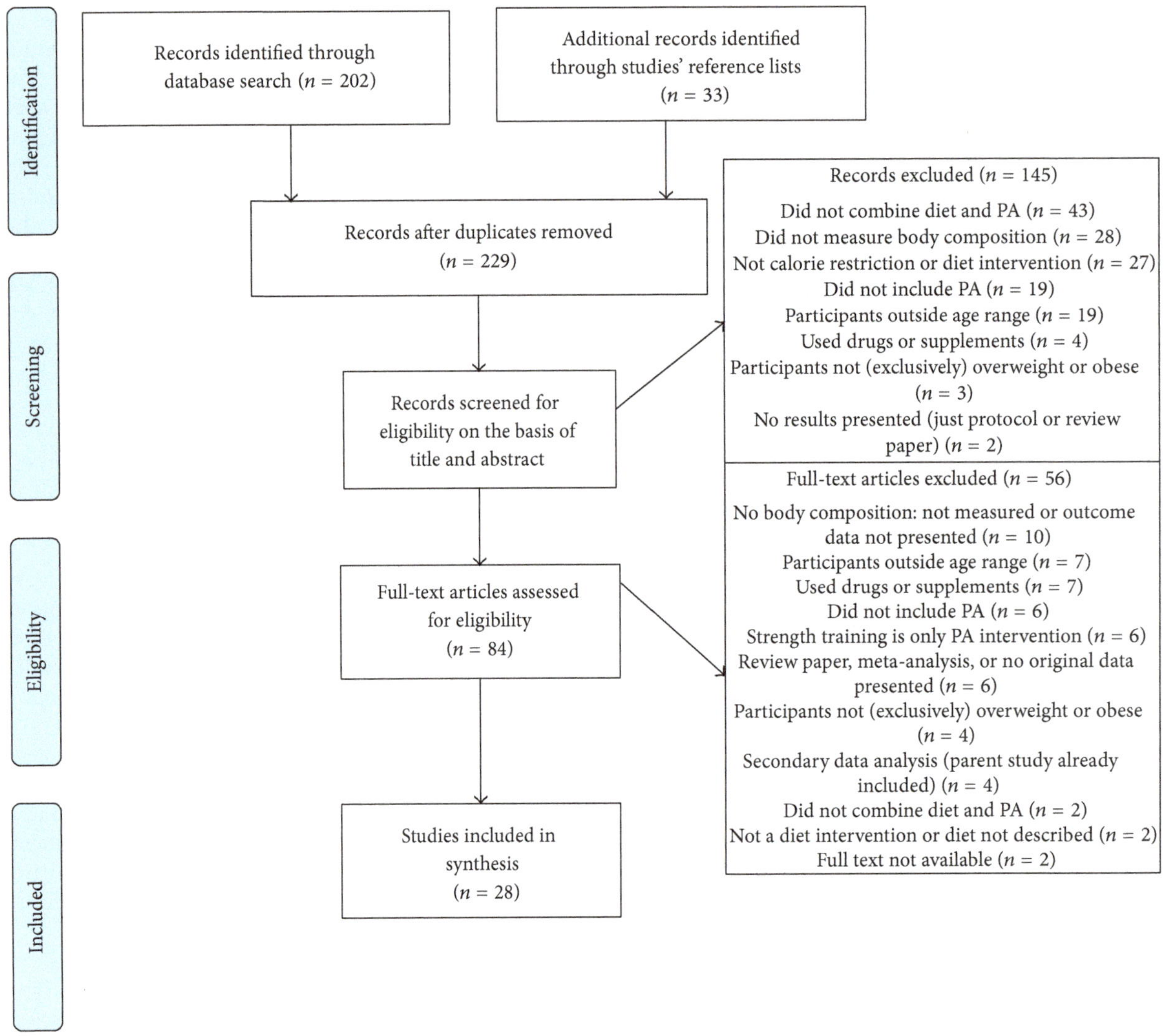

Figure 1: PRISMA flow diagram of article screening and eligibility.

the average weight change from the baseline to follow-up weight).

3. Results

3.1. Study Selection. The flowchart describing study screening, exclusion, eligibility, and selection is shown in Figure 1. Two hundred and thirty-five studies were identified through the various search sources. Of the nonduplicate studies, 145 were excluded for the reasons listed in the flowchart; the main reasons were that the intervention did not include both diet and PA, that the study did not include or measure body composition, and that the intervention did not involve a calorie restricted diet (e.g., manipulating macronutrient composition in isocaloric weight maintenance diets). Of the 56 excluded full-text articles, the main exclusion reasons were that the studies did not report body composition data, included older adults, and used drugs or supplements to aid weight loss. Ultimately, 28 studies met all inclusion criteria and were included in this review.

3.2. Study Characteristics. The study characteristics are presented in Table 1. There was much heterogeneity across studies, with respect to other characteristics. Sample sizes ranged from 5 to 111, and a majority of studies (18 studies = 64%) used only women participants. The studies included and focused on a range of adult ages, with baseline ages ranging from 28 to 54.7. No studies reported participants' socioeconomic status (SES). Of the few studies that reported racial composition of samples, most participants were white. The country of origin could perhaps be used to infer racial composition of samples when otherwise not indicated (e.g., primarily white participants in Belgium, The Netherlands, and Australia). Most studies were conducted in the United States, with fewer from Europe and Australia. Three main body composition measures were used (categories were not mutually exclusive, as some studies used multiple measures): DEXA was used in 13 studies, BIA was used in eight, and underwater weighing was used in six. Two studies used skinfold thickness (bi- or triceps), one used air displacement, and one used doubly labeled water with

Table 1: Descriptive and technical components of the included studies.

Study (reference number)	Participants*	Baseline age (years) (mean ± SD or range)	Country and SES/race (if reported)	Degree of overweight for inclusion and comorbidities	Intervention length and setting	Intervention components	Body composition measure
[27]	19 women	43.2 ± 9.1	US: white, Hispanic, black	≥15 kg above ideal MetLife table weight, no comorbidities	16 week intervention, 1 year follow-up, university clinical research center	1200 cal/day, structured aerobic PA: 3x 45 min step aerobics classes/week	DEXA
[28]	111 women	35 ± 11.2 (19–50)	Cyprus	BMI >25, no comorbidities	18-week intervention plus 18-week follow-up, university clinical research center	1,500 ± 200 cal/day (50% carb, 30% fat, 20% protein) mod intensity PA 30–60 min/day; behavioral modification consult 1x/week	BIA
[29]	32 men and women	37.6 ± 4.4 (30–45)	Australia	BMI 27–32, no comorbidities	32 weeks, university clinical research center	Diet and PA general advice, cal counting resources, heart rate monitor, personalized goals	DEXA
[30]	44 women	54.7 ± 7.9 (postmenopausal)	US	BMI >30, postmenopausal, no comorbidities	24 weeks, university clinical research center	LEARN program: lifestyle nutrition, PA skills, weekly coaching, self-control training	BIA
[31] (aerobic treatment arm)	46 women	35.2 ± 7	US: white and black	BMI 27–30, no comorbidities	Hospital inpatient and outpatient; intervention until BMI <25: ~5.5 mo.	800 cal/day (meals provided), 40 min aerobic PA 3x/week	DEXA
[32]	105 men and women	44.9 ± 10.2 (18–65)	US: white and black	BMI 30–40, no comorbidities	2 years, multisite: university clinical research center/academic medical centers	1200–1500 cal/day (women), 1500–1800 cal/day (men), 20–50 min PA/walking 4x/week, behavioral counseling 1x/week	DEXA
[33] (aerobic treatment arm)	20 men and women	36 ± 7 (19–48)	US	Body weight >20% above "desired amount," no comorbidities	8 weeks, university hospital GCRC	Liquid-formula diet: 1286 ± 281 cal/day, 30 min cycling (arms and legs) 3x/week, weekly nutrition counseling	Underwater weighing, BIA, and bi-/triceps skinfold
[34]	43 men and women	43 ± 12	Switzerland	BMI >30, no comorbidities	6 weeks, hospital inpatient	1000 cal/day low carb (15%) or moderate carb (45%), 1 h aerobic PA and 1 h underwater PA/day, nutrition education, "standard behavioral techniques"	BIA and triceps skinfold thickness
[35]	6 women	33 ± 8	US	>30% body fat, no comorbidities	16 weeks, university clinical research center	800 cal/day (low-fat diet), about 3 miles brisk walking/jogging 5 d/week, weekly diet education classes	Underwater weighing and triceps skinfold thickness
[36]	12 men and women	36 ± 6 (28–45)	US: mostly white	BMI 25–30, no comorbidities	24 weeks, university clinical research center	12.5% cal restriction (all food provided), 12.5% increase in energy expenditure structured PA (45–50 min cardio 5x/wk), weekly CBT group	DEXA
[37]	5 women	35 ± 4	US	37–50% body fat, no comorbidities	6 weeks, university hospital inpatient	800 cal/day (all food provided), about 4 miles daily walking	Underwater weighing

Table 1: Continued.

Study (reference number)	Participants*	Baseline age (years) (mean ± SD or range)	Country and SES/race (if reported)	Degree of overweight for inclusion and comorbidities	Intervention length and setting	Intervention components	Body composition measure
[38]	18 women	35 ± 7	US	130–160% "ideal body weight," no comorbidities	12 weeks, university clinical research center	1200 cal/day constant or rotating 600–1800 cal/day, 5 d/week walking, behavior modification program	Underwater weighing
[39]	81 women	28 ± 1 (19–45)	Canada: multiracial (white, Indian, Asian primarily)	BMI 27–40, no comorbidities	16 weeks, university clinical research center	Individual weight maintenance cal level −500 cal/day (varying dairy levels), 5 d/week supervised aerobic PA (to burn 250 cal), 2 d/week strength training	DEXA
[40]	10 women	39.3 ± 5.4 (25–50)	The Netherlands	BMI >30, no comorbidities	8 weeks, no location noted	Diet: 3.5 mJ/day plus 1.4 mJ/day formula, 90 min each aerobics, fitness/strength 2 d/week	Underwater weighing
[41]	126 women	38.5 ± 8.5	US	BMI ≥30, no comorbidities	14 weeks, university clinical research center	1200 cal/day f/both: very low carb, high protein (63 : 7 : 30%), low carb, moderate protein (50 : 20 : 30%), high carb, low protein (55 : 15 : 30%), curves fitness program: 30 min circuit (strength and aero.) 3 d/week	DEXA
[42]	118 women	38.7 ± 7.5	US	BMI ≥30, no comorbidities	14 weeks, university clinical research center	1200 cal/day (phase 1) then 1600 cal (phase 2) diets varying macronutrients (carb/protein), curves fitness program: 30 min circuit (strength and aero.) 3 d/week	DEXA
[43]	34 men and women	18+	Ireland	BMI ≥28, any comorbidities except eye diseases and pregnancy	12 months, no location noted	1500 cal/day (women), 1800 cal/day (men) low-fat, 1 h/day PA classes, 1x/month motivational seminars	DEXA
[44] (diet + aerobic ex.)	9 women	Premenopausal (exact ages unknown)	US	>120% above MetLife table ideal weight or BMI ≥27, no comorbidities	12 weeks, university clinical research center	Matola food products provided, 30–50 min aerobic PA 3x/wk at 70–80% max HR, 1 h/wk group education	Underwater weighing
[44] (diet + aerobic + strength)	8 women	Same as above	Same as above	Same as above	Same as above	Diet and aerobic PA as above, plus 3x/week strength circuit	Same as above
[45]	90 women	18–55	US	BMI 27–40, no comorbidities	10 weeks, university clinical research center	Week 1: 1200 cal/day, weeks 2–10: 1600 cal/day, curves supervised PA program: 30 min circuit 3 days/week, met w/RD every 2 weeks	BIA

Table 1: Continued.

Study (reference number)	Participants*	Baseline age (years) (mean ± SD or range)	Country and SES/race (if reported)	Degree of overweight for inclusion and comorbidities	Intervention length and setting	Intervention components	Body composition measure
[46]	24 women	47.2 ± 1.3 (40–56)	US	BMI ≥26, no comorbidities	16 weeks, no location noted	Low (15%) or high (30%), pro: carb ratio, both: 1700 cal/day, ≥5 d/week walking, 2 d/week strength training, weekly nutritionist counsel.	DEXA
[47]	40 men and women	41 ± 7.7	US	Body fat ≥25% (men), 30% (wom), no comorbidities	12 weeks, university clinical research center	OPTIFAST: 420 cal/day, walking (60% HR) 3x/week to reduce 300 cal	BIA
[48]	22 men and women	43.0 ± 5.3 (29–50)	US	BMI 27–35, no comorbidities	24 weeks, independent outpatient clinical research center	500 cal deficit diet (meal replacements of 25–40% cals, plus supplemental foods), plus walking ≥5 d/week moderate intensity (300–500 cal), RD, PA couns. 1x/week	Air displacement plethysmography
[49]	100 women	20–65	Australia	BMI 27–40, no comorbidities	12 weeks, hospital outpatient GCRC	Food provided: 5600 kJ/day high protein (34%) or high carb (64%), ≥3x 30 min PA/week, nutritionist 1x/mo.	DEXA
[50]	10 women	38 ± 4.5 (21–47)	US	140%–180% weight f/height based on MetLife tables and body fat ≥35%, no comorbidities	12 weeks, university clinical research center	75% individuals' RMR: low-fat or low-carb diets, 45 min aerobic PA (60–60% max HR) 3x/week, nutrition and education meetings 1x/week	Doubly labeled water and appropriate calculations
[51]	48 women	53.8 ± 2.5 (postmenopausal)	Denmark	BMI ≥25, no comorbidities	12 weeks, university clinical research center	Up to 10 260 kJ portions of formula diet, small food supplementation, 1–1.5 h each aerobic and strength PA 3x/week	DEXA
[52]	20 men	25–50	The Netherlands	BMI ≥30, no comorbidities noted	12 weeks, outpatient site (no details noted)	Very low cal protein-enriched formula diet (Modifast): around 5 mJ/day), 1 hour low-intensity aerobic PA (40% HR) 4x/week, weekly meetings w/nutrition and exercise pros	underwater weighing
[53]	20 men and women	44.7 ± 13.0	Belgium	BMI 25–40, no comorbidities	24 week intervention + 24 week observation, outpatient hospital	600 cal deficit, 1 hour aerobic and strength training 2x/week, biweekly dietician	BIA
[54]	34 men and women	45.6 ± 9.0 (18–65)	US: mostly white, minority black	BMI 30–60, elevated lipids, no other comorbidities	24 weeks, outpatient clinical research center	Low-fat diet: 500–1000 cal/day deficit, ≥30 min aerobic PA 3 d/week, biweekly nutritionist meetings	BIA and DEXA (on a subset)

Note: university clinical research center: outpatient unless indicated otherwise.
Abbreviations. GCRC: General Clinical Research Center, BIA: bioelectrical impedance, DEXA: dual-energy X-ray absorptiometry, cal: calories, d/week: days per week, and h/day: hours per day.
*Only reporting participant data for the diet + aerobic PA intervention arm.

appropriate body composition calculations. Fifteen studies reported attrition rates, with values ranging from 2.5 to 48.8% loss.

The studies also demonstrated heterogeneity in factors such as intervention emphases, diet types, and length (Table 1). Most studies took place at outpatient facilities (university or hospital clinical research centers), two studies took place in inpatient clinics, and three studies did not give details on intervention setting. The interventions ranged from six weeks to two years in length, with half (13 studies) in the 12- to 16-week range, and five studies lasted 24 weeks. There were varied emphases and strategies for the different components of the interventions. For the diet component of the interventions, strategies included visits with a dietician, calorie restriction (ranges: 420–1800 cal/day), specific macronutrient proportions (e.g., % calories from fat and carbohydrates), liquid diets with and without solid food supplementation, food provided by programs, and nutrition education. For the calorie restriction, the most common target value was 1200 calories/day or a range including 1200, but the liquid and inpatient diets were much lower (e.g., 420 calories/day), and when men were included in studies, the calorie ranges were higher (1500–1800 calories/day). Several studies did not report a specific calorie target but rather gave each participant an individualized goal based on their resting metabolic rate and subtracting 250–1000 calories from that as the daily goal. For the PA components of the interventions, the strategies included varied amounts of activity per day and week, structured and supervised aerobic activity (commonly walking and indoor cycling), circuit classes, skills-based and noncompetitive activity classes, individualized heart rate training goals, and emphases on lifestyle activity. Most of the studies included weekly or biweekly individual or group meetings with a nutritionist, exercise counselor, and/or psychologist for support, education, and behavior modification strategies.

3.3. Outcome Measures. The outcome measures of baseline and follow-up BMI, body fat, and fat mass, plus the average changes, summary of agreement, and attrition are presented in Table 2. The follow-up outcomes are reported for the intervention group and for the longest follow-up period reported in the paper.

Weight and BMI Outcomes. Averages are presented as overall values, unadjusted for length or intensity of intervention and unweighted for sample size. All 28 studies reported weight as an outcome, and all reported losses with wide ranging values (−2.9 to −17.3 kg). The average weight loss at the longest reported follow-up time point was −8.2 kg. While most studies measured baseline BMI, only nine reported follow-up BMI values as an outcome. The range of BMI losses was −1.1 to −5.1 kg/m^2, with an average loss of −3.1 kg/m^2.

Body Composition Outcomes. Twenty studies measured % body fat, all finding losses (−0.7 to −10.2%), with an average decrease of −5.1%. Twenty-two of the studies measured fat mass, all finding losses. The range of fat mass lost was −0.9 to −14.9 kg, and the average decrease was −6.6 kg.

3.4. Agreement between Weight and Body Composition Measures. All of the studies found a decrease in weight, and those that measured BMI showed decreases that were in agreement with the weight losses. That is, there was no discrepancy between the interpretations of weight and BMI changes, though weight loss showed greater changes than BMI. All of the studies found a decrease in the body composition measure(s) used. Most of the weight lost was accounted for by fat loss. In control groups that included diet only interventions, the percentage of fat lost was significantly greater in the diet + PA groups, while the diet + PA groups preserved or increased their fat-free mass more so than control (diet only) groups. Overall, all of the studies had agreement between the weight or BMI and body composition measure(s). Two studies were discrepant between the significance of the measures: one had a borderline significant reduction in fat mass, with a significant weight decrease, and the other had a more significant reduction in fat mass and % body fat than weight. The details and significance of the agreement across outcomes are presented in Table 2.

Overall agreement did not describe the variability among weight loss measures completely. Across studies, the losses of fat mass and % body fat were proportionally greater than losses of BMI or weight and, in one instance, fat mass or % body fat losses were more significantly different before and after intervention than the overall weight or BMI losses. While there is some redundancy across weight loss measures, body composition was a consistent measure of success in these studies, with all of the studies that reported it finding decreases after intervention in parallel or more so than weight or BMI decreases.

4. Discussion

This review focused on evaluating measures of success in diet and PA weight loss interventions for adults. There was heterogeneity across studies in terms of designs, settings, participants, and types of outcome measures. Despite the differences, several themes emerged. All of the studies demonstrated agreement across measures (weight or BMI and % body fat or fat mass). However, nuances were present as well. Specifically and as expected, fat accounted for most of weight lost. While not a primary focus of this review, several studies reported that fat-free mass was preserved or increased following the interventions that included PA. The conclusions from measures of weight or BMI and body fat % or fat mass were largely in agreement and body fat is more metabolically informative than overall weight. Thus, it can be proposed that measuring body fat should be considered a primary outcome of weight loss studies. In aggregate, the data presented here support the conclusion that measuring % body fat or fat mass before and after weight loss interventions that include diet and PA may be the most efficient and informative measure of success or change.

Body fat measures are highly accurate, though each type has pros and cons. Underwater weighing is the gold-standard, but it can be unpleasant and inconvenient and often not feasible for obese participants [17]. It is rarely used in more recent studies, given the technological advances. DEXA,

TABLE 2: Outcome data from the included studies: weight or BMI, body fat (BF) % or fat mass (FM), summary of weight or BMI and body fat measures, and attrition.

Study (reference number)	Baseline average weight (kg) ± SD and/or BMI (kg/m^2) ± SD (range)	Baseline average % fat and/or fat mass (kg) ± SD (range)	Follow-up average weight (kg) ± SD and/or BMI (kg/m^2) ± SD (range)	Follow-up average % fat and/or fat mass (kg) ± SD (range)	Average weight (kg) ± SD and/or BMI (kg/m^2) change ± SD (range)	Average % fat or fat mass (kg) change ± SD (range or CI)	Summary/agreement of measures	Attrition (if reported)
[27]	Weight: 83.6 ± 8.6 BMI: 31.4 ± 3.7	% BF: 46.9 ± 3.6 FM: 39.3 ± 6.1	Not reported	% BF: 41.9 ± 4.3	Weight loss: −8.3 ± 3.1	% BF: −5.0 ± 3.9 FM: −7.4 ± 3.7	Weight and fat measures decreased significantly. Weight loss mostly accounted for by FM.	
[28]	Weight: 79.8 ± 11.8 BMI: 29.1 ± 4.8	% BF: 39.3 ± 7.5	Weight: 62.5 ± 8.3 BMI: ±24 ± 3.5	% BF: 29.3 ± 7.0	Weight loss: −17.3 (CI: 15.7–18.9) (=−22%) BMI: −5.1 (CI: −3.8–−6.4) (=−17.5%)	% BF: −10.0 (CI: 9–11)	All weight or BMI and fat measures decreased significantly.	
[29]	Weight: 87.2 ± 12.6 BMI: 29.3 ± 1.6	% BF: 35.5 ± 6.0	Not reported	Not reported	Weight loss: −6.1 ± 0.6 (=−7.1%)	FM: −5.9 ± 0.6	Weight and fat measures decreased significantly. Weight loss mostly accounted for by FM.	20%
[30]	Weight: 96.4 ± 16.0 BMI: 36.4 ± 5.5	% BF: 46.2 ± 5.6 FM: 45.8 ± 12.3	Weight: 88.6 ± 15.7 BMI: 33.6 ± 5.3	% BF: 44.5 ± 6.1 FM: 39.6 ± 11.8	Weight loss: −7.8 ± 15.9 BMI: −2.8 ± 5.4	% BF: −2.4 ± 5.8 FM: −3.6 ± 12.0	Weight or BMI and fat measures decreased significantly.	25%
[31]	Weight: 76.9 ± 6.6 BMI: 28.5 ± 1.5	% BF: ±44.1 ± 3.8 FM: 34.1 ± 5.0	Weight: 64.3 ± 6.1 BMI: 23.8 ± 1.1	% BF: 33.9 ± 4.5 FM: 22.0 ± 4.6	Weight loss: −12.5 ± 2.2 BMI: −4.7 ± 1.3	% BF: −10.2 ± 4.1 FM: −12.1 ± 4.8	Weight or BMI and fat measures decreased significantly. Weight loss mostly accounted for by FM.	
[32]	Weight: 103.5 ± 14.4 BMI: 36.1 ± 3.5	FM: 40.4 ± 7.8	Not reported	Not reported	Weight loss: −7.4 (CI: −9.1–−5.6)	FM: −3.8 (−5.0–−2.6)	Weight and fat measures decreased (not noted whether differences were significant).	32%
[33] (aerobic treatment arm)	Weight: 96.0 ± 23.0	FM: 38.4 ± 12.5	Weight: 86.4 ± 19.8	FM: 31.2 ± 11.0	Weight loss: −9.6 ± 4.5	FM: −7.2 ± 3.0	Weight or BMI and fat measures decreased significantly. Weight loss mostly accounted for by FM.	
[34]	Weight: 104.5 ± 5.5 BMI: 39.5 ± 7.0	FM: 44.0 ± 2.5	Weight: 97.0 ± 3.5	FM: 36.0 ± 2.0	Weight loss: −7.5 ± 4.5	FM: −8.0 ± 2.2 (=−17.2%)	Weight or BMI and fat measures decreased significantly. Weight loss mostly accounted for by FM.	
[35]	Weight: 87.7 ± 22.6 BMI: 32.2 ± 7.8	% BF: 38.4 ± 6.0	Weight: 74.8 ± 21.4	% BF: 30.3 ± 7.1	Weight loss: −8.2 ± 0.7	% BF: −8.1 ± 6.5	Weight and fat measures decreased significantly. Weight loss mostly accounted for by FM (91% loss was fat).	28%
[36]	Weight: 81.9 ± 10.5 BMI: 27.5 ± 1.6	% BF: 32.6 ± 7.6	Not reported	Not reported	Weight loss: −8.0 ± 2.0 (=−10.0%)	% BF: −25% change in % BF ± 3	Weight and fat measures decreased significantly.	4%
[37]	Weight: 100.2	% BF: 45.0%	Weight: 92.6	% BF: 41.8	Weight loss: −7.6	% BF: −3.2	Weight and fat measures decreased significantly. Diet and PA group lost significantly more fat and significantly less fat-free mass than non-PA subjects.	

TABLE 2: Continued.

Study (reference number)	Baseline average weight (kg) ± SD and/or BMI (kg/m^2) ± SD (range)	Baseline average % fat and/or fat mass (kg) ± SD (range)	Follow-up average weight (kg) ± SD and/or BMI (kg/m^2) ± SD (range)	Follow-up average % fat and/or fat mass (kg) ± SD (range)	Average weight (kg) ± SD and/or BMI (kg/m^2) change ± SD (range)	Average % fat or fat mass (kg) change ± SD (range or CI)	Summary/agreement of measures	Attrition (if reported)
[38]	BMI: 30.5 ± 3	% BF: 43.9 ± 1.1	Not reported	% BF: 38.5 ± 1.3	Weight loss: −8.6 ± 0.9	% BF: −4.5% FM: −7.0 ± 0.6	Diet + PA group lost significantly more weight than non-PA group; diet + PA group had significantly greater decrease in % BF than non-PA group. PA: 86% of weight loss from fat versus 73% f/non-PA group	20%
[39]	Weight: 85.2 kg ± 2.1 BMI: 31.6 ± 0.6	% BF: 40.2 ± 0.7 FM: 34.5 ± 1.3	Not reported	Not reported	Weight loss: −4.3 ± 0.7 BMI decrease: −1.8 ± 0.3	% BF: −1.3 ± 0.2 FM: −1.7 ± 0.5	Weight or BMI and fat measures decreased significantly. Fat loss more significant than weight loss.	10% at halfway point of study
[40]	Weight: 90.4 ± 2.9 BMI: 32.4 ± 1.3	% BF: 41.6 ± 1.5 FM: 38.0 ± 2.9	Not reported	Not reported	Weight loss: −9.1 ± 1.1	FM: −7.8 ± 0.82	Weight and fat measures decreased significantly. Weight loss mostly accounted for by FM: percentage of weight lost as fat significantly different for diet + PA group (88.6 ± 5.1%), versus diet only group (77.1 ± 4.0%).	
[41]	Weight: 95.7 ± 16.6 BMI: 35.3 ± 5.7	FM: 39.3 ± 10	Not reported	Not reported	Weight loss: −5.4 (−7.0–3.0)	% BF: −1.9% (−2.6–−1.3) FM: −3.3 (−4.0–−2.0)	Weight and fat measures decreased significantly.	33%
[42]	Weight: 94.3 kg ± 16.0 BMI: 35.7 ± 6.0	% BF: 45.1 ± 4.2 FM: 40.0 ± 9.7	Weight: 90.1 ± 15.6	% BF: 43.2 ± 4.7 FM: 36.6 ± 9.2	Weight loss: −4.2 ± 15.8	% BF: −1.9 ± 4.5 FM: −3.4 ± 9.5	Weight and fat measures decreased significantly. Weight loss mostly accounted for by FM.	35%
[43]	Weight: 96.5 ± 20.2 BMI: 35.3 ± 6.5	% BF: 41.9 ± 9.4 FM: 40.3 ± 12.8	Weight: 93.3 ± 21.2 BMI: 34.2 ± 6.6	% BF: 41.2 ± 8.9 FM: 38.6 ± 12.6	Weight loss: −2.9 ± 20.7 BMI: −1.1 ± 6.5	% BF: −0.7 ± 9.2 FM: −0.9 ± 12.7 (=−2.3%)	Weight decreased significantly; body FM was "borderline significant" ($P = 0.053$). BMI and % BF declined more in intervention group than control, no significant differences between groups.	
[44] (diet + aerobic ex.)	Not reported	Not reported	Not reported	Not reported	Weight loss: −6.8	% BF: −8.0	Weight and fat measures decreased significantly.	
[44] (diet + aerobic + strength)	Not reported	Not reported	Not reported	Not reported	Weight loss: −7.0	% BF: −4.3	Results same as above. Weight loss through diet was not altered by aerobic or aerobic + strength training, but diet + PA increased muscle during weight loss.	

Table 2: Continued.

Study (reference number)	Baseline average weight (kg) ± SD and/or BMI (kg/m^2) ± SD (range)	Baseline average % fat and/or fat mass (kg) ± SD (range)	Follow-up average weight (kg) ± SD and/or BMI (kg/m^2) ± SD (range)	Follow-up average % fat and/or fat mass (kg) ± SD (range)	Average weight (kg) ± SD and/or BMI (kg/m^2) change ± SD (range)	Average % fat or fat mass (kg) change ± SD (range or CI)	Summary/agreement of measures	Attrition (if reported)
[45]	Weight: 89.2 ± 12.0 BMI: 33.1 ± 4.0	% BF: 44.1 ± 4.0 FM: 36.8 ± 8.0	Weight: 86.1 ± 11.0 BMI: 32.0 ± 4.0	% BF: 43.1 ± 5.0 FM: 34.5 ± 7.0	Weight loss: −3.1 ± 11.5 BMI: −1.1 ± 4.0	% BF: −1.0 ± 4.5 FM: −2.3 ± 7.5	Weight or BMI and fat measures decreased significantly. Weight loss mostly accounted for by FM.	
[46]	Weight: 82.9 ± 3.6 BMI: 30.8 ± 1.5	FM: 38.6 ± 2.9	Weight: 74.7 ± 3.3	FM: 31.4 ± 2.6	Weight loss: −8.2 ± 3.4	FM: −7.2 ± 0.7	Weight or BMI and fat measures decreased significantly. Weight loss mostly accounted for by FM. More fat loss associated with high protein diet + PA	
[47]	Weight: 106.0 ± 25.6	% BF: 44.4 ± 5.6 FM: 47.6 ± 15.2	Weight: 90.7 ± 21.6	% BF: 35.3 ± 7.3 FM: 32.7 ± 12.7	Weight loss: −15.3 ± 1.1	% BF: −9.1 ± 0.5 FM: −14.9 ± 0.8	Weight and fat measures decreased significantly. Weight loss mostly accounted for by FM.	
[48]	Weight: 88.8 ± 3.3 BMI: 31.8 ± 0.7	% BF: 45.1 ± 1.3 FM: 40.2 ± 2.1	Weight: 81.7 ± 2.9 BMI: 28.0 ± 1.5	% BF: 39.7 ± 1.6 FM: 32.6 ± 1.9	Weight loss: −7.1 ± 3.1 (=−8%) BMI: −3.8 ± 1.1 (=−12%)	% BF: −5.4 ± 1.4 FM: −7.6 ± 2.0	Weight and fat measures decreased significantly. Weight loss mostly accounted for by FM. Diet + PA group lost more fat than PA only group.	48.8% in diet + PA group
[49]	Weight: 86.5 ± 12.0 BMI: 32.5 ± 5.0	FM: 42.0 ± 1.1	Not reported	FM: 36.8 ± 1.1	Weight loss: −7.6 ± 0.4	FM: −5.1 ± 0.5	Weight and fat measures decreased significantly. Weight loss mostly accounted for by FM.	16%
[50]	Weight: 92.1 ± 8.8	% BF: 44.0 ± 3.2	Not reported	Not reported	Weight loss: −10.5 ± 3.3	FM: −8.8 ± 2.1	Weight and fat measures decreased significantly. Weight loss mostly accounted for by FM.	25%
[51]	Weight: 78.1 ± 10.3	FM: 31.9 ± 6.2	Not reported	Not reported	Weight loss: −10.3 ± 3.0	FM: −9.6 ± 2.7	Weight and fat measures decreased significantly. Weight loss mostly accounted for by FM. Diet + PA group lost no lean tissue mass.	2.5%
[52]	Weight: 101.9 ± 11.2 BMI: 32.6 ± 2.5	% BF: 33.5 ± 4.2 FM: 34.2 ± 6.1	Weight: 86.7 ± 9.3 BMI: 27.8 ± 2.5	% BF: 25.0 ± 5.0 FM: 21.7 ± 5.2	Weight loss: −15.2 ± 6.3 BMI: −4.8 ± 2.5	% BF: −8.5 ± 4.7 FM: −12.5 ± 5.6	Weight or BMI and fat measures decreased significantly. Weight loss mostly accounted for by FM.	
[53]	Weight: 94.5 ± 11.7 BMI: 33.1 ± 3.4	Not reported	Not reported	Not reported	Weight loss: −6.6 ± 6.4 BMI: −2.3 ± 2.1	% BF: −4.0 ± 4.1	Weight or BMI and fat measures decreased significantly.	23%
[54]	Weight: 95.7 ± 18.0 BMI: 33.9 ± 5.3	Not reported	Not reported	Not reported	Weight loss: −6.5 (CI: −4.6−−8.4)	% BF: −2.8 (CI: −1.9−−3.9) FM: −4.8 (CI:−3.2−−6.3)	Weight or BMI and fat measures decreased significantly. Weight loss mostly accounted for by FM.	24–43% (different groups)

Note: outcome data are presented for the longest follow-up time available. CI: 95% confidence interval. When body weight or BMI and/or body composition outcome measures were not reported, the data were not imputed as long as the change measures were intact and reported; in those cases "not reported" is presented. Significance indicates differences compared to baseline values and to diet only/control group(s) when indicated.

which was commonly used in the studies reviewed here, is highly accurate but expensive [17, 21, 22]. Cost is the most common barrier against using DEXA. However, given the increased precision of measurement, the cost of DEXA scans may be justifiable for research groups, with long-term use. BIA is also commonly used to measure fat, but its accuracy is more questionable and likely depends on different equations used to estimate body composition and the quality of the equipment [17, 23]. Skinfold thickness is an inexpensive and useful method for large surveillance studies, but its accuracy and reliability are variable, depending on rater training and precision of caliper location [17, 23]. Body fat also is the most metabolically harmful tissue type, so it makes sense to promote its measurement over others [11, 12, 15, 16]. Further, just as weight and BMI do not provide nuanced measures of health, body fat mass is a similarly gross measure. Body fat % may be a more indicative measure of health, as it allows more specificity by accounting for other tissue types' contributions to weight and body composition.

As BMI and weight are ubiquitous in weight loss studies [5], it is not likely that a paradigm shift will occur quickly, in which measurement shifts to focus on % body fat. Further, most people do not know their body fat percentage or have a context for its interpretation the way people do for weight and, increasingly, BMI. However, healthy body fat % ranges do exist for different ages and genders [13, 24], and these values could become more commonly evaluated and discussed. The evidence presented herein suggests that % body fat should become more of a primary measure of health and weight loss success, as it provides a succinct and meaningful indication of a person's body composition, and likely disease risk, than weight, BMI, and fat mass.

This review had several limitations. First, only studies available in English were included. While this may introduce bias, most of the countries that bear the largest burden of adult obesity are economically developed and English-speaking. So it is unlikely that many contradictory or critical studies have been published in other languages. As with all reviews, this one encompassed studies with considerable heterogeneity of study/intervention and outcome components. While, at this stage of research, this problem is mostly inevitable, future studies may become more homogenous in measurement and reporting of outcomes if they follow the CONSORT and EQUATOR Network guidelines [25]. Another limitation is that most studies did not describe their attrition rates. Of those that did, many had high loss, over the 20% considered acceptable for weight loss studies [26]. Attrition is an important consideration for generalizing the results of this and other studies, so more consistent reporting is necessary, along with improved strategies for retaining participants in weight loss studies.

The risk of bias across studies merited attention. There is a concern that authors and journal editors typically prefer to show weight/fat loss, and so there is a risk of positive publication bias in this field [26]. Indeed, while we know that when properly carried out, diet and PA studies do promote fat loss, many interventions suffer from attrition, loss of participant motivation, and weight/fat regain over time; most studies presented here do not include long-term follow-up data. It is likely that many studies with negative findings do not make it into the published literature. Unfortunately, there is no way to tell how many of those studies existed (particularly before the NIH clinical trials registry: http://clinicaltrials.gov/), but everyone in the weight loss research field should consider results with this caveat in mind [26]. Risk of bias in individual studies also merited attention. While, ideally, this would have been assessed, most studies did not provide enough information to make consistent or relevant judgments of bias (i.e., blinding is not possible in weight loss trials and attrition and funding sources were not always reported). Four out of the 28 studies (8.5%) reported funding conflicts of interest, indicating a low risk of bias. The risk of bias in the present review is also minimal, as the author had no sources of financial support in its creation.

In conclusion, % body fat in addition to or along with weight, BMI, and fat mass appears to be a useful, consistent, and meaningful measure of success in adults weight loss studies. It is recommended that researchers include it as a primary outcome measure in future studies.

Conflict of Interests

The author has no potential conflict of interests to declare.

Acknowledgments

Drs. Linda Gallo, Kerri Boutelle, Karen Calfas, and James Sallis, all from the SDSU/UCSD Joint Doctoral Program in Clinical Psychology, are gratefully acknowledged, for their reviews of this paper and input into its development and structure.

References

[1] A. Avenell, T. J. Brown, M. A. McGee et al., "What interventions should we add to weight reducing diets in adults with obesity? A systematic review of randomized controlled trials of adding drug therapy, exercise, behaviour therapy or combinations of these interventions," *Journal of Human Nutrition and Dietetics*, vol. 17, no. 4, pp. 293–316, 2004.

[2] K. M. Flegal, M. D. Carroll, B. K. Kit, and C. L. Ogden, "Prevalence of obesity and trends in the distribution of body mass index among US adults, 1999–2010," *The Journal of the American Medical Association*, vol. 307, no. 5, pp. 491–497, 2012.

[3] A.-M. Glenny, S. O'Meara, A. Melville, T. A. Sheldon, and C. Wilson, "The treatment and prevention of obesity: a systematic review of the literature," *International Journal of Obesity*, vol. 21, no. 9, pp. 715–737, 1997.

[4] J. D. Douketis, C. Macie, L. Thabane, and D. F. Williamson, "Systematic review of long-term weight loss studies in obese adults: clinical significance and applicability to clinical practice," *International Journal of Obesity*, vol. 29, no. 10, pp. 1153–1167, 2005.

[5] M. J. Franz, J. J. VanWormer, A. L. Crain et al., "Weight-loss outcomes: a systematic review and meta-analysis of weight-loss clinical trials with a minimum 1-year follow-up," *Journal of the American Dietetic Association*, vol. 107, no. 10, pp. 1755–1767, 2007.

[6] R. N. Baumgartner, S. B. Heymsfield, and A. F. Roche, "Human body composition and the epidemiology of chronic disease," *Obesity Research*, vol. 3, no. 1, pp. 73–95, 1995.

[7] Z. Mei, L. M. Grummer-Strawn, A. Pietrobelli, A. Goulding, M. I. Goran, and W. H. Dietz, "Validity of body mass index compared with other body-composition screening indexes for the assessment of body fatness in children and adolescents," *American Journal of Clinical Nutrition*, vol. 75, no. 6, pp. 978–985, 2002.

[8] M. Dalton, A. J. Cameron, P. Z. Zimmet et al., "Waist circumference, waist-hip ratio and body mass index and their correlation with cardiovascular disease risk factors in Australian adults," *Journal of Internal Medicine*, vol. 254, no. 6, pp. 555–563, 2003.

[9] World Health Organization (WHO), *Obesity—Preventing and Managing the Global Epidemic: Report of a Who Consultation on Obesity*, World Health Organization, Geneva, Switzerland, 1998.

[10] S. R. Daniels, J. A. Morrison, D. L. Sprecher, P. Khoury, and T. R. Kimball, "Association of body fat distribution and cardiovascular risk factors in children and adolescents," *Circulation*, vol. 99, no. 4, pp. 541–545, 1999.

[11] A. R. Folsom, S. A. Kaye, T. A. Sellers et al., "Body fat distribution and 5-year risk of death in older women," *The Journal of the American Medical Association*, vol. 269, no. 4, pp. 483–487, 1993.

[12] S. C. Ho, Y. M. Chen, J. L. F. Woo, S. S. F. Leung, T. H. Lam, and E. D. Janus, "Association between simple anthropometric indices and cardiovascular risk factors," *International Journal of Obesity*, vol. 25, no. 11, pp. 1689–1697, 2001.

[13] C. L. Ogden, Y. Li, D. S. Freedman, L. G. Borrud, and K. M. Flegal, "Smoothed percentage body fat percentiles for U.S. children and adolescents, 1999–2004," *National Health Statistics Report*, no. 43, pp. 1–7, 2011.

[14] D. T. Villareal, C. M. Apovian, R. F. Kushner, and S. Klein, "Obesity in older adults: Technical review and position statement of the American Society for Nutrition and NAASO, The Obesity Society," *The American Journal of Clinical Nutrition*, vol. 82, no. 5, pp. 923–934, 2005.

[15] E. W. Gregg, Y. J. Cheng, B. L. Cadwell et al., "Secular trends in cardiovascular disease risk factors according to body mass index in US adults," *Journal of the American Medical Association*, vol. 293, no. 15, pp. 1868–1874, 2005.

[16] P. W. Wilson, R. B. D'Agostino, L. Sullivan, H. Parise, and W. B. Kannel, "Overweight and obesity as determinants of cardiovascular risk: the Framingham experience," *Archives of Internal Medicine*, vol. 162, no. 16, pp. 1867–1872, 2002.

[17] M. Davis, "Measuring adiposity in health research," in *Handbook of Physiological Research Methods in Health Psychology*, L. C. Luecken and L. J. Gallo, Eds., pp. 259–275, Sage, Thousand Oaks, Calif, USA, 2007.

[18] W. C. Miller, D. M. Koceja, and E. J. Hamilton, "A meta-analysis of the past 25 years of weight loss research using diet, exercise or diet plus exercise intervention," *International Journal of Obesity*, vol. 21, no. 10, pp. 941–947, 1997.

[19] P. Stiegler and A. Cunliffe, "The role of diet and exercise for the maintenance of fat-free mass and resting metabolic rate during weight loss," *Sports Medicine*, vol. 36, no. 3, pp. 239–262, 2006.

[20] D. L. Ballor and E. T. Poehlman, "Exercise-training enhances fat-free mass preservation during diet-induced weight loss: a meta-analytical finding," *International Journal of Obesity*, vol. 18, no. 1, pp. 35–40, 1994.

[21] J. C. Eisenmann, K. A. Heelan, and G. J. Welk, "Assessing body composition among 3- to 8-year-old children: anthropometry, BIA, and DXA," *Obesity Research*, vol. 12, no. 10, pp. 1633–1640, 2004.

[22] J. Steinberger, D. R. Jacobs Jr., S. Raatz, A. Moran, C.-P. Hong, and A. R. Sinaiko, "Comparison of body fatness measurements by BMI and skinfolds vs dual energy X-ray absorptiometry and their relation to cardiovascular risk factors in adolescents," *International Journal of Obesity*, vol. 29, no. 11, pp. 1346–1352, 2005.

[23] N. Wattanapenpaiboon, W. Lukito, B. J. G. Strauss, B. H.-H. Hsu-Hage, M. L. Wahlqvist, and D. B. Stroud, "Agreement of skinfold measurement and bioelectrical impedance analysis (BIA) methods with dual energy X-ray absorptiometry (DEXA) in estimating total body fat in Anglo-Celtic Australians," *International Journal of Obesity and Related Metabolic Disorders*, vol. 22, no. 9, pp. 854–860, 1998.

[24] D. Gallagher, S. B. Heymsfield, M. Heo, S. A. Jebb, P. R. Murgatroyd, and Y. Sakamoto, "Healthy percentage body fat ranges: an approach for developing guidelines based on body mass index," *American Journal of Clinical Nutrition*, vol. 72, no. 3, pp. 694–701, 2000.

[25] EQUATOR Network, Library for health research reporting, 2011, http://www.equator-network.org/reporting-guidelines/.

[26] R. M. Kaplan and C. J. Atkins, "Selective attrition causes overestimates of treatment effects in studies of weight loss," *Addictive Behaviors*, vol. 12, no. 3, pp. 297–302, 1987.

[27] R. E. Andersen, T. A. Wadden, S. J. Bartlett, B. Zemel, T. J. Verde, and S. C. Franckowiak, "Effects of lifestyle activity vs structured aerobic exercise in obese women: a randomized trial," *The Journal of the American Medical Association*, vol. 281, no. 4, pp. 335–340, 1999.

[28] E. Andreou, C. Philippou, and D. Papandreou, "Effects of an intervention and maintenance weight loss diet with and without exercise on anthropometric indices in overweight and obese healthy women," *Annals of Nutrition and Metabolism*, vol. 59, no. 2–4, pp. 187–192, 2011.

[29] N. M. Byrne, J. D. Meerkin, R. Laukkanen, R. Ross, M. Fogelholm, and A. P. Hills, "Weight loss strategies for obese adults: personalized weight management program vs. standard care," *Obesity*, vol. 14, no. 10, pp. 1777–1788, 2006.

[30] R. A. Carels, L. A. Darby, H. M. Cacciapaglia, and O. M. Douglass, "Reducing cardiovascular risk factors in postmenopausal women through a lifestyle change intervention," *Journal of Women's Health*, vol. 13, no. 4, pp. 412–426, 2004.

[31] P. del Corral, P. C. Chandler-Laney, K. Casazza, B. A. Gower, and G. R. Hunter, "Effect of dietary adherence with or without exercise on weight loss: a mechanistic approach to a global problem," *The Journal of Clinical Endocrinology & Metabolism*, vol. 94, no. 5, pp. 1602–1607, 2009.

[32] G. D. Foster, H. R. Wyatt, J. O. Hill et al., "Weight and metabolic outcomes after 2 years on a low-carbohydrate versus low-fat diet: a randomized trial," *Annals of Internal Medicine*, vol. 153, no. 3, pp. 147–157, 2010.

[33] A. Geliebter, M. M. Maher, L. Gerace, B. Gutin, S. B. Heymsfield, and S. A. Hashim, "Effects of strength or aerobic training on body composition, resting metabolic rate, and peak oxygen consumption in obese dieting subjects," *The American Journal of Clinical Nutrition*, vol. 66, no. 3, pp. 557–563, 1997.

[34] A. Golay, A.-F. Allaz, Y. Morel, N. de Tonnac, S. Tankova, and G. Reaven, "Similar weight loss with low- or high-carbohydrate

diets," *American Journal of Clinical Nutrition*, vol. 63, no. 2, pp. 174–178, 1996.

[35] R. L. Hammer, C. A. Barrier, E. S. Roundy, J. M. Bradford, and A. G. Fisher, "Calorie-restricted low-fat diet and exercise in obese women," *The American Journal of Clinical Nutrition*, vol. 49, no. 1, pp. 77–85, 1989.

[36] L. K. Heilbronn, L. de Jonge, M. I. Frisard et al., "Effect of 6-month calorie restriction on biomarkers of longevity, metabolic adaptation, and oxidative stress in overweight individuals: a randomized controlled trial," *The Journal of the American Medical Association*, vol. 295, no. 13, pp. 1539–1548, 2006.

[37] J. O. Hill, P. B. Sparling, T. W. Shields, and P. A. Heller, "Effects of exercise and food restriction on body composition and metabolic rate in obese women," *The American Journal of Clinical Nutrition*, vol. 46, no. 4, pp. 622–630, 1987.

[38] J. O. Hill, D. G. Schlundt, T. Sbrocco et al., "Evaluation of an alternating-calorie diet with and without exercise in the treatment of obesity," *The American Journal of Clinical Nutrition*, vol. 50, no. 2, pp. 248–254, 1989.

[39] A. R. Josse, S. A. Atkinson, M. A. Tarnopolsky, and S. M. Phillips, "Increased consumption of dairy foods and protein during diet- and exercise-induced weight loss promotes fat mass loss and lean mass gain in overweight and obese premenopausal women," *Journal of Nutrition*, vol. 141, no. 9, pp. 1626–1634, 2011.

[40] K. P. G. Kempen, W. H. M. Saris, and K. R. Westerterp, "Energy balance during an 8-wk energy-restricted diet with and without exercise in obese women," *American Journal of Clinical Nutrition*, vol. 62, no. 4, pp. 722–729, 1995.

[41] C. Kerksick, A. Thomas, B. Campbell et al., "Effects of a popular exercise and weight loss program on weight loss, body composition, energy expenditure and health in obese women," *Nutrition and Metabolism*, vol. 6, article 023, 2009.

[42] C. M. Kerksick, J. Wismann-Bunn, D. Fogt et al., "Changes in weight loss, body composition and cardiovascular disease risk after altering macronutrient distributions during a regular exercise program in obese women," *Nutrition Journal*, vol. 9, no. 1, article 59, 2010.

[43] M. L. Kirby, S. Beatty, J. Stack et al., "Changes in macular pigment optical density and serum concentrations of lutein and zeaxanthin in response to weight loss," *British Journal of Nutrition*, vol. 105, no. 7, pp. 1036–1046, 2011.

[44] W. J. Kraemer, J. S. Volek, K. L. Clark et al., "Physiological adaptations to a weight-loss dietary regimen and exercise programs in women," *Journal of Applied Physiology*, vol. 83, no. 1, pp. 270–279, 1997.

[45] R. B. Kreider, M. Serra, K. M. Beavers et al., "A structured diet and exercise program promotes favorable changes in weight loss, body composition, and weight maintenance," *Journal of the American Dietetic Association*, vol. 111, no. 6, pp. 828–843, 2011.

[46] D. K. Layman, E. Evans, J. I. Baum, J. Seyler, D. J. Erickson, and R. A. Boileau, "Dietary protein and exercise have additive effects on body composition during weight loss in adult women," *Journal of Nutrition*, vol. 135, no. 8, pp. 1903–1910, 2005.

[47] B. C. Leutholtz, R. E. Keyser, W. W. Heusner, V. E. Wendt, and L. Rosen, "Exercise training and severe caloric restriction: effect on lean body mass in the obese," *Archives of Physical Medicine and Rehabilitation*, vol. 76, no. 1, pp. 65–70, 1995.

[48] K. J. Melanson, J. Dell'Olio, M. R. Carpenter, and T. J. Angelopoulos, "Changes in multiple health outcomes at 12 and 24 weeks resulting from 12 weeks of exercise counseling with or without dietary counseling in obese adults," *Nutrition*, vol. 20, no. 10, pp. 849–856, 2004.

[49] M. Noakes, J. B. Keogh, P. R. Foster, and P. M. Clifton, "Effect of an energy-restricted, high-protein, low-fat diet relative to a conventional high-carbohydrate, low-fat diet on weight loss, body composition, nutritional status, and markers of cardiovascular health in obese women," *American Journal of Clinical Nutrition*, vol. 81, no. 6, pp. 1298–1306, 2005.

[50] S. B. Racette, D. A. Schoeller, R. F. Kushner, K. M. Neil, and K. Herling-Iaffaldano, "Effects of aerobic exercise and dietary carbohydrate on energy expenditure and body composition during weight reduction in obese women," *The American Journal of Clinical Nutrition*, vol. 61, no. 3, pp. 486–494, 1995.

[51] O. L. Svendsen, C. Hassager, and C. Christiansen, "Effect of an energy-restrictive diet, with or without exercise, on lean tissue mass, resting metabolic rate, cardiovascular risk factors, and bone in overweight postmenopausal women," *The American Journal of Medicine*, vol. 95, no. 2, pp. 131–140, 1993.

[52] D. P. C. Van Aggel-Leijssen, W. H. M. Saris, G. B. Hul, and M. A. van Baak, "Short-term effects of weight loss with or without low-intensity exercise training on fat metabolism in obese men," *American Journal of Clinical Nutrition*, vol. 73, no. 3, pp. 523–531, 2001.

[53] D. Vissers, A. Verrijken, I. Mertens et al., "Effect of long-term whole body vibration training on visceral adipose tissue: a preliminary report," *Obesity Facts*, vol. 3, no. 2, pp. 93–100, 2010.

[54] W. S. Yancy Jr., M. K. Olsen, J. R. Guyton, R. P. Bakst, and E. C. Westman, "A low-carbohydrate, ketogenic diet versus a low-fat diet to treat obesity and hyperlipidemia: a randomized, controlled trial," *Annals of Internal Medicine*, vol. 140, no. 10, pp. 769–777, 2004.

9

The Relationship between the Neighborhood Safety and Nutritional Status of Children in Baghdad City, Iraq

Hasanain Faisal Ghazi, Zaleha Md. Isa, Shamsul Azhar Shah, Mohammed A. Abdal Qader, Tiba Nezar Hasan, and AL-abed Ali AL-abed

Department of Community Health, Universiti Kebangsaan Malaysia Medical Centre, Bandar Tun Razak, Cheras, 56000 Kuala Lumpur, Malaysia

Correspondence should be addressed to Hasanain Faisal Ghazi; dr.hasanainhabasha@gmail.com

Academic Editor: Maurizio Muscaritoli

Introduction. The neighborhoods where the children live play an important role in their development physically and mentally. The objective of this study was to assess the relationship between neighborhood safety and child nutritional status in Baghdad city, Iraq. *Methods.* A cross-sectional study was carried out in Baghdad city, Iraq, among 400 primary school children from 4 schools. BMI-for-age Z score was used to assess the nutritional status of the children. Newly developed questionnaires on neighborhood safety were distributed to the parents to answer them. *Results.* In this study, males were more predominant than females with 215 participants compared to 185 females. A total of 49% were normal weight, 38.8% either overweight or obese, and only 12.2% underweight. There was a significant relationship between father education, father and mother working status, family income, and children nutritional status ($P = 0.10, 0.009, <0.001, 0.37$), respectively. The association between neighborhood safety variables and child nutritional status was significant except for worrying about child safety and thinking of leaving the neighborhood ($P = 0.082, 0.084$), respectively. *Conclusion.* Nutritional status of school children continues to be a public health issue in Iraq especially Baghdad city. There was a significant association between neighborhood safety and children nutritional status.

1. Introduction

Child nutrition remains a serious challenging issue not only in Iraq but also worldwide. As a developing country located in the Middle East, Iraq faced several conflicts starting from 1980 followed by 1991 war then 13 years of economic sanctions and last war in 2003. The problem for Iraqis and especially children is not the war itself but the following years of unstable situation that can impact their daily life and activities.

Environmental factors can include various physical and social elements in people's surroundings that influence their choices for diet and physical activities. School and nearby neighborhood environments are of interest because children spend at least 5–7 hours daily within school boundaries during weekdays and make various health choices while at school [1].

Iraq is a developing country experiencing constraints on economic and social development and most of the environmental factors that affect the physical growth of children before puberty, including poor food consumption patterns, illness, lack of sanitation, poor hygiene practices, and poor health care coverage and resources, are present. Almost half of Iraq's total populations of 27 million are children. United Nations agencies estimate that 1 out of 8 Iraqi children dies before the age of 5 years; one-third are undernourished; one-quarter are born underweight and one-quarter do not have access to safe water. Rates for stunting among children under the age of 5 years in southern and central Iraq peaked in 1996 at 32.0% and then declined to 23.1% in 2002 [2].

ALDoori et al. [3] did a study in Basrah city in south of Iraq following the 1991 war and found that 24% of the children were stunted. Stunting and low weight-for-age were significantly higher among children of low socioeconomic

Table 1: Sociodemographic characteristics of children.

Variables	Nutritional status Underweight N (%)	Normal N (%)	Overweight and obese N (%)	P value
Gender				
Male	23 (10.7)	109 (50.7)	83 (38.6)	0.551[a]
Female	26 (14.1)	87 (47.0)	72 (38.9)	
Mother education				
Low	10 (16.9)	28 (47.5)	21 (35.6)	0.448[a]
High	39 (11.5)	166 (49.0)	134 (39.5)	
Father education				
Low	4 (8.7)	32 (69.6)	10 (21.7)	**0.010**[a]
High	36 (10.7)	155 (46.3)	144 (43.0)	
Mother working status				
Housewife	35 (16.8)	101 (48.6)	72 (34.6)	**0.009**[a]
Working	14 (7.4)	93 (48.9)	83 (43.7)	
Father working status				
Not working	6 (42.9)	6 (42.9)	2 (14.2)	**<0.001**[a]
Working	34 (9.3)	181 (49.3)	152 (41.4)	
Family income				
Mean ± SD	615,673.46 ± 361,872.29	763,418.36 ± 546,916.34	810,225.80 ± 356,521.04	**0.037**[b]

[a]Chi Square test was performed; [b]Student t-test was performed, level of significance at P value <0.05.

households. Comparison of these data with an earlier nutritional survey in the area showed the nutritional status of children in Basrah city has deteriorated as a result of successive armed conflicts.

Another study done by Guerrero-serdan [4] in Iraq assumed the war has affected the physical growth of children, where the young born after the war in high-intensity conflict areas had lower height-for-age Z-scores than children born in less violent areas. Also the weight-forage Z-scores increased in 2004 but decreased in 2006. These results suggest that children are not losing weight but rather are not growing properly in length in high violence affected areas.

The links between neighborhood and nutritional status can be because of many reasons such as safety concerns (crime; unstable political situation), the environment (lack of parks, playgrounds, and walkable destinations such as restaurants and supermarkets), and access to and affordability of healthy foods [5–9].

The objective of this study was to assess the effects of unstable neighborhood in the last 8 years in Baghdad city on child nutritional status.

2. Methods

A cross-sectional study was conducted among 400 respondents aged between 7 and 8 years from four different primary schools in Baghdad city in 2011. Baghdad is divided into two sides by Tigris river, so 2 schools from each side were chosen randomly. Subsequently, a complete list of student names in each selected school was obtained. A total of 100 children from each school were then identified by stratified random sampling according to grade. Questionnaires on neighborhood safety and standardized measurements for weight and height were used in this study. Self-administered questionnaires were distributed to respondents' parents during the monthly meeting at schools.

BMI for age Z-score was used to assess the nutritional status of the children based on WHO 2007 growth references cut-off points [10]. Weight and height of the respondents were measured by using calibrated weighing machine to be compared to WHO cut-off points. Weight was measured using digital weighing scale manufactured by Beurer Company in Germany. Children were weighed after taking off their shoes.

Height was measured by using tape measure to the nearest 0.5 cm by standing with their back touching the wall and their head in upright position. Then BMI-for-Age Z score is recategorized into three groups, underweight, normal weight, and overweight or obese children in one group.

Neighborhood safety in this study was defined as the place where the child's family is living. The questionnaire consists of 10 items, each with three response options: never (2), sometimes (1), and a lot of times (0). The maximum total score was 20. A score of 10 was taken as the cut-point because if the respondents were to answer "sometimes" for all questions, the total score is 10. Any score less than 10 indicated unsafe neighborhood, whereas a score equal to or above 10 indicated safe neighborhood.

A questionnaire validation was carried out using face validity, factor analysis, and back-to-back translation. After that a pretest was done by distributing the questionnaires to 50 parents to ensure that the questions are easily understandable and validated. Living environment scale has a good internal consistency, with Cronbach's alpha coefficient of 0.820.

This study was approved by the Research and Ethics Committee of Universiti Kebangsaan Malaysia Medical Center.

Table 2: Relationship between neighbourhood safety variables and children nutritional status.

Variables	Nutritional status Underweight *N* (%)	Normal *N* (%)	Overweight and obese *N* (%)	*P* value[a]
Did that unstable security situation affect your daily life?				
A lot of times	15 (19.7)	30 (39.5)	31 (40.8)	
Sometimes	29 (10.8)	131 (48.7)	109 (40.5)	**0.034**
Never	5 (9.1)	35 (63.6)	15 (27.3)	
Did you hear gun shots in your neighborhood during the last month?				
A lot of times	8 (13.1)	26 (42.6)	27 (44.3)	
Sometimes	37 (14.0)	146 (55.3)	81 (30.7)	**<0.001**
Never	4 (5.3)	24 (32.0)	47 (62.7)	
Did you hear explosions in your neighborhood during the last month?				
A lot of times	8 (13.1)	29 (47.5)	24 (39.4)	
Sometimes	38 (15.2)	125 (50.0)	87 (34.8)	**0.023**
Never	3 (3.4)	42 (47.2)	44 (49.4)	
Did you see any dead bodies in your neighborhood during the last month?				
A lot of times	2 (33.3)	4 (66.7)	0 (0.0)	
Sometimes	27 (30.7)	37 (42.0)	24 (27.3)	**<0.001**
Never	20 (6.5)	155 (50.7)	131 (42.8)	
Did your child witness any explosions in your neighborhood?				
A lot of times	4 (40.0)	4 (40.0)	2 (20.0)	
Sometimes	27 (14.8)	84 (45.9)	72 (39.3)	**0.024**
Never	18 (8.7)	108 (52.2)	81 (39.1)	
Did your child face any problems like closed roads in going to school during the last month?				
A lot of times	12 (54.5)	6 (27.3)	4 (18.2)	
Sometimes	25 (13.2)	82 (43.1)	83 (43.7)	**<0.001**
Never	12 (6.4)	108 (57.4)	68 (36.2)	
Do you worry about child safety when they go to school?				
A lot of times	23 (18.4)	57 (45.6)	45 (36.0)	
Sometimes	17 (8.5)	99 (49.2)	85 (42.3)	0.082
Never	9 (12.2)	40 (54.0)	25 (33.8)	
Did your child miss any school days because of neighborhood safety issues?				
A lot of times	6 (37.4)	5 (31.3)	5 (31.3)	
Sometimes	27 (11.0)	119 (48.4)	100 (40.6)	**0.030**
Never	16 (11.6)	72 (52.2)	50 (36.2)	
Do you ever think of leaving your neighborhood?				
A lot of times	9 (20.5)	22 (50.0)	13 (29.5)	
Sometimes	26 (12.7)	90 (44.1)	88 (43.2)	0.084
Never	14 (9.2)	84 (55.3)	54 (35.5)	
Do you think that your neighborhood area is insecure?				
A lot of times	10 (37.0)	10 (37.0)	7 (26.0)	
Sometimes	21 (10.1)	88 (42.3)	99 (47.6)	**<0.001**
Never	18 (10.9)	98 (59.4)	49 (29.7)	

[a]Pearson Chi Square test was performed; level of significance is at $P < 0.05$.

3. Results

The response rate in this study was 85.1% (400/470). The results show that the mean body weight of the children was 25.20 ± 5.6 kg. The average standing height of the children was 121.00 ± 8.8 cm. The most common nutritional status was normal (49.0%) followed by overweight and obese (38.8%) and underweight (12.2%).

There was a significant relationship between father education, father and mother working status, family income, and children nutritional status (P = 0.10, 0.009, <0.001, 0.37), respectively, as shown in Table 1.

For neighborhood safety variables, the relationship between thinking of unstable situation, hearing gun shots, hearing explosions, saying dead bodies, child witnessing explosions, difficulties in going to school, thinking neighborhood is unsafe, child missing school days, and nutritional status was significant (P = 0.034, <0.001, 0.023, <0.001, 0.024, <0.001, 0.030, <0.001), respectively, as shown in Table 2.

Table 3: The relationship between neighborhood safety status and children nutritional status.

Neighborhood safety	Nutritional status: Underweight N (%)	Nutritional status: Normal N (%)	Nutritional status: Overweight and obese N (%)	P value
Unsafe	23 (27.1)	39 (45.8)	23 (27.1)	**<0.001**[a]
Safe	26 (8.3)	157 (49.8)	132 (41.9)	

[a]Pearson Chi Square test was performed; level of significance is at $P < 0.05$.

There was a significant association between neighborhood safety status and nutritional status of the children ($P < 0.001$) as shown in Table 3, which means children living in unsafe neighborhood have more risk to be overweight or obese as they tend to be physically inactive.

4. Discussion

In this study, the neighborhood safety status reflects the daily life of Iraqi families and their children. There was some improvement in the living environment due to changes in the security situation in the capital Baghdad in the last 2 years as more than two-thirds of the children's parents were found to live in a good environment according to our cut-off point.

The nutritional status of the children was affected by the unstable living environment in Baghdad city in the last 10 years after the end of 2003 war. The association between neighborhood safety status and child nutritional status was significant, which means children who live in bad living environment have more risk to be overweight and obese compared to those living in acceptable living environment and this may be due to lack of physical activity as the children cannot play outside the house in the playground and also cannot walk to school because parents are concerned about their safety.

Tol et al. [11] noted that exposure to violence is a risk factor for adverse outcomes of child development in low-income settings, and that childhood mental health problems are difficult to address within the contexts of ongoing poverty and political instability. Parental perception of neighborhood safety likely operates more strongly than that of the child in altering lifestyle because parents exert significant control over young children's activities [12, 13].

The results of this study was supported by similar studies done in armed conflicts zones elsewhere in the world, such as one study done following the conflict in Guinea-Bissau [14]. They found that in a noncamp setting, residents may be more malnourished and have higher mortality than refugees. Major improvements in nutritional status and a reduction in mortality occurred in resident and refugee children as soon as refugees returned home despite the fact that there was no improvement in food availability.

Previous study done in Basrah city in south of Iraq following the 1991 war [3] found that 24% of the children were stunted. Stunting and low weight-for-age were significantly higher among children of low socioeconomic households. Comparison of these data with an earlier nutritional survey in the area showed the nutritional status of children in Basrah city has worsened as a result of successive armed conflicts.

Latest study done to assess the effects of the 2003 war in Iraq on nutrition and health [4] concluded that the war had affected the physical growth of children, where young cohorts born after the war in high-intensity conflict areas had lower height-for-age Z-scores than children born in less violent areas. Also the weight-forage Z-scores increased in 2004 but decreased in 2006. These results suggest that children are not losing weight but rather are not growing properly in length in high violence affected areas.

There were some limitations in this study. First of all, its cross-sectional design only measured the prevalence at specific point of time in specified areas. Second, is the recall bias of the respondents' parents, especially in their evaluation of the security situation in their living neighborhood. lastly, is the logistic issues because of unstable security situation.

5. Conclusion

Nutritional status of school children continues to be a public health issue in Iraq especially Baghdad city. There was a significant association between neighborhood safety and children nutritional status. There is a transition from undernutrition among children during the economic sanctions in the 90s to rapid increase in overweight and obesity prevalence in Iraq. Children living in safe area tend to be more overweight and obese.

Conflict of Interests

Authors declare no conflict of interests.

Acknowledgment

This study was funded by the Universiti Kebangsaan Malaysia Medical Centre Fundamental Research Grant, Code no. FF-180-2011, without which the study would not have been possible.

References

[1] S. Park, B. Y. Choi, Y. Wang, E. Colantuoni, and J. Gittelsohn, "School and neighborhood nutrition environment and their association with students'nutrition behaviors and weight status in seoul, South Korea," *Journal of Adolescent Health*, vol. 53, pp. 655–662, 2013.

[2] A. J. Al-Saffar, "Stunting among primary-school children: A sample from Baghdad, Iraq," *Eastern Mediterranean Health Journal*, vol. 15, no. 2, pp. 322–329, 2009.

[3] W. AlDoori, N. Armijo-Hussein, W. W. Fawzi, and M. G. Herrera, "Child nutrition and armed conflicts in Iraq," *Journal of Tropical Pediatrics*, vol. 40, no. 1, pp. 32–36, 1994.

[4] G. Guerrero-Serdan, "The effects of the war in Iraq on nutrition and health: an analysis using anthropometric outcomes of children," Royal Holloway College, University of London, 2009, http://mpra.ub.uni-muenchen.de/14056.

[5] D. Rose and R. Richards, "Food store access and household fruit and vegetable use among participants in the US Food Stamp Program," *Public Health Nutrition*, vol. 7, no. 8, pp. 1081–1088, 2004.

[6] B. Giles-Corti and R. J. Donovan, "Socioeconomic status differences in recreational physical activity levels and real and perceived access to a supportive physical environment," *Preventive Medicine*, vol. 35, no. 6, pp. 601–611, 2002.

[7] N. Humpel, N. Owen, and E. Leslie, "Environmental factors associated with adults' participation in physical activity: a review," *The American Journal of Preventive Medicine*, vol. 22, no. 3, pp. 188–199, 2002.

[8] I. Kawachi and L. Berkman, "Social cohesion, social capital, and health," in *Social Epidemiology*, L. Berkman and I. Kawachi, Eds., pp. 174–190, Oxford University Press, New York, NY, USA, 2000.

[9] L. V. Moore, A. V. Diez Roux, K. R. Evenson, A. P. McGinn, and S. J. Brines, "Availability of recreational resources in minority and low socioeconomic status areas," *The American Journal of Preventive Medicine*, vol. 34, no. 1, pp. 16–22, 2008.

[10] World Health Organization, *Development of a WHO Growth Reference for School-Aged Children and Adolescents*, WHO, Geneva, Switzerland, 2007.

[11] W. A. Tol, I. H. Komproe, D. Susanty, M. J. D. Jordans, R. D. Macy, and J. T. V. M. De Jong, "School-based mental health intervention for children affected by political violence in Indonesia: a cluster randomized trial," *Journal of the American Medical Association*, vol. 300, no. 6, pp. 655–662, 2008.

[12] R. O'Neil, R. D. Parke, and D. J. McDowell, "Objective and subjective features of children's neighborhoods: relations to parental regulatory strategies and children's social competence," *Journal of Applied Developmental Psychology*, vol. 22, no. 2, pp. 135–155, 2001.

[13] N. E. Hill and M. A. Herman-Stahl, "Neighborhood safety and social involvement: associations with parenting behaviors and depressive symptoms among African American and Euro-American mothers," *Journal of Family Psychology*, vol. 16, no. 2, pp. 209–219, 2002.

[14] P. Aaby, J. Gomes, M. Fernandes, Q. Djana, I. Lisse, and H. Jensen, "Nutritional status and mortality of refugee and resident children in a non-camp setting during conflict: follow up study in Guinea-Bissau," *British Medical Journal*, vol. 319, no. 7214, pp. 878–881, 1999.

10

Effects of High Phosphorus Diet on Bone Metabolism-Related Gene Expression in Young and Aged Mice

Shinichi Katsumata,[1] Hiroshi Matsuzaki,[1] Rie Katsumata-Tsuboi,[1] Mariko Uehara,[2] and Kazuharu Suzuki[1]

[1] *Department of Nutritional Science, Faculty of Applied Bioscience, Tokyo University of Agriculture, 1-1-1 Sakuragaoka, Setagaya-ku, Tokyo 156-8502, Japan*
[2] *Department of Nutritional Science and Food Safety, Faculty of Applied Bioscience, Tokyo University of Agriculture, 1-1-1 Sakuragaoka, Setagaya-ku, Tokyo 156-8502, Japan*

Correspondence should be addressed to Shinichi Katsumata; s1katsum@nodai.ac.jp

Academic Editor: Michael B. Zemel

In this study, the effects of high phosphorus (P) diet on bone metabolism-related gene expression were investigated in young and aged mice. Twelve- and 80-week-old ddY male mice were divided into two groups, respectively, and fed a control diet containing 0.3% P or a high P diet containing 1.2% P. After 4 weeks of treatment, serum parathyroid hormone (PTH) concentration was significantly higher in the high P groups than in the control groups in both young and aged mice and was significantly higher in aged mice than in young mice fed the high P diet. High P diet significantly increased receptor activator of NF-κB ligand (RANKL) mRNA in the femur of both young and aged mice and significantly increased the RANKL/osteoprotegerin (OPG) mRNA ratio only in aged mice. High P diet significantly increased mRNA expression of transient receptor potential vanilloid type 6, calbindin-D9k, and plasma membrane Ca^{2+}-ATPase 1b in the duodenum of both young and aged mice. These results suggest that high P diet increased RANKL mRNA expression in the femur and calcium absorption-related gene expression in the duodenum regardless of age. Furthermore, the high P diet-induced increase in PTH secretion might increase the RANKL/OPG mRNA ratio in aged mice.

1. Introduction

High phosphorus (P) intake is known as one of the risk factors for impaired bone health. Several researchers have investigated the adverse effects of a high P diet on bone metabolism. In human adults, a diet containing P additives increases urinary hydroxyproline excretion, a bone resorption marker [1]. In growing male rats, high P intake decreases bone mineral density (BMD) and bone strength [2]. Elevation in parathyroid hormone (PTH) secretion is considered one of the mechanisms by which a high P diet impairs bone metabolism [3].

Bone metabolism results from the balance between osteoblastic bone formation and osteoclastic bone resorption. Chronic PTH stimulation is known to induce osteoclastogenesis. PTH stimulates osteoblasts, which produce mediators of osteoclastic bone resorption such as macrophage colony-stimulating factor (M-CSF), interleukin-6 (IL-6), or receptor activator of NF-κB ligand (RANKL) [4–6]. We previously reported that a high P diet increased RANKL mRNA expression and the osteoclast number in rats [3]. It could therefore be deduced that high P diet-induced elevated PTH secretion leads to an increase in RANKL expression, which enhances osteoclastic bone resorption.

Aging is also one of the risk factors of bone loss in elderly individuals. Overton and Basu suggested that bone loss occurs with increasing age at a rate of approximately 1% per year averaged over the age range of 29–76 y [7]. In a previous mouse study by Ferguson et al., bone mass and mechanical properties approached mature levels by 12 weeks of age while age-related osteopenia was observed after 42 weeks of age [8]. We hypothesized that a high P diet would accelerate age-related bone loss. We previously reported the effects of a high P diet on mechanical properties of

TABLE 1: Composition of the experimental diets.

	Control	High P
	g/kg diet	
Casein	200.0	200.0
Corn starch	529.486	489.938
Sucrose	100.0	100.0
Soybean oil	70.0	70.0
Cellulose powder	50.0	50.0
AIN-93G mineral mixture	35.0	35.0
AIN-93 vitamin mixture	10.0	10.0
L-Cysteine	3.0	3.0
Choline bitartrate	2.5	2.5
tert-Butylhydroquinone	0.014	0.014
KH_2PO_4	—	39.548

the femur in 4-, 12-, 24-, and 80-week-old mice [9]. The results showed that a high P diet decreased the breaking force of the femur in 80-week-old mice and the stiffness of the femur in 24- and 80-week-old mice. We also found that a high P diet increased serum PTH concentration in 12-, 24-, and 80-week-old mice, and 80-week-old mice had a higher serum PTH concentration than mice at other ages. Therefore, it was thought that a high P diet strongly influences aged mice in terms of PTH response. However, the mechanisms are not fully understood.

The purpose of this study was to clarify the mechanism by which a high P diet affects bone metabolism in older mice. We assessed the changes in bone metabolism in young and aged mice fed a high P diet by measuring the mRNA expression of bone metabolism mediators using real-time polymerase chain reaction (PCR).

2. Methods

2.1. Experimental Design. This study was approved by the Tokyo University of Agriculture Animal Use Committee. The mice were maintained in accordance with the university guidelines for the care and use of laboratory animals. The experimental diets were based on the AIN-93G diet (Table 1) [10]. A control diet containing 0.3% P and a high P diet containing 1.2% P were prepared. Each experimental diet contained 0.5% calcium (Ca). Twenty-four 10-week-old ddY male mice were purchased from SLC (Shizuoka, Japan) and housed individually in stainless cages in a room maintained at 22°C with a 12-hour light/dark cycle. Half of the mice were fed a commercial diet (CE-2, CLEA Japan, Tokyo, Japan) until 78 weeks of age. All mice were fed the control diet for 2 weeks of acclimatization period. After the acclimatization period, 12 young (12-week-old) and 12 aged (80-week-old) mice were randomly divided into two experimental groups and fed the control diet or the high P diet for 4 weeks. They were given free access to the diets and distilled water. Their urine samples were collected during the 5 days prior to euthanasia for the further analyses. At the end of the experimental period, all mice were euthanized under anesthesia and blood, bone, and duodenum samples were collected for analyses. The blood samples were centrifuged and the supernatants were used as serum samples. The femur samples were removed and cleaned of all soft tissues. All samples were stored at −80°C until further analyses.

2.2. Serum and Urine Analyses. The serum Ca level was analyzed by atomic absorption spectrophotometry (Hitachi A-2000; Hitachi, Tokyo, Japan) according to the method of Gimblet et al. [11]. The serum P level was analyzed by colorimetry using Phospha C-test Wako (Wako Pure Chemical Industries, Osaka, Japan). Serum PTH concentration was measured using the mouse intact PTH enzyme-linked immunosorbent assay (ELISA) kit (ALPCO, NH, USA). Serum intact osteocalcin (OC) concentration was measured using the mouse osteocalcin EIA kit (Biomedical Technologies, MA, USA). Urinary C-terminal telopeptide of type I collagen (CTx) level was measured using RatLaps EIA kit (Immunodiagnostic Systems, Boldon, UK). Urinary creatinine level was measured using the Jaffe reaction, as described by Lustgarten and Wenk [12]. Urinary CTx level was normalized to the urinary creatinine level.

2.3. Isolation of Total RNA and Real-Time PCR. Total RNA was isolated from the homogenized femurs or duodenum by using TRIzol reagent (Life Technologies, CA, USA) according to the manufacturer's specifications. The amount and purity of the RNA were assessed using a NanoDrop 2000c (Thermo Fisher Scientific, MA, USA). Complementary DNA (cDNA) was synthesized using the High-Capacity RNA-to-cDNA Kit (Applied Biosystems, CA, USA). For real-time PCR, the reaction mixture was prepared using the TaqMan Gene Expression Master Mix (Applied Biosystems) with TaqMan gene expression assays (Applied Biosystems) for mouse PTH receptor (Assay ID: Mm00441046_m1), mouse RANKL (Assay ID: Mm00441906_m1), mouse osteoprotegerin (OPG) (Assay ID: Mm01205928_m1), mouse tartrate resistant acid phosphatase (TRAP) (Assay ID: Mm00475698_m1), mouse runt related transcription factor 2 (Runx2) (Assay ID: Mm00501580_m1), mouse Osterix (Assay ID: Mm04209856_m1), mouse alkaline phosphatase (ALP) (Assay ID: Mm00475834_m1), mouse osteopontin (OPN) (Assay ID: Mm00436767_m1), mouse OC (Assay ID: Mm03413826_mH), mouse type I collagen (Col1a1) (Assay ID: Mm00801666_g1), mouse transient receptor potential vanilloid type 6 (TRPV6) (Assay ID: Mm00499069_m1), mouse calbindin-D9k (Assay ID: Mm00486654_m1), mouse plasma membrane Ca^{2+}-ATPase 1b (PMCA1b) (Assay ID: Mm01245805_m1), and mouse glyceraldehyde-3-phosphate dehydrogenase (GAPDH) (Assay ID: Mm99999915_g1). Real-time PCR was performed using a StepOne Real-Time PCR System (Applied Biosystems). The mRNA expression was normalized to GAPDH mRNA as a housekeeping gene. The value of the young mice fed the control diet was considered to be 1.00.

2.4. Statistical Analysis. Results are expressed as the mean ± SEM for each group of six mice. After two-way analysis of variance (ANOVA), Fisher's protected least significant

TABLE 2: Body weight, serum Ca and P, serum PTH, and markers of bone turnover.

	Young		Aged		Two-way ANOVA[1]
	Control	High P	Control	High P	
Initial body weight (g)	44.10 ± 0.79^A	44.20 ± 0.65^A	52.71 ± 2.17^B	53.84 ± 2.56^B	A
Final body weight (g)	48.66 ± 1.43	46.18 ± 0.67	51.47 ± 2.28	50.69 ± 2.19	
Serum Ca (mg/dL)	9.15 ± 0.17	8.67 ± 0.20	8.81 ± 0.27	8.60 ± 0.12	
Serum P (mg/dL)	9.23 ± 0.41	9.49 ± 0.47	9.07 ± 0.68	9.08 ± 0.60	
Serum PTH (pg/mL)	52.8 ± 12.5^A	181.4 ± 33.0^B	76.7 ± 13.8^A	306.1 ± 51.1^C	P, A
Serum OC (ng/mL)	26.58 ± 0.83^A	37.53 ± 2.94^B	7.31 ± 0.48^C	13.44 ± 0.78^D	P, A
Urine CTx (μg/mmol creatinine)	5.98 ± 0.91^A	16.73 ± 1.22^B	5.62 ± 1.02^A	9.61 ± 1.13^C	P, A, P × A

The data indicate the mean ± SEM of 6 mice.
[A,B,C,D]The different superscript letters denote significant differences ($P < 0.05$).
[1]Significant effect ($P < 0.05$): P = effect of high P diet; A = effect of age; P × A = effect of interaction.

difference (PLSD) test was used to determine significant differences among the groups. The homogeneity of variance was analyzed with Levene's test. Differences were considered to be significant when the P value was less than 0.05.

3. Results

3.1. Body Weight. In both young and aged mice, there was no significant difference in the initial body weight between mice fed the control and high P diets (Table 2). The initial body weight of aged mice was significantly higher than that of young mice. There was no significant difference in the final body weight among groups.

3.2. Serum Ca, P, and PTH Concentrations. There were no significant differences in serum Ca and P concentrations among the groups (Table 2). In both young and aged mice, the high P diet significantly increased serum PTH concentration. Although there was no significant difference in serum PTH concentration between young and aged mice fed the control diet, serum PTH concentration was significantly higher in aged mice than in young mice fed the high P diet.

3.3. Markers of Bone Turnover. In both young and aged mice, the high P diet significantly increased serum OC concentration (Table 2). In mice fed both the control and high P diets, serum OC concentration was significantly lower in aged mice than in young mice. In both young and aged mice, the high P diet significantly increased urinary excretion of CTx. Although there was no significant difference in urinary excretion of CTx between young and aged mice fed the control diet, urinary excretion of CTx was significantly lower in aged mice than in young mice fed the high P diet.

3.4. mRNA Expression in the Femur. In both young and aged mice, the high P diet significantly increased mRNA expression of PTH receptor, RANKL, TRAP, Runx2, Osterix, ALP, OPN, OC, and Col1a1 (Figure 1) compared to the control diet. In mice fed the control and high P diets, mRNA expression of PTH receptor, RANKL, TRAP, Runx2, Osterix, ALP, OPN, OC, and Col1a1 was significantly lower in aged mice than in young mice. There was no significant difference in OPG mRNA expression among the groups. The high P diet significantly increased RANKL/OPG ratio in aged mice but did not in young mice.

3.5. mRNA Expression in the Duodenum. In both young and aged mice, high P diet significantly increased mRNA expression of TRPV6, CaBP9k, and PMCA1b (Figure 2) compared to the control diet. In mice fed the control and high P diets, mRNA expression of TRPV6 and CaBP9k was significantly lower in aged mice than in young mice.

4. Discussion

In humans, bone loss occurs with increasing age at a rate of approximately 1% per year averaged over the ages of 29–76 years [7]. Therefore, aging is one of the risk factors for bone loss. In a previous mouse study, bone mass and mechanical properties were shown to approach mature levels by 12 weeks of age, and age-related osteopenia was observed after 42 weeks [8]. In this study, we investigated bone metabolism by measuring markers of bone formation and resorption, serum OC concentration [13], and urinary CTx excretion [14]. Serum OC concentration was significantly lower in aged mice than in young mice fed the control diet, whereas there is no difference in urinary excretion of CTx between young and aged mice fed the control diet. These results showed that aged mice present a decrease in bone formation, and it appears that the balance between bone formation and resorption may be disrupted in aged mice.

Bone formation is mediated by osteoblasts. Using gene-deficient mouse models, Runx2 and Osterix were shown to be essential transcription factors for osteoblast differentiation and bone formation [15, 16]. In this study, Runx2 and Osterix mRNA expression were significantly lower in aged mice than in young mice. These results suggest that aging leads to a reduction in osteoblast differentiation and that Runx2 and Osterix mRNA expression changes are associated with a decrease in bone formation in aged mice. Consequently, decreases in mRNA expression of ALP and bone matrix proteins such as OPN, OC, and Col1a1 also occurred in aged mice. Ikeda et al. showed that the mRNA expression of OPN, OC, and Col1a1 decreased in both cortical and

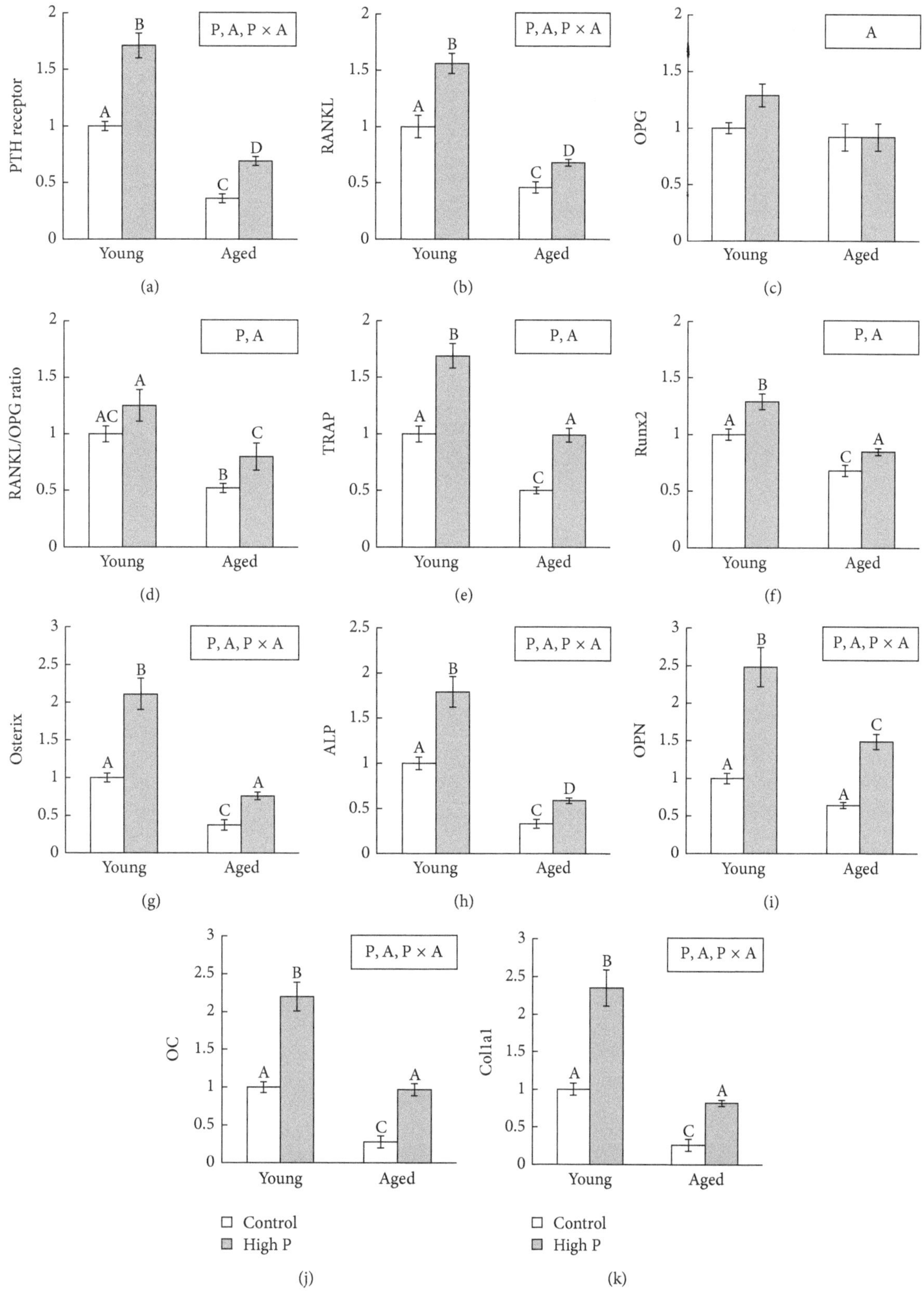

FIGURE 1: Bone metabolism-related gene expression in the femur. (a) PTH receptor; (b) RANKL; (c) OPG; (d) RANKL/OPG ratio; (e) TRAP; (f) Runx2; (g) Osterix; (h) ALP; (i) OPN; (j) OC; (k) Col1a1. The data indicate the mean ± SEM of 6 mice. [A,B,C,D]The different letters denote significant differences ($P < 0.05$). Significant effect ($P < 0.05$): P = effect of high P diet; A = effect of age; P × A = effect of interaction. The value of young mice fed a control diet is considered to be 1.00.

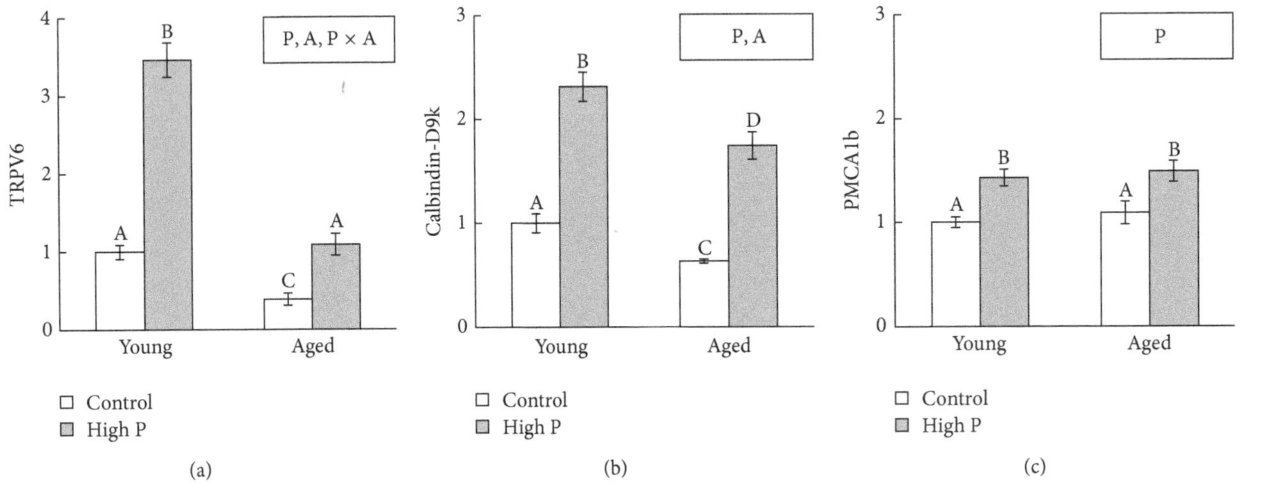

FIGURE 2: Ca absorption-related gene expression in the duodenum. (a) TRPV6; (b) calbindin-D9k; (c) PMCA1b. The data indicate the mean ± SEM of 6 mice. A,B,C,D The different letters denote significant differences ($P < 0.05$). Significant effect ($P < 0.05$): P = effect of high P diet; A = effect of age; P × A = effect of interaction. The value of young mice fed a control diet is considered to be 1.00.

trabecular bones in aged rats compared to young animals [17]. In addition, Cao et al. reported that ALP and Col1a1 expression declined in aged mice compared to young mice [18]. Thus, aging results in a decrease in bone formation with declined osteoblast-related gene expression. With regard to bone resorption, this study showed that RANKL and TRAP mRNA expression were decreased in aged mice compared to young mice, despite unchanging serum PTH concentration. Since PTH stimulates RANKL [6], the result of RANKL mRNA expression seems to contradict that of serum PTH concentration. However, PTH receptor mRNA expression was decreased in aged mice compared to young mice in this study. This result suggested that PTH action was suppressed, which decreased RANKL mRNA expression in aged mice. Though urinary excretion of CTx was unchanged, expression of bone resorption-rerated genes was decreased in aged mice compared to young mice. However, it is generally known that age-related increase in serum PTH contributes to the increase in bone resorption [19]. Therefore, results of serum PTH concentration and bone resorption between young and aged mice in this study are contradictory. Further studies with the increase in number of mice per group are needed to address these discrepancies.

A decline in intestinal Ca absorption is also one of the causes of age-related bone loss. The transfer of Ca via the intestine occurs through both transcellular and paracellular pathways [20]. The transcellular Ca pathway, which is affected by 1,25-dihydroxyvitamin D ($1,25(OH)_2D$), has been proposed to involve Ca entry via TRPV6, intracellular diffusion of Ca by calbindin-D9k, and basolateral extrusion of Ca by PMCA1b [21]. In this study, TRPV6 and calbindin-D9k mRNA expression in the duodenum were decreased in aged mice compared to young mice. Wood et al. demonstrated that plasma $1,25(OH)_2D$, duodenal calbindin D protein, and Ca absorption decreased with age in rats [22]. From this study, we could deduce that similar findings would be found in aged mice. Therefore, a decrease in serum $1,25(OH)_2D$ might reflect our results on TRPV6 and calbindin-D9k mRNA expression.

Many studies have reported that high P intake induces an increase in serum PTH concentration in humans [1, 23] and animals [2, 3, 24]. In this study, the high P diet increased serum PTH concentration in both young and aged mice and was greater in aged mice than in young mice fed the high P diet. Similar to our previous study [9], these results suggest that the response to a high P diet in terms of PTH secretion might be different and greater with aging. Kidney function decreases with age [25], and declining kidney function causes an increase in PTH secretion [26]. Furthermore, we previously reported that high P diet decreased kidney function in rats [27]. Therefore, the combination of aging and high P diet might be one of the reasons that higher serum PTH concentration was observed in aged mice fed the high P diet.

We previously reported that a high P diet increased RANKL mRNA expression and bone resorption in growing rats [3]. RANKL is a mediator of osteoclastic bone resorption and is stimulated by PTH [6]. Therefore, it was suggested that elevated PTH secretion induced by the high P diet led to an increase in RANKL expression, which increased bone resorption. In this study, the high P diet significantly increased urinary excretion of CTx in both young and aged mice. Regarding mRNA expression of bone resorption-related molecules in the femora, the high P diet significantly increased RANKL and TRAP mRNA expressions in both young and aged mice. These results suggested that the high P diet enhanced bone resorption independently of age. RANKL actions are inhibited by OPG, which acts as a decoy receptor by blocking RANKL binding to its receptor [28]. In this study, the high P diet significantly increased the RANKL/OPG ratio in aged mice, whereas the ratio was unchanged in young mice. Many reports have supported the assertion that the increase in RANKL/OPG ratio promotes osteoclastogenesis, accelerates bone resorption, and induces bone loss [29]. Our previous study showed that a high P diet decreased the breaking force and stiffness of the femur in aged mice compared to young mice [9]. Our findings show that the high P diet in the aged mice leads to increased PTH secretion and consequent increases in the RANKL/OPG ratio accelerating osteoclastogenesis.

Previous studies showed that PTH regulates Runx2 and Osterix mRNA expression [30, 31]. In this study, the high P diet significantly increased serum OC concentration and mRNA expression of Runx2, Osterix, ALP, OPN, OC, and Col1a1 in both young and aged mice. From the results of bone resorption and formation markers, high bone turnover with resorption exceeding formation was observed. Since high bone turnover is a risk factor for bone fracture and osteoporosis [32], a high P diet might be a risk factor for bone loss not only in young mice but also in aged mice.

PTH secretion might reflect a decrease in Ca absorption in the high P diet group. Although the mechanism underlying the high P diet-induced decreased Ca absorption remains unknown, it is thought that the formation of insoluble Ca and P salts in the intestinal lumen is an important factor [33]. Our previous study also showed that high P diet decreased Ca absorption in female rats [34]. While we did not evaluate Ca absorption in this study, it is possible to estimate the decrease in Ca absorption by high P diet in young and aged mice. However, this study showed that the high P diet significantly increased mRNA expression of TRPV6, CaBP9k, and PMCA1b in both young and aged mice. Previous study demonstrated that Ca restriction during lactation stimulated Ca-binding protein and active Ca transport in jejunum and ileum [35]. Furthermore, Ca deficient diet resulted in an increase in duodenal PMCA mRNA in chickens [36]. These studies suggested that $1,25(OH)_2D$-regulated Ca transporters might be stimulated by low Ca status in the intestinal lumen. Thus, a decrease in soluble Ca induced by high P diet might lead to TRPV6, CaBP9k, and PMCA1b mRNA expression, and these increases in $1,25(OH)_2D$-regulated gene expression seem to compensate for a decrease in Ca absorption by the high P diet. In brief, our results suggest that high P diet accelerates the transcellular Ca pathway, though absorbed amount of Ca was insufficient to maintain serum PTH concentration. It is known that fibroblast growth factor 23 (FGF23) and $1,25(OH)_2D$ as well as PTH are key factors for Ca and P metabolism. FGF23 inhibits P reabsorption and $1,25(OH)_2D$ synthesis in the kidney [37]. Furthermore, high P diet increased serum FGF23 concentration in mice [38]. Therefore, measuring serum FGF23 and $1,25(OH)_2D$ is important to fully elucidate the mechanisms by which high P diet changes Ca and P metabolism, and further studies are needed to clarify the details.

5. Conclusion

In conclusion, we demonstrated that the high P diet increased bone metabolism-related gene expression in both young and aged mice. Furthermore, the high P diet affected PTH secretion differently in young and aged mice, leading to an increase in the RANKL/OPG mRNA ratio in aged mice.

Conflict of Interests

The authors declare that there is no conflict of interests regarding the publication of this paper.

Acknowledgment

This work was supported by KAKENHI (Grant-in-Aid for Young Scientists (B), 20700598).

References

[1] R. R. Bell, H. H. Draper, D. Y. M. Tzeng, H. K. Shin, and G. R. Schmidt, "Physiological responses of human adults to foods containing phosphate additives," *Journal of Nutrition*, vol. 107, no. 1, pp. 42–50, 1977.

[2] M. M. Huttunen, I. Tillman, H. T. Viljakainen et al., "High dietary phosphate intake reduces bone strength in the growing rat skeleton," *Journal of Bone and Mineral Research*, vol. 22, no. 1, pp. 83–92, 2007.

[3] S.-I. Katsumata, R. Masuyama, M. Uehara, and K. Suzuki, "High-phosphorus diet stimulates receptor activator of nuclear factor-κB ligand mRNA expression by increasing parathyroid hormone secretion in rats," *British Journal of Nutrition*, vol. 94, no. 5, pp. 666–674, 2005.

[4] E. C. Weir, C. W. G. M. Lowik, I. Paliwal, and K. L. Insogna, "Colony stimulating factor-1 plays a role in osteoclast formation and function in bone resorption induced by parathyroid hormone and parathyroid hormone-related protein," *Journal of Bone and Mineral Research*, vol. 11, no. 10, pp. 1474–1481, 1996.

[5] A. Grey, M.-A. Mitnick, U. Masiukiewicz et al., "A role for interleukin-6 in parathyroid hormone-induced bone resorption *in vivo*," *Endocrinology*, vol. 140, no. 10, pp. 4683–4690, 1999.

[6] S.-K. Lee and J. A. Lorenzo, "Parathyroid hormone stimulates TRANCE and inhibits osteoprotegerin messenger ribonucleic acid expression in murine bone marrow cultures: correlation with osteoclast-like cell formation," *Endocrinology*, vol. 140, no. 8, pp. 3552–3561, 1999.

[7] T. R. Overton and T. K. Basu, "Longitudinal changes in radial bone density in older men," *European Journal of Clinical Nutrition*, vol. 53, no. 3, pp. 211–215, 1999.

[8] V. L. Ferguson, R. A. Ayers, T. A. Bateman, and S. J. Simske, "Bone development and age-related bone loss in male C57BL/6J mice," *Bone*, vol. 33, no. 3, pp. 387–398, 2003.

[9] S. Katsumata, H. Matsuzaki, R. Tsuboi, M. Uehara, and K. Suzuki, "Effects of aging and a high-phosphorus diet on bone metabolism in mice," *Japanese Journal of Nutrition and Dietetics*, vol. 64, no. 1, pp. 55–60, 2006 (Japanese).

[10] P. G. Reeves, F. H. Nielsen, and G. C. Fahey Jr., "AIN-93 purified diets for laboratory rodents: final report of the American Institute of Nutrition ad hoc writing committee on the reformulation of the AIN-76A rodent diet," *Journal of Nutrition*, vol. 123, no. 11, pp. 1939–1951, 1993.

[11] E. G. Gimblet, A. F. Marney, and R. W. Bonsnes, "Determination of calcium and magnesium in serum, urine, diet, and stool by atomic absorption spectrophotometry," *Clinical Chemistry*, vol. 13, no. 3, pp. 204–214, 1967.

[12] J. A. Lustgarten and R. E. Wenk, "Simple, rapid, kinetic method for serum creatinine measurement," *Clinical Chemistry*, vol. 18, no. 11, pp. 1419–1422, 1972.

[13] J. P. Brown, P. D. Delmas, L. Malaval, C. Edouard, M. C. Chapuy, and P. J. Meunier, "Serum bone Gla-protein: a specific marker for bone formation in postmenopausal osteoporosis," *The Lancet*, vol. 1, no. 8386, pp. 1091–1093, 1984.

[14] M. Bonde, P. Qvist, C. Fledelius, B. J. Riis, and C. Christiansen, "Immunoassay for quantifying type I collagen degradation products in urine evaluated," *Clinical Chemistry*, vol. 40, no. 11, pp. 2022–2025, 1994.

[15] T. Komori, H. Yagi, S. Nomura et al., "Targeted disruption of Cbfa1 results in a complete lack of bone formation owing to maturational arrest of osteoblasts," *Cell*, vol. 89, no. 5, pp. 755–764, 1997.

[16] K. Nakashima, X. Zhou, G. Kunkel et al., "The novel zinc finger-containing transcription factor Osterix is required for osteoblast differentiation and bone formation," *Cell*, vol. 108, no. 1, pp. 17–29, 2002.

[17] T. Ikeda, Y. Nagai, A. Yamaguchi, S. Yokose, and S. Yoshiki, "Age-related reduction in bone matrix protein mRNA expression in rat bone tissues: application of histomorphometry to in situ hybridization," *Bone*, vol. 16, no. 1, pp. 17–23, 1995.

[18] J. Cao, L. Venton, T. Sakata, and B. P. Halloran, "Expression of RANKL and OPG correlates with age-related bone loss in male C57BL/6 mice," *Journal of Bone and Mineral Research*, vol. 18, no. 2, pp. 270–277, 2003.

[19] K. A. Kennel, B. L. Riggs, S. J. Achenbach, A. L. Oberg, and S. Khosla, "Role of parathyroid hormone in mediating age-related changes in bone resorption in men," *Osteoporosis International*, vol. 14, no. 8, pp. 631–636, 2003.

[20] D. Pansu, C. Bellaton, and F. Bronner, "Effect of Ca intake on saturable and nonsaturable components of duodenal Ca transport," *The American Journal of Physiology*, vol. 240, no. 1, pp. G32–G37, 1981.

[21] J. C. Fleet and R. D. Schoch, "Molecular mechanisms for regulation of intestinal calcium absorption by vitamin D and other factors," *Critical Reviews in Clinical Laboratory Sciences*, vol. 47, no. 4, pp. 181–195, 2010.

[22] R. J. Wood, J. C. Fleet, K. Cashman, M. E. Bruns, and H. F. Deluca, "Intestinal calcium absorption in the aged rat: evidence of intestinal resistance to 1,25$(OH)_2$ vitamin D," *Endocrinology*, vol. 139, no. 9, pp. 3843–3848, 1998.

[23] E. Reiss, J. M. Canterbury, M. A. Bercovitz, and E. L. Kaplan, "The role of phosphate in the secretion of parathyroid hormone in man," *The Journal of Clinical Investigation*, vol. 49, no. 11, pp. 2146–2149, 1970.

[24] Y. Tani, T. Sato, H. Yamanaka-Okumura et al., "Effects of prolonged high phosphorus diet on phosphorus and calcium balance in rats," *Journal of Clinical Biochemistry and Nutrition*, vol. 40, no. 3, pp. 221–228, 2007.

[25] R. D. Lindeman and R. Goldman, "Anatomic and physiologic age changes in the kidney," *Experimental Gerontology*, vol. 21, no. 4-5, pp. 379–406, 1986.

[26] S. Schindler, M. Mannstadt, P. Urena, G. V. Segre, and G. Stein, "PTH secretion in patients with chronic renal failure assessed by a modified CiCa clamp method: effects of 1-year calcitriol therapy," *Clinical Nephrology*, vol. 61, no. 4, pp. 253–260, 2004.

[27] H. Matsuzaki, T. Kikuchi, Y. Kajita et al., "Comparison of various phosphate salts as the dietary phosphorus source on nephrocalcinosis and kidney function in rats," *Journal of Nutritional Science and Vitaminology*, vol. 45, no. 5, pp. 595–608, 1999.

[28] H. Yasuda, N. Shima, N. Nakagawa et al., "Osteoclast differentiation factor is a ligand for osteoprotegerin/osteoclastogenesis-inhibitory factor and is identical to TRANCE/RANKL," *Proceedings of the National Academy of Sciences of the United States of America*, vol. 95, no. 7, pp. 3597–3602, 1998.

[29] L. C. Hofbauer and M. Schoppet, "Clinical implications of the osteoprotegerin/RANKL/RANK system for bone and vascular diseases," *Journal of the American Medical Association*, vol. 292, no. 4, pp. 490–495, 2004.

[30] V. Krishnan, T. L. Moore, Y. L. Ma et al., "Parathyroid hormone bone anabolic action requires cbfa1/runx2-dependent signaling," *Molecular Endocrinology*, vol. 17, no. 3, pp. 423–435, 2003.

[31] B. L. Wang, C. L. Dai, J. X. Quan et al., "Parathyroid hormone regulates osterix and Runx2 mRNA expression predominantly through protein kinase A signaling in osteoblast-like cells," *Journal of Endocrinological Investigation*, vol. 29, no. 2, pp. 101–108, 2006.

[32] P. Ravn, M. Rix, H. Andreassen, B. Clemmesen, M. Bidstrup, and M. Gunnes, "High bone turnover is associated with low bone mass and spinal fracture in postmenopausal women," *Calcified Tissue International*, vol. 60, no. 3, pp. 255–260, 1997.

[33] N. Brautbar, B. S. Levine, M. W. Walling, and J. W. Coburn, "Intestinal absorption of calcium: role of dietary phosphate and vitamin D," *The American Journal of Physiology*, vol. 241, no. 1, pp. G49–G53, 1981.

[34] M. Koshihara, S.-I. Katsumata, M. Uehara, and K. Suzuki, "Effects of dietary phosphorus intake on bone mineralization and calcium absorption in adult female rats," *Bioscience, Biotechnology and Biochemistry*, vol. 69, no. 5, pp. 1025–1028, 2005.

[35] M. E. Bruns, A. Boass, and S. U. Toverud, "Regulation by dietary calcium of vitamin D-dependent calcium-binding protein and active calcium transport in the small intestine of lactating rats," *Endocrinology*, vol. 121, no. 1, pp. 278–283, 1987.

[36] Q. Cai, J. S. Chandler, R. H. Wasserman, R. Kumar, and J. T. Penniston, "Vitamin D and adaptation to dietary calcium and phosphate deficiencies increase intestinal plasma membrane calcium pump gene expression," *Proceedings of the National Academy of Sciences of the United States of America*, vol. 90, no. 4, pp. 1345–1349, 1993.

[37] T. Shimada, H. Hasegawa, Y. Yamazaki et al., "FGF-23 is a potent regulator of vitamin D metabolism and phosphate homeostasis," *Journal of Bone and Mineral Research*, vol. 19, no. 3, pp. 429–435, 2004.

[38] F. Perwad, N. Azam, M. Y. H. Zhang, T. Yamashita, H. S. Tenenhouse, and A. A. Portale, "Dietary and serum phosphorus regulate fibroblast growth factor 23 expression and 1,25-dihydroxyvitamin D metabolism in mice," *Endocrinology*, vol. 146, no. 12, pp. 5358–5364, 2005.

11

Vitamin D Supplementation in Australia: Implications for the Development of Supplementation Guidelines

Kellie Bilinski[1] and Peter Talbot[2]

[1] *Westmead Breast Cancer Institute, Westmead Hospital and The University of Sydney, P.O. Box 143, Westmead, NSW 2145, Australia*
[2] *Department of Dietetics & Nutrition, Westmead Hospital, Westmead, NSW 2145, Australia*

Correspondence should be addressed to Kellie Bilinski; kellie.bilinski@me.com

Academic Editor: Cindy Davis

High rates of vitamin D deficiency and testing have been reported in Australia, yet there are few reports regarding vitamin D supplement use. Australian wholesale sales data was obtained for vitamin D supplements for the period 2000–2011. There has been a threefold increase in supplement sales over the past decade, whereby over A$94 million supplements containing vitamin D in Australia were sold during the year 2010. There were eighty-nine manufacturers that produce a variety of 195 vitamin D products. The amount of vitamin D in these products varies considerably, from 40 to 1000 IU per unit, although supplements containing only vitamin D had the highest amount of vitamin D. There was a trend for sales to increase in winter months. Given the potential public health benefits of vitamin D, there is an urgent need for a better understanding of supplementation use and for the development of supplementation.

1. Introduction

A high prevalence of vitamin D deficiency (25-hydroxyvitamin D (25(OH)D) ≤50 nmol/L) in Australians has been reported [1]. The Australian Health Survey showed that one-quarter (23.5%) of Australians are vitamin D deficient, 17% of whom had mild deficiency (20–40 nmol/L); 6% had moderate deficiency (13–29 nmol/L), and <1% had severe deficiency (<13 nmol/L). Serum vitamin D levels varied by season whereby deficiency rates ranged from 36% in winter to 14% in summer. The highest prevalence of vitamin D deficiency occurred in the south eastern states of Victoria (49%), ACT (49%), and Tasmania (43%), whereas, in the northern states of Queensland and the Northern Territory, the prevalence of vitamin D deficiency was 15% and 17%, respectively.

This may be driving the 95-fold increase in testing of 25(OH)D concentration in Australia reported over the last decade [2]. In addition to osteoporosis, achieving adequate vitamin D has been linked with protection from numerous chronic diseases including diabetes and cancer [3]. Exposure of the skin to ultraviolet-B (UVB) radiation from sunlight provides the body's principal source of vitamin D [4], although it can also be obtained by diet and supplements.

It has been assumed that most individuals achieve adequate sunlight exposure to meet their requirements [5]. However, adherence to sun protection messages [6] dramatically reduces 25(OH)D synthesis. We have previously shown that there are few opportunities in Australia to obtain the equivalent of 1000 IU 25(OH)D within a realistic duration while adhering to sun smart messages [7].

It is difficult to meet vitamin D requirements from the diet as few foods contain adequate amounts of vitamin D [8]. It has been estimated that Australians consume approximately 80 and 176 IU vitamin D from foods and supplements daily [9, 10]. This is despite the mandatory fortification of vitamin D in Australia of edible oils and spreads, whereby these foods must contain a minimum of 220 IU vitamin D per 100 g of food and voluntary fortification of modified and skim milk and powdered milk, yoghurts, and table confections (as well as cheese) [11]. This estimated intake is considerably lower than the estimated average requirements publicized by the US Institute of Medicine (IOM) of 400 IU/d for all adults [12] and the current adequate intake (AI) values for Australia of

Table 1: Annual vitamin D containing supplement sales in thousands of units[1] and thousands of dollars according to year and product type.

Year	Units by product type (1000s) Vitamin D (plain)	Vitamins A and D combination	Calcium plus vitamin D	Total units	Value by product type (A$1000) Vitamin D (plain)	Vitamins A and D combination	Calcium plus vitamin D	Total value
2000	5.0	2.1	13.4	20.5	233.9	14.1	93.3	341.2
2001	5.0	1.9	14.7	21.6	216.3	12.1	104.0	332.4
2002	4.8	1.8	16.6	23.2	170.7	11.0	118.6	300.4
2003	5.1	3.1	18.3	26.5	144.2	30.2	132.8	307.2
2004	6.3	3.7	18.9	28.9	140.7	45.3	141.8	327.9
2005	6.9	3.5	20.9	31.4	141.6	44.7	162.0	348.2
2006	8.1	4.1	20.2	32.4	148.2	58.6	166.5	373.3
2007	10.6	3.7	24.4	38.8	177.3	51.8	210.2	439.3
2008	15.2	3.3	28.8	47.3	270.2	41.4	263.5	575.1
2009	19.6	3.8	31.9	55.3	408.3	48.1	316.5	772.9
2010	24.6	4.0	32.4	61.0	534.1	57.5	348.7	940.3

[1]Units represent individual items as purchased; source: IMS Health, Australia, 2013.

200 IU for adults up to age of 50 years, 400 IU for those aged 50 to 70 years, and 600 IU for those aged 70 years [13].

Supplementation has been suggested as a safe and effective way of meeting 25(OH)D requirements [14], although only one report that we are aware of has documented vitamin D supplement use in Australians. In that report, 5% of Australian adults reported taking vitamin D supplements and 18% reported taking multivitamin supplements in 2011-2012 [1]. As expected, those who took vitamin D supplements had lower levels of vitamin D deficiency than those who did not take supplements (7% compared with 23%).

The aim of this study was to investigate the patterns of vitamin D supplementation in Australians between 2000 and 2010.

2. Methods

IMS Health is a leading provider of information to the health care industry. Data representing unit sales (as individual bottles) and dollar value in Australian dollars (Aus$1 ≈ US0.93) from manufacturers and wholesalers to pharmacies and hospitals for all vitamin D containing supplements was requested directly from IMS. IMS captures approximately 95% of total pharmacy sales. A panel of pharmacies is used to supplement any sales that are not captured or projected by imputation to give a national view to account for 100% of the market.

Data was obtained in Excel (Microsoft) for all manufacturers, including private label manufacturers and supplement brand name for each month between January 2000 and December 2010 for the following categories of vitamin D containing supplements: vitamin D plain, 100% vitamin D; vitamins A and D combination, containing some vitamin D; and calcium and vitamin D. Multivitamin supplements were not included as these generally were less potent than vitamin D or vitamin D and calcium supplements in Australia (personal observation).

In Australia, seasons are defined as follows: summer: December–February; autumn: March–May; winter: June–August; and spring: September–November. Descriptive analyses were used to describe the data in absolute values. The effect of season on vitamin D sales was investigated by ANOVA with Bonferroni *post hoc* analysis. A P value of 0.05 was used to determine statistical significance. Descriptive analyses and ANOVA were conducted in SPSS version 19.0 (IBM SPSS, USA, version 19.0).

3. Results

Eighty-nine manufacturers produced 195 different vitamin D containing products, with 26 (15.9%) of which containing vitamin D alone, 34 (17.4%) containing a combination of vitamins A and D, and 135 (69.2%) predominately containing calcium with or without vitamin D.

In terms of units sold, the calcium with vitamin D supplement market comprised the largest proportion, which was then followed by plain vitamin D supplements (32.4 million and 24.6 million units in 2010, resp.) (Table 1). However, when sales according to dollar value were compared for each product type, plain vitamin D supplements comprised the largest value, followed by calcium units (A$54.4 million and A$34.9 million units in 2010, resp.).

There was a consistent increase in wholesale vitamin D supplement sales from 2000 to 2010. During this period, the volume of vitamin D containing supplements sold almost tripled, increasing from 2.1 million to 6.10 million units, representing a value of A$34.1 to A$94.0 million in wholesale sales (Table 1).

As expected, a nonsignificant trend for increased vitamin D supplement sales from summer through spring was observed (Figure 1); however, the difference in sales did not reach statistical significance for units or value (data not shown).

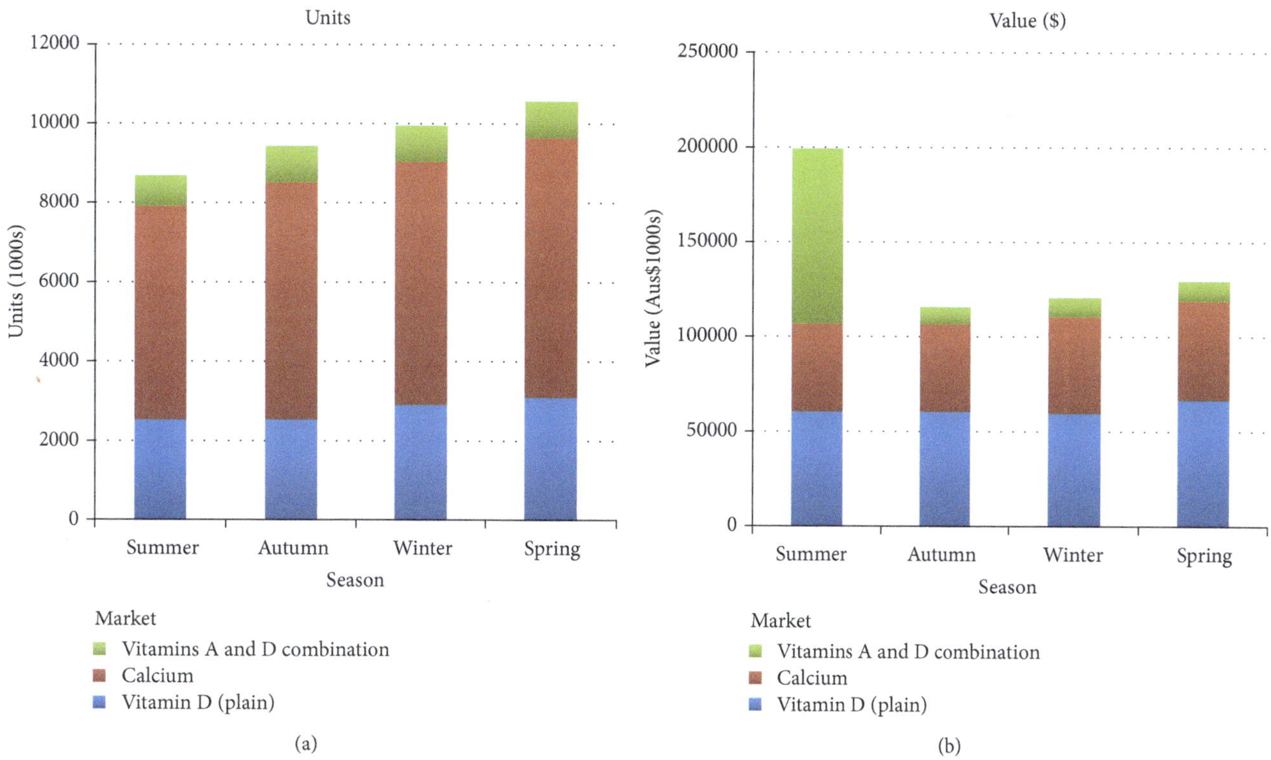

FIGURE 1: Total vitamin D supplement sales in thousands of units (units represent individual package as purchased from 2000 to 2010, source: IMS Health, Australia, 2012) and thousands of dollars ($Aus) from 2000 to 2010 according to season and product type.

4. Discussion

There has been an incremental increase in the use of vitamin D supplements as evident by the rise in vitamin D containing supplement sales. In 2010, more than A$94 million wholesale of vitamin D containing supplements was sold, of which over half contained only vitamin D. Based on a retail mark-up from anywhere between 50% and 75%, this equates to A$140–$165 million spent by consumers. Similar findings have been shown internationally. In an analysis of the NHANES III study, Gahche showed that the use of dietary supplements containing vitamin D increased from 29.7% of individuals in 1988–1994 to as high as 56.3% in women aged 60 and over in 2003–2006, whereas, in individuals aged 20–39, the rate remained around 22–26% for males and 30–34% for females [15].

The vitamin D content of supplements sold in Australia varies widely, with vitamin D only supplements containing the highest content of vitamin D per supplement (1000–2000 IU 25(OH)D per capsule). The content of vitamin D in vitamins A and D combination supplements also varied between approximately 100 and 400 IU. Calcium products contain a mix of supplements containing calcium alone or calcium plus vitamin D. If present, the vitamin D content in calcium products also ranges between approximately 40 and 1000 IU vitamin D. Our data do not include multivitamin supplements, which also provide vitamin D, and, as such, underrepresent vitamin D containing supplement use in Australians. However, multivitamin supplements generally contained lesser amounts of vitamin D than vitamin D only supplements (personal observation) and were, therefore, not included in this dataset. Individuals who take supplements have been shown to lead healthier lifestyles and consume better quality diets [16]. For instance, supplement users have been shown to have better dietary patterns, exercise regularly, maintain a healthy body weight, avoid tobacco products [16], and, therefore, obtain additional vitamin D from being out in the sun and eating foods containing vitamin D compared to those who do not use supplements.

Although a higher proportion of units of calcium supplements were sold (32.4 million units in 2010), vitamin D supplements compromised the highest proportion in regard to dollar sales (A$53.4 million). This is likely to be explained by the higher cost of vitamin D supplements relative to calcium supplements.

As previous reports have shown, the highest prevalence of vitamin D deficiency occurs in winter whereby 36% of Australians have been shown to be vitamin D deficient [1]. It would be expected that, following a diagnosis of vitamin D deficiency, individuals are more likely to purchase vitamin D supplement. A seasonal trend was observed where supplement sales increased linearly from summer to spring (Figure 1); however, the difference did not reach statistical significance ($P > 0.05$) for all supplements combined or by each product type. A drop in sales from summer would be expected on the basis that many Australians have access to adequate sunlight in the summer months. These figures represent absolute wholesale sales in units and Australian dollars to pharmacies and hospitals and not direct dates of purchase by individuals which may occur several weeks afterwards; thus, a trend may have been more evident if we had been able to investigate the lag in retail sales to consumers.

It is concerning that large amounts are spent on vitamin D supplements in Australia. Sunlight and food sources are alternative sources of vitamin D. Wide scale vitamin D testing of the Australian population in order to determine the need for supplementation would be expensive, as would universal vitamin D supplementation; therefore, neither of these is viable option. A simple tool to calculate the risk of vitamin D deficiency, which could then be followed up by a blood test if indicated, would potentially save the healthcare system a sizable amount of money and avoid potentially unnecessary testing and supplementation.

There has been an enormous increase in vitamin D awareness, testing, and supplementation in Australia since 2000. Supplementation may be the most feasible option for many individuals to reverse the high prevalence of vitamin D deficiency reported in Australians. However, it is possible that some individuals are unnecessarily taking supplements or, conversely, the dose of vitamin D in their supplement is inadequate and unlikely to assist individuals in achieving adequate vitamin D status.

5. Conclusion

Australians are spending enormous amounts of money on vitamin D supplements, a proportion of which may potentially be unnecessary. Further information is needed regarding the reasons individuals are taking supplements and the effect of such supplementation on serum vitamin D levels. Additionally, it would be useful to include multivitamin supplements in our analysis. Vitamin D supplementation guidelines are clearly warranted in Australia in effort to prevent unnecessary use of vitamin D containing supplements; and, conversely, for those who require vitamin D supplements to correct vitamin D deficiency, these guidelines would provide details on the correct dose required to lead to vitamin D repletion.

Conflict of Interests

The authors have no conflict of interests.

Acknowledgment

This study was internally funded by Westmead Breast Cancer Institute.

References

[1] Australian Bureau of Statistics, *Australian Health Survey: Biomedical Results for Nutrients*, 43640DO011_20112012, Australian Bureau of Statistics, 2011.

[2] K. L. Bilinski and S. C. Boyages, "The rising cost of vitamin D testing in Australia: time to establish guidelines for testing," *Medical Journal of Australia*, vol. 197, article 90, no. 2, 2012.

[3] M. F. Holick, "Vitamin D: A D-Lightful health perspective," *Nutrition Reviews*, vol. 66, supplement 2, pp. S182–S194, 2008.

[4] A. F. McKinlay, "A reference action spectrum for ultraviolet induced erythema in human skin," *CIE-Journal*, vol. 61, no. 1, pp. 17–22, 1987.

[5] M. F. Holick, "Vitamin D and bone health," *Journal of Nutrition*, vol. 126, supplement 4, pp. 1159S–1164S, 1996.

[6] Cancer Council of Victoria, SunSmart Campaign, 2008, http://www.sunsmart.com.au/.

[7] K. L. Stalgis-Bilinski, J. Boyages, E. L. Salisbury, C. R. Dunstan, S. I. Henderson, and P. L. Talbot, "Burning daylight: balancing vitamin D requirements with sensible sun exposure," *Medical Journal of Australia*, vol. 194, no. 7, pp. 345–348, 2011.

[8] M. F. Holick, "Sunligth "D"ilemma: risk of skin cancer or bone disease and muscle weakness," *The Lancet*, vol. 357, no. 9249, pp. 4–6, 2001.

[9] C. A. Nowson and C. Margerison, "Vitamin D intake and vitamin D status of Australians," *Medical Journal of Australia*, vol. 177, no. 3, pp. 149–152, 2002.

[10] N. Jayaratne, M. C. B. Hughes, T. I. Ibiebele, S. van den Akker, and J. C. van der Pols, "Vitamin D intake in Australian adults and the modeled effects of milk and breakfast cereal fortification," *Nutrition*, vol. 29, no. 7-8, pp. 1048–1053, 2013.

[11] "Australia New Zealand Food Standards Code—Standard 2.4.2—Edible Oil Spreads—F2011C00574," ComLaw, 2011.

[12] Institute of Medicine, *Dietary Reference Intakes for Calcium and Vitamin D*, The National Academies Press, Washington, DC, USA, 2011.

[13] NHMRC, "Nutrient reference values for Australia and New Zealand including recommended dietary intakes," Report No. 1864962372, Australian Government Department of Health and Ageing, National Health and Medical Research Council, 2005.

[14] W. Shrapnel and S. Truswell, "Vitamin D deficiency in Australia and New Zealand: what are the dietary options?" *Nutrition and Dietetics*, vol. 63, no. 4, pp. 206–212, 2006.

[15] J. Gahche, R. Bailey, V. Burt et al., "Dietary supplement use among U.S. adults has increased since NHANES III (1988–1994)," *NCHS Data Brief*, no. 61, pp. 1–8, 2011.

[16] A. Dickinson and D. MacKay, "Health habits and other characteristics of dietary supplement users: a review," *Nutrition Journal*, vol. 13, article 14, 2014.

12

Food Insecurity and Not Dietary Diversity Is a Predictor of Nutrition Status in Children within Semiarid Agro-Ecological Zones in Eastern Kenya

Zipporah N. Bukania,[1] Moses Mwangi,[1] Robert M. Karanja,[2] Richard Mutisya,[1] Yeri Kombe,[1] Lydia U. Kaduka,[1] and Timothy Johns[3]

[1] *Centre for Public Health Research, Kenya Medical Research Institute, P.O. Box 20752, Nairobi 00200, Kenya*
[2] *Centre for Biotechnology Research and Development, Kenya Medical Research Institute, P.O. Box 54840, Nairobi 00200, Kenya*
[3] *School of Dietetics and Human Nutrition, McGill University, Sainte Anne de Bellevue, QC, Canada H9X 3V9*

Correspondence should be addressed to Zipporah N. Bukania; zbukania@gmail.com

Academic Editor: Christel Lamberg-Allardt

Machakos and Makueni counties in Kenya are associated with historical land degradation, climate change, and food insecurity. Both counties lie in lower midland (LM) lower humidity to semiarid (LM4), and semiarid (LM5) agroecological zones (AEZ). We assessed food security, dietary diversity, and nutritional status of children and women. *Materials and Methods.* A total of 277 woman-child pairs aged 15–46 years and 6–36 months respectively, were recruited from farmer households. Food security and dietary diversity were assessed using standard tools. Weight and height, or length in children, were used for computation of nutritional status. *Findings.* No significant difference ($P > 0.05$) was observed in food security and dietary diversity score (DDS) between LM4 and LM5. Stunting, wasting, and underweight levels among children in LM4 and LM5 were comparable as were BMI scores among women. However, significant associations ($P = 0.023$) were found between severe food insecurity and nutritional status of children but not of their caregivers. Stunting was significantly higher in older children (>2 years) and among children whose caregivers were older. *Conclusion.* Differences in AEZ may not affect dietary diversity and nutritional status of farmer households. Consequently use of DDS may lead to underestimation of food insecurity in semiarid settings.

1. Introduction

Demonstration of the potential of food-based approaches that draw on local agriculture resources to improve food and nutrition security of small-holder farmers in sub-Saharan Africa is challenged by the lack of valid assessment tools able to recognize meaningful nutritional changes within short time frames [1–3]. Nutrition-sensitive and sustainable agriculture has found growing conceptual support [4, 5]; however, agricultural interventions within complex and dynamic ecological and socioeconomic environments typically fail to demonstrate significant improvement in direct measures of nutrient status [3].

Dietary diversity as a practical indicator of nutrient adequacy may be the most satisfactory proxy for nutritional quality both in comparing differences in available food resources and in response to dietary change [3, 5] but requires evaluation in relation to agricultural interventions. Individual dietary diversity score (IDDS) is useful as a proxy measure of the nutritional quality of an individual's diet [6] and as a reflection of nutrient adequacy [5, 7, 8]. Household dietary diversity score (HDDS) on the other hand is meant to reflect, in a snapshot form, the economic ability of a household access to variety of foods [5].

Nutrient intake, food choices, and dietary diversity are key determinants of nutritional status, especially in areas of high food insecurity such as the arid and semiarid lands (ASAL) of Eastern Kenya. Food and nutrition insecurity of public health importance affects the urban and rural poor in both developed and developing countries [9] while undernutrition contributes to increased mortality and morbidity in developing countries, with critical impacts on children's cognitive and physical development, quality of life, and lifetime productivity [10].

Diets recognized as monotonous, cereal-based, and lacking diversity that are characteristic of most developing countries, especially in Africa, are comprised of foods low in energy with few animal products, fruits, and vegetables [9, 11]. Consequently, inadequate quantities and unbalanced distribution of food types consumed by the household often result in nutritional deficiencies. Women of childbearing age and children under 5 years are particularly at risk of poor health; children have higher nutrient requirements for growth and are susceptible to infectious diseases such as diarrhoea and respiratory infections which can inhibit nutrient absorption and decrease appetite [12, 13].

2. Project Background

This paper presents baseline findings of dietary diversity and nutritional status of a study population participating in a 42-month (Canadian International Food Security Research Fund (CIFSRF) project). The project was implemented by Kenya Agricultural Research Institute (KARI) and McGill University, Canada. The two institutions partnered with Kenya Medical Research Institute, Ministry of Agriculture, Kenya, Cascade Development Organization, and FreshCo Seed Company to support the project in nutrition and health, agricultural extension, marketing dynamics, and seed production and supply. The general objective of the proposed project was to contribute to improved food security among women and men in hunger-prone communities. This was to be achieved by facilitating farmers to adopt proven agricultural technologies that enhance ecological resilience in the face of climate change.

The overarching theme of the study was to measure and evaluate any differences in the adoption, uptake, and performance of introduced agricultural technologies in the different agroecological zones (AEZs) and to understand factors that drive these differences. The agricultural productivity potential of land is one such factor that is inherently linked to nutritional and health status of rural communities that rely on subsistence farming. The two counties of Machakos and Makueni in Eastern Kenya, where this study was conducted, predominantly occupy semiarid lower midland (LM) agroecological zones (AEZs) 4 and 5 where farmers typically combine crop and livestock production under conditions of moderate intensity of land use. LM4 is a transitional zone between lower humidity and semiarid zone with a mean annual rainfall of 800 mm, whereas LM5 is semiarid with mean annual rainfall of 500 mm [14, 15]. In both zones, rain-fed dependent agriculture is the mainstay system, and thus directly affecting food production and subsistence-sourced dietary diversity.

We explored DDS as a predictor of nutritional status within the framework of baseline assessment and methods development for an agricultural technology-transfer project. While we predict correlations of DDS in this population characterized by high rates of childhood and adult malnutrition with measures of stunting, wasting, and underweight, as seen in published studies [16], we further tested the hypothesis that, in comparing agro-ecological zones that differ in agricultural potential and production, dietary diversity would reflect resultant differences in nutritional status. Specifically, we examined the impact of agro-ecological zones (AEZs) relative to known determinants on the dietary diversity of women and children in farming households.

3. Materials and Methods

A descriptive study was carried out to assess the dietary diversity and nutritional status of women and children. Participants included children aged from 6 to 36 months and nonpregnant women of child-bearing age (15–46 years) from households of two counties of Machakos and Makueni in Eastern Kenya. At the outset, district agricultural officers were consulted in order to compile a list of registered farmer groups (FG) in the two counties. 72 farmers' groups from 119 villages met the inclusion criteria of being active in the last 6 months. Households were listed and data on age and sex of household members were collected to create a sampling frame of eligible households to recruit into the study. The proportion of FG membership whose households met the inclusion criteria of a woman-child pair where the woman was not pregnant but of child-bearing age (15–46 years) and the child of 6–36 months was determined. A simple random sampling technique was used to sample the households for participation in the study.

Lastly, 324 households were randomly selected from the sampling frame for inclusion in the survey with 307 successfully recruited and assessed. The nonresponses were due to absence. Thirty (30) households were excluded from the analysis due to missing or incomplete anthropometric measurements. Two hundred and seventy-seven, 134 child-woman pairs in Makueni and 143 in Machakos, were included in the study carried out in May and June 2012, a season described as "plenty" following harvest. Data collection and assessments were undertaken by a trained research team.

3.1. Sociodemographic Data. Demographic data included gender of the children, household size, and child's date of birth. Parents or guardians were asked to provide information on child's age which was confirmed using child immunization cards. Where cards were unavailable, the caretakers/guardians were asked to recall or use references to calendar events. Furthermore, information on household composition was collected.

3.2. Nutritional Status. Anthropometric measurements undertaken in all eligible respondents in the selected households included height/length and weight for children and women. Standard categories of nutritional status are reported according to WHO classification of anthropometric measurements cut-offs [17, 18]. Height and length were measured to the nearest 0.1 cm using the UNICEF wooden height and length boards while weight was assessed to the nearest 0.1 kg using a UNICEF Seca 762 classic mechanical medical weighing scale.

The nutritional status of the women of reproductive age (WRA) was assessed by body mass index (BMI) which was categorized according to World Health Organization standards [18].

3.3. Dietary Data. Household food consumption and dietary diversity data were collected with reference to both the child and the mother/caretaker/guardian. Respondents were requested to list all the foods consumed both at home and out of the home in the 24 hours preceding the interview. According to FAO, there are no established cut-offs points in terms of number of food groups to indicate adequate or inadequate dietary diversity for HDDS or IDDS [5]. Dietary diversity score (DDS) was calculated based on the number of different food groups consumed over a given reference period [19]. Mean scores or distribution of scores is, therefore, used for analysis [5].

Dietary diversity consisted of nine food groups for older children (24–36 months) and women. These foods included (1) starch staples, (2) grains and tubers, (3) dark green leafy vegetables, (4) other vitamin A rich fruits and vegetables, (5) other fruits and vegetables, (6) organ meats, (7) meat and fish, (8) eggs, and (9) legumes, nuts, and seeds. For the younger children 6–23 months, seven food groups are used [20]. Starchy staples, roots, and tubers were combined to form the grains, roots, and tubers food group, while meats, fish, and organ meats were combined and described as flesh foods group. Details about children's dietary intake were collected from their mothers or caregivers. Since the emphasis of this paper is on micronutrient intake as opposed to economic access, information on use of fats and oils has not been included in the scores.

3.4. Ethical Considerations. The study was approved by the Kenya Medical Research Institute's/National Ethical Review Committee. Prior to undertaking interviews, written consent was obtained from all adults, whereas for enrolment of children permission was obtained from their parents or guardians.

3.5. Statistical Analysis. Data entry was performed with Microsoft Access 2007. Data quality was maintained by quality checks during both data collection and entry (double entry) and further cleaning. All statistical analyses were carried out using IBM SPSS Statistics (version 20.0; IBM Corporation Software Group, NY, USA). Exploratory data analysis techniques were used to uncover the distribution structure of the study variables and identify outliers or unusually entered values. Distribution of continuous variables was tested for normality using Shapiro-Wilk test. All were found to follow normal distribution; therefore, independent t-test was used to test for mean differences between ecological zones. Chi-square test or Fisher exact test was used to test for independence in distribution of categorical variables (demographic characteristics, categorized nutritional variables, and dietary intake) between ecological zones. The same technique (Chi-square test or Fisher exact test) was used to assess the effect of various explanatory variables on nutritional status (stunting for children, underweight for WRA), with one explanatory variable assessed at a time. Odds ratios (OR) with corresponding 95% confidence intervals were estimated. In order to control for confounders, all explanatory variables that were associated with the nutritional status at $P < 0.1$ were considered together in a multiple logistic regression. Backward conditional method with removal at $P < 0.05$ was specified. Adjusted odds ratios (OR) with corresponding 95% confidence intervals were estimated.

4. Results

Table 1 presents characteristics of participating children and women in the survey. A total of 277 households were interviewed constituting 138 woman-child pairs from LM4 and 139 woman-child pairs from LM 5 AEZs. Mean age for WRA was 29.5 ± 0.4 (SE) years and 21.2 ± 0.5 (SE) months for the children (Table 1). Among the children, 49.8% were males and 50.2% were females.

The education levels and marital status of the WRAs were comparable in the two ecological zones. Less than 20% of the respondents reported having attained education above secondary levels but 81.2% had completed primary school. 87.3% of the WRA were married with household sizes averaging 7.0 members. This was also comparable in LM4 (mean household size 7.1) and LM5 (mean household size 6.8). Analyses of household food insecurity revealed that majority of the households (86.6%) were severely insecure. All variables were not significantly different between LM4 and LM5 ($P > 0.05$).

4.1. Nutritional Status and Dietary Diversity Score (DDS) in Women. The nutritional status of women between AEZs LM4 and LM5 showed no statistically significant difference ($P > 0.05$) (Table 2). The mean (SE) BMI for the study population was 21.9(0.2) Kg/m^2 with thinness and overweight observed in 14.4% and 17.7% of the women, respectively. Interestingly, obesity levels exceeded those of wasting in both AEZs although these differences between zones were not significant ($P = 0.05$) (LM4: 18.8% versus LM5: 16.5%).

For the classification of dietary diversity, three categories were used in ranking DD score for women: low (1–3 food groups), medium (4-5 food groups), and high (≥6 food groups) [5, 6]. Women's dietary diversity score (DDS) was similar between LM4 and LM5 with high (≥6 food groups), medium (4-5 food groups), and low (1–3 food groups) percentages of 8.7%, 61.6%, and 29.7% compared to 5.8%, 63.3%, and 30.9%, respectively (Table 2).

4.2. Nutritional Status and Dietary Diversity (DDS) in Children. Children in AEZs LM4 and LM5 did not differ with statistical significance ($P > 0.05$) in mean z-scores (HAZ, WAZ, and WHZ) (Table 2). The mean Z scores (SE) for stunting, underweight, and wasting were −1.51(0.08), 0.78(0.07), and 0.02(0.07), respectively. Overall 33.8%, 11.6%, and 2.5% of

TABLE 1: Characteristics of women and children in farmer households in lower midland ASAL regions of Eastern Kenya.

Characteristic WRA	Total (n = 277)	LM4 (n = 138)	LM5 (n = 139)	P value
Mean (SE) years	29.5 (0.4)	29.3 (0.5)	29.6 (0.6)	0.675
15–29 years %	56.7	58.7	54.7	
30–39 years %	33.9	34.1	33.8	0.463
40–46 years %	9.4	7.2	11.5	
Marital status				
Never married %	9.6	9.0	10.2	
Married %	86.0	87.3	84.7	0.795
Separated/widowed/divorced %	4.4	3.7	5.1	
Education level				
Primary level and below %	81.2	81.9	80.4	0.878
Secondary level and below %	18.8	18.1	19.6	
Characteristics, children 6–36 months				
Gender (%)				
Girls %	50.2	53.6	46.8	0.254
Boys %	49.8	46.4	53.2	
6–23 months %	57.4	59.4	55.4	0.498
24–36 months %	42.6	40.6	44.6	
Age in months				
Mean (SE) months	21.2 (0.5)	20.8 (0.7)	21.5 (0.7)	0.446
Mean (SE) 6–23 months	15.0 (0.4)	15.0 (0.5)	15.0 (0.5)	0.986
Mean (SE) 24–36 months	29.4 (0.3)	29.2 (0.4)	29.6 (0.4)	0.525
Household characteristics				
Family size				
Mean household size (SE)	7.0 (0.1)	7.1 (0.2)	6.8 (0.2)	0.274
3-4 people %	11.4	9.6	13.1	
5-6 people %	37	40.4	33.6	0.410
≥7 people %	51.6	50.0	53.3	
Household food security				
Food secure %	6.5	5.1	8.0	
Moderately insecure %	6.9	9.4	4.3	0.173
Food insecure %	86.6	85.5	87.7	

SE: standard error; LM: lower midlands agroecological zones.

the children aged 6–36 months were stunted, underweight, and wasted, respectively.

Stunting was slightly more prevalent in the semiarid LM5 (32.5%) compared to transitional LM4 (26.8%). However, proportions of underweight (LM4: 9.8% versus LM5: 10.4%) and wasting (LM4: 2.2% versus LM5: 2.9%) were nearly similar as shown in Table 2.

When disaggregated into age groups, significantly more older children (24–36 months) were stunted (39.7%) as compared to children younger than two years (29.6%; $P < 0.05$). Similarly, there was a greater burden of underweight and wasting in the older children (13.6% and 2.6%, resp.) than in the younger category (10.1% and 2.5%, resp.) although this was not statistically significant ($P > 0.05$).

Only 27.7% of children aged 6–23 months had adequate dietary diversity (a minimum of 4 food groups or more in the previous 24 hours). Within this same age group, more (76.6%) children in LM5 had inadequate dietary diversity compared to 68.3% in LM4. When DDS was calculated for the older children, upto 78.8% had low dietary diversity that was higher in LM4 (89.3%) than in LM5 (69.4%). A statistically significant difference in DDS between children aged 6–23 months and 24–36 months was observed (P = 0.008).

Using Pearson's correlation, a direct significant relationship was found between the DDS of the child and DDS of the caregiver/mother (r = 0.487; P = 0.001); as the DDS of the mother increases, the DDS of the child increases.

4.3. Nutritional Status of Children and Women in relation to Selected Sociodemographic Characteristics. Table 3 presents the nutritional status in children in relation to selected sociodemographic characteristics. Age 31–46 years among women was found to be significantly associated with stunting in children. A higher proportion of the older children (24–36 month) were stunted (39.7%) compared to their younger counterparts (29.6%) (OR = 1.57; 95% CI: (0.95–2.59); P = 0.081). For children whose caregivers were older

TABLE 2: Nutritional status and dietary diversity score in women and children by agroecological zones.

	Total (n = 277)	LM4 (n = 138)	LM5 (n = 139)	P value
Nutritional status of women				
Mean (SE) weight (Kg)	54.3 (0.6)	54.2 (0.9)	54.4 (0.8)	0.863
Mean (SE) height (cm)	157.4 (0.4)	157.4 (0.5)	157.3 (0.6)	0.912
Mean (SE) BMI (Kg/m^2)	21.9 (0.2)	21.8 (0.3)	22.0 (0.3)	0.738
Wasted %	14.4	16.7	12.2	
Overweight %	17.7	18.8	16.5	0.447
Normal %	67.9	64.5	71.2	
Dietary diversity in women				
Dietary diversity score (DDS)				
Low DDS %	30.3	29.7	30.9	
Medium DDS %	62.5	61.6	63.3	0.639
High DDS %	7.2	8.7	5.8	
Nutritional status in children				
Mean Z scores in children				
Mean (SE) HAZ Z-score	−1.51 (0.08)	−1.44 (0.12)	−1.59 (0.12)	0.392
Mean (SE)WAZ Z-score	−0.78 (0.07)	−0.76 (0.09)	−0.80 (0.10)	0.783
Mean (SE) WHZ Z-score	0.02 (0.07)	−0.02 (0.09)	0.05 (0.10)	0.596
Stunting %	33.8	31.6	36.0	0.445
Underweight %	11.6	10.9	12.2	0.723
Wasting %	2.5	2.2	2.9	1.000
6–23 months				
Stunting %	29.6	26.8	32.5	0.436
Underweight %	10.1	9.8	10.4	0.894
Wasting %	2.5	2.4	2.6	1.000
24–36 months				
Stunting %	39.7	38.9	40.3	0.875
Underweight %	13.6	12.5	14.5	0.749
Wasting %	2.6	1.9	3.2	1.000
Dietary diversity score in children				
6–23 months				
Inadequate %	72.3	68.3**	76.6	0.241
Adequate %	27.7	31.7	23.4	
24–36 months				
Low %	78.8	89.3**	69.4	0.008
Medium %	21.2	10.7	30.6	

*Difference in DDS between 6–23 and 24–36 months is significant ($P < 0.05$).

(31–46 years) a significantly higher proportion was stunted compared to children whose caregivers were relatively young (15–30 years) (OR = 1.81; 95% CI: (1.08–3.02); P = 0.024). No significant associations were found with age of child, sex of child, education level of the woman/caregiver, marital status of the woman/caregiver, ecological zones, DDS of child, household size, and nutritional status of the woman. Upon controlling for confounders, binary logistic regression revealed older age in woman/caregiver (AOR = 1.77; 95% CI: (1.05–2.97); P = 0.032) and severe food insecurity (AOR = 2.90; 95% CI: (1.16–7.25); P = 0.023) to be strongly associated with stunting in children.

Table 4 presents the nutritional status of women respondents in relation to selected sociodemographic. Contrary to expectation, no significant association was found between nutritional status of the mother and dietary diversity or food security. Low education level among women could have an implication on their own nutritional status since our findings show a marginal significant association (OR = 3.23; 95% CI: (0.96–10.9) (P = 0.05)).

5. Discussion

Agricultural potential as defined by the ecological zones did not affect the dietary diversity (DDSs) and nutritional status of households. Household characteristics that posed as confounders did not differ significantly between LM4 and LM5 (Table 1). However, the overall mean household size of 7.0 was found to be much higher than the national mean of 4.2

TABLE 3: Nutritional status of children in relation to selected sociodemographic and economic characteristics.

Characteristics	Stunting status: Stunted n	Stunted %	Normal n	Normal %	OR	Bivariate analysis 95% CI Lower	Upper	P value	Multivariate analysis AOR	95% CI Lower	Upper	P value
Age of the child in months												
6–23	47	29.6	112	70.4	1.00							
24–36	46	39.7	70	60.3	1.57	0.95	2.59	0.081				
Sex of the child												
Male	49	36.0	87	64.0	1.22	0.74	2.01	0.443				
Female	44	31.7	95	68.3	1.00							
Age in years of the woman												
15–30	51	29.0	125	71.0	1.00				1.00			
3–46	42	42.4	57	57.6	1.81	1.08	3.02	**0.024**	1.77	1.05	2.97	**0.032**
Education level of the woman												
Primary and below	78	35.1	144	64.9	1.47	0.75	2.88	0.259				
Secondary and above	14	26.9	38	73.1	1.00							
Marital status of the woman												
Never married	9	34.6	17	65.4	0.74	0.18	3.02	0.676				
Married	76	32.9	155	67.1	0.69	0.21	2.23	0.532				
Separated/divorced/widowed	5	41.7	7	58.3	1.00							
Ecological zone												
LM4	43	31.6	93	68.4	1.00							
LM5	50	36.0	89	64.0	1.22	0.74	2.00	0.445				
Household with severe food insecurity												
Yes	87	36.6%	151	63.4%	2.98	1.19	7.42	**0.015**	2.90	1.16	7.25	**0.023**
No	6	16.2%	31	83.8%	1.00				1.00			
DDS of children												
Inadequate	75	36.4	131	63.6	1.62	0.88	2.98	0.117				
Adequate	18	26.1	51	73.9								
Household size												
3-4 people	13	39.4	20	60.6	1.16	0.53	2.52	0.709				
5-6 people	29	29.0	71	71.0	0.73	0.42	1.27	0.261				
≥7 people	51	35.9	91	64.1	1.00							
*NS of the woman												
Underweight	12	30.0	28	70.0	0.88	0.36	2.18	0.789				
Normal	65	34.9	121	65.1	1.11	0.57	2.16	0.764				
Overweight	16	32.7	33	67.3	1.00							

*Age of woman in years, nutritional status significant at $P < 0.05$.

as reported by the Kenya Demographic Health Survey [16]. A household was defined as members of a community that live as a cohesive unit and eat from the same "pot." Machakos and Makueni counties are historically food insecure and characterized by land degradation, cycles of drought and famine, and reliance on food aid [20, 21]. The large household size reported may, therefore, represent a coping strategy in the face of persistent food insecurity through leveraging food access by increasing the ratio of number of persons per pot [22].

5.1. Nutritional Status. Individual dietary diversity has been described [6] as a proxy measure of nutrition/dietary quality and can be a good indicator of overall household food security and positively associated with nutritional status in children [23]; this would, therefore, explain why long term poor dietary diversity is likely to be reflected in stunting. We observed an association between severe food insecurity and stunting (AOR = 2.90; 95% CI: (1.16–7.25); $P = 0.023$) but not with wasting and underweight. Stunting or chronic malnutrition is usually an indication of long term deprivation

TABLE 4: Nutritional status of women respondent in relation to selected sociodemographic and economic characteristics.

Characteristics	Underweight		Normal/overweight		OR	95% CI		*P* value
	n	%	*n*	%		Lower	Upper	
Age in years								
15–30	28	15.8	149	84.2	1.38	0.67	2.85	0.385
31–46	12	12.0	88	88.0	1.00			
Education level*								
Primary and below	37	16.5	187	83.5	3.23	0.96	10.9	0.050
Secondary and above	3	5.77	49	94.2	1.00			
Marital status								
Never married	3	11.5	23	88.5	1.43	0.13	15.4	0.766
Married	36	15.5	197	84.5	2.01	0.25	16.1	0.510
Separated/divorced/widowed	1	8.33	11	91.7	1.00			
Ecological zone								
LM4	23	16.7	115	83.3	1.44	0.73	2.82	0.294
LM5	17	12.2	122	87.8	1.00			
Dietary diversity score								
Low	12	14.3	72	85.7	0.94	0.24	3.72	0.935
Medium	25	14.5	148	85.5	0.96	0.26	3.51	0.947
High	3	15.0	17	85.0	1.00			
Household food security								
Food secure	1	5.56	17	94.4	1.00			
Moderately food insecure	3	15.8	16	84.2	3.19	0.30	33.89	0.336
Food insecure	36	15.1	203	84.9	3.01	0.39	23.36	0.291
Household size								
3-4 people	4	11.8	30	88.2	1.00			
5-6 people	13	13.0	87	87.0	1.12	0.34	3.70	0.852
≥7 people	23	16.1	120	83.9	1.44	0.46	4.47	0.531

*Education level of woman.

of nutrients in children. Stunting remains a problem of greater magnitude than underweight or wasting, and it more accurately reflects nutritional deficiencies and illness that occur during the most critical periods for growth and development in early life [24].

The consequences of seasonal changes on the nutritional status of adults have been well documented although dietary diversity scores are usually measured during a single period of the year and their seasonal variations remain largely unknown [23]. Our findings showed no association between DDS and nutritional status of the women in this study; however, almost two thirds of the women (62.3%) had medium and less than 8% had high DDS at the time of this survey. This could be attributed to seasonality since our survey was undertaken in May/June at a season dubbed "season of plenty" immediately after harvest. It would be useful to reconsider measuring dietary diversity during the food shortage season, to understand the seasonal differences.

The lower the education level, the higher the prevalence of underweight among women. These findings are similar in many populations where the level of education is directly related with obesity especially in women [25]. Education, therefore, appears as a protective factor for the underweight women but as a risk factor for the overweight and obese women [26]. More immediately, at least, increasing the options of a household as mediated by factors, such as women's education, to source for other preferred food choices may have greater impact on nutritional status. FAO analysed data from 48 low income countries and found that primary education is a key determinant of food security in low income communities. FAO further emphasizes that female education has a direct additional benefit to nutritional status [27].

5.2. Dietary Diversity. In addition to improved outcomes in child birth weights, anthropometric status and improved haemoglobin status, dietary diversity is highly correlated with caloric and protein adequacy [6]. Our findings indicate moderate DDS among the women however, not statistically significant ($P = 0.639$). Most of the children on the other hand had low diversity 72.3% and 78.8% 6–23 and 24–36 months old respectively, mostly consuming less than 4 food groups in a day. Statistically significant differences in DD in children were identified ($P = 0.008$) between the ecological zones. However, it has been found that although there is a definite and strong association between mother and child DD, mother child agreement in intake of foods from different food groups may not necessary allow using mothers DD for calculating the child's DD [3].

5.3. Strengths and Limitations. The use of experienced research assistants ensured accuracy in data collection. Moreover, the customization of the data collection tool to suit the study population ensured that all the relevant foods were listed to guide research assistants in probing for all foods possibly consumed in the household including alternative sources foraged from the wild with the potential of augmenting or negating differences in DD between the AEZs. Careful field supervision also allowed us to have accurate data collection with minimal missing variables.

One limitation of our study was that dietary data was collected for only one recall. This can potentially limit the true dietary diversity on different days of the week. Self-reporting of dietary data in a food insecure population where respondents might assume that underreporting of consumption offered a possible opportunity for food support could introduce a potential bias. An improvement on the dietary diversity data tool could help collect information on food varieties specific to food groups. This is an aspect that may need consideration in future surveys. Additionally, the survey was undertaken in one season and at a time of plenty. More DDS done in different seasons may better explain the true dietary diversity of the study population.

6. Conclusion

Child nutritional status was dependent on household food security and not dietary diversity. Moreover, differences in AEZ did not affect dietary diversity and nutritional status of farmer households. This may be attributed to seasonal variation that has been shown to significantly impact the estimation of food security status. DDS may therefore not be the preferred predictor for nutritional status, and further interrogation of its applicability in diverse settings is necessary. Conversely, the caregivers' education levels strongly correlated with the nutritional status of the children. In addressing food insecurity, enhancing women's nutrition knowledge in combination with nutrition-sensitive agricultural interventions, can be expected to significantly improve food security and nutritional status of children and women.

Certainly nutrition-sensitive agriculture that focuses on increasing locally-sourced diversity is more likely to achieve impact in conjunction with population wide nutrition education directed especially to caregivers that emphasizes the need for diversified diets. Increased household income is a major contributor to improved nutrition, although on its own it is not enough; it may be inefficient or ineffective if women have no level of control. Women are more likely to spend the income they control on health, food, and education of their children [28]. Additionally, improvements in mechanization focused on agricultural production in the study area can free women of heavy workloads, thus creating more time to care for their children and families.

Conflict of Interests

All the authors declare that there is no conflict of interests regarding the publication of this paper.

Authors' Contribution

Zipporah N. Bukania analysed and interpreted the data and wrote the paper. Moses Mwangi guided the statistical analysis of the data and participated in revision of the paper; Robert M. Karanja, Lydia U. Kaduka, Richard Mutisya, Yeri Kombe, and Timothy Johns contributed to data interpretation and writing of the paper. All authors approved the final paper.

Acknowledgments

This work was carried out with the aid of a grant from the Canadian International Food Security Research Fund (CIFSRF), a program of Canada's International Development Research Centre (IDRC) undertaken with the financial support of the Government of Canada provided through Foreign Affairs, Trade and Development Canada (DFATD). The authors acknowledge the support of Lutta W. Muhammad and Gordon M. Hickey, Principal investigators of this CIFSRF project for the Kenya Agricultural Research Institute (KARI) and McGill University, respectively. The authors also thank all administrative personnel in the research area and all the respondents who accepted and consented to take part in the study. Further gratitude goes out to all the research assistants who tirelessly participated in this survey during data collection. The authors also thank the Director of KEMRI for granting permission to undertake this study.

References

[1] E. Masset, L. Haddad, A. Cornelius, and J. Isaza-Castro, "A systematic review of agricultural interventions that aim to improve nutritional status of children," http://agris.fao.org/agris-search/search/display.

[2] P. Pinstrup-Andersen, "Can agriculture meet future nutrition challenges?" *European Journal of Development Research*, vol. 25, no. 1, pp. 5–12, 2013.

[3] M. T. Ruel and H. Alderman, "Nutrition-sensitive interventions and programmes: how can they help to accelerate progress in improving maternal and child nutrition?" *The Lancet*, vol. 382, no. 9891, pp. 536–551, 2013.

[4] A. Herforth, A. Jones, and P. Pinstrup-Andersen, "Prioritizing nutrition in agriculture and rural development: guiding principles for operational investments," Working Paper, Health Nutrition and Population, The World Bank, 2012.

[5] FAO, *Guidelines for Measuring Household and Individual Dietary Diversity*, 2013.

[6] A. Swindale and P. Bilinsky, *Household Dietary Diversity Score (HDDS) for Measurement of Household Food Access: Indicator Guide*, Version 2, 2006.

[7] P. Mirmiran, L. Azadbakht, and F. Azizi, "Dietary diversity within food groups: an indicator of specific nutrient adequacy in Tehranian women," *Journal of the American College of Nutrition*, vol. 25, no. 4, pp. 354–361, 2006.

[8] T. Belachew, D. Lindstrom, A. Gebremariam et al., "Food insecurity, food based coping strategies and suboptimal dietary practices of adolescents in Jimma zone Southwest Ethiopia," *PLoS ONE*, vol. 8, no. 3, Article ID e57643, 2013.

[9] G. H. Leyna, E. J. Mmbaga, K. S. Mnyika, A. Hussain, and K. Klepp, "Food insecurity is associated with food consumption

patterns and anthropometric measures but not serum micronutrient levels in adults in rural Tanzania," *Public Health Nutrition*, vol. 13, no. 9, pp. 1438–1444, 2010.

[10] N. S. Scrimshaw and J. P. SanGiovanni, "Synergism of nutrition, infection, and immunity: an overview," *The American Journal of Clinical Nutrition*, vol. 66, no. 2, pp. 464S–477S, 1997.

[11] G. L. Kennedy, M. R. Pedro, C. Seghieri, G. Nantel, and I. Brouwer, "Dietary diversity score is a useful indicator of micronutrient intake in non-breast-feeding Filipino children," *The Journal of Nutrition*, vol. 137, no. 2, pp. 472–477, 2007.

[12] L. C. Ivers and K. A. Cullen, "Food insecurity: special considerations for women," *The American Journal of Clinical Nutrition*, vol. 94, no. 6, pp. 1740–1744, 2011.

[13] N. P. Steyn, J. H. Nel, G. Nantel, G. Kennedy, and D. Labadarios, "Food variety and dietary diversity scores in children: are they good indicators of dietary adequacy?" *Public Health Nutrition*, vol. 9, no. 5, pp. 644–650, 2006.

[14] G. Ndeng'e, C. Opiyo, J. Mistiaen, and P. Kristjanson, *Geographic Dimensions of Well-Being in Kenya: Where are the Poor? From Districts to Locations*, vol. 1, 2003.

[15] R. Jaetzold, H. Schmidt, B. Hornetz, and C. Shisanya, *Farm Management Handbook of Kenya: Part A. West Kenya; Subpart A2. Nyanza Province*, Ministry of Agriculture, in Cooperation with the German Agency for Technical Cooperation (GTZ), Nairobi, Kenya, 2nd edition, 2006.

[16] KDHS: Kenya. Kenya National Bureau of Statistics (KNBS) and ICF Macro, *Kenya Demographic and Health Survey 2008-09*, KNBS and ICF Macro, Calverton, Md, USA, 2010.

[17] WHO Child Growth Standards, "Length/height-for-age, weight-for-age, weight-for-length, weight-forheight and body mass index-for-age: methods and development," 2006.

[18] World Health Organization(WHO), "Obestiy, preventing and managing the global epidemic. Report of a WHO Consultation," WHO Technical Report Series 894, World Health Organization, Geneva, Switzerland, 2000.

[19] K. M. Rathnayake, P. Madushani, and K. Silva, "Use of dietary diversity score as a proxy indicator of nutrient adequacy of rural elderly people in Sri Lanka," *BMC Research Notes*, vol. 5, article 469, 2012.

[20] World Health Organization, "Indicators for assessing infant and young child feeding practices," 2008.

[21] D. Rocheleau and P. Benjami, "The Ukambani region of Kenya," in *Regions at Risk: Comparisons of Threatened Environments*, J. X. Kasperson, R. E. Kasperson, and B. L. Turner, Eds., United Nations University Press, Tokyo, Japan, 1995.

[22] A. Ehrensperger, O. Grimm, and B. Kiteme, "Spatial analysis of food insecurity drivers and potential impacts of biofuels cultivation: a contribution to sustainable regional development and national biofuel policies in Kenya," *Research Design and Methodology*, vol. 11, 2012.

[23] M. Savy, Y. Martin-Prével, P. Traissac, S. Eymard-Duvernay, and F. Delpeuch, "Dietary diversity scores and nutritional status of women change during the seasonal food shortage in rural Burkina Faso," *Journal of Nutrition*, vol. 136, no. 10, pp. 2625–2632, 2006.

[24] UNICEF, *Tracking Progress on Child and Maternal Nutrition*, UNICEF, New York, NY, USA, 2009.

[25] A. T. Ali and N. J. Crowther, "Factors predisposing to obesity: a review of the literature," *Journal of Endocrinology, Metabolism and Diabetes of South Africa*, vol. 14, no. 2, pp. 81–84, 2009.

[26] M. M. H. Khan and A. Kraemer, "Factors associated with being underweight, overweight and obese among ever-married non-pregnant urban women in Bangladesh," *Singapore Medical Journal*, vol. 50, no. 8, pp. 804–813, 2009.

[27] Food Agricutural Organization (FAO), *Education for Rural People and Food Security. A Cross Country Analysis*, 2007.

[28] R. Meinzen-Dick, J. Behrman, P. Menon, and A. Quisumbing, in *Gender: A Key Dimension Linking Agricultural Programs to Improved Nutrition and Health*, C. Shenggen, Ed., vol. 16, chapter 16, pp. 135–144, International Food Policy Research Institute, Washington, DC, USA.

13

Anthropometry and Dietary Intake before and during a Competition in Mountain Runners

Anja Carlsohn[1] and Wolfram Müller[2]

[1] *University of Education Schwäbisch Gmünd, Institute of Health Sciences, Oberbettringer Straße, 200 73525 Schwäbisch Gmünd, Germany*
[2] *Medical University of Graz, Institute of Biophysics, Harrachgasse 21, 8010 Graz, Austria*

Correspondence should be addressed to Anja Carlsohn; anja.carlsohn@arcor.de

Academic Editor: C. S. Johnston

Mountain running is a non-Olympic sport consisting of uphill or up- and downhill races at moderate-to-high altitude. Special nutritional requirements are anticipated, but no nutritional data of mountain runners are available. In three studies, physique of elite and recreational athletes ($N = 62$), maximum oxygen uptake ($N = 3$), and prerace and race day dietary intake ($N = 6$) were measured (mean ± SD). Mean oxygen uptake was 68.7 ± 5.2 mL/kg/min. Energy and carbohydrate intake before a race (29 ± 15 km, 1596 ± 556 m HD) was 3199 ± 701 kcal/d (13.4 ± 2.9 MJ/d) and 497 ± 128 g/d (8.3 ± 1.8 g/kg/d) in German national team members. Fluid intake was calculated as 2783 ± 1543 mL/d. During the race, athletes consumed 336 ± 364 kcal and 927 ± 705 mL of fluids. Substrate intake per hour was calculated as 23 ± 22 g of carbohydrates and 4.0 ± 3.2 g of proteins. In conclusion, anthropometric and oxygen uptake characteristics of mountain runners were similar to those reported for elite distance runners. Carbohydrate intake before and during the race was below recommendations for endurance athletes. This is of concern when considering the increased reliance on carbohydrates at altitude.

1. Introduction

Mountain running is a non-Olympic track and field sport. Under the auspices of the International Association of Athletics Federation (IAAF) and the World Mountain Running Association (WMRA, association under IAAF patronage), international championships in Africa, Europe, North America, Central America, and South America are conducted.

There are two main types of mountain running competitions: uphill races and up- and downhill events. Although the racetracks are of natural origin and are not artificially constructed, there are some course requirements stated by the WMRA. The length of the uphill tracks should approximate 8 km with an ascent of 800 m for females or 12 km with an ascent of 1200 m for males (10% ascent in average for both genders). The up- and downhill courses approximate a length of 8 km for females and 12 km for males, with ascents/descents of 500 m and 750 m, respectively.

Uphill races often finish at or above an altitude of 2500 m above sea level (a.s.l.), whereas up- and downhill courses most often start and finish at the same point at altitudes of 500 m a.s.l. or higher. Due to the shifting running conditions (surface, elevation, and direction) and increase in altitude, average running velocity in mountain races is much slower than during flat races. As a raw estimation, running time during a mountain race is 50% to 75% longer compared to the expected running time during flat races of the same distance. Drinking or eating is almost impossible at steep or rocky passages, where athletes breathe heavily and/or full concentration is required. According to the WMRA technical regulations, water and other "suitable" refreshments shall be made available at the finish and at intervals of 2-3 kilometres. Converted into running time, refreshment stations are available approximately every 10–30 minutes.

Mountain runners do not necessarily live and train in alpine landscapes all the year but divide their competition preparation into training periods at sea level and others at moderate or high altitude (3-4 weeks at 1800–3000 m a.s.l. for several times a year). Elite mountain runners conduct 10–14 training sessions per week and cover a weekly running distance of 120–200 km. During altitude training camps, up to twenty training sessions per week are conducted with a focus on repeated uphill runs and long-distance sessions with a cumulative gain in altitude of 500–1000 m (personal communication to German national team members).

The unique physiological strain on mountain runners during training and competition suggests special nutritional demands. However, little is known about the nutrition and physique of mountain runners. The purpose was to collect a set of data (1) on the physique of mountain runners and (2) on the dietary intake at the day before and during mountain running competitions of elite and recreational mountain runners and to compare the results from this sample to nutritional habits in other endurance sports. We hypothesized that elite mountain runners have a low body mass, as this might be a competitive advantage. We also hypothesized that the dietary intake during the race does not meet the recommendations for endurance runners, as the nutritional supply of the athletes might be difficult in the mountains.

2. Materials and Methods

Three independent pilot studies were conducted to collect preliminary data about physique and endurance capacity, prerace, and race day dietary intake of mountain runners. All volunteers gave written informed consent before participating in the studies. The assessment of anthropometric characteristics, dietary intake, and maximum oxygen uptake of athletes was approved by the Ethical Committee of the University of Potsdam. The studies were carried out according to the Declaration of Helsinki.

2.1. Anthropometry. The day before an international mountain running competition that served as qualification race for the European Championships (Feuerkogel Mountain Race, Austria: length 10 km, 1250 gain in altitude), anthropometric data of 62 elite and recreational athletes were collected. Here, age, body mass, and body height were measured with the barefoot subjects standing in an upright position [1]. Dressed body mass was corrected by 1 kg to account for the clothes worn by the athletes, and body-mass index (BMI) was calculated using the corrected body mass and the body height.

2.2. Maximum Oxygen Uptake. Maximum oxygen uptake ($\dot{V}O_2$max) as a surrogate parameter of endurance capacity was measured in three elite athletes from the German national team (2 females, 1 male). Here, a stepwise incremental running test on a treadmill was conducted with a starting velocity of 6 km/h for females and 8 km/h for the male athlete and an increment of 2 km/h every 3 minutes until exhaustion as described elsewhere [2]. Oxygen uptake and carbon dioxide expiration were measured continuously using a face mask with a stationary breath-by-breath system (ZAN 600 metabolic device, nSpire, Health, Oberthulba, Germany).

2.3. Prerace and Race Day Nutrition. Dietary intake was measured at the day before a mountain running competition in six elite mountain runners of the German national team (5 males, 1 female) using a 24-hour diet recall. Athletes were asked to collect their nutritional data before and during their upcoming competitions. Thus, data were collected at four different competitions: Dachstein Mountain race (12,1 km with 884 m gain in altitude, Dachstein, Austria), Inferno Half Marathon (21,1 km with 2175 m gain in altitude, Lauterbrunnen, Switzerland), LGT Alpin Marathon (42,2 km with 1870 m gain altitude, Malbun, Liechtenstein), and finally Schauinsland Mountain race (13 km with 905 m gain in altitude, Freiburg, Germany). In all four races, a regular supply approximately every 5 km was ensured.

To analyze the dietary intake during the race, athletes were asked to complete a diet recall at the same day for the race only. Athletes were instructed to use household measures (i.e., cups, packages, etc.) to record the amounts of food and fluids consumed. Energy and macronutrient intake was analysed based on the German food database BLS II.3 using PRODI expert software (PRODI NutriScience, Hausach, Germany). For foods not contained in the database such as energy bars or carbohydrate gels, standardized recipes according to the company's food labeling were created. For items not listed in the food database as well as for composite dishes a coding list was established to standardize the data input. To avoid interexaminer set-offs when transferring the food records into nutrient calculation [3], data input into the database was performed by one examiner only. All data are given as means ± standard deviations and 95% confidence intervals (CI). Student's *t*-test was conducted to test for gender differences where appropriate (alpha = 0.05).

3. Results

3.1. Age and Anthropometry. Mean age of elite and recreational participants of a qualification race for the European Championships was 37.3 ± 11.5 years (95% CI: 34.1–40.5 years). Body height was 1.76 ± 0.07 m (95% CI: 1.74–1.78 m), and body weight was measured as 65.3 ± 8.8 kg (95% CI: 63.0–67.5 kg). BMI was calculated as 21.0 ± 1.8 kg/m^2 (95% CI: 20.5–21.4 kg/m^2), with significantly higher values in males than females (21.4 ± 1.7 kg/m^2 versus 19.6 ± 1.7 kg/m^2, $P = .009$). Anthropometric data for males and females are detailed in Table 1.

3.2. Maximum Oxygen Uptake. The $\dot{V}O_2$max measured in three members of the German national mountain running team was 64 and 66 mL/kg/min for the two females and 76 mL/kg/min for the male athlete.

3.3. Prerace and Race Day Nutrition. The distance covered during the mountain running competitions, where prerace and race diets were observed, was 28.8 ± 15.0 km with a gain in altitude of 1596 ± 556 m. Running time was 2 hours 25

TABLE 1: Anthropometric characteristics of elite and recreational mountain runners participating at a qualifying for the European Championships.

	General cohort ($N = 62$)	Females ($N = 14$)	Males ($N = 48$)
Age (years)	37.3 ± 11.5	35.8 ± 9.3	37.7 ± 12.1
Height (cm)	176 ± 7	167 ± 4[a]	178 ± 6[a]
Body mass (kg)	65.3 ± 9.8	54.8 ± 6.09[b]	68.4 ± 6.9[b]
BMI (kg/m^2)	21.0 ± 1.8	19.6 ± 1.7[c]	21.4 ± 1.7[c]

Data are given as means ± SD.
a, b, c indicate significant differences between the genders ($P > 0.05$).

minutes ± 1 hour 7 minutes. Total energy intake at the day before competition was 3199 ± 701 kcal/d (95% CI: 2463–3935 kcal/d). At the prerace day, athletes ingested 497±128 g/d of carbohydrates (95% CI: 362–632 g/d), which corresponds to 8.3 ± 1.8 g/kg body mass/d (95% CI: 6.4–10.2 g/kg/d, Table 2). A protein intake of 99.9 ± 29.2 g/d of protein (95% CI: 69.1–130.4 g/d) was documented, which corresponds to 1.6±0.3 g/kg body mass/d (95% CI: 1.3–2.0 g/kg/d). Fat intake was calculated as 85.8 ± 30.8 g/d (95% CI: 53.6–118.1 g/d). Total fluid intake was 2783 ± 1409 mL/d at the day before competition, with a 95% CI of 1164–4403 mL/d.

During the race, athletes preferred consuming liquid energy sources such as sports drinks, caffeine containing soft drinks, and/or commercial carbohydrate gels. Altogether, athletes ingested 336 ± 364 kcal while running (95% CI: −46–717 kcal). Mean carbohydrate intake was 68 ± 76 g, which corresponds to 23 ± 22 g carbohydrate per hour competition (95% CI: 0–46 g/h carbohydrates). Total fluid intake during the race was calculated 927 ± 704 mL/race (449 ± 376 mL per hour competition). Athletes consumed 4.0 ± 3.2 g of protein per hour (95% CI: 0.7–7.3 g/h) and 0.3 ± 0.5 g of fat per hour competition (95% CI: −0.2–0.8 g/h).

4. Discussion

Considering the few available data [4, 5], elite mountain runners seem to have similar anthropometric and physiological characteristics compared to distance runners. In participants of a WMRA Masters World Championship (46 ± 9 years) a body mass of 64.5 ± 7.9 kg and a height of 177 ± 8 cm were measured [4], which is comparable to elite middle-distance runners [6]. In most endurance sports, low body fat is of advantage, as this improves the power-to-weight ratio [7]. Mountain running athletes also had a low body mass which indicates that low body weight may increase running efficiency, at least as long as it is above critically low body weight, which may decrease $\dot{V}O_2$max relative to body mass and increase running time [8]. In an experimental setting, an additional 7.5% of weight resulted in a 30% reduction of running performance in male runners [9, 10].

Furthermore, energy cost for running was shown to considerably increase with steepness on both ascents and descents compared to running or walking at level [11]. In a single case analysis, a remarkable increase in energy expenditure was observed during uphill walking when either weight, speed, slope, or altitude was augmented [12]. Thus, a low body mass may be of special importance for mountain runners who compete on steep trails at altitude. This issue might explain the relatively low body mass in both male and female runners. However, as body mass or body mass index is of limited value for the use in athletes, information regarding body composition of mountain runners is required for interpreting data on a finer scale.

Regarding aerobic capacity, $\dot{V}O_2$max measured in three internationally elite mountain runners was quite similar to values observed in master mountain runners with a $\dot{V}O_2$max of 67.8 ± 6.9 mL/kg/min [4]. The values of $\dot{V}O_2$max measured in the current study in elite mountain runners are close to results reported for elite marathoners. Typically, elite marathon runners achieve values of 70 mL/kg/min for males and 60 mL/kg/min for females, respectively [13].

The dietary intake at the day before competition might be considered generally adequate, although the interindividual variation in energy, carbohydrate, and fluid intake was high.

The mean observed carbohydrate intake of 8.5 g/kg/d at the day before competition is lower than recommended to optimize glycogen stores before a marathon [14]. However, as no information is available about the nutrition and tapering strategies of the mountain runners during the last 3–5 days before the competition, conclusions about the adequacy of carbohydrate intake in mountain runners before the race cannot be drawn.

Carbohydrate utilization during exercise and postexercise recovery is higher in hypoxia than at sea level, increasing the athletes' reliance on carbohydrate-rich foods [15]. This is a special challenge for endurance athletes with high training volumes at moderate-to-high intensities, as the important fuel source glycogen rapidly becomes a limiting factor. However, this effect is more pronounced in men than in women, who utilize fat to a greater extent at altitude than men [16]. In addition to a relatively low carbohydrate intake before the race, carbohydrate consumption of elite mountain runners during the race was found to be lower than recommended. Currently, for endurance events a carbohydrate intake of 30–60 g/hour is recommended [17], and there is growing evidence for beneficial effects of even higher dosages during competitions lasting longer than 2.5 hours (for review, see [18]).

However, due to the increased carbohydrate reliance at altitude, mountain runners competing at moderate to high altitudes may require even more carbohydrates than athletes competing at sea level. At very high altitude (>5000 m a.s.l.), carbohydrate supplementation was shown to improve time trial performance by 17% and reduce ratings of perceived exertion by 18% [19]. Thus, increasing carbohydrate ingestion during the race might be recommended to the athletes.

Interestingly, elite mountain runners consumed relatively large amounts of protein at the prerace day, pretty close to what is recommended for endurance athletes during heavy training periods. This observation is found consistently in all six elite mountain runners, although nutritionists might recommend reducing the protein intake in favor of carbohydrates.

TABLE 2: Prerace day nutrition and dietary intake during mountain races in $N = 6$ elite athletes.

	Prerace day diet	Race diet
Energy intake	3199 ± 701 kcal/day (2463; 3935)	336 ± 364 kcal/race (−46; 717)
Fluid intake	2783 ± 1543 mL/day (1164; 4403)	927 ± 705 mL/race (187; 1665)
Relative water intake	—	449 ± 376 mL/hour (55; 844)
Carbohydrate intake	497 ± 128 g/d (362; 632)	68 ± 76 g/race (−11; 148)
Relative carbohydrate intake	8.3 ± 1.8 g/kg/d (6.4; 10.2)	23 ± 22 g/hour (0; 46)
Protein intake	99.8 ± 29.2 g/d (69; 130)	11 ± 11 g/race (0; 22)
Relative protein intake	1.65 ± 0.31 g/kg/d (1.32; 1.97)	4.0 ± 3.2 g/hour (0; 46)
Fat intake	85.8 ± 30.8 g/d (53.6; 118.1)	0.7 ± 1.6 g/race (−1.0; 2.4)
Relative fat intake	1.4 ± 0.5 g/kg/d (0.85; 1.95)	0.3 ± 0.5 g/hour (−0.2; 0.8)

Data are given as mean ± SD (upper; lower 95% CI).

Water intake during the race was highly variable. As no data are available on sweat rates and water loss during mountain running in the altitude, it remains elusive whether the ingested amount of ~400 mL per hour is sufficient to prevent dehydration higher than 2% of body mass.

To our knowledge, this is the first time that data on anthropometry, $\dot{V}O_2$max, and prerace and race day nutrition in WMRA mountain runners are presented. However, the study has some limitations that need to be considered when interpreting the data. Firstly, the data were collected in three different pilot studies, which means that anthropometric, nutritional, and $\dot{V}O_2$max data were in part collected in different subjects. Secondly, the sample sizes in the two pilot studies with elite athletes only ($\dot{V}O_2$max measurements and nutritional data) are very limited with three and six athletes and represent a case series rather than a study. Therefore, this first set of data presented here is preliminary and should be extended in further studies including larger number of subjects and coherent data sets in all subgroups. However, it is difficult to motivate international elite athletes to participate in studies, where data are collected shortly before or even during the race. In addition, the number of mountain runners is much smaller than in Olympic sports or in sports that are commonly exerted in the general population such as cycling.

Thirdly, there are some issues about the data collection of the race day nutrition. Here, a recall for the race time only was used, where athletes had to record their estimated intake as soon as possible after the race. A recall of the dietary intake during the race might be biased by the remembrance of the individuals.

5. Conclusions

In conclusion, little data are available on nutrition and anthropometry of elite mountain runners. Preliminary date of this study shows that carbohydrate intake of mountain runners before and during the race is low when compared to distance runners. Therefore, increase of carbohydrate intake may increase performance in mountain running, but this can only be hypothesized on the basis of data presented here. Energy intake was rather low, which might reflect the challenge to find a balance between low body mass and adequately fueled glycogen stores. More studies are needed to create reliable nutrition recommendations in terms of performance optimization for elite mountain runners.

Conflict of Interests

The authors declare that there is no conflict of interests regarding the publication of this paper.

Acknowledgment

The authors thank Juliane Heydenreich (Technical University of Munich, Germany) for her valuable assistance during the analysis of the dietary intake data.

References

[1] T. R. Ackland, T. G. Lohman, J. Sundgot-Borgen et al., “Current status of body composition assessment in sport: review and position statement on behalf of the Ad Hoc research working group on body composition health and performance, under the auspices of the I.O.C. medical commission,” *Sports Medicine*, vol. 42, no. 3, pp. 227–249, 2012.

[2] F. Scharhag-Rosenberger, A. Carlsohn, M. Cassel, F. Mayer, and J. Scharhag, “How to test maximal oxygen uptake: a study on timing and testing procedure of a supramaximal verification test,” *Applied Physiology, Nutrition and Metabolism*, vol. 36, no. 1, pp. 153–160, 2011.

[3] A. J. Braakhuis, K. Meredith, G. R. Cox, W. G. Hopkins, and L. M. Burke, “Variability in estimation of self-reported dietary intake data from elite athletes resulting from coding by different sports dietitians,” *International Journal of Sport Nutrition and Exercise Metabolism*, vol. 13, no. 2, pp. 152–165, 2003.

[4] M. Burtscher, H. Förster, and J. Burtscher, “Superior endurance performance in aging mountain runners,” *Gerontology*, vol. 54, no. 5, pp. 268–271, 2008.

[5] B. Knechtle, P. Knechtle, and T. Rosemann, “Race performance in male mountain ultra-marathoners: anthropometry or training?” *Perceptual and Motor Skills*, vol. 110, no. 3, pp. 721–735, 2010.

[6] A. L. Arrese and E. S. Ostáriz, “Skinfold thicknesses associated with distance running performance in highly trained runners,” *Journal of Sports Sciences*, vol. 24, no. 1, pp. 69–76, 2006.

[7] H. O’Connor, T. Olds, and R. J. Maughan, “Physique and performance for track and field events,” *Journal of Sports Sciences*, vol. 25, supplement 1, pp. S49–S60, 2007.

[8] K. J. Cureton, P. B. Sparling, B. W. Evans, S. M. Johnson, U. D. Kong, and J. W. Purvis, "Effect of experimental alterations in excess weight on aerobic capacity and distance running performance," *Medicine and Science in Sports and Exercise*, vol. 10, no. 3, pp. 194–199, 1978.

[9] K. J. Cureton and P. B. Sparling, "Distance running performance and metabolic responses to running in men and women with excess weight experimentally equated," *Medicine and Science in Sports and Exercise*, vol. 12, no. 4, pp. 288–294, 1980.

[10] R. M. Malina, "Anthropometric correlates of strength and motor performance," *Exercise and Sport Sciences Reviews*, vol. 3, pp. 249–274, 1975.

[11] A. E. Minetti, C. Moia, G. S. Roi, D. Susta, and G. Ferretti, "Energy cost of walking and running at extreme uphill and downhill slopes," *Journal of Applied Physiology*, vol. 93, no. 3, pp. 1039–1046, 2002.

[12] K. Koehler, F. Huelsemann, M. De Marees, B. Braunstein, H. Braun, and W. Schaenzer, "Case study: simulated and real-life energy expenditure during a 3-week expedition," *International Journal of Sport Nutrition and Exercise Metabolism*, vol. 21, no. 6, pp. 520–526, 2011.

[13] C. T. M. Davies and M. W. Thompson, "Aerobic performance of female marathon and male ultramarathon athletes," *European Journal of Applied Physiology and Occupational Physiology*, vol. 41, no. 4, pp. 233–245, 1979.

[14] L. M. Burke, "Nutrition strategies for the marathon: fuel for training and racing," *Sports Medicine*, vol. 37, no. 4-5, pp. 344–347, 2007.

[15] K. Katayama, K. Goto, K. Ishida, and F. Ogita, "Substrate utilization during exercise and recovery at moderate altitude," *Metabolism*, vol. 59, no. 7, pp. 959–966, 2010.

[16] B. Braun, G. E. Butterfield, S. B. Dominick et al., "Women at altitude: changes in carbohydrate metabolism at 4,300-m elevation and across the menstrual cycle," *Journal of Applied Physiology*, vol. 85, no. 5, pp. 1966–1973, 1998.

[17] N. R. Rodriguez, N. M. DiMarco, and S. Langley, "Position of the American dietetic association, dietitians of Canada, and the american college of sports medicine: nutrition and athletic performance," *Journal of the American Dietetic Association*, vol. 109, no. 3, pp. 509–527, 2009.

[18] L. M. Burke, J. A. Hawley, S. H. S. Wong, and A. E. Jeukendrup, "Carbohydrates for training and competition," *Journal of Sports Sciences*, vol. 29, no. 1, pp. S17–S27, 2011.

[19] S. J. Oliver, P. Golja, and J. H. MacDonald, "Carbohydrate supplementation and exercise performance at high altitude: a randomized controlled trial," *High Altitude Medicine and Biology*, vol. 13, no. 1, pp. 22–31, 2012.

14

Subclinical Iodine Deficiency among Pregnant Women in Haramaya District, Eastern Ethiopia: A Community-Based Study

Haji Kedir,[1] Yemane Berhane,[2] and Alemayehu Worku[3]

[1] *College of Health and Medical Sciences, Haramaya University, P.O. Box 235, Harar, Ethiopia*
[2] *Addis Continental Institute of Public Health, Addis Ababa, Ethiopia*
[3] *School of Public Health, Addis Ababa University, P.O. Box 26751/1000, Addis Ababa, Ethiopia*

Correspondence should be addressed to Haji Kedir; hajikedir2007@yahoo.com

Academic Editor: C. S. Johnston

Background. Iodine deficiency in pregnancy is a worldwide problem. This study aimed to assess prevalence and predictors of subclinical iodine deficiency among pregnant women in Haramaya district, eastern Ethiopia. *Methods.* A cross-sectional, community-based study was conducted on 435 pregnant women existing in ten randomly selected rural kebeles (kebele is the smallest administrative unit in Ethiopia). Data on the study subjects' background characteristics, dietary habits, and gynecological/obstetric histories were collected via a structured questionnaire. UIC of <150 μg/L defined subclinical iodine deficiency. Data were analyzed by Stata 11. A multivariable logistic regression was used to identify the predictors of subclinical iodine deficiency. *Results.* The median urinary iodine concentration (MUIC) was 58.1 μg/L and 82.8% of the women who had subclinical iodine deficiency. The risk of subclinical iodine deficiency was reduced by the use of iodized salt (AOR = 0.13) and by intake of milk twice a month or more (AOR = 0.50), but it was increased by maternal illiteracy (AOR = 3.52). *Conclusion.* Iodine nutritional status of the pregnant women was poor. This shows that women and their children are exposed to iodine deficiency and its adverse effects. Thus, they need urgent supplementation with iodine and improved access to and intake of iodized salt and milk during pregnancy.

1. Background

Iodine is an essential nutrient which is used in the synthesis of thyroid hormones that are important for the normal development of the nervous system [1]. Its inadequate intake brings iodine deficiency disorders (IDD) and hypothyroidism, either clinical or subclinical, is one of them [2]. Previous researchers have reported that during pregnancy, mild iodine deficiency causes maternal hypothyroxinemia that is related to mild and subclinical cognitive and psychomotor deficit in neonates, infants, and children [3]. Several others have shown that severe iodine deficiency during pregnancy increases the risk of spontaneous abortion, stillbirth, reduced intelligence, neurological cretinism, poor cognitive functions, and delayed psychomotor developments [1, 2, 4, 5].

Nearly two billion (28%) of the world's population, of whom more than 321 million (39%) are Africans, are at risk of insufficient iodine intake [6]. Based on the global data of iodine nutrition, Ethiopia is categorized among moderately iodine deficient countries [6]. Similarly, Ethiopia is one of the African countries with the highest prevalence of IDD and with the weakest program to prevent IDD [7]. From 34.5% to 37% of the childbearing women in the country have goiter [8].

Because of the higher production and fetal transfer of maternal thyroid hormone, fetal iodine transfer, and raised renal iodine clearance [9], pregnant women need more iodine than their counterparts do. They are required to consume 250 μg of iodine a day [10]. However, they hardly meet their requirement solely from the usual diets [10, 11].

Universal salt iodization (USI) is the most cost effective strategy recommended to eliminate IDD [12, 13]. However, more than one-third of the African households do not have access to iodized salt [14]. Only 20% of the Ethiopian households use adequately iodized salt [15]. Thus, iodine deficiency has been identified as a serious public health problem for the Ethiopian population. As a result a national regulatory provision for mandatory universal salt iodization was offered in 2011.

Still, hardly any of the previous studies done on the problem in Ethiopia have considered subclinical iodine deficiency in pregnant women. Therefore, this study aimed to determine median urinary iodine concentration, prevalence of subclinical iodine deficiency, and factors associated with iodine deficiency among the pregnant women in rural communities of Haramaya district of eastern Ethiopia.

2. Methods

2.1. Study Setting, Design, and Period. A community-based cross-sectional study was done in Haramaya district, eastern Ethiopia, from March 16 to 29, 2012. The district is 1400–2340 meters above sea level. It is divided into 4 semiurban kebeles and 33 rural kebeles (the lowest administrative units) and has a population of about 271,018 people, of whom more than 96% are Oromo in ethnicity and Muslim. We have described the study area in more detail in the previously published article [16].

2.2. Study Population, Sample Size, and Sampling. Pregnant women in the randomly selected rural kebeles of the district were the study population. Pregnant women were identified by self-report, from the report of health extension workers in the kebeles and urine testing when pregnancy was doubtful. The sample size was estimated by assuming a 50% subclinical iodine deficiency, a 95% confidence interval, a 5% margin of error, and a 15% nonresponse rate and this yielded a sample size of 443. According to Andersen et al., a sample size of 500 subjects is adequate to assess iodine nutrition of the population from spot samples [17].

This study is a subsample of a prospective cohort study of the effects of maternal nutrition on birth outcomes. It was based on random subsample of ten of the twelve kebeles selected for the cohort study. To select the kebeles (primary sampling units), all the rural kebeles in the district (n = 33) were listed and each was assigned a unique number. Then, a simple random sampling technique was applied to select the twelve kebeles for the cohort study. Finally, the list of the twelve kebeles was considered as a sampling frame and ten kebeles were randomly selected from it. Cognizing the little difference between the larger estimated sample size (n = 500) and the number of the pregnant women in the ten selected kebeles (n = 525), we recruited all of them.

2.3. Data Collection Methods and Procedures. Eight college graduates collected the data and two public health professionals and the principal investigator supervised the fieldwork. The respondents' sociodemographic, reproductive history and dietary characteristics were obtained via a pretested and interviewer-administered questionnaire. The items included in the questionnaire were taken from the Ethiopian Health and Demographic Survey (EDHS) and were adapted to suit the study context.

A teaspoon of salt sample was taken from each household of the respondent and tested for contents of iodine, using a rapid test kit distributed by UNICEF for the purpose of assessing household salt iodine content. The rapid test kit comprises test solution and a color chart. We put a drop of the test solution on each salt sample. Then, salt samples that immediately turned into a purple blue color of any intensity as shown on the color chart after putting a drop of test solution were classified as containing iodine and those not remaining unchanged after putting a drop of test solution were classified as not containing iodine.

A 10 mL urine sample was taken from each study subject in wide-opened plastic caps covered by an opaque paper bag and transferred into a labeled, clean tightly sealed plastic tubes that were free from iodine or any other chemical to avoid leakage and cross-contaminations with iodine from other sources. The urine samples were kept in a cold box and transported to iodine laboratory at Ethiopian Health and Nutrition Research Institute (EHNRI) under the cold chain. In the national laboratory of food and nutrition research at ENHRI, duplicates of each urine sample brought from each pregnant women were prepared and the determination of urinary iodine concentration was made using the duplicate samples. The Sandell-Kolthoff reaction method which is recommended by World Health Organization (WHO), United Nations Children's Fund (UNICEF), and International Council for the Control of Iodine Deficiency Disorders (ICCIDD) was used to determine the UIC [18]. The method is described in further detail by the publication of WHO [18].

2.4. Statistical Analysis. Data were double-entered and validated by EpiData Version 3.1. Stata 11 and SPSS V. 16 were used to analyze the data. Using Mann-Whitney U test or Kruskal-Wallis test, we compared the median and the UIC between groups of categorical independent variables. Through bivariate and multivariable logistic regressions, the predictors of subclinical iodine deficiency (UIC <150 μg/L) were identified. Variables were included into the multivariable model based on the existing literature about their suspected effects on subclinical iodine deficiency and its predictors. In order to avoid confounding factors, the enter method of logistic regression was used and the risk estimates were adjusted for all the variables entered. A two-sided P value of <0.05 was considered to declare statistical significance.

WHO recommends urinary iodine concentrations of <150, 150–249, 250–499, and ≥500 μg/L that should be used to indicate insufficient, adequate, more than adequate, and excessive levels of iodine intake in population of pregnant women [19]. In this study, UIC <150 μg/L defined subclinical iodine deficiency and value ≥150 μg/L absence of subclinical iodine deficiency. Median UIC and subclinical iodine deficiency were the dependent variables. The group with <150 μg/L were further categorized into <20 μg/L (severe iodine deficiency), 20–49 μg/L (moderate iodine deficiency), and 50–149 μg/L (mild iodine deficiency) [20]. The independent variables and covariates examined for their association with subclinical iodine deficiency were maternal age, educational status, family size, possession of milk cows, consumption of milk, consumption of cabbage, household use of iodized salt, prenatal visit, trimester of pregnancy, and number of pregnancies.

2.5. Ethics. The Institutional Review Board (IRB) of Haramaya University and the National Research Ethics Review

TABLE 1: Characteristics of pregnant women in Haramaya, eastern Ethiopia, April, 2012.

Variable	Total Frequency (%)	No iodine deficiency Frequency (%)	Iodine deficiency Frequency (%)	P value*
Age group				0.100
18–24 years	126 (28.97)	18 (14.29)	108 (85.71)	
25–34 years	241 (55.40)	39 (16.18)	202 (83.82)	
35–49 years	68 (15.63)	18 (26.47)	50 (73.53)	
Educational status				0.213
Literate	31 (7.13)	8 (25.81)	23 (74.19)	
Illiterate	404 (92.87)	67 (16.58)	337 (83.42)	
Family size				0.636
<5 persons	41 (18.06)	41 (18.06)	186 (81.94)	
≥5 persons	34 (16.35)	34 (16.35)	174 (83.65)	
Previous pregnancies				0.281
0	87 (20.00)	12 (13.79)	75 (86.21)	
1–3	280 (64.37)	47 (16.79)	233 (83.21)	
≥4	68 (15.63)	16 (23.53)	52 (76.47)	
Pregnancy trimester				0.694
First	52 (11.95)	9 (17.31)	43 (82.69)	
Second	169 (38.85)	26 (15.38)	143 (84.62)	
Third	214 (49.20)	40 (18.69)	174 (81.31)	
Possessed milk cows				0.189
No cow	171 (39.31)	26 (15.20)	145 (84.80)	
One cow	124 (28.51)	28 (22.58)	96 (77.42)	
Two or more cows	140 (32.18)	21 (15.00)	119 (85.00)	
Prenatal visit				0.440
No	112 (25.75)	22 (19.64)	90 (80.36)	
Yes	323 (74.25)	53 (16.41)	270 (83.59)	
Cabbage consumption				0.019
<1 serving/week	334 (76.78)	65 (19.46)	269 (80.54)	
≥1 serving/week	101 (23.22)	10 (9.90)	91 (90.10)	
Household salt				<0.0001
Not iodized	406 (93.33)	62 (15.27)	344 (84.73)	
Iodized	29 (6.67)	13 (44.83)	16 (55.17)	
Consumption of milk				0.500
<3 times/month	230 (52.87)	37 (16.09)	193 (83.91)	
≥3 times/month	205 (47.13)	38 (18.54)	167 (81.46)	

* P values are based on likelihood ratio chi-square test.

Committee of Ethiopia reviewed and approved the protocol. All the study participants provided their written informed consent. The purposes, the data collection procedures, the risks, and the benefits of the research were explained to the eligible respondents before obtaining their informed consent.

3. Results

3.1. Sociodemographic Characteristics. Of the 525 pregnant women recruited, 435 were included in the analysis and the response rate was 82.8%. The remaining 90 pregnant women did not give their urine and they were not different from the women included in the analysis in their sociodemographic characteristics. Nearly all of the respondents were Oromo (99.5%) and Muslim (99.3%). Their mean age and duration of pregnancy were 27.0 (SD ± 5.9) years and 25.8 (SD ± 5.8) weeks, respectively. The characteristics of women participating in this study are detailed in Table 1.

Urinary Iodine Concentration and Associated Factors. The median UIC was 58.1 μg/L (interquartile range (IQR) = 21.4–111.1). The women in the households which used iodized salt more than those in the households that did not and women who consumed milk at least twice a month more than those who never did so had significantly higher median UIC (Mann-Whitney U test, $P < 0.05$) (Table 2). However, the median UIC did not show significant associations with the respondents' educational status, parity levels, trimester of pregnancy, and other sociodemographic characteristics studied (Table 2).

TABLE 2: Distribution of median urinary iodine concentration of pregnant women with selected variables, Haramaya, April 2012.

Variable	Percent	Median (μg/L)	Q1 (μg/L)	Q3 (μg/L)	P value
Age groups					0.355**
18–24 years	28.97	54.47	19.30	97.97	
25–34 years	55.40	68.66	25.56	151.89	
35 years or older	15.63	68.66	25.56	151.89	
Education					0.776*
Literate	7.13	64.13	23.79	150.75	
Illiterate	92.87	57.80	20.47	109.68	
Number of births					0.168**
0	20.00	49.37	18.84	96.79	
1–3	64.37	56.78	20.34	111.84	
>4	15.63	68.89	31.84	138.72	
Trimester					0.743**
First	11.95	61.87	24.84	114.25	
Second	38.85	58.07	27.33	94.00	
Third	49.20	55.91	13.86	113.79	
Prenatal visit					0.348*
No	25.75	42.84	18.99	124.32	
Yes	74.25	58.78	23.27	110.87	
Milk cows					0.971**
0	39.31	58.34	21.70	103.04	
1	28.51	54.20	20.92	126.87	
2+	32.18	57.95	19.47	113.18	
Milk intake					0.023*
<3 times/month	52.87	47.76	15.70	104.22	
≥3 times/month	47.13	69.14	27.11	113.98	
Cabbage use					0.051*
<1/week	76.78	58.42	23.35	114.28	
≥1/week	23.22	52.10	8.23	89.67	
Household salt					0.018*
Not iodized	93.33	56.57	20.37	103.56	
Iodized	6.67	114.08	28.40	195.64	
Family size					0.669*
<5 persons	52.18	58.07	19.45	103.31	
≥5 persons	47.82	57.93	23.46	113.45	

*P values are based on Mann-Whitney U test; **P values are based on Kruskal-Wallis test; Q1: first quartile (25th percentile); Q3: third quartile (75th percentile).

Subclinical Iodine Deficiency and Its Predictors. In this study, 360 (82.8%) respondents had subclinical iodine deficiency. Of the total respondents, 40 (9.2%), 35 (8.0%), and none had adequate, more than adequate, and excessive iodine intakes, respectively. When the levels of subclinical iodine deficiency were further categorized, 105 (24.1%), 91 (20.9%), and 164 (37.7%) of all respondents had UIC of <20 μg/L (severe iodine deficiency), 20–49 μg/L (moderate iodine deficiency), and 50–149 μg/L (mild iodine deficiency), respectively.

In the bivariate analysis, the risk of subclinical iodine deficiency was significantly low among the women who were 35 to 49 years of age, who consumed milk at least three times a month, who eat cabbage once a week or more frequently, and who were from households which used iodized salt (Table 3).

In a multivariable analysis, the factors significantly associated with subclinical iodine deficiency at the bivariate still remained significant. Thus, the respondents who used iodized salt had 87% lower risk of subclinical iodine deficiency than their counterparts who did not and those who consumed milk at least twice a month had a 50% lower risk of the deficiency (Table 3). The risk of subclinical iodine deficiency increased by more than threefold in illiterate women than literate ones and in women within a family size of five or more than in those within family size below five, while the magnitude of subclinical iodine deficiency was significantly reduced with increased parity level (Table 3).

4. Discussion

The study showed a high prevalence of subclinical iodine deficiency, a low median UIC, and a low proportion of households with iodized salt, while consuming iodized salt

TABLE 3: Factors associated with subclinical iodine deficiency in logistic regression analysis among pregnant women, Haramaya, April 2012.

Variable	Crude OR (95% CI)	*P* value*	Adjusted OR (95% CI)†	*P* value*
Age group				
18–24 years	1		1	
25–34 years	0.86 (0.47, 1.58)	0.634	0.59 (0.29, 1.19)	0.138
35–49 years	0.46 (0.22, 0.96)	0.040	0.30 (0.12, 0.75)	0.010
Educational status				
Literate	1		1	
Illiterate	1.75 (0.75, 4.08)	0.195	3.52 (1.30, 9.55)	0.013
Family size				
<5 persons	1		1	
≥5 persons	1.13 (0.68, 1.86)	0.636	3.22 (1.57, 6.58)	0.001
Previous pregnancies				
0	1		1	
1–3	0.79 (0.40, 1.57)	0.508	0.36 (0.15, 0.82)	0.016
≥4	0.52 (0.23, 1.19)	0.122	0.23 (0.07, 0.74)	0.013
Trimester of pregnancy				
First	1		1	
Second	1.15 (0.50, 2.64)	0.740	0.97 (0.38, 2.46)	0.942
Third	0.91 (0.41, 2.02)	0.817	0.78 (0.32, 1.89)	0.582
Possessed milk cows				
No cow	1		1	
One cow	0.61 (0.34, 1.11)	0.108	0.41 (0.21, 0.81)	0.010
Two or more cows	1.02 (0.54, 1.90)	0.960	0.78 (0.38, 1.61)	0.503
Prenatal visit				
No	1		1	
Yes	1.25 (0.72, 2.16)	0.436	1.85 (0.92, 3.70)	0.083
Cabbage consumption				
<3 times/month	1		1	
≥3 times/month	2.20 (1.08, 4.46)	0.029	3.04 (1.37, 6.77)	0.006
Household salt				
Not iodized	1		1	
Iodized	0.22 (0.10, 0.48)	<0.001	0.13 (0.05, 0.32)	<0.0001
Consumption of milk				
<3 times/month	1		1	
≥3 times/month	0.84 (0.51, 1.39)	0.500	0.50 (0.27, 0.93)	0.029

* *P* values are based on likelihood chi-square test; †adjusted for all variables shown in the table; OR: odds ratio; CI: confidence interval.

and milk that were predictors of higher UIC and were associated with reduced prevalence of subclinical iodine deficiency. Lack of education among pregnant women was associated with a threefold increased risk of subclinical iodine deficiency.

Recently, other researchers working in the Sidama Zone of southern Ethiopia on 172 pregnant women have witnessed that iodized salt is not used in higher than 90% of the households and 60% of the respondents had UIC lower than 20 μg/L and the median UIC of 15 μg/L [21]. Compared to the study from southern Ethiopia, the present study observed lower proportion of pregnant women with UIC less than 20 μg/L and a higher median UIC [21]. Nevertheless, both studies indicate that iodine deficiency is unacceptably high in Ethiopian pregnant women. The differences between the findings of our study conducted in eastern Ethiopia and the study from the southern Ethiopia could be explained by sociocultural and ecological differences and the compositions of the study subjects enrolled in the two studies. The variations could also indicate the need of regional studies with detailed investigation of risk factors specific to particular locality or a nationwide study with representative coverage of different regions and contextual risk factors.

High level of iodine deficiency has been witnessed by previous clinical and subclinical assessments made on samples of school-aged children [22, 23]. For example, a median UIC of 56 μg/L and 34.2 μg/L is reported from southwestern and southern Ethiopia, respectively [22, 23]. Despite the differences in the study subjects, the findings from the present study were confirmations of the degree of iodine deficiency in Ethiopia.

Like other authors, we observed that household salt iodization significantly reduces the risk of iodine deficiency [24, 25]. However, low proportion of iodized salt utilization at the household level reported in our study calls for urgent need of coordinated efforts. Our study and other previous studies demonstrated that milk is an important source of iodine [25, 26], and this strengthens the promotion of the intake of milk and other dairy products during or before prenatal period and of the importance of possessing milk cows at the household level in the communities like ours that do not have ease of access to fish and other sea foods. In our study, the illiterate study subjects were more likely to be iodine deficient than the literate ones. It has also been demonstrated by other authors that better educated pregnant women are more likely to consume food with higher dietary quality index [27]. The decrease of the UIC in advanced stages of pregnancy that have been demonstrated by several others [28–30] was not confirmed by the present study. This could be due to the small number of respondents with the first trimester of pregnancy in the present study.

Unlike the understanding of the negative effect of repeated pregnancy on maternal nutritional status, in this study the risk of subclinical iodine deficiency decreased with an increase in the number of pregnancy experiences. This could be explained by the possibility that pregnant women with more pregnancy experiences could receive nutrition related information from various sources during their previous pregnancy experience and as a result they may have higher iodine intake [31].

Our study might have had some limitations. The questionnaire based assessment of dietary iodine intake and other data might not be as accurate as it should be. Moreover, as UIC can only indicate the present iodine nutrition, the findings might not reflect levels of chronic iodine deficiency. To get an insight on long standing iodine deficiency, it would have been better if the prevalence of goiter had also been assessed. Furthermore, UIC data cannot tell thyroid function in the group of women studied, which could only be possible with determination of thyroid hormones. The data on thyroid hormones and prevalence of goiter were not possible to obtain with the resource available for the present study. Nevertheless, the data are important for better understanding of the iodine deficiency in this community. Another limitation is that this study is a cross-sectional study conducted during the hot season which might lead to the underestimation of iodine deficiency and overestimation of median urinary iodine concentration as a result of seasonal variation in urinary iodine concentration that increases during the dry hot season [32]. In addition, the household salt iodization test was qualitative; it did not provide iodine concentration. Even though it has been recommended that salt samples identified as containing iodine based on the color change should be further examined through iodiometric titration for the quantification of iodine content and evaluation of its adequacy [18], this was not done in the present study, because only a few samples were found to contain iodine based on the rapid test and further classification of such a small number would make further analysis impossible. Had it not been for the resource limitation, we could have considered design effect to account for the cluster based sampling. However, the effect of this problem could be minimal due to the inclusion of all pregnant women that were identified in each selected kebele.

Regardless of its limitations, the present study contributed to the global body of knowledge by establishing the Ethiopian data on UIC and iodine nutrition among pregnant women. It is a contribution of Ethiopian context to the global body of knowledge regarding iodine deficiency during pregnancy [33, 34].

This study also revealed high prevalence of subclinical iodine deficiency and the scarcity of iodized salt usage among the households in eastern Ethiopia using the two simple and practical methods recommended [35]. Therefore, the findings have implications for public policy, health service provision, self-care, and future research work.

The legislation of Universal Salt Iodization (USI) law is an indication of political commitment and is a corner stone in the prevention of iodine deficiency, but it may not be adequate enough as a self-standing strategy to meet the goal. A number of previous studies have reported persistent occurrence of IDD in pocket geographical locations of countries, regardless of the USI law in place [36]. The same authors witnessed that public awareness about the importance of iodized salt and appropriate handlings of salt at household level are also important factors [36].

The level of subclinical iodine deficiency reported here raises policy concerns for effective interventions. This study could serve the policy on salt iodization to monitor progress and implementation. The health care workers can use the findings to initiate appropriate interventions during prenatal services and to counsel clients to improve their iodized salt and milk intake. Based on international criteria for intervention [37], the findings of the present study strongly indicate the need of iodine supplementation to the pregnant women in the area.

5. Conclusion

There was moderate to mild iodine deficiency among the pregnant women in the study area. The poor iodine nutritional status of the women resulted from low iodine intake and this in turn was caused by the scarcity of iodized salt utilization in this community. Therefore, there is an urgent need for iodine supplementation to the pregnant women in the setting. Enhancing access to iodized salt and the promotion of its health benefits are priority interventions that need serious attention. Interventions should also focus on young women, on women with no education, and on those with first time pregnancy. Encouraging milk consumption and improving household livestock raring, with particular emphasis to milk cows is also vital.

Abbreviations

AOR: Adjusted odds ratio
CI: Confidence interval
EDHS: Ethiopian Health and Demographic Survey
EHNRI: Ethiopian Health and Nutrition Research Institute
ICCIDD: International Council for Control of Iodine Deficiency Disorders

IDD: Iodine deficiency disorders
UIC: Urinary iodine concentration
UNICEF: United Nations Children's Fund
USI: Universal salt iodization
WHO: World Health Organization.

Conflict of Interests

The authors did not receive payments, funding, or salary from any organization in relation to the work and publication of this paper in the past five years and there will also be no financial support to be received in the future. There is not any organization affected positively or negatively by the publication of this paper. There are not any competing interests related with patents of the content of the paper. The authors also declare that there are not any other financial or nonfinancial competing interests.

Authors' Contribution

All authors, Haji Kedir, Yamane Berhane, and Alemayehu Worku, conceptualized the research question, monitored and managed fieldwork, analyzed and interpreted the findings, and written the paper.

Acknowledgments

Haramaya University provided financial and other supports for this research project. The authors thank EHNRI for laboratory facility. The authors are grateful to the pregnant women who sacrificed their time. The authors acknowledge Eastern Hararghe Zone Health Department for its cooperation and for providing them with cold boxes and icepacks used for keeping and transporting the urine samples. The authors got enabling cooperation from Haramaya District Administration, District Health Office, and the Administrations of the Selected Kebeles during the fieldwork.

References

[1] C. Yarrington and E. N. Pearce, "EN: iodine and pregnancy," *Journal of Thyroid Research*, vol. 2011, Article ID 934104, 8 pages, 2011.

[2] M. B. Zimmermann, "Iodine deficiency," *Endocrine Reviews*, vol. 30, pp. 376–408, 2009.

[3] C. Trumpff, J. De Schepper, J. Tafforeau, H. Van Oyen, J. Vanderfaeillie, and S. Vandevijvere, "Mild iodine deficiency in pregnancy in Europe and its consequences for cognitive and psychomotor development of children: a review," *Journal of Trace Elements in Medicine and Biology*, vol. 27, no. 3, pp. 174–183, 2013.

[4] M. B. Zimmermann, "The effects of iodine deficiency in pregnancy and infancy," *Paediatric and Perinatal Epidemiology*, vol. 26, supplement 1, pp. 108–117, 2012.

[5] A. Melse-Boonstra and N. Jaiswal, "Iodine deficiency in pregnancy, infancy and childhood and its consequences for brain development," *Best Practice & Research Clinical Endocrinology & Metabolism*, vol. 24, no. 1, pp. 29–38, 2010.

[6] M. Andersson, V. Karumbunathan, and M. B. Zimmermann, "Global iodine status in 2011 and trends over the past decade," *Journal of Nutrition*, vol. 142, no. 4, pp. 744–750, 2011.

[7] I. Hussein and V. Assey, *Elimination of Iodine Deficiency through Salt Iodization in Ethiopia*, ICCIDD, Addis Ababa, Ethiopia, 2012.

[8] C. Abuye and Y. Berhane, "The goitre rate, its association with reproductive failure, and the knowledge of iodine deficiency disorders (IDD) among women in Ethiopia: cross-section community based study," *BMC Public Health*, vol. 7, article 316, 2007.

[9] M. B. Zimmermann, "The role of iodine in human growth and development," *Seminars in Cell and Developmental Biology*, vol. 22, pp. 645–652, 2011.

[10] M. Andersson, B. D. Benoist, F. Delange, and J. Zupan, "WHO Secretariat on behalf of the participants to the Consultation: Prevention and control of iodine deficiency in pregnant and lactating women and in children less than 2-years-old: conclusions and recommendations of the Technical Consultation," *Public Health Nutrition*, vol. 10, no. 12, pp. 1606–1611, 2007.

[11] M. B. Zimmermann, "Iodine deficiency in pregnancy and the effects of maternal iodine supplementation on the offspring: a review," *The American Journal of Clinical Nutrition*, vol. 89, supplement, pp. 668S–672S, 2009.

[12] *UNICEF: The State of the World's Children 2009: Maternal and Newborn Health*, United Nations Children's Fund, New York, NY, USA, 2008.

[13] V. D. Assey, S. Peterson, S. Kimboka et al., "Tanzania national survey on iodine deficiency: impact after twelve years of salt iodation," *BMC Public Health*, vol. 9, article 319, 2009.

[14] M. Andersson, Bd. Benoist, and L. Rogers, "Epidemiology of iodine deficiency: salt iodisation and iodine status," *Best Practice & Research: Clinical Endocrinology*, vol. 24, pp. 1–11, 2010.

[15] Central Statistical Agency [Ethiopia], "Ethiopia Demographic and Health Survey 2005," Addis Ababa, Ethiopia and Calverton, Md, USA: Central Statistical Agency and ORC Macro. 2006.

[16] H. Kedir, Y. Berhane, and A. Worku, "Khat chewing and restrictive dietary behaviors are associated with anemia among pregnant women in high prevalence rural communities in Eastern Ethiopia," *PLoS ONE*, vol. 8, no. 11, Article ID e78601, 2013.

[17] S. Andersen, J. Karmisholt, K. M. Pedersen, and P. Laurberg, "Reliability of studies of iodine intake and recommendations for number of samples in groups and in individuals," *British Journal of Nutrition*, vol. 99, no. 4, pp. 813–818, 2008.

[18] WHO, UNICEF, and ICCIDD, *Assessment of Iodine Deficiency Disorders and Monitoring their Elimination: A Guide for Programme Managers*, WHO, Paris, France, 3rd edition, 2007.

[19] WHO, *Urinary Iodine Concentrations for Determining Iodine Status Deficiency in Populations*, VMNIS, World Health Organization, Geneva, Switzerland, 2013.

[20] K. E. Charlton, L. Gemming, H. Yeatman, and G. Ma, "Suboptimal iodine status of Australian pregnant women reflects poor knowledge and practices related to iodine nutrition," *Nutrition*, vol. 26, no. 10, pp. 963–968, 2010.

[21] G. Ersino, H. Tadele, A. Bogale, C. Abuye, and B. J. Stoecker, "Clinical assessment of goiter and low urinary iodine concentration depict presence of severe iodine deficiency in pregnant Ethiopian women: a cross-sectional study in rural Sidama," *Ethiopian Medical Journal*, vol. 51, no. 2, pp. 133–141, 2013.

[22] Y. Mezgebu, A. Mossie, P. N. Rajesh, and G. Beyene, "Prevalence and severity of iodine deficiency disorder among children 6-12 years of age in shebe senbo district, jimma zone, southwest ethiopia," *Ethiopian Journal of Health Sciences*, vol. 22, no. 3, pp. 196–204, 2012.

[23] M. Girma, E. Loha, A. Bogale, N. Teyikie, and C. Abuye, "Iodine deficiency in primary school children and knowledge of iodine deficiency and iodized salt among caretakers in Hawassa Town: Southern Ethiopia," *Ethiopian Journal of Health Development*, vol. 26, no. 1, pp. 30–35, 2012.

[24] M. Murcia, M. Rebagliato, M. Espad et al., "Iodine intake in a population of pregnant women: INMA mother and child cohort study, Spain," *Journal of Epidemiology & Community Health*, vol. 64, no. 12, pp. 1094–1099, 2010.

[25] C. Mian, P. Vitaliano, D. Pozza et al., "Iodine status in pregnancy: role of dietary habits and geographical origin," *Clinical Endocrinology*, vol. 70, pp. 776–780, 2009.

[26] M. Alvarez-Pedrerol, N. Ribas-Fitó, R. García-Esteban et al., "Iodine sources and iodine levels in pregnant women from an area without known iodine deficiency," *Clinical Endocrinology*, vol. 72, no. 1, pp. 81–86, 2010.

[27] L. M. Bodnar and A. M. Siega-Riz, "A diet quality index for pregnancy detects variation in diet and differences by sociodemographic factors," *Public Health Nutrition*, vol. 5, no. 6, pp. 801–809, 2002.

[28] G. Stilwell, P. J. Reynolds, V. Parameswaran, L. Blizzard, T. M. Greenaway, and J. R. Burgess., "The influence of gestational stage on urinary iodine excretion in pregnancy," *Journal of Clinical Endocrinology and Metabolism*, vol. 93, pp. 1737–1742, 2008.

[29] E. Ainy, A. Ordookhani, M. Hedayati, and F. Azizi, "Assessment of intertrimester and seasonal variations of urinary iodine concentration during pregnancy in an iodine-replete area," *Clinical Endocrinology*, vol. 67, pp. 577–581, 2007.

[30] L. Habimana, K. E. Twite, P. Wallemacq et al., "Iodine and iron status of pregnant women in Lubumbashi, Democratic Republic of Congo," *Public Health Nutrition*, vol. 16, no. 8, pp. 1362–1370, 2013.

[31] A. L. Brantsæter, M. H. Abel, M. Haugen, and H. M. Meltzer, "Risk of suboptimal iodine intake in pregnant norwegian women," *Nutrients*, vol. 5, no. 2, pp. 424–440, 2013.

[32] K. J. Schulze, K. P. West Jr., L. A. Gautschi et al., "Seasonality in urinary and household salt iodine content among pregnant and lactating women of the plains of Nepal," *European Journal of Clinical Nutrition*, vol. 57, pp. 969–976, 2003.

[33] A. O. Kutlu and C. Kara, "Iodine deficiency in pregnant women in the apparently iodine-sufficient capital city of Turkey," *Clinical Endocrinology*, vol. 77, pp. 615–620, 2012.

[34] J. Fisher, T. Tran, B. Biggs et al., "Iodine status in late pregnancy and psychosocial determinants of iodized salt use in rural northern Viet Nam," *Bulletin of the World Health Organization*, vol. 89, pp. 813–820, 2011.

[35] B. S. Hetzel, "The development of a global program for the elimination of brain damage due to iodine deficiency," *Asia Pacific Journal of Clinical Nutrition*, vol. 21, no. 2, pp. 164–170, 2012.

[36] S. Yadav, S. K. Gupta, M. M. Godbole et al., "Persistence of severe iodine-deficiency disorders despite universal salt iodization in an iodine-deficient area in northern India," *Public Health Nutrition*, vol. 13, no. 3, pp. 424–429, 2009.

[37] M. Andersson, B. De Benoist, F. Delange, J. Zupan, and WHO Secretariat, "Prevention and control of iodine deficiency in pregnant and lactating women and in children less than 2-years-old: conclusions and recommendations of the technical consultation," *Public Health Nutrition*, vol. 10, no. 12, pp. 1606–1611, 2007.

15

Exercise and Dietary-Mediated Reductions in Postprandial Lipemia

Eric P. Plaisance and Gordon Fisher

Department of Human Studies, Exercise and Nutritional Physiology Laboratory, University of Alabama at Birmingham, Birmingham, AL 35294, USA

Correspondence should be addressed to Eric P. Plaisance; plaisep@uab.edu

Academic Editor: C. S. Johnston

Postprandial hyperlipemia produces long-term derangements in lipid/lipoprotein metabolism, vascular endothelial dysfunction, hypercoagulability, and sympathetic hyperactivity which are strongly linked to atherogenesis. The purpose of this review is to (1) provide a qualitative analysis of the available literature examining the dysregulation of postprandial lipid metabolism in the presence of obesity, (2) inspect the role of adiposity distribution and sex on postprandial lipid metabolism, and (3) examine the role of energy deficit (exercise- and/or energy restriction-mediated), isoenergetic low-carbohydrate diets, and omega-3 (n-3) fatty acid supplementation on postprandial lipid metabolism. We conclude from the literature that central adiposity primarily accounts for sex-related differences in postprandial lipemia and that aerobic exercise attenuates this response in obese or lean men and women to a similar extent through potentially unique mechanisms. In contrast, energy restriction produces only mild reductions in postprandial lipemia suggesting that exercise may be superior to energy restriction alone as a strategy for lowering postprandial lipemia. However, isoenergetic very low-carbohydrate diets and n-3 fatty acid supplementation reduce postprandial lipemia indicating that macronutrient manipulations reduce postprandial lipemia in the absence of energy restriction. Therefore, interactions between exercise/energy restriction and alterations in macronutrient content remain top priorities for the field to identify optimal behavioral treatments to reduce postprandial lipemia.

1. Introduction

Reductions in vocational physical activity and the availability and consumption of energy-dense foods are often cited as primary culprits of the rising incidence of obesity observed throughout the world [1–6]. The increased incidence of obesity has presented both health-related and economic challenges which exceed $110 billion dollars per year in the US alone [7], a level which surpasses any other country in the world. One of the conventional consequences of obesity (especially centrally distributed adiposity) is an increase in triglyceride (TG)-rich lipoproteins (TRLs) and associated hypertriglyceridemia. TRLs (very low-density lipoprotein (VLDL) and chylomicrons (CM)) reduce cholesterol content of high-density lipoproteins (HDL) and decrease the size of low-density lipoproteins (LDL) which increase the propensity for vascular endothelial infiltration and oxidation [8]. Fasting HDL-C concentrations often account for a greater amount of variance in the risk of cardiovascular disease (CVD) compared to TGs leading to their dismissal as a primary risk factor for atherosclerosis [9–11].

While the contribution of fasting TGs and HDL-C to the overall risk of CVD remains controversial [12, 13], accumulating evidence suggests that exaggerated postprandial lipemia produces rapid derangements of lipid/lipoprotein metabolism, vascular endothelial dysfunction, hypercoagulability, and sympathetic hyperactivity that is strongly linked to atherogenesis [10, 14–20]. A randomized clinical trial of 602 men and women illustrated that asymptomatic carotid atherosclerosis was positively correlated with postprandial lipemia [21]. Despite comparable fasting lipids, male offspring of men with CVD had significantly higher serum TGs up to 12 hours following a high-fat meal, indicative of a delayed clearance of TGs [22]. Furthermore, postprandial hyperlipemia increased the number of myocardial infarctions by 40% for every 100 mg/dL increase in Physician's Health Study and Multiple Risk Factor Intervention Trials [17, 23].

The relationship between postprandial lipid metabolism and atherosclerosis is not surprising when one considers that the majority of individuals following a typical Western Diet consume 3 to 5 meals per day. Since each meal requires 6–12 hours to fully dissipate TGs in circulation, the implication is that most individuals spend over two-thirds of the day in a postprandial state with elevated TRLs (postprandial hyperlipidemia) [24, 25]. The capacity to regulate incoming chylomicrons from exogenous sources, tasked with counterregulation of the endogenous production and secretion of VLDL-TG, may offer a more valid investigation of the capacity of tissues to appropriately regulate lipid metabolism. Therefore, interventions which improve the capacity to regulate tissue and blood lipid metabolism following a meal would be expected to lower CVD risk.

The purpose of the current review is to (1) provide a qualitative analysis of the available literature examining the dysregulation of postprandial lipid metabolism in the presence of obesity, (2) inspect the role of adiposity distribution and sex on postprandial lipid metabolism, and (3) examine the role and mechanisms by which energy deficit produced via exercise and/or energy restriction, isoenergetic low-carbohydrate diets, and n-3 fatty acid supplementation improves postprandial lipid metabolism.

2. Obesity and Postprandial Lipemia

Systematic abnormalities in fasting and postprandial blood lipid and lipoprotein metabolism have been well documented in obesity and other conditions which produce hyperinsulinemia/insulin resistance [26]. Insulin resistance produces de novo lipogenesis, increases microsomal triglyceride transfer protein, and enhances intracellular apolipoprotein B48 stability in the intestine [27, 28]. Along with increased free fatty acid loading, increases in intestinal de novo lipogenesis increase the secretion of apolipoprotein B48 from enterocytes which increases the absorption and circulating concentrations of chylomicrons [28]. The increase in secretion is accompanied by a reduction in the fractional catabolic rate of apolipoprotein B48. Reductions in adipose tissue lipoprotein lipase (LPL) mass and activity [26, 29, 30] have also been observed in the presence of obesity-mediated insulin resistance. Furthermore, downregulation of LDL-receptor expression [31] leads to reductions in remnant lipolysis and removal [28]. Concurrently, obesity associated hepatic insulin resistance dampens the repression of VLDL-TG secretion normally observed in the presence of hepatic insulin sensitivity. The additional VLDL-TG competes with chylomicrons for hydrolysis by LPL and hepatic receptors, thereby reducing the overall clearance of serum TGs [28, 29].

2.1. Adiposity Distribution and Sex. Subsequent studies have attempted to address the precise role of adiposity distribution on postprandial lipemia. Overall, visceral adipose tissue distribution appears to be a better predictor of postprandial TG responses than body mass index (BMI) alone or gynoid distribution of adipose tissue [32–35]. Since the distribution and magnitude of visceral adiposity is greater in men than in women, some have proposed that these differences may be responsible for gender dimorphisms in postprandial lipemia and rates of CVD [34, 36].

In agreement with this interpretation, variation in postprandial TG responses between men and women was abrogated after controlling for visceral adiposity suggesting that the distribution of body fat is responsible for postprandial TG responses as opposed to sex differences [34, 37]. Furthermore, Mekki et al. [35] showed that android obese women had greater postprandial lipemia than women who were lean or exhibited a gynoid distribution of excess adiposity despite the presence of comparable BMI and fasting TGs. In addition, postprandial lipemia was similar between normolipidemic lean and gynoid obese groups, suggesting that excess adipose tissue distribution away from the viscera is insufficient to alter postprandial lipid metabolism. When men with impaired or normal glucose tolerance were individually matched for visceral adipose tissue accumulation, no significant differences were found in postprandial responses of all TRL-TG fractions between groups [38]. Additional studies [39, 40] showed similar results suggesting that visceral adipose tissue accumulation may be a more important determinant of postprandial lipemia than impaired glucose tolerance. These findings suggest that insulin resistance/hyperinsulinemia-mediated increases in visceral adiposity may be responsible for postprandial hyperlipemia. Future studies are needed to address cause and effect between visceral adiposity and insulin resistance *per se*. Taken together, abnormal postprandial lipemia can be detected in the presence of excess visceral adiposity with comparable BMI and in the absence of fasting hypertriglyceridemia, providing further support that postprandial lipemia is a more sensitive predictor of metabolic and CVD risk than fasting parameters.

3. Exercise and Postprandial Lipemia

3.1. Exercise Training Produces Effects on Postprandial Lipemia That Are Primarily Associated with Energy Expended during the Most Recent Bout of Exercise Performed. Cross-sectional and longitudinal studies show that regular aerobic exercise reduces postprandial lipemia in the presence or absence of weight loss [41–51]. However, the effects of aerobic exercise training on postprandial lipemia are diminished when blood sampling occurs 48 hours after the most recent bout of exercise suggesting that exercise produces acute effects that are rapidly reversed [49, 50]. Detraining studies provide further support for the concept that exercise performed in the hours prior to a mixed or high-fat meal is responsible for generating a metabolic environment which leads to postprandial lowering of TGs. Indeed, endurance-trained individuals who discontinued training for more than 60 hours had postprandial circulating TGs that were over 35% higher than levels following the last exercise training session [51]. Direct evidence for this phenomenon has been observed in multiple populations following single sessions of exercise to provide unequivocal evidence that exercise training produces important acute metabolic responses that reduce postprandial lipemia.

3.2. Timing and Composition of Test Meals. Although the nutrient composition of test meals, populations studied, and exercise conditions/modalities employed are highly variable throughout the literature, several conclusions can be drawn that support the therapeutic benefits of consistent exercise participation as a means to improve metabolic health in the postprandial state. For instance, significant disparities exist between studies regarding the type of test meals employed to evaluate the impact of exercise on postprandial lipemia. The literature is split regarding the use of mixed high-fat meals and exclusively high-fat meals to explore these effects. Although the validity of high-fat diets exclusively as a test meal has come into question, what is important is that, in the majority of cases, a single session of aerobic exercise robustly lowers postprandial lipemia regardless of the composition of the test meal.

It is important to point out that the amount of dietary fat in the test meal required to increase plasma TG concentrations appears to be dose-dependent [22]. Absolute doses of 5–15 g do not increase postprandial lipemia (29–32); whereas doses of 30–50 g increase postprandial lipemia by 75 to 110 mg/dL above baseline. Doses above 80 g exaggerate postprandial lipemia but are similar in magnitude to the increase with 50 g of dietary TG suggesting that a threshold is achieved beyond a certain dose of TGs consumed.

Another important aspect regarding test meal design and administration that should be considered is the fact that although the single test meal provides a powerful laboratory design, it may not reflect a real-world situation that occurs as additional meals are consumed. Farah and colleagues [52] addressed the question of how subsequent meals impact the capacity of exercise to lower postprandial lipemia and reported that prior exercise lowered the overall TG response to 3 meals. These results indicate that aerobic exercise reduces the postprandial TG response to multiple meals suggesting that the observed effects are not limited to effects on a single meal or the laboratory setting.

3.3. Timing of Exercise Interventions. The majority of investigations which have examined the effects of exercise on postprandial lipemia have been designed where a single session of aerobic exercise is conducted 11–20 hours prior to the test meal in a 2-day design. On day 1, participants perform a prescribed exercise modality and on day 2 they receive a test meal followed by temporal blood sampling for up to 8 hours [16, 53–64]. In contrast, others have used exercise sessions conducted 30–60 minutes before [65–67], 4 hours before [68], or 60–90 minutes after a meal [65, 66] to examine the effects of a single bout of exercise on postprandial lipemia. Aerobic exercise conducted 11–20 hours before the test meal appears to produce the greatest magnitude of reduction in postprandial lipemia compared to aerobic exercise conducted 30 minutes to 4 hours before a test meal [65, 66]. Finally, aerobic exercise conducted 1 hour following a high-fat meal produced only modest reductions in postprandial lipemia suggesting that prior exercise is superior to exercise conducted after a meal [65].

3.4. Effects of Exercise on Postprandial Lipemia Are Dose- and Intensity-Related. Early studies demonstrate that total energy expenditure achieved through isoenergetic low- or moderate-intensity aerobic exercise produces similar reductions in postprandial lipemia [54, 55], with exceptions [69]. Tsetsonis [54] reported that normolipidemic young men and women had lower postprandial lipemia after walking on a treadmill for 90 minutes at 61% of $VO_{2\,max}$ and that 90 minutes of walking at 31% of $VO_{2\,max}$ elicited little effect providing initial evidence that the effects of aerobic exercise on postprandial lipemia were due to either greater intensity or to greater energy expenditure at a higher intensity. In an attempt to answer this question, the authors conducted a follow-up study where they compared the effects of isoenergetic low- and moderate-intensity walking on postprandial lipemia [55]. The authors reported that 3 hours of walking at low-intensity (32% $VO_{2\,max}$) or walking for 1.5 hours at moderate-intensity (63%) (~1000 kcal energy expenditure for both) decreased postprandial lipemia to a similar extent. Further evidence that the beneficial effects of aerobic exercise on postprandial lipemia are related to the magnitude of energy expended come from studies which show that the accumulation of aerobic exercise throughout the day in multiple bouts produces similar benefits as a single session of isoenergetic aerobic exercise conducted at the same intensity [56, 58, 61–63, 70–74]. In contrast, a single session of aerobic exercise at 65% $VO_{2\,max}$ designed to expend ~1100 kcals reduced postprandial TG area under the curve (AUC) by 39% whereas isoenergetic exercise at 25% $VO_{2\,max}$ produced a statistically insignificant 9% decrease [69]. A recent study by Kim and colleagues [75] showed that isoenergetic low (25% $VO_{2\,max}$) and moderate (65% $VO_{2\,max}$) intensity aerobic exercise both lowered the TG AUC_I, but moderate intensity was over 17% more effective than low-intensity exercise. Available evidence suggests that energy deficit may be the most important determinant of the magnitude of reduction in postprandial lipemia with exercise. However, multiple investigations suggest that single or accumulated bouts of moderate-intensity exercise lower postprandial lipemia to a greater extent than isoenergetic exercise of low intensity [61, 69, 75].

3.5. High-Intensity Interval Exercise. Although total energy expenditure achieved during and after an exercise bout may outweigh the role of intensity in low- to moderate-intensity aerobic exercise, an interesting observation from emerging studies is that high-intensity, short-duration interval exercise may yield results which are at least as effective as low- to moderate-intensity exercise of longer duration despite lower energy expenditure [76]. High intensity interval (HII) exercise consists of multiple short-term (30 s–240 s) bouts at >85% $VO_{2\,max}$ followed by 60 s–180 s of active or passive recovery periods. Indeed, moderate-intensity walking for 30 minutes at an energy expenditure of 240 kcals produced no effect on postprandial lipemia whereas HII exercise expending 103 kcals produced reductions in the TG AUC [76]. Trombold et al. [77] found that moderate-intensity and HII exercise both decreased the AUC_I, but reported that HII exercise was more effective than moderate-intensity exercise for lowering postprandial lipemia despite identical energy expenditure during the exercise session. In contrast, isoenergetic (500 kcal) moderate-intensity and HII exercise decreased TG concentrations to a similar extent [73]. These differences may be due to disparities in the timing of test-meals as the meals were delivered 30 minutes after exercise in the latter study [73] versus 12–16 hours in the former study [77]. Additional studies show that sprint interval cycling produced no effects on postprandial lipemia [78, 79]. While additional studies are needed to sort out the role of HII exercise on postprandial lipemia, the notion that prior HII exercise can lower postprandial lipemia to a similar extent as moderate-intensity exercise, despite lower energy expenditure and time commitment, has important practical implications for individuals with limited access to exercise opportunities.

3.6. Is Resistance Training an Effective Modality to Reduce Postprandial Lipemia? A growing number of investigations have examined the effects of resistance training on postprandial lipemia and its efficacy in comparison to traditional aerobic exercise training. Available evidence suggests that resistance exercise is an effective strategy to reduce postprandial lipemia [80–84], although exceptions exist [85–87]. In fact, Burns and colleagues [87] showed that resistance training actually increased postprandial lipemia following a test meal. Studies which show a reduction in postprandial lipemia with resistance exercise observe these changes in recreationally active normal weight, normolipidemic males and females [80–82] and sedentary obese females [83]. Parallel comparisons of aerobic and resistance exercise have been somewhat disappointing on the surface in the sense that each of the studies conducted showed that aerobic exercise produced no effect on postprandial lipemia (which is inconsistent with the aerobic exercise literature). Some have suggested that although energy expenditure was similar in magnitude to levels which have previously shown reductions, it is possible that an intensity threshold was not met (~30% VO_{2peak}). However, it is important to note that reductions in postprandial lipemia were observed at a similar energy expenditure as that conducted for aerobic exercise suggesting that resistance training may be more effective than aerobic exercise at lowering postprandial lipemia. Support for this comes from a recent study in obese premenopausal women where the authors had the same group of women complete 2 separate 60 minute trials (aerobic exercise at 60–65% VO_{2peak} and resistance training exercise) [83]. Although they did not measure energy expenditure, it would be predicted that energy expenditure in the resistance training condition was approximately 50% lower than the aerobic exercise group. Taken together, these results suggest that resistance training exercises that work the upper and lower body may be more efficacious than aerobic exercise at lowering postprandial lipemia.

Overall, the literature suggests that the energy expenditure requirements to produce reductions in postprandial lipemia occur at a minimum of 250 kcals, but occur more consistently at 400 kcals and greater during low- to moderate-intensity aerobic exercise [88]. Zhang et al. [59] showed that 400 kcals produced similar results to that of EE over 700 kcals suggesting that an upper capacity may be achieved where only limited returns are produced beyond a certain level of energy expenditure. Future studies will be required to examine the effects of HII exercise (>85%) and resistance training to provide additional evidence that energy expenditure requirements are different with higher intensity exercise.

3.7. What Are the Mechanisms Responsible for Exercise-Mediated Reductions in Postprandial Lipemia? Prior exercise reduces postprandial lipemia by three possible mechanisms acting alone or in combination: (1) decreased appearance of chylomicron-TG concentrations from the gut, (2) increased clearance of TRLs (VLDL and/or CM) via exercise-mediated increases in skeletal muscle and/or adipose tissue LPL activity, and (3) decreased hepatic VLDL-TG secretion [89] (Figure 1). Although studies are limited, available evidence indicates that aerobic exercise does not reduce gastric emptying, as determined by paracetamol administration [89–91] and does not delay the time to achieve peak TG concentrations compared to control conditions [89, 91, 92] which would be expected to occur if exercise altered the rate of gastric emptying.

In contrast, numerous investigations have reported that exercise produces a delayed and transient increase in LPL protein expression and activity that occurs within 4 to 8 hours after exercise and remains elevated for up to 24 hours [93–100]. These results are metabolically relevant as skeletal muscle LPL activity increases the hydrolysis of TG and clearance of free fatty acids in the postabsorptive state when serum insulin concentrations are relatively low. Therefore, an increase in skeletal muscle LPL activity produced by exercise would be expected to increase the clearance of circulating TG in the postabsorptive state. There is a strong association between the elevation in LPL activity and reductions in fasting serum TGs in most [96, 97, 100–102] but not all investigations [97, 103, 104].

Studies which have examined the effects of aerobic exercise on postprandial lipemia when exercise was performed 12–18 hours before a meal consistently show significant reductions in TGs. However, the effects of exercise on LPL and TG clearance are unclear. For example, 90 minutes of walking reduced postprandial lipemia in men, but the same amount of exercise did not increase clearance of an intravenous lipid emulsion [90]. Herd and colleagues [105] showed that 90 minutes of moderate-intensity aerobic exercise conducted 18 h before a meal reduced postprandial lipemia but did not produce statistically significant increases in LPL activity (although the individuals who experienced increases in LPL activity after exercise had the greatest reduction in TGs). An additional study showed that 2 hours of moderate-intensity exercise did not significantly increase TG clearance across the leg 18 hours after exercise despite lowering postprandial TG concentrations [92].

While a role for skeletal muscle LPL in the reduction of postprandial lipemia cannot be ruled out, multiple factors suggest that other mechanisms must be involved.

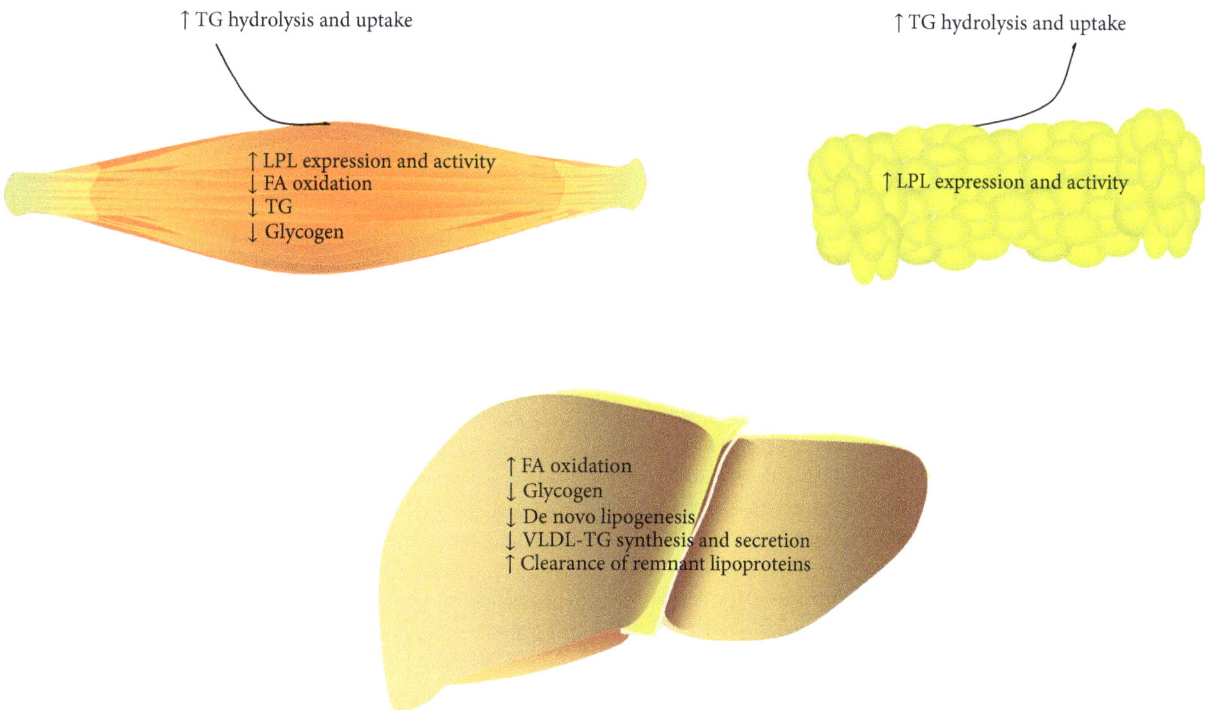

FIGURE 1: Potential mechanisms by which prior exercise reduces postprandial lipemia. Aerobic exercise has been shown to increase FA substrate utilization during metabolic studies presumably in skeletal muscle and liver. Increased hepatic FA oxidation and decreased de novo lipogenesis are thought to decrease VLDL-TG synthesis and secretion. Increased clearance of postprandial triglycerides through upregulation of LPL expression and activity has also been proposed in skeletal muscle and to a lesser extent in adipose tissue.

For example, it is well known that postprandial hyperinsulinemia promotes the storage of TG in adipose tissue by increasing LPL activity [106, 107] and that skeletal muscle LPL activity and TG storage are minimal [105, 108, 109]. The question is whether exercise-mediated increases in skeletal muscle LPL activity are maintained in the presence of hyperinsulinemia as in the postabsorptive state or if exercise increases adipose tissue LPL activity. Furthermore, the increase in skeletal muscle LPL activity is likely to have dissipated in most protocols where exercise is conducted the day before the test meal.

Early studies in rodents revealed that aerobic exercise training reduced the secretion of hepatic VLDL-TG by decreasing serum NEFAs and de novo lipogenesis [110–112]. While direct evidence in humans is limited, emerging studies indicate that reductions in serum TG produced by aerobic and resistance exercise are attenuated by reductions in either hepatic VLDL-TG secretion or clearance. Indeed, 50–70% of the reduction in postprandial lipemia produced by aerobic exercise was accounted for by reductions in hepatic VLDL-TG secretion [105, 113, 114]. In one of the most elegant investigations conducted to date, Davitt and colleagues [83] provided evidence using stable isotopes that the reduction in postprandial lipemia with aerobic and resistance training was not achieved by enhanced clearance of dietary fat, but instead by reduced abundance of endogenous fatty acids in circulating TGs. Others [115] reported that a single session of aerobic exercise reduced postprandial lipemia by decreasing hepatic secretion and increasing clearance in women whereas in men, the reduction in postprandial lipemia was due to increased clearance alone. From the available literature, it is difficult to draw strong conclusions about the specific roles and magnitude of reduction in postprandial lipemia produced by hepatic TG secretion and clearance. Nonetheless, it seems plausible that reductions in VLDL-TG secretion may be the primary mechanism by which exercise reduces postprandial lipemia. A role for postprandial TG clearance may exist, especially with HII exercise. For example, beta-hydroxybutyrate (BOHB) concentrations were not different from control following a single HII exercise bout [76] suggesting that the reduction in TG was not due to reductions in hepatic production of VLDL. While this conclusion should be reached with caution due to the equivocal role of BOHB as a valid marker of VLDL production [116, 117], it does suggest that reductions in postprandial lipemia may be attributed, at least in part, to increased clearance. It is possible that HII exercise may more effectively alter LPL mass and activity than low- to moderate-intensity exercise.

The mechanisms by which aerobic and resistance exercise decrease VLDL-TG secretion and postprandial lipemia is unclear. We propose the possibility that aerobic exercise-mediated reductions in hepatic and skeletal muscle glycogen content is part of a metabolic program produced by exercise which preferentially shuttles fatty acids to oxidation as a strategy to spare glucose for storage and to reestablish intracellular ATP concentrations, thereby lowering the packaging/secretion of VLDL-TG and potentially increasing clearance. Evidence for this hypothesis is supported by the finding that a prior bout of endurance or resistance exercise increases whole body fatty acid oxidation [80, 91, 118, 119] and increases glycogen synthase activity and protein expression in skeletal muscle [120]. This hypothesis is also supported by the findings of some studies which show that moderate-intensity exercise produces greater reductions in postprandial lipemia than low-intensity exercise (where carbohydrate oxidation would presumably be lower than moderate intensity exercise) [61, 69, 75]. In contrast, if moderate-intensity exercise produces similar reductions in postprandial lipemia as isoenergetic low-intensity exercise as previously reported [55] this would suggest an alternative possibility. This possibility is further suggested by the observation that aerobic exercise produced similar reductions in postprandial lipemia in the presence or absence of pharmacological inhibition of adipose tissue lipolysis by acipimox [121]. If the increase in glycogen oxidation is correct, then it would be expected that exercise in the presence of acipimox would produce a greater reduction in postprandial lipemia, but this was not the case. Overall, these results suggest that regardless of the substrate used, the hepatic and skeletal muscle program following exercise is to preferentially use fatty acids from TRLs to replenish ATP thus sparing glucose for glycogen storage as part of the training adaptation for subsequent exercise bouts. This is in agreement with studies which show that exercise increases glycogen synthase activity and protein expression in skeletal muscle [120].

3.8. Does Acute Energy Deficit Produced by Diet Yield Similar Effects on Postprandial Lipemia as Energy Deficit Produced by Exercise? An important question arising from acute exercise studies is whether the accompanying energy deficit is responsible for the reduction in postprandial lipemia. To address this question, Gill and Hardman [16] compared an energy deficit of 500 kcals produced by exercise with an equivalent deficit produced by energy restriction. Exercise decreased total and incremental TG concentrations while decreasing energy intake an equivalent amount produced only mild reductions in serum TG concentrations that were 3-fold lower than that produced by exercise. In a similar fashion, Maraki and colleagues [122] reported that isoenergetic deficits created by energy restriction or aerobic exercise decreased postprandial lipemia to a similar extent statistically, with a numerically stronger effect elicited by exercise compared to energy restriction. The interpretation of the results is complicated by the fact that Gill and Hardman [16] reported that the energy deficit caused by energy restriction was 17% lower than that induced by exercise. Based on limited evidence, it appears that greater amounts of energy restriction are required to produce reductions in postprandial lipemia compared to the energy deficit created by exercise.

Using a different approach, Burton et al. [123] showed that moderate-intensity aerobic exercise producing an energy deficit of ~668 kcal lowered postprandial lipemia as expected. However, when an isoenergetic mixed meal was provided shortly after exercise to achieve energy balance, the effects of aerobic exercise on postprandial lipemia were dramatically attenuated. Additional evidence in men using a combination of moderate- and high-intensity exercise showed that aerobic exercise produced a profound reduction in postprandial lipemia which was ameliorated when glucose was consumed shortly after exercise to reestablish energy balance [124].

The results of these studies suggest that dietary and exercise-induced reductions in postprandial lipemia may be mediated through both common and different pathways [16, 123]. These differences may be related to the fact that moderate-intensity exercise produces quantitatively larger deficits in skeletal muscle and liver glycogen than energy restriction alone [123]. As described above, the reduction in glycogen content in both tissues would be expected to preferentially partition intracellular glucose to storage as glycogen and increase the hydrolysis and uptake of circulating TG to provide fatty acids as a substrate for oxidation. In contrast, energy restriction-mediated deficits would likely use a greater proportion of adipose tissue TGs which would not generate an equivalent reduction in skeletal muscle and liver glycogen content. While the energy deficit created by exercise appears to be a primary mediator of the exercise-induced reduction in postprandial lipemia, energy balance created by carbohydrate replacement alone abolished the reduction in postprandial lipemia by exercise providing further support that glycogen resynthesis and energy balance are driving forces in this process. Additional studies are required to further evaluate the role of carbohydrate and other dietary macronutrients such as protein or fat to determine the interaction between macronutrient replacement, energy balance, and exercise-mediated reductions in postprandial lipemia.

4. Energy and Carbohydrate Restriction Effects on Postprandial Lipemia

Accumulating evidence indicates that negative energy balance, achieved through increases in energy expenditure or decreases in energy intake, decrease postprandial lipemia [89, 125]. While studies to date have focused on assessing the absolute energy deficit on postprandial lipemia, it has been suggested that the improvements from either diet or exercise may be greater when there is a larger carbohydrate deficit [124]. Indeed, this is supported by dietary studies that demonstrate greater improvements in fasting serum lipids and significantly greater reductions in postprandial lipemia following a very low-carbohydrate diet as compared to a low-fat diet, and exercise studies that have shown that acute improvements in postprandial TG metabolism is abolished if a carbohydrate rich postexercise meal is consumed [124, 126]. Thus, it is possible that a low-carbohydrate diet may be more beneficial than a low-fat diet for improving both fasting plasma lipids and postprandial lipemia.

A reduction in the intake of dietary fat has long been recommended as a means to reduce risk factors associated with metabolic and CVD [127]. However, it is generally thought that a significant amount of weight loss must occur to appreciate the beneficial effects of a low-fat diet. Furthermore, it has been shown that low-fat diets lower HDL-C concentrations and increase fasting plasma TG concentrations [128, 129], which is associated with an increased risk of development and mortality from CVD [130]. While it has been suggested that hypertriglyceridemia is a relatively short-term adaptive response to the increase in dietary carbohydrate [131], results from epidemiological studies show that this response may actually be a longer lasting phenomenon [132]. Thus, it remains to be determined whether or not carbohydrate-induced hypertriglyceridemia is transient in nature or can be avoided. An overview of the literature suggests that the carbohydrate-induced hypertriglyceridemic response can only be partially normalized if there is a significant (at least 10%) amount of weight loss that accompanies the low-fat high-carbohydrate diet [133]. Given these observations, more recent studies have begun to focus on the potential benefit of reducing carbohydrate content in the diet as a means to improve fasting and postprandial lipid metabolism [125, 126, 129, 134].

Carbohydrate restriction has been adopted by a large number of people; however, because carbohydrate restriction can increase the production of ketone bodies, there has been caution issued by the scientific community in regard to the safety of these diets. This concern has prompted the USDA to call for further research into the safety and efficacy of low-carbohydrate diets. Thus, in recent years a number of well-designed studies have tested the effects and safety of a low-carbohydrate diet on reducing biomarkers for CVD risk, and improving fasting and postprandial lipid metabolism. There have been several well-controlled randomized studies comparing the effects of a carbohydrate-restricted diet with a fat-restricted diet on weight loss, blood lipids, and other CVD risk markers. Given that responses may differ between genders, weight status, and blood lipid profile, there have been several investigations assessing a number of different cohorts, including men and women with atherogenic dyslipidemia [135], normal weight-normolipidemic men [136, 137], normal weight-normolipidemic women [126], and overweight men [125].

In a cohort of normal weight men [136] and women [126] assigned to an isoenergetic diet comprised of a low-carbohydrate diet or low-fat diet, there was a reduction in fasting TG, postprandial lipemia, and fasting insulin compared to low-fat diet in men and a significant increase in postprandial lipemia compared to low-fat diet in women. However, in a cohort of overweight men that incorporated a hypocaloric diet, both a low-carbohydrate and a low-fat diet led to a reduction in total cholesterol, fasting insulin, and HOMA-IR. In contrast, fasting TAG, fasting glucose, and increases in mean and peak LDL particle size were only reduced by the low-carbohydrate diet [125]. Overall, the authors concluded that the hypocaloric low-carbohydrate diet had a similar or better effect on overall blood lipids compared to the low-fat diet. In a more recent study in men and women with MetS, Volek et al. [135] found that carbohydrate restriction led to a two-fold greater weight loss as compared to the low-fat control. Additionally, carbohydrate restriction resulted in a significantly greater reduction in fasting glucose and HOMA-IR, lower total postprandial TG AUC following an oral fat load, reduction in fasting TG and total cholesterol, and increased circulating HDL and LDL particle size as compared to a fat restricted diet [135]. Thus, carbohydrate restriction provided a more comprehensive improvement in clinical risk factors associated with MetS than fat restriction at a reduced caloric intake. In summary, it appears that a very low-carbohydrate diet may represent an alternative and safe strategy for metabolic and cardiovascular health that extends beyond weight regulation.

In addition to the theory that insulin resistance may exacerbate both fasting and PPL, there is also accumulating evidence that the type of sugar may differentially regulate these effects [138]. Fructose, in particular, has been given a great deal of attention due to its potential role for improving glucose tolerance and attenuating the postprandial insulin response [139, 140]. However, these responses have been shown in smaller doses of dietary fructose (7.5 g), whereas higher doses have been shown to increase postprandial lipemia [141]. Thus, while fructose may attenuate the postprandial glucose response it may potentiate postprandial hyperlipemia [138, 142]. While the exact mechanism in which fructose potentiates postprandial lipemia remains to be determined, current data in humans suggests a decrease in activation of adipose tissue LPL activity due to a lower postprandial insulin excursion [138]. It is also possible that fructose-mediated increases in hepatic de novo lipogenesis [143, 144] increase postprandial VLDL synthesis and secretion. Additional studies in rodents suggest a reduction in hepatic clearance of VLDL-TG [145]. The role of fructose consumption on postprandial lipemia is an area that warrants further investigation and may explain why high-carbohydrate diets can exaggerate postprandial lipemia in individuals that are insulin sensitive.

4.1. Potential Mechanisms in Which Carbohydrate Restriction Improves Fasting and Postprandial Lipemia. In order to discuss potential mechanisms in which carbohydrate restriction improves fasting and postprandial lipemia, it is important to first discuss potential ways in which carbohydrate consumption can induce lipemia and increase fasting TG. Simply put, an increase in the fasting TG and postprandial lipemia response is a result of either an increase in hepatic de novo lipogenesis or a reduction in skeletal muscle and/or adipose tissue TG clearance. Carbohydrate-induced lipemia was first recognized in the early 1950s, where low-fat diets were prescribed to lower blood cholesterol levels [146, 147]. Paradoxically, when investigators prescribed this diet, patients exhibited postprandial lipemia [146]. The proposed mechanisms for postprandial hyperlipemia in early studies of low-fat/higher carbohydrate diets were that higher carbohydrate promoted hepatic TG synthesis and a net decrease in TG removal rates leading to a more pronounced lipemia [148]. Shortly thereafter, it was recognized that the carbohydrate-induced hypertriglyceridemic response was exacerbated in individuals that were insulin resistant [149]. Reaven et al. [150] performed a number of studies demonstrating that even mild peripheral insulin resistance could result in failure of an inhibition of lipolysis when insulin is elevated, and also increases hepatic TG secretion due to a reduction of insulin's ability to inhibit liver TG secretion when hepatic insulin resistance was present [151]. Given these observations, peripheral insulin resistance has become one of the prevailing theories in which high-carbohydrate diets increase fasting and postprandial lipemia. Over the ensuing years, a number of mechanisms have been identified that may explain carbohydrate-induced lipemia, including increased secretion of hepatic VLDL particles and/or upregulation of apolipoprotein synthesis and TG packaging per VLDL particle, or a reduction in clearance due to a decrease in LPL activity [133]. Additionally, while the lipemic response to high carbohydrate has been clearly demonstrated, it is important to note that evidence exists that reveals lower lipemia when weight loss occurs or less fat composition is included in the diet [133]. Thus, hyperinsulinemia, hepatic and peripheral insulin resistance, and body weight all appear to be critical factors associated with the magnitude of carbohydrate-induced lipemia.

While the paradoxical postprandial lipemia response following a high-carbohydrate/low-fat diet was observed over 50 years ago, the idea of lowering carbohydrate and increasing fat composition in the diet has only begun to gain traction over the last 5–10 years. The precise mechanisms in which a low-carbohydrate diet improves postprandial lipemia remain to be determined. However, it is currently thought that reducing carbohydrate intake reduces fasting and postprandial TG by reducing VLDL production rate and increasing TG removal by increasing both plasma and skeletal muscle LPL activity [135, 152]. It has also been shown that carbohydrate restriction can decrease fasting and postprandial insulin concentrations [135]. This is important since insulin suppresses skeletal muscle and adipose tissue lipolysis and increases de novo lipogenesis. In the presence of insulin resistance, an increase in skeletal muscle lipolysis and a decrease in hepatic lipid storage would be expected, leading to increased production of larger TG-enriched LDL particles, an increased formation of small LDL particles, and a decrease in HDL-C. In addition to the effects of insulin on hepatic lipid production, it is also possible that there is greater lipid clearance due to a shift towards fat oxidation and a lower reliance on insulin for skeletal muscle glucose uptake when fat content is increased in the diet. The role of insulin resistance in determining lipemia has been shown by Petersen et al. [153]. Dietary carbohydrate substrate partitioning, liver and muscle TG and glycogen synthesis, and de novo lipogenesis were determined using ^{1}H and ^{13}C NMR spectroscopy and deuterium enrichment. They found that insulin resistant men had impaired skeletal muscle and hepatic glycogen formation following carbohydrate intake, directed dietary carbohydrate toward hepatic de novo lipogenesis, and TG synthesis and produced an overall increase in plasma TG concentrations [153]. Thus, it is likely that the benefits of carbohydrate restriction would be greater in individuals with insulin resistance. Indeed, it has been shown that carbohydrate restriction can decrease malonyl-CoA concentrations, removing the disinhibition of carnitine acetyltransferase and enabling greater fatty acid transport and fatty acid oxidation [152]. Furthermore, lowering carbohydrate intake for as few as three days has been shown to upregulate genes associated with fatty acid oxidation [154].

5. Omega-3-Fatty Acids

The omega-3 (n-3) fatty acids, docosahexaenoic acid (DHA), and eicosapentaenoic acid (EPA) beneficially modify fasting/postprandial blood lipid and lipoprotein metabolism and independently decrease mortality due to myocardial infarction and sudden death [155]. An analysis of 72 placebo-controlled human studies of at least 2 weeks in length providing 2 to 7 grams of n-3 fatty acids per day found that fish oils dose-dependently decrease serum fasting TG concentrations in normo- and hypertriglyceridemic individuals by 4% to 40% [155, 156].

The effects of n-3 fatty acids have also been explored to determine their impact on postprandial lipid/lipoprotein metabolism. Tinker and colleagues showed that a liquid fish oil supplement containing 5.2 g of EPA and DHA decreased postprandial TGs in hypertriglyceridemia participants [157]. Similar results have been observed by others in both normo- and hypertriglyceridemic participants [158–160] with exceptions [161, 162]. Although the precise mechanisms are unknown, fish oil supplementation has been shown to decrease hepatic triglyceride synthesis via inhibition of diacylglycerol transferase (DGAT), fatty acid synthase, and acetyl coenzyme A carboxylase ACC activities [163, 164]. Fish oils also enhance fatty acid oxidation by stimulating peroxisome proliferator-activated receptor α (PPARα) [165]. Each of these mechanisms would be expected to reduce hepatic apolipoprotein B-100 synthesis and reduce VLDL synthesis and secretion. Indeed, a number of studies show that reductions in postprandial lipemia with n-3 fatty acids are associated with lower synthesis of apolipoprotein B-100 [160, 166–168]. A more recent study showed that n-3 fatty acids decreased postprandial apolipoprotein B-48 concentrations by 22% suggesting that the reduction in postprandial lipemia with n-3 fatty acids are also related to improvements in chylomicron metabolism [28].

5.1. n-3 Fatty Acids and Exercise. Potentially overlapping mechanisms have prompted some to explore the combined effects of n-3 fatty acids and aerobic exercise on postprandial lipemia. An early investigation by Thomas et al. [161] showed that a combination of short-term n-3 fatty acids (4 g/d) and a single bout of exercise failed to reduce postprandial lipemia in sedentary individuals. The authors suggested that the reduction in fasting TGs by n-3 fatty acids reduced the postprandial TG-lowering effects of exercise. In a follow-up study by the same group, aerobic exercise and n-3 fatty acids reduced postprandial lipemia in recreationally active individuals in an additive fashion [169]. It was hypothesized that trained and untrained individuals may respond differently to n-3 fatty acid and aerobic exercise interventions and that if sedentary individuals were to become trained, that they might benefit from the combination of exercise and n-3 supplementation. To test this hypothesis, the authors examined the effects of 4 weeks of training in previously sedentary overweight men and women [170]. n-3 fatty acids reduced fasting and postprandial TG concentrations, but the addition of exercise training produced no additional benefits. In a more recent study, n-3 fatty acid supplementation for 16 weeks with an exercise intervention (which did not produce weight loss) provided significantly greater reductions in TG AUC_I compared to n-3 fatty acids alone in viscerally obese individuals [158]. From the available literature, it is difficult to form a straightforward conclusion about the effects of combining n-3 fatty acids and exercise on postprandial lipemia. Future studies will be required to examine whether combination therapy is superior in hypertriglyceridemic versus normotriglyceridemic individuals and/or if the disparity in results are due to differences in the duration of treatment.

6. Quantifying Postprandial Lipemia

Assessment of postprandial lipemia has traditionally occurred over the course of 6–8 hours in laboratory settings. The extensive length of time and variety of macronutrient manipulations used has hindered the transition of oral fat tolerance tests to the clinic. Furthermore, the optimal mathematical representation for reporting postprandial lipemia with multiple time points continues to be debated. The traditional approach has been to report the absolute increase in TG concentrations that occur over the course of 6–8 hours with measurements conducted at 1-2 hour intervals. This technique is complemented by using a summary measurement referred to as the AUC which uses the trapezoidal rule [171] to calculate the average area under the absolute curve. While this strategy provides a valid assessment of the absolute response to the test meal, it does not factor in effects of interventions which influence baseline fasting TG concentrations. Since fasting TG concentrations are highly associated with the absolute hourly response and total TG AUC, it can be argued that absolute reductions in the postprandial response to a meal challenge are due to the reduction in fasting TGs as opposed to effects of the intervention during the postprandial period. To account for the effects of an intervention on fasting concentrations prior to a meal, the incremental AUC has been used as a method to control for reductions in fasting TG concentrations. The incremental AUC factors out the fasting TG concentration from the hourly responses and produces a curve that strictly reflects the success of the intervention on postprandial serum TG concentrations. However, it should be cautioned that one study reported that the reproducibility of the incremental TG AUC is relatively low when compared to the total TG AUC suggesting this as a limitation for incremental AUC [172].

While it seems clear that the most effective mathematical strategy at this point is to use both the total and incremental AUC, we still have not adequately dealt with the clinical utility of these tests as it relates to the length of time required to conduct the tests. It is well known that TG concentrations peak in most individuals within a 4-hour period [173, 174] (although it may occur sooner in women than men) and that the 4-hour peak is highly related to the total 8 h postprandial lipemia response [175], which can be used for accurate estimation of the postprandial response in healthy and at-risk individuals. Additional studies provide evidence that an abbreviated 4 h test is appropriate and is highly correlated with the results of full length tests [172, 175, 176] with the caveat that it may be a better predictor in healthy lean or obese populations but not in individuals with hypertriglyceridemia [176].

6.1. Other Markers of Postprandial Lipid/Lipoprotein Metabolism. While the effects of a lifestyle intervention on the postprandial response to a high fat or mixed meal is most often measured by the magnitude and duration of the TG response, these responses provide little information about the source of lipids (exogenous or endogenous). As described above, a primary objective of current research programs is to determine the mechanisms by which aerobic exercise and dietary interventions reduce postprandial lipemia. One of the earliest methods used to evaluate the source of TGs following a meal is retinyl palmitate [177]. Retinyl palmitate is the ester of retinol (vitamin A) and palmitate which is given in concert with a test meal. In the intestinal mucosa, retinyl palmitate is incorporated into the chylomicron core where it is thought to remain during triglyceride hydrolysis [35, 178–180]. This method assumes that the retinyl ester remains associated with apolipoprotein B-48 and does not exchange with other lipoproteins. Numerous challenges [181–184] to this assumption have been observed suggesting that retinyl palmitate may not be an ideal approach to evaluate the contribution of exogenous and endogenous sources of lipoproteins following a meal.

A more accurate approach to quantifying the source of TRLs may lie in the direct measurement of apolipoprotein B-48. Apolipoprotein B-48 has a molecular weight (MW) of 264 kDa and is approximately 48% the mass of apolipoprotein B-100 with a MW of 550 kDa [185]. In this technique, TRLs are separated from plasma by ultracentrifugation and the concentrations of apolipoprotein B-48 and B-100 are quantified by HPLC or densitometry [8, 35, 186]. The postprandial apolipoprotein concentrations are subsequently reported in relative units or considered as AUC responses in a similar fashion to that used for TGs.

More sophisticated approaches to apolipoprotein quantification and kinetics have been employed using stable isotopes to examine whether defects in secretion and/or catabolism of apolipoproteins are responsible for hypertriglyceridemia following a meal. For example, Wong and colleagues [28] used a bolus of d_3-leucine (5 mg/kg BW) with a high-fat meal in obese and lean individuals to evaluate chylomicron metabolism. Using this approach, the authors were able to show that central obesity results in an overproduction and impaired catabolism of apolipoprotein B-48 containing lipoproteins. Additional studies have used stable isotopes to label test meal fatty acid content to differentiate between exogenous and endogenous TGs where [U-^{13}C] palmitate (5 mg/kg FFM) was administered in a liquid test meal. Lipids were isolated using a heptane/isopropanol extraction procedure and analyzed by LC/MS [83].

7. Conclusions and Future Directions

Moderate-high intensity aerobic and resistance exercise produce consistent reductions in postprandial lipemia when performed 30 minutes to 20 hours prior to mixed or high-fat meals. Energy deficits created by moderate-intensity exercise appear to be responsible for a considerable proportion of the decrease in postprandial TGs with exercise. However, studies using resistance and HII exercise suggest that other factors may be involved. Future studies are needed to compare the effects of resistance exercise and HII in men and women. Furthermore, studies are required to determine the impact of weight loss (produced by energy restriction and/or increased energy expenditure) on the acute postprandial lipemia response to exercise. On a practical level, the minimum threshold of exercise energy expenditure to lower postprandial lipemia in sedentary and active populations is crucial. Additional studies using innovative approaches are also needed to more thoroughly evaluate the mechanisms by which exercise lowers postprandial TGs and if these mechanisms are different between obese and nonobese populations, sex, and race/ethnicity.

Reductions in dietary carbohydrate without reducing energy intake also decrease postprandial lipemia. Future studies will be needed to determine the mechanisms by which low-carbohydrate (glucose and/or fructose) diets work. Studies using deuterated water could be used to evaluate de novo lipogenesis between tissues.

Abbreviations

TRLs:	Triglyceride rich lipoproteins
TG:	Triglyceride
LPL:	Lipoprotein lipase
HTG:	Hypertriglyceridemia
HOMA-IR:	Homeostasis model assessment-insulin resistance.

Conflict of Interests

The authors have no conflict of interests to disclose.

Acknowledgments

The authors wish to thank Kelsey Miller for providing editorial support. Dr. Eric P. Plaisance was responsible for generating the concept of the paper and was primarily responsible for writing and editing the exercise-related sections. Dr. Gordon Fisher was responsible for writing the sections on low-carbohydrate diets and provided editorial support for all sections of the paper.

References

[1] B. A. Swinburn, I. Caterson, J. C. Seidell, and W. P. James, "Diet, nutrition and the prevention of excess weight gain and obesity," *Public Health Nutrition*, vol. 7, no. 1, pp. 123–146, 2004.

[2] S. A. French, M. Story, and R. W. Jeffery, "Environmental influences on eating and physical activity," *Annual Review of Public Health*, vol. 22, pp. 309–335, 2001.

[3] S. D. Poppitt and A. M. Prentice, "Energy density and its role in the control of food intake: evidence from metabolic and community studies," *Appetite*, vol. 26, no. 2, pp. 153–174, 1996.

[4] S. M. Grundy, G. Blackburn, M. Higgins, R. Lauer, M. G. Perri, and D. Ryan, "Physical activity in the prevention and treatment of obesity and its comorbidities," *Medicine and Science in Sports and Exercise*, vol. 31, supplement 11, pp. S502–S508, 1999.

[5] D. F. Williamson, J. Madans, R. F. Anda, J. C. Kleinman, H. S. Kahn, and T. Byers, "Recreational physical activity and ten-year weight change in a US national cohort," *International Journal of Obesity and Related Metabolic Disorders*, vol. 17, no. 5, pp. 279–286, 1993.

[6] R. Ross and I. Janssen, "Physical activity, total and regional obesity: dose-response considerations," *Medicine & Science in Sports & Exercise*, vol. 33, no. 6, pp. S521–S527, 2001.

[7] A. G. Tsai, D. F. Williamson, and H. A. Glick, "Direct medical cost of overweight and obesity in the USA: a quantitative systematic review," *Obesity Reviews*, vol. 12, no. 1, pp. 50–61, 2011.

[8] J. C. Mamo, G. F. Watts, P. H. Barrett, D. Smith, A. P. James, and S. Pal, "Postprandial dyslipidemia in men with visceral obesity: an effect of reduced LDL receptor expression?" *American Journal of Physiology: Endocrinology and Metabolism*, vol. 281, no. 3, pp. E626–E632, 2001.

[9] H. N. Hodis and W. J. Mack, "Triglyceride-rich lipoproteins and the progression of coronary artery disease," *Current Opinion in Lipidology*, vol. 6, no. 4, pp. 209–214, 1995.

[10] S. Bansal, J. E. Buring, N. Rifai, S. Mora, F. M. Sacks, and P. M. Ridker, "Fasting compared with nonfasting triglycerides and risk of cardiovascular events in women," *The Journal of the American Medical Association*, vol. 298, no. 3, pp. 309–316, 2007.

[11] N. Sarwar, J. Danesh, G. Eiriksdottir et al., "Triglycerides and the risk of coronary heart disease: 10 158 incident cases among 262 525 participants in 29 Western prospective studies," *Circulation*, vol. 115, no. 4, pp. 450–458, 2007.

[12] L. Wideman, L. A. Kaminsky, and M. H. Whaley, "Postprandial lipemia in obese men with abdominal fat patterning," *The Journal of Sports Medicine and Physical Fitness*, vol. 36, no. 3, pp. 204–210, 1996.

[13] R. McBride, "The role of niacin in raising high-density lipoprotein cholesterol to reduce cardiovascular events in patients with atherosclerotic cardiovascular disease and optimally treated low-density lipoprotein cholesterol: rationale and study design. the Atherothrombosis Intervention in Metabolic syndrome with low HDL/high triglycerides: Impact on Global Health outcomes (AIM-HIGH)," *American Heart Journal*, vol. 161, no. 3, article e2, pp. 471–477, 2011.

[14] D. B. Zilversmit, "Atherogenesis: a postprandial phenomenon," *Circulation*, vol. 60, no. 3, pp. 473–485, 1979.

[15] B. Nordestgaard, M. Benn, P. Schnohr, and A. Tybjaerg-Hansen, "Nonfasting triglycerides and risk of myocardial infarction, ischemic heart disease, and death in men and women," *The journal of the American Medical Association*, vol. 298, no. 3, pp. 299–308, 2007.

[16] J. M. R. Gill and A. E. Hardman, "Postprandial lipemia: effects of exercise and restriction of energy intake compared," *The American Journal of Clinical Nutrition*, vol. 71, no. 2, pp. 465–471, 2000.

[17] J. H. O'Keefe and D. S. Bell, "Postprandial hyperglycemia/hyperlipidemia (postprandial dysmetabolism) is a cardiovascular risk factor," *The American Journal of Cardiology*, vol. 100, no. 5, pp. 899–904, 2007.

[18] Y. Nakashima, E. W. Raines, A. S. Plump, J. L. Breslow, and R. Ross, "Upregulation of VCAM-1 and ICAM-1 at atherosclerosis-prone sites on the endothelium in the ApoE-deficient mouse," *Arteriosclerosis, Thrombosis, and Vascular Biology*, vol. 18, no. 5, pp. 842–851, 1998.

[19] W. C. Tsai, Y. H. Li, C. C. Lin, T. H. Chao, and J. H. Chen, "Effects of oxidative stress on endothelial function after a high-fat meal," *Clinical Science*, vol. 106, no. 3, pp. 315–319, 2004.

[20] H. Gaenzer, W. Sturm, G. Neumayr et al., "Pronounced postprandial lipemia impairs endothelium-dependent dilation of the brachial artery in men," *Cardiovascular Research*, vol. 52, no. 3, pp. 509–516, 2001.

[21] A. R. Sharrett, G. Heiss, L. E. Chambless et al., "Metabolic and lifestyle determinants of postprandial lipemia differ from those of fasting triglycerides the atherosclerosis risk in communities (ARIC) study," *Arteriosclerosis, Thrombosis, and Vascular Biology*, vol. 21, no. 2, pp. 275–281, 2001.

[22] J. Lopez-Miranda, C. Williams, and D. Larion, "Dietary, physiological, genetic and pathological influences on postprandial lipid metabolism," *The British Journal of Nutrition*, vol. 98, no. 3, pp. 458–473, 2007.

[23] J. M. Gaziano, C. H. Hennekens, C. J. O'Donnell, J. L. Breslow, and J. E. Buring, "Fasting triglycerides, high-density lipoprotein, and risk of myocardial infarction," *Circulation*, vol. 96, no. 8, pp. 2520–2525, 1997.

[24] L. A. Moreno, I. Quintela, J. Fleta et al., "Postprandial triglyceridemia in obese and non-obese adolescents. Importance of body composition and fat distribution," *Journal of Pediatric Endocrinology & Metabolism*, vol. 14, no. 2, pp. 193–202, 2001.

[25] D. Hyson, J. C. Rutledge, and L. Berglund, "Postprandial lipemia and cardiovascular disease," *Current Atherosclerosis Reports*, vol. 5, no. 6, pp. 437–444, 2003.

[26] J. L. Potts, S. W. Coppack, R. M. Fisher, S. M. Humphreys, G. F. Gibbons, and K. N. Frayn, "Impaired postprandial clearance of triacylglycerol-rich lipoproteins in adipose tissue in obese subjects," *The American Journal of Physiology: Endocrinology and Metabolism*, vol. 268, no. 4, part 1, pp. E588–E594, 1995.

[27] K. Adeli and G. F. Lewis, "Intestinal lipoprotein overproduction in insulin-resistant states," *Current Opinion in Lipidology*, vol. 19, no. 3, pp. 221–228, 2008.

[28] A. T. Wong, D. C. Chan, J. Pang, G. F. Watts, and P. H. Barrett, "Plasma apolipoprotein B-48 transport in obese men: a new tracer kinetic study in the postprandial state," *The Journal of Clinical Endocrinology and Metabolism*, vol. 99, no. 1, pp. E122–E126, 2014.

[29] J. Kobayashi, J. Tashiro, S. Murano, N. Morisaki, and Y. Saito, "Lipoprotein lipase mass and activity in post-heparin plasma from subjects with intra-abdominal visceral fat accumulation," *Clinical Endocrinology*, vol. 48, no. 4, pp. 515–520, 1998.

[30] D. Panarotto, P. Rémillard, L. Bouffard, and P. Maheux, "Insulin resistance affects the regulation of lipoprotein lipase in the postprandial period and in an adipose tissue-specific manner," *European Journal of Clinical Investigation*, vol. 32, no. 2, pp. 84–92, 2002.

[31] T. Mazzone, D. Foster, and A. Chait, "In vivo stimulation of low-density lipoprotein degradation by insulin," *Diabetes*, vol. 33, no. 4, pp. 333–338, 1984.

[32] T. Nakamura, K. Tokunaga, I. Shimomura et al., "Contribution of visceral fat accumulation to the development of coronary artery disease in non-obese men," *Atherosclerosis*, vol. 107, no. 2, pp. 239–246, 1994.

[33] B. Larsson, K. Svardsudd, L. Welin, L. Wilhelmsen, P. Bjorntorp, and G. Tibblin, "Abdominal adipose tissue distribution, obesity, and risk of cardiovascular disease and death: 13 year follow up of participants in the study of men born in 1913," *British Medical Journal*, vol. 288, no. 6428, pp. 1401–1404, 1984.

[34] C. Couillard, N. Bergeron, D. Prud'homme et al., "Gender difference in postprandial lipemia: importance of visceral adipose tissue accumulation," *Arteriosclerosis, Thrombosis, and Vascular Biology*, vol. 19, no. 10, pp. 2448–2455, 1999.

[35] N. Mekki, M. A. Christofilis, M. Charbonnier et al., "Influence of obesity and body fat distribution on postprandial lipemia and triglyceride-rich lipoproteins in adult women," *Journal of Clinical Endocrinology and Metabolism*, vol. 84, no. 1, pp. 184–191, 1999.

[36] S. Lemieux, J. P. Després, S. Moorjani et al., "Are gender differences in cardiovascular disease risk factors explained by the level of visceral adipose tissue?" *Diabetologia*, vol. 37, no. 8, pp. 757–764, 1994.

[37] K. Taira, M. Hikita, J. Kobayashi et al., "Delayed post-prandial lipid metabolism in subjects with intra-abdominal visceral fat accumulation," *European Journal of Clinical Investigation*, vol. 29, no. 4, pp. 301–308, 1999.

[38] P. Blackburn, B. Lamarche, C. Couillard et al., "Contribution of visceral adiposity to the exaggerated postprandial lipemia of men with impaired glucose tolerance," *Diabetes Care*, vol. 26, no. 12, pp. 3303–3309, 2003.

[39] E. Henkel, T. Temelkova-Kurktschiev, C. Koehler, J. Pietzsch, W. Leonhardt, and M. Hanefeld, "Impaired glucose tolerance is not associated with lipid intolerance," *Diabetes, Nutrition and Metabolism—Clinical and Experimental*, vol. 15, no. 2, pp. 84–90, 2002.

[40] K. Higashi, H. Shige, T. Ito et al., "Impaired glucose tolerance without hypertriglyceridemia does not enhance postprandial lipemia," *Hormone and Metabolic Research*, vol. 33, no. 2, pp. 101–105, 2001.

[41] G. G. Ziogas, T. R. Thomas, and W. S. Harris, "Exercise training, postprandial hypertriglyceridemia, and LDL subfraction distribution," *Medicine and Science in Sports and Exercise*, vol. 29, no. 8, pp. 986–991, 1997.

[42] G. H. Hartung, S. J. Lawrence, R. S. Reeves, and J. P. Foreyt, "Effect of alcohol and exercise on postprandial lipemia and triglyceride clearance in men," *Atherosclerosis*, vol. 100, no. 1, pp. 33–40, 1993.

[43] J. R. Merrill, R. G. Holly, R. L. Anderson, N. Rifai, M. E. King, and R. DeMeersman, "Hyperlipemic response of young trained and untrained men after a high fat meal," *Arteriosclerosis*, vol. 9, no. 2, pp. 217–223, 1989.

[44] J. C. Cohen, T. D. Noakes, and A. J. S. Benade, "Postprandial lipemia and chylomicron clearance in athletes and in sedentary men," *The American Journal of Clinical Nutrition*, vol. 49, no. 3, pp. 443–447, 1989.

[45] J. D. Brunzell, W. R. Hazzard, D. Porte Jr., and E. L. Bierman, "Evidence for a common, saturable, triglyceride removal mechanism for chylomicrons and very low density lipoproteins in man," *The Journal of Clinical Investigation*, vol. 52, no. 7, pp. 1578–1585, 1973.

[46] S. M. Grundy and H. Y. I. Mok, "Chylomicron clearance in normal and hyperlipidemic man," *Metabolism: Clinical and Experimental*, vol. 25, no. 11, pp. 1225–1239, 1976.

[47] G. Schlierf, A. Dinsenbacher, H. Kather, M. Kohlmeier, and W. Haberbosch, "Mitigation of alimentary lipemia by postprandial

exercise-Phenomena and mechanisms," *Metabolism*, vol. 36, no. 8, pp. 726–730, 1987.

[48] G. Annuzzi, E. Jansson, L. Kaijser, L. Holmquist, and L. A. Carlson, "Increased removal rate of exogenous triglycerides after prolonged exercise in man: time course and effect of exercise duration," *Metabolism*, vol. 36, no. 5, pp. 438–443, 1987.

[49] S. L. Herd, J. E. M. Lawrence, D. Malkova, M. H. Murphy, S. Mastana, and A. E. Hardman, "Postprandial lipemia in young men and women of contrasting training status," *Journal of Applied Physiology*, vol. 89, no. 5, pp. 2049–2056, 2000.

[50] N. V. Tsetsonis, A. E. Hardman, and S. S. Mastana, "Acute effects of exercise on postprandial lipemia: a comparative study in trained and untrained middle-aged women," *American Journal of Clinical Nutrition*, vol. 65, no. 2, pp. 525–533, 1997.

[51] A. E. Hardman, J. E. M. Lawrence, and S. L. Herd, "Postprandial lipemia in endurance-trained people during a short interruption to training," *Journal of Applied Physiology*, vol. 84, no. 6, pp. 1895–1901, 1998.

[52] N. M. F. Farah, D. Malkova, and J. M. R. Gill, "Effects of exercise on postprandial responses to ad libitum feeding in overweight men," *Medicine and Science in Sports and Exercise*, vol. 42, no. 11, pp. 2015–2022, 2010.

[53] H. E. Aldred, I. C. Perry, and A. E. Hardman, "The effect of a single bout of brisk walking on postprandial lipemia in normolipidemic young adults," *Metabolism: Clinical and Experimental*, vol. 43, no. 7, pp. 836–841, 1994.

[54] N. V. Tsetsonis, "Effects of low and moderate intensity treadmill walking on postprandial lipaemia in healthy young adults," *European Journal of Applied Physiology and Occupational Physiology*, vol. 73, no. 5, pp. 419–426, 1996.

[55] N. V. Tsetsonis and A. E. Hardman, "Reduction in postprandial lipemia after walking: influence of exercise intensity," *Medicine and Science in Sports and Exercise*, vol. 28, no. 10, pp. 1235–1242, 1996.

[56] J. M. R. Gill, M. H. Murphy, and A. E. Hardman, "Postprandial lipemia: effects of intermittent versus continuous exercise," *Medicine and Science in Sports and Exercise*, vol. 30, no. 10, pp. 1515–1520, 1998.

[57] M. Kolifa, A. Petridou, and V. Mougios, "Effect of prior exercise on lipemia after a meal of moderate fat content," *European Journal of Clinical Nutrition*, vol. 58, no. 10, pp. 1327–1335, 2004.

[58] M. Miyashita, S. F. Burns, and D. J. Stensel, "Exercise and postprandial lipemia: effect of continuous compared with intermittent activity patterns," *The American Journal of Clinical Nutrition*, vol. 83, no. 1, pp. 24–29, 2006.

[59] J. Q. Zhang, L. L. Ji, V. S. Fretwell, and G. Nunez, "Effect of exercise on postprandial lipemia in men with hypertriglyceridemia," *European Journal of Applied Physiology*, vol. 98, no. 6, pp. 575–582, 2006.

[60] J. Q. Zhang, L. L. Ji, D. L. Fogt, and V. S. Fretwell, "Effect of exercise duration on postprandial hypertriglyceridemia in men with metabolic syndrome," *Journal of Applied Physiology*, vol. 103, no. 4, pp. 1339–1345, 2007.

[61] M. L. Mestek, E. P. Plaisance, L. A. Ratcliff, J. K. Taylor, S. Wee, and P. W. Grandjean, "Aerobic exercise and postprandial lipemia in men with the metabolic syndrome," *Medicine and Science in Sports and Exercise*, vol. 40, no. 12, pp. 2105–2111, 2008.

[62] M. Miyashita, S. F. Burns, and D. J. Stensel, "Accumulating short bouts of brisk walking reduces postprandial plasma triacylglycerol concentrations and resting blood pressure in healthy young men," *The American Journal of Clinical Nutrition*, vol. 88, no. 5, pp. 1225–1231, 2008.

[63] M. Miyashita, "Effects of continuous versus accumulated activity patterns on postprandial triacylglycerol concentrations in obese men," *International Journal of Obesity*, vol. 32, no. 8, pp. 1271–1278, 2008.

[64] K. A. Shannon, R. M. Shannon, J. N. Clore, C. Gennings, B. J. Warren, and J. A. Potteiger, "Aerobic exercise alters postprandial lipemia in African American versus White women," *International Journal of Sport Nutrition and Exercise Metabolism*, vol. 18, no. 1, pp. 37–48, 2008.

[65] J. Q. Zhang, T. R. Thomas, and S. D. Ball, "Effect of exercise timing on postprandial lipemia and HDL cholesterol subfractions," *Journal of Applied Physiology*, vol. 85, no. 4, pp. 1516–1522, 1998.

[66] C. S. Katsanos and R. J. Moffatt, "Acute effects of premeal versus postmeal exercise on postprandial hypertriglyceridemia," *Clinical Journal of Sport Medicine*, vol. 14, no. 1, pp. 33–39, 2004.

[67] E. P. Plaisance, M. L. Mestek, A. J. Mahurin, J. K. Taylor, J. Moncada-Jimenez, and P. W. Grandjean, "Postprandial triglyceride responses to aerobic exercise and extended-release niacin," *American Journal of Clinical Nutrition*, vol. 88, no. 1, pp. 30–37, 2008.

[68] R. Silvestre, W. J. Kraemer, E. E. Quann et al., "Effects of exercise at different times on postprandial lipemia and endothelial function," *Medicine and Science in Sports and Exercise*, vol. 40, no. 2, pp. 264–274, 2008.

[69] C. S. Katsanos, P. W. Grandjean, and R. J. Moffatt, "Effects of low and moderate exercise intensity on postprandial lipemia and postheparin plasma lipoprotein lipase activity in physically active men," *Journal of Applied Physiology*, vol. 96, no. 1, pp. 181–188, 2004.

[70] M. Miyashita, S. F. Burns, and D. J. Stensel, "An update on accumulating exercise and postprandial lipaemia: translating theory into practice," *Journal of Preventive Medicine and Public Health*, vol. 46, supplement 1, pp. S3–S11, 2013.

[71] M. H. Murphy, A. M. Nevill, and A. E. Hardman, "Different patterns of brisk walking are equally effective in decreasing postprandial lipaemia," *International Journal of Obesity and Related Metabolic Disorders*, vol. 24, no. 10, pp. 1303–1309, 2000.

[72] M. Miyashita, S. F. Burns, and D. J. Stensel, "Acute effects of accumulating exercise on postprandial lipemia and C-reactive protein concentrations in young men," *International Journal of Sport Nutrition and Exercise Metabolism*, vol. 19, no. 6, pp. 569–582, 2009.

[73] A. P. Ferreira, C. B. Ferreira, V. C. de Souza et al., "The influence of intense intermittent versus moderate continuous exercise on postprandial lipemia," *Clinics*, vol. 66, no. 4, pp. 535–541, 2011.

[74] T. S. Altena, J. L. Michaelson, S. D. Ball, and T. R. Thomas, "Single sessions of intermittent and continuous exercise and postprandial lipemia," *Medicine and Science in Sports and Exercise*, vol. 36, no. 8, pp. 1364–1371, 2004.

[75] I. Y. Kim, S. Park, J. R. Trombold, and E. F. Coyle, "Effects of moderate- and intermittent low-intensity exercise on postprandial lipemia," *Medicine and Science in Sports and Exercise*, 2014.

[76] B. Gabriel, A. Ratkevicius, P. Gray, M. P. Frenneaux, and S. R. Gray, "High-intensity exercise attenuates postprandial lipaemia and markers of oxidative stress," *Clinical Science*, vol. 123, no. 5, pp. 313–321, 2012.

[77] J. R. Trombold, K. M. Christmas, D. R. MacHin, I. Kim, and E. F. Coyle, "Acute high-intensity endurance exercise is more effective than moderate-intensity exercise for attenuation of postprandial triglyceride elevation," *Journal of Applied Physiology*, vol. 114, no. 6, pp. 792–800, 2013.

[78] E. Allen, P. Gray, A. Kollias-Pearson et al., "The effect of short-duration sprint interval exercise on plasma postprandial triacylglycerol levels in young men," *Journal of Sports Sciences*, vol. 32, no. 10, pp. 911–916, 2014.

[79] M. S. Tan, A. Mok, M. C. Yap, and S. F. Burns, "Effect of sprint interval versus continuous cycling on postprandial lipaemia," *Journal of Sports Sciences*, vol. 31, no. 9, pp. 989–995, 2013.

[80] D. S. Petitt, S. Á. Arngrímsson, and K. J. Cureton, "Effect of resistance exercise on postprandial lipemia," *Journal of Applied Physiology*, vol. 94, no. 2, pp. 694–700, 2003.

[81] A. Zafeiridis, E. Goloi, A. Petridou, K. Dipla, V. Mougios, and S. Kellis, "Effects of low- and high-volume resistance exercise on postprandial lipaemia," *The British Journal of Nutrition*, vol. 97, no. 3, pp. 471–477, 2007.

[82] A. Singhal, J. L. Trilk, N. T. Jenkins, K. A. Bigelman, and K. J. Cureton, "Effect of intensity of resistance exercise on postprandial lipemia," *Journal of Applied Physiology*, vol. 106, no. 3, pp. 823–829, 2009.

[83] P. M. Davitt, S. M. Arent, M. A. Tuazon, D. L. Golem, and G. C. Henderson, "Postprandial triglyceride and free fatty acid metabolism in obese women after either endurance or resistance exercise," *Journal of Applied Physiology*, vol. 114, no. 12, pp. 1743–1754, 2013.

[84] Z. K. Pafili, G. C. Bogdanis, N. V. Tsetsonis, and M. Maridaki, "Postprandial lipemia 16 and 40 hours after low-volume eccentric resistance exercise," *Medicine and Science in Sports and Exercise*, vol. 41, no. 2, pp. 375–382, 2009.

[85] K. A. Shannon, R. M. Shannon, J. N. Clore, C. Gennings, B. J. Warren, and J. A. Potteiger, "Resistance exercise and postprandial lipemia: the dose effect of differing volumes of acute resistance exercise bouts," *Metabolism: Clinical and Experimental*, vol. 54, no. 6, pp. 756–763, 2005.

[86] S. F. Burns, H. Corrie, E. Holder, T. Nightingale, and D. J. Stensel, "A single session of resistance exercise does not reduce postprandial lipaemia," *Journal of Sports Sciences*, vol. 23, no. 3, pp. 251–260, 2005.

[87] S. F. Burns, D. R. Broom, M. Miyashita, C. Ueda, and D. J. Stensel, "Increased postprandial triacylglycerol concentrations following resistance exercise," *Medicine and Science in Sports and Exercise*, vol. 38, no. 3, pp. 527–533, 2006.

[88] M. Pfeiffer, C. Wenk, and P. C. Colombani, "The influence of 30 minutes of light to moderate intensity cycling on postprandial lipemia," *European Journal of Cardiovascular Prevention and Rehabilitation*, vol. 13, no. 3, pp. 363–368, 2006.

[89] J. M. R. Gill and A. E. Hardman, "Exercise and postprandial lipid metabolism: an update on potential mechanisms and interactions with high-carbohydrate diets (review)," *The Journal of Nutritional Biochemistry*, vol. 14, no. 3, pp. 122–132, 2003.

[90] J. M. R. Gill, G. P. Mees, K. N. Frayn, and A. E. Hardman, "Moderate exercise, postprandial lipaemia and triacylglycerol clearance," *European Journal of Clinical Investigation*, vol. 31, no. 3, pp. 201–207, 2001.

[91] J. M. R. Gill, K. N. Frayn, S. A. Wootton, G. J. Miller, and A. E. Hardman, "Effects of prior moderate exercise on exogenous and endogenous lipid metabolism and plasma factor VII activity," *Clinical Science*, vol. 100, no. 5, pp. 517–527, 2001.

[92] D. Malkova, R. D. Evans, K. N. Frayn, S. M. Humphreys, P. R. Jones, and A. E. Hardman, "Prior exercise and postprandial substrate extraction across the human leg," *American Journal of Physiology: Endocrinology and Metabolism*, vol. 279, no. 5, pp. E1020–E1028, 2000.

[93] R. L. Seip, K. Mair, T. G. Cole, and C. F. Semenkovich, "Induction of human skeletal muscle lipoprotein lipase gene expression by short-term exercise is transient," *The American Journal of Physiology—Endocrinology and Metabolism*, vol. 272, no. 2, part 1, pp. E255–E261, 1997.

[94] R. L. Seip and C. F. Semenkovich, "Skeletal muscle lipoprotein lipase: molecular regulation and physiological effects in relation to exercise," *Exercise and Sport Sciences Reviews*, vol. 26, pp. 191–218, 1998.

[95] B. Kiens, H. Lithell, K. J. Mikines, and E. A. Richter, "Effects of insulin and exercise on muscle lipoprotein lipase activity in man and its relation to insulin action," *Journal of Clinical Investigation*, vol. 84, no. 4, pp. 1124–1129, 1989.

[96] M. A. Kantor, E. M. Cullinane, P. N. Herbert, and P. D. Thompson, "Acute increase in lipoprotein lipase following prolonged exercise," *Metabolism: Clinical and Experimental*, vol. 33, no. 5, pp. 454–457, 1984.

[97] M. A. Kantor, E. M. Cullinane, S. P. Sady, P. N. Herbert, and P. D. Thompson, "Exercise acutely increases high density lipoprotein-cholesterol and lipoprotein lipase activity in trained and untrained men," *Metabolism: Clinical and Experimental*, vol. 36, no. 2, pp. 188–192, 1987.

[98] S. P. Sady, P. D. Thompson, E. M. Cullinane, M. A. Kantor, E. Domagala, and P. N. Herbert, "Prolonged exercise augments plasma triglyceride clearance," *Journal of the American Medical Association*, vol. 256, no. 18, pp. 2552–2555, 1986.

[99] R. L. Seip, T. J. Angelopoulos, and C. F. Semenkovich, "Exercise induces human lipoprotein lipase gene expression in skeletal muscle but not adipose tissue," *American Journal of Physiology: Endocrinology and Metabolism*, vol. 268, no. 2, part 1, pp. E229–E236, 1995.

[100] P. W. Grandjean, S. F. Crouse, and J. J. Rohack, "Influence of cholesterol status on blood lipid and lipoprotein enzyme responses to aerobic exercise," *Journal of Applied Physiology*, vol. 89, no. 2, pp. 472–480, 2000.

[101] P. M. Gordon, F. L. Goss, P. S. Visich et al., "The acute effects of exercise intensity on HDL-C metabolism," *Medicine and Science in Sports and Exercise*, vol. 26, no. 6, pp. 671–677, 1994.

[102] M. A. Ferguson, N. L. Alderson, S. G. Trost, D. A. Essig, J. R. Burke, and J. L. Durstine, "Effects of four different single exercise sessions on lipids, lipoproteins, and lipoprotein lipase," *Journal of Applied Physiology*, vol. 85, no. 3, pp. 1169–1174, 1998.

[103] J. M. R. Gill, S. L. Herd, V. Vora, and A. E. Hardman, "Effects of a brisk walk on lipoprotein lipase activity and plasma triglyceride concentrations in the fasted and postprandial states," *European Journal of Applied Physiology*, vol. 89, no. 2, pp. 184–190, 2003.

[104] S. D. Weise, P. W. Grandjean, J. J. Rohack, J. W. Womack, and S. F. Crouse, "Acute changes in blood lipids and enzymes in postmenopausal women after exercise," *Journal of Applied Physiology*, vol. 99, no. 2, pp. 609–615, 2005.

[105] S. L. Herd, B. Kiens, L. H. Boobis, and A. E. Hardman, "Moderate exercise, postprandial lipemia, and skeletal muscle lipoprotein lipase activity," *Metabolism: Clinical and Experimental*, vol. 50, no. 7, pp. 756–762, 2001.

[106] K. N. Frayn, S. Shadid, R. Hamlani et al., "Regulation of fatty acid movement in human adipose tissue in the postabsorptive-to-postprandial transition," *The American Journal of Physiology: Endocrinology and Metabolism*, vol. 266, no. 3, pp. E308–E317, 1994.

[107] T. J. Yost, D. R. Jensen, B. R. Haugen, and R. H. Eckel, "Effect of dietary macronutrient composition on tissue-specific

lipoprotein lipase activity and insulin action in normal-weight subjects," *The American Journal of Clinical Nutrition*, vol. 68, no. 2, pp. 296–302, 1998.

[108] J. L. Potts, R. M. Fisher, S. M. Humphreys, S. W. Coppack, G. F. Gibbons, and K. N. Frayn, "Peripheral triacylglycerol extraction in the fasting and post-prandial states," *Clinical Science*, vol. 81, no. 5, pp. 621–626, 1991.

[109] T. T. Nguyen, A. H. Mijares, C. M. Johnson, and M. D. Jensen, "Postprandial leg and splanchnic fatty acid metabolism in nonobese men and women," *American Journal of Physiology: Endocrinology and Metabolism*, vol. 271, no. 6, part 1, pp. E965–E972, 1996.

[110] G. M. Reaven and E. P. Reaven, "Prevention of age-related hypertriglyceridemia by caloric restriction and exercise training in the rat," *Metabolism: Clinical and Experimental*, vol. 30, no. 10, pp. 982–986, 1981.

[111] C. Simonelli and R. P. Eaton, "Reduced triglyceride secretion: a metabolic consequence of chronic exercise," *The American Journal of Physiology—Endocrinology and Metabolism*, vol. 234, no. 3, pp. E221–E227, 1978.

[112] I. Zavaroni, Y. I. Chen, C. E. Mondon, and G. M. Reaven, "Ability of exercise to inhibit carbohydrate-induced hypertriglyceridemia in rats," *Metabolism*, vol. 30, no. 5, pp. 476–480, 1981.

[113] A. D. Sniderman and K. Cianflone, "Substrate delivery as a determinant of hepatic apoB secretion," *Arteriosclerosis and Thrombosis: A Journal of Vascular Biology/American Heart Association*, vol. 13, no. 5, pp. 629–635, 1993.

[114] M. J. Dekker, T. E. Graham, T. C. Ooi, and L. E. Robinson, "Exercise prior to fat ingestion lowers fasting and postprandial VLDL and decreases adipose tissue IL-6 and GIP receptor mRNA in hypertriacylglycerolemic men," *The Journal of Nutritional Biochemistry*, vol. 21, no. 10, pp. 983–990, 2010.

[115] E. Bellou, A. Siopi, M. Galani et al., "Acute effects of exercise and calorie restriction on triglyceride metabolism in women," *Medicine and Science in Sports and Exercise*, vol. 45, no. 3, pp. 455–461, 2013.

[116] P. A. Mayes and J. M. Felts, "Regulation of fat metabolism in the liver," *Nature*, vol. 215, no. 5102, pp. 716–718, 1967.

[117] M. E. Laker and P. A. Mayes, "Regulation of 3-hydroxybutyrate formation and secretion of very low density lipoprotein triacylglycerol by perfused livers from fed and starved rats," *Biochemical Journal*, vol. 206, no. 2, pp. 427–430, 1982.

[118] F. Magkos, D. C. Wright, B. W. Patterson, B. S. Mohammed, and B. Mittendorfer, "Lipid metabolism response to a single, prolonged bout of endurance exercise in healthy young men," *American Journal of Physiology—Endocrinology and Metabolism*, vol. 290, no. 2, pp. E355–E362, 2006.

[119] N. V. Tsetsonis, A. E. Hardman, and S. S. Mastana, "Acute effects of exercise on postprandial lipemia: a comparative study in trained and untrained middle-aged women," *The American Journal of Clinical Nutrition*, vol. 65, no. 2, pp. 525–533, 1997.

[120] Y. Manabe, K. S. C. Gollisch, L. Holton et al., "Exercise training-induced adaptations associated with increases in skeletal muscle glycogen content," *The FEBS Journal*, vol. 280, no. 3, pp. 916–926, 2013.

[121] D. Malkova, A. E. Hardman, R. J. Bowness, and I. A. Macdonald, "The reduction in postprandial lipemia after exercise is independent of the relative contributions of fat and carbohydrate to energy metabolism during exercise," *Metabolism: Clinical and Experimental*, vol. 48, no. 2, pp. 245–251, 1999.

[122] M. Maraki, F. Magkos, N. Christodoulou et al., "One day of moderate energy deficit reduces fasting and postprandial triacylglycerolemia in women: the role of calorie restriction and exercise," *Clinical Nutrition*, vol. 29, no. 4, pp. 459–463, 2010.

[123] F. L. Burton, D. Malkova, M. J. Caslake, and J. M. Gill, "Energy replacement attenuates the effects of prior moderate exercise on postprandial metabolism in overweight/obese men," *International Journal of Obesity*, vol. 32, no. 3, pp. 481–489, 2008.

[124] M. Harrison, D. J. O'Gorman, N. McCaffrey et al., "Influence of acute exercise with and without carbohydrate replacement on postprandial lipid metabolism," *Journal of Applied Physiology*, vol. 106, no. 3, pp. 943–949, 2009.

[125] M. J. Sharman, A. L. Gómez, W. J. Kraemer, and J. S. Volek, "Very low-carbohydrate and low-fat diets affect fasting lipids and postprandial lipemia differently in overweight men," *The Journal of Nutrition*, vol. 134, no. 4, pp. 880–885, 2004.

[126] J. S. Volek, M. J. Sharman, A. L. Gómez, T. P. Scheett, and W. J. Kraemer, "An isoenergetic very low carbohydrate diet improves serum HDL cholesterol and triacylglycerol concentrations, the total cholesterol to HDL cholesterol ratio and postprandial lipemic responses compared with a low fat diet in normal weight, normolipidemic women," *Journal of Nutrition*, vol. 133, no. 9, pp. 2756–2761, 2003.

[127] R. M. Krauss, R. H. Eckel, B. Howard et al., "AHA dietary guidelines revision 2000: a statement for healthcare professionals from the nutrition committee of the American heart association," *Circulation*, vol. 102, no. 18, pp. 2284–2299, 2000.

[128] R. P. Mensink and M. B. Katan, "Effect of dietary fatty acids on serum lipids and lipoproteins: a meta- analysis of 27 trials," *Arteriosclerosis and Thrombosis*, vol. 12, no. 8, pp. 911–919, 1992.

[129] R. D. Feinman and J. S. Volek, "Low carbohydrate diets improve atherogenic dyslipidemia even in the absence of weight loss," *Nutrition & Metabolism*, vol. 3, article 24, 2006.

[130] M. A. Austin, "Epidemiology of hypertriglyceridemia and cardiovascular disease," *The American Journal of Cardiology*, vol. 83, no. 9B, pp. 13F–16F, 1999.

[131] A. S. Truswell, "Food carbohydrates and plasma lipids—an update," *The American Journal of Clinical Nutrition*, vol. 59, supplement 3, pp. 710S–718S, 1994.

[132] C. E. West, D. R. Sullivan, M. B. Katan, I. L. Halferkamps, and H. W. van der Torre, "Boys from populations with high-carbohydrate intake have higher fasting triglyceride levels than boys from populations with high-fat intake," *American Journal of Epidemiology*, vol. 131, no. 2, pp. 271–282, 1990.

[133] E. J. Parks and M. K. Hellerstein, "Carbohydrate-induced hypertriacylglycerolemia: historical perspective and review of biological mechanisms," *American Journal of Clinical Nutrition*, vol. 71, no. 2, pp. 412–433, 2000.

[134] J. Volek, M. Sharman, A. Gomez et al., "Comparison of energy-restricted very low-carbohydrate and low-fat diets on weight loss and body composition in overweight men and women," *Nutrition & Metabolism*, vol. 1, no. 1, article 13, 2004.

[135] J. S. Volek, S. D. Phinney, C. E. Forsythe et al., "Carbohydrate restriction has a more favorable impact on the metabolic syndrome than a low fat diet," *Lipids*, vol. 44, no. 4, pp. 297–309, 2009.

[136] M. J. Sharman, W. J. Kraemer, D. M. Love et al., "A ketogenic diet favorably affects serum biomarkers for cardiovascular disease in normal-weight men," *The Journal of Nutrition*, vol. 132, no. 7, pp. 1879–1885, 2002.

[137] C. Koutsari, D. Malkova, and A. E. Hardman, "Postprandial lipemia after short-term variation in dietary fat and carbohydrate," *Metabolism: Clinical and Experimental*, vol. 49, no. 9, pp. 1150–1155, 2000.

[138] M. F. Chong, B. A. Fielding, and K. N. Frayn, "Mechanisms for the acute effect of fructose on postprandial lipemia," *The American Journal of Clinical Nutrition*, vol. 85, no. 6, pp. 1511–1520, 2007.

[139] M. C. Moore, S. N. Davis, S. L. Mann, and A. D. Cherrington, "Acute fructose administration improves oral glucose tolerance in adults with type 2 diabetes," *Diabetes Care*, vol. 24, no. 11, pp. 1882–1887, 2001.

[140] C. M. Donmoyer, J. Ejiofor, D. B. Lacy, S. Chen, and O. P. McGuinness, "Fructose augments infection-impaired net hepatic glucose uptake during TPN administration," *American Journal of Physiology: Endocrinology and Metabolism*, vol. 280, no. 5, pp. E703–E711, 2001.

[141] S. S. Elliott, N. L. Keim, J. S. Stern, K. Teff, and P. J. Havel, "Fructose, weight gain, and the insulin resistance syndrome," *The American Journal of Clinical Nutrition*, vol. 76, no. 5, pp. 911–922, 2002.

[142] K. L. Teff, S. S. Elliott, M. Tschöp et al., "Dietary fructose reduces circulating insulin and leptin, attenuates postprandial suppression of ghrelin, and increases triglycerides in women," *Journal of Clinical Endocrinology and Metabolism*, vol. 89, no. 6, pp. 2963–2972, 2004.

[143] D. Faeh, K. Minehira, J. Schwarz, R. Periasami, P. Seongsu, and L. Tappy, "Effect of fructose overfeeding and fish oil administration on hepatic de novo lipogenesis and insulin sensitivity in healthy men," *Diabetes*, vol. 54, no. 7, pp. 1907–1913, 2005.

[144] K. L. Stanhope, J. M. Schwarz, N. L. Keim et al., "Consuming fructose-sweetened, not glucose-sweetened, beverages increases visceral adiposity and lipids and decreases insulin sensitivity in overweight/obese humans," *The Journal of Clinical Investigation*, vol. 119, no. 5, pp. 1322–1334, 2009.

[145] J. C. L. Mamo, T. Hirano, L. James, L. Szeto, and G. Steiner, "Partial characterization of the fructose-induced defect in very-low-density lipoprotein triglyceride metabolism," *Metabolism: Clinical and Experimental*, vol. 40, no. 9, pp. 888–893, 1991.

[146] D. M. Watkin, H. F. Froeb, F. T. Hatch, and A. B. Gutman, "Effects of diet in essential hypertension. II. Results with unmodified kempner rice diet in fifty hospitalized patients," *The American Journal of Medicine*, vol. 9, no. 4, pp. 441–493, 1950.

[147] E. Ahrens Jr., W. Insull Jr., R. Blomstrand, J. Hirsch, T. Tsaltas, and M. Peterson, "The influence of dietary fats on serum-lipid levels in man," *The Lancet*, vol. 269, no. 6976, pp. 943–953, 1957.

[148] J. W. Farquhar, G. M. Reaven, R. Gross, and R. Wagner, "Rate of plasma triglyceride synthesis in carbohydrate-induced lipaemia," *Journal of Clinical Investigation*, vol. 42, p. 930, 1963.

[149] R. J. Havel, "Conversion of plasma free fatty acids into triglycerides of plasma lipoprotein fractions in man," *Metabolism: Clinical and Experimental*, vol. 10, pp. 1031–1034, 1961.

[150] G. M. Reaven, D. B. Hill, R. C. Gross, and J. W. Farquhar, "Kinetics of triglyceride turnover of very low density lipoproteins of human plasma.," *The Journal of Clinical Investigation*, vol. 44, no. 11, pp. 1826–1833, 1965.

[151] G. M. Reaven, R. L. Lerner, M. P. Stern, and J. W. Farquhar, "Role of insulin in endogenous hypertriglyceridemia," *The Journal of Clinical Investigation*, vol. 46, no. 11, pp. 1756–1767, 1967.

[152] J. S. Volek, M. L. Fernandez, R. D. Feinman, and S. D. Phinney, "Dietary carbohydrate restriction induces a unique metabolic state positively affecting atherogenic dyslipidemia, fatty acid partitioning, and metabolic syndrome," *Progress in Lipid Research*, vol. 47, no. 5, pp. 307–318, 2008.

[153] K. F. Petersen, S. Dufour, D. B. Savage et al., "The role of skeletal muscle insulin resistance in the pathogenesis of the metabolic syndrome," *Proceedings of the National Academy of Sciences of the United States of America*, vol. 104, no. 31, pp. 12587–12594, 2007.

[154] L. M. Sparks, H. Xie, R. A. Koza, R. Mynatt, G. A. Bray, and S. R. Smith, "High-fat/low-carbohydrate diets regulate glucose metabolism via a long-term transcriptional loop," *Metabolism: Clinical and Experimental*, vol. 55, no. 11, pp. 1457–1463, 2006.

[155] H. C. Bucher, P. Hengstler, C. Schindler, and G. Meier, "N-3 polyunsaturated fatty acids in coronary heart disease: a meta-analysis of randomized controlled trials," *The American Journal of Medicine*, vol. 112, no. 4, pp. 298–304, 2002.

[156] E. P. Plaisance, P. W. Grandjean, and A. J. Mahurin, "Independent and combined effects of aerobic exercise and pharmacological strategies on serum triglyceride concentrations: a qualitative review," *The Physician and Sportsmedicine*, vol. 37, no. 1, pp. 11–19, 2009.

[157] L. F. Tinker, E. J. Parks, S. R. Behr, B. O. Schneeman, and P. A. Davis, "(n-3) Fatty acid supplementation in moderately hypertriglyceridemic adults changes postprandial lipid and apolipoprotein B responses to a standardized test meal," *The Journal of Nutrition*, vol. 129, no. 6, pp. 1126–1134, 1999.

[158] K. M. Slivkoff-Clark, A. P. James, and J. C. L. Mamo, "The chronic effects of fish oil with exercise on postprandial lipaemia and chylomicron homeostasis in insulin resistant viscerally obese men," *Nutrition & Metabolism*, vol. 9, article 9, 2012.

[159] D. S. Kelley, D. Siegel, M. Vemuri, and B. E. Mackey, "Docosahexaenoic acid supplementation improves fasting and postprandial lipid profiles in hypertriglyceridemic men," *The American Journal of Clinical Nutrition*, vol. 86, no. 2, pp. 324–333, 2007.

[160] H. M. Roche and M. J. Gibney, "Long-chain $n-3$ polyunsaturated fatty acids and triacylglycerol metabolism in the postprandial state," *Lipids*, vol. 34, no. 6, pp. S259–S265, 1999.

[161] T. R. Thomas, B. A. Fischer, W. B. Kist, K. E. Horner, and R. H. Cox, "Effects of exercise and n-3 fatty acids on postprandial lipemia," *Journal of Applied Physiology*, vol. 88, no. 6, pp. 2199–2204, 2000.

[162] H. E. Hanwell, C. D. Kay, J. W. Lampe, B. J. Holub, and A. M. Duncan, "Acute fish oil and soy isoflavone supplementation increase postprandial serum (n-3) polyunsaturated fatty acids and isoflavones but do not affect triacylglycerols or biomarkers of oxidative stress in overweight and obese hypertriglyceridemic men," *The Journal of Nutrition*, vol. 139, no. 6, pp. 1128–1134, 2009.

[163] R. A. Baillie, R. Takada, M. Nakamura, and S. D. Clarke, "Coordinate induction of peroxisomal acyl-CoA oxidase and UCP-3 by dietary fish oil: a mechanism for decreased body fat deposition," *Prostaglandins Leukotrienes and Essential Fatty Acids*, vol. 60, no. 5-6, pp. 351–356, 1999.

[164] D. B. Jump, "The biochemistry of n-3 polyunsaturated fatty acids," *The Journal of Biological Chemistry*, vol. 277, no. 11, pp. 8755–8758, 2002.

[165] B. M. Forman, J. Chen, and R. M. Evans, "Hypolipidemic drugs, polyunsaturated fatty acids, and eicosanoids are ligands for peroxisome proliferator-activated receptors α and δ," *Proceedings of the National Academy of Sciences of the United States of America*, vol. 94, no. 9, pp. 4312–4317, 1997.

[166] A. T. Wong, D. C. Chan, P.H. Barrett, L. A. Adams, and G. F. Watts, "Effect of omega-3 fatty acid ethyl esters on apolipoprotein B-48 kinetics in obese subjects on a weight loss diet: a new tracer kinetic study in the postprandial state," *The Journal of Clinical Endocrinology and Metabolism*, 2014.

[167] W. S. Harris, W. E. Connor, N. Alam, and D. R. Illingworth, "Reduction of postprandial triglyceridemia in humans by dietary n-3 fatty acids," *Journal of Lipid Research*, vol. 29, no. 11, pp. 1451–1460, 1988.

[168] W. S. Harris and F. Muzio, "Fish oil reduces postprandial triglyceride concentrations without accelerating lipid-emulsion removal rates," *The American Journal of Clinical Nutrition*, vol. 58, no. 1, pp. 68–74, 1993.

[169] B. K. Smith, G. Y. Sun, O. M. Donahue, and T. R. Thomas, "Exercise plus n-3 fatty acids: additive effect on postprandial lipemia," *Metabolism: Clinical and Experimental*, vol. 53, no. 10, pp. 1365–1371, 2004.

[170] T. R. Thomas, Y. Liu, M. A. Linden, and R. S. Rector, "Interaction of exercise training and *n*-3 fatty acid supplementation on postprandial lipemia," *Applied Physiology, Nutrition and Metabolism*, vol. 32, no. 3, pp. 473–480, 2007.

[171] J. N. S. Matthews, D. G. Altman, M. J. Campbell, and P. Royston, "Analysis of serial measurements in medical research," *British Medical Journal*, vol. 300, no. 6719, pp. 230–235, 1990.

[172] E. P. Weiss, D. A. Fields, B. Mittendorfer, M. A. D. Haverkort, and S. Klein, "Reproducibility of postprandial lipemia tests and validity of an abbreviated 4-hour test," *Metabolism: Clinical and Experimental*, vol. 57, no. 10, pp. 1479–1485, 2008.

[173] H. H. Ditschuneit, M. Flechtner-Mors, E. Hagel, and H. Ditschuneit, "Postprandial lipoprotein metabolism in obese patients with moderate hypertriglyceridaemia: effects of gemfibrozil," *The Journal of International Medical Research*, vol. 20, no. 3, pp. 197–210, 1992.

[174] M. Miller, P. O. Kwiterovich Jr., P. S. Bachorik, and A. Georgopoulos, "Decreased postprandial response to a fat meal in normotriglyceridemic men with hypoalphalipoproteinemia," *Arteriosclerosis, Thrombosis, and Vascular Biology*, vol. 13, no. 3, pp. 385–392, 1993.

[175] R. Scott Rector, M. A. Linden, J. Q. Zhang et al., "Predicting postprandial lipemia in healthy adults and in at-risk individuals with components of the cardiometabolic syndrome," *Journal of Clinical Hypertension*, vol. 11, no. 11, pp. 663–671, 2009.

[176] M. Maraki, N. Aggelopoulou, N. Christodoulou et al., "Validity of abbreviated oral fat tolerance tests for assessing postprandial lipemia," *Clinical Nutrition*, vol. 30, no. 6, pp. 852–857, 2011.

[177] F. Berr and F. Kern Jr., "Plasma clearance of chylomicrons labeled with retinyl palmitate in healthy human subjects," *Journal of Lipid Research*, vol. 25, no. 8, pp. 805–812, 1984.

[178] D. S. Goodman, "Vitamin A metabolism," *Federation Proceedings*, vol. 39, no. 10, pp. 2716–2722, 1980.

[179] D. L. Sprecher, S. L. Knauer, D. M. Black et al., "Chylomicron-retinyl palmitate clearance in type I hyperlipidemic families," *Journal of Clinical Investigation*, vol. 88, no. 3, pp. 985–994, 1991.

[180] G. F. Lewis, N. M. O'Meara, P. A. Soltys et al., "Postprandial lipoprotein metabolism in normal and obese subjects: comparison after the vitamin A fat-loading test," *The Journal of Clinical Endocrinology and Metabolism*, vol. 71, no. 4, pp. 1041–1050, 1990.

[181] J. S. Cohn, E. J. Johnson, J. S. Millar et al., "Contribution of apoB-48 and apoB-100 triglyceride-rich lipoproteins (TRL) to postprandial increases in the plasma concentration of TRL triglycerides and retinyl esters," *Journal of Lipid Research*, vol. 34, no. 12, pp. 2033–2040, 1993.

[182] S. D. Krasinski, J. S. Cohn, R. M. Russell, and E. J. Schaefer, "Postprandial plasma vitamin A metabolism in humans: a reassessment of the use of plasma retinyl esters as markers for intestinally derived chylomicrons and their remnants," *Metabolism: Clinical and Experimental*, vol. 39, no. 4, pp. 357–365, 1990.

[183] S. Lemieux, R. Fontani, K. D. Uffelman, G. F. Lewis, and G. Steiner, "Apolipoprotein B-48 and retinyl palmitate are not equivalent markers of postprandial intestinal lipoproteins," *Journal of Lipid Research*, vol. 39, no. 10, pp. 1964–1971, 1998.

[184] E. J. Campos, N. Qureshi, and H. P. Blaschek, "Production of acetone butanol ethanol from degermed corn using Clostridium beijerinckii BA101," *Applied Biochemistry and Biotechnology A: Enzyme Engineering and Biotechnology*, vol. 98–100, pp. 553–561, 2002.

[185] S. Otokozawa, M. Ai, M. R. Diffenderfer et al., "Fasting and postprandial apolipoprotein B-48 levels in healthy, obese, and hyperlipidemic subjects," *Metabolism: Clinical and Experimental*, vol. 58, no. 11, pp. 1536–1542, 2009.

[186] C. Couillard, J. Gagnon, J. Bergeron et al., "Contribution of body fatness and adipose tissue distribution to the age variation in plasma steroid hormone concentrations in men: the HERITAGE family study," *The Journal of Clinical Endocrinology & Metabolism*, vol. 85, no. 3, pp. 1026–1031, 2000.

16

The Intake of Energy and Selected Nutrients by Thai Urban Sedentary Workers: An Evaluation of Adherence to Dietary Recommendations

Katiya Ivanovitch,[1] Jeeranun Klaewkla,[1] Rewadee Chongsuwat,[1] Chukiat Viwatwongkasem,[2] and Wanicha Kitvorapat[3]

[1] *Department of Nutrition, Faculty of Public Health, Mahidol University, 420/1 Ratchawithi Road, Ratchathewi, Bangkok 10400, Thailand*
[2] *Department of Biostatistics, Faculty of Public Health, Mahidol University, 420/1 Ratchawithi Road, Ratchathewi, Bangkok 10400, Thailand*
[3] *Regional Health Promotion Center 8, Nakhon Sawan, 157 Moo 1, Phahonyothin Road, Muang District, Nakhon Sawan 6000, Thailand*

Correspondence should be addressed to Jeeranun Klaewkla; jeeranun.kla@mahidol.ac.th

Academic Editor: Joel B. Mason

Rapid changes in Thailand's nutrition and lifestyles have led to increasing diet-related pathologies among people with sedentary occupations. This study examines the extent to which the dietary intake of nutrients and energy by a sample of Thai sedentary workers conforms to the Thai Dietary Reference Intakes (Thai DRIs). The nutrients and energy intake estimates were based on self-reported information collected with a single 24-hour dietary recall and nonweighed 2-day food record. The study participants were Thai adults aged 20–50 years employed in sedentary occupations. A convenience sample of 215 healthy individuals (75 males and 140 females) was based on four randomly selected worksites in the Bangkok metropolitan area. For male participants, the study found a median energy intake of 1,485 kcal/day, with 54.4% of energy coming from carbohydrate, 15.9% from protein, and 29.6% from fat. Females' median energy intake was 1,428 kcal/day, 56% of which came from carbohydrate, 16.2% from protein, and 28.6% from fat. Both genders showed insufficient intake of fiber and most micronutrients. This study provides the material for preventive public health interventions focusing on nutrition-related diseases affecting Thailand's rapidly growing sedentary workforce.

1. Introduction

The objective of this study is to examine nutrition-related risks of overweight/obesity for a group of sedentary workers in the Bangkok metropolitan area. The main methodological step in that process is a comparison of selected nutrients and energy intakes of a sample of 215 sedentary workers with the Thai Dietary Reference Intakes (Thai DRIs). Epidemiological studies typically use the term sedentary to describe physically inactive individuals [1]. A somewhat more nuanced view has also been adopted to describe sedentary behavior as a prolonged sitting rather than a severe lack of physical activity. Sedentary lifestyles are characterized by low energy expenditure levels, usually less than 1.5 METs (metabolic equivalents) [2]. Evidence also shows that sedentary behaviors are associated with growing health risks [3, 4].

A prolonged occupational sitting is a fact of daily life in both developed and developing economies. In Thailand, for example, automation, urbanization, and mass transit travel have led to increasingly sedentary lifestyles [5, 6]. As a result, 13 million people, or 33% of the active civilian labor force, are becoming seriously affected by overweight and obesity [7, 8]. The acute nature of that public health problem is highlighted by the fact that 47.3% of men and 57.5% of women in adult Thai population have been classified as overweight/obese [9], based on the Asian BMI criteria (BMI $\geq 23\,\mathrm{kg/m^2}$) [10]. This growing public health problem has negative implications for Thailand's productive capacity, because excessive body weight leads to increased mortality risks [11, 12] caused by type 2 diabetes, cardiovascular diseases, and various forms of cancer [13, 14]. In addition to their negative impact on productivity and workplace safety, these illnesses also require a broad range of medical treatments, creating an increasing burden on public health services and constant upward pressures on healthcare costs [15, 16].

As in many other countries, Thailand's public health authorities have used dietary guidelines to inform and assist consumers in making healthy nutrition choices and in following healthy lifestyles. However, no specific public health intervention has been conducted with regard to sedentary workers, partly because of the lack of adequate data on their nutrient and energy intakes. A research was, therefore, needed to examine the extent to which estimated energy and nutrients intakes of adults with sedentary occupations conform to the current Thai DRIs [17]. This paper is a first attempt to fill that important gap in public health research based on evidence collected in Thailand.

2. Materials and Methods

2.1. Workplaces and Participants. The data collection phase for this study commenced in February 2012 and was completed in April 2012. The study's research protocol was reviewed and approved by the Institutional Review Research in Human Subjects Committee, Faculty of Public Health, Mahidol University (COA. number: MUPH 2012-001). At each worksite, data were collected for two weeks. Letters were sent to each worksite to explain the study's purpose and objectives. Informed consent forms were read and signed by all study subjects. The participants in this study were working adults aged 20–50 years. They were part of a convenience sample based on four randomly selected worksites in the Bangkok metropolitan area. All the participants satisfied sedentary job characteristics: sitting occupation/desk-bound jobs, involving at least 4 hours per day of sitting during work hours. Most of the sedentary jobs were clerical, administrative, or data processing jobs. Excluded from the study were (1) people taking medication to control blood pressure or plasma lipids, (2) athletes/sports persons, (3) individuals on diets, (4) pregnant and/or lactating women, (5) people with self-reported chronic diseases, (6) persons with diseases that induce weight loss or weight gain, and (7) persons with physical or psychological disabilities.

The sample size of the study was determined by a priori power statistical analysis using the G*Power, Version 3.1.9.2, developed by the University of Düsseldorf, Germany [18]. Following this statistical procedure, a small-to-medium effect size of 0.4 [19], the α-error = 0.05 and the power = 0.8, gave a sample size of 66 participants for comparison of three male and female nutrition status groups. After a 15% adjustment to account for likely dropouts, a total sample size of 75 participants was obtained for each gender group. The final result was that a sample size of at least 150 (75 men and 75 women) study subjects would be required. For this study, 470 subjects from four worksites in the Bangkok metropolitan area were invited to participate. But since only 300 subjects (100 men and 200 women) agreed to participate, the response rate was 64%. The data collection was done by means of self-administered questionnaires handed out to study participants.

2.2. Anthropometric Measurements. The body weight (BW), body mass index (BMI), body fat percentage (%BF), and basal metabolic rate (BMR) were measured by Body Composition Analyzer (Tanita TBF 410, Tanita Corp., Tokyo, Japan) with the bioelectrical impedance (BIA) technology for accurate analysis. Waist circumference (WC) was also measured.

(1) WC was measured using a nonstretch, spring weighted anthropometric tape with a tension meter attached in the horizontal plane. The measure is taken midway between the inferior margin of the ribs and the superior border of the iliac crest at the end of normal expiration, as recommended by the World Health Organization and the International Diabetes Federation [20]. Abdominal measurements in this study were done in accordance with Asian-Pacific recommendations for WC cut-off points [21], where WC < 90 cm for males and WC < 80 cm for females are considered normal, while WC ≥ 90 cm for males and WC ≥ 80 cm for females indicate abdominal obesity.

(2) BW, BMI, and %BF were measured using the Tanita body composition digital scale. The Tanita scale used in this study had the "foot-to-foot" BIA technology. The bioelectrical impedance analysis (BIA) is performed with a nonharmful, low dose, conduction of electrical waves to measure weight and body composition. The Tanita scale was placed on a flat surface, and participants were asked to remove their shoes and heavy objects (such as jackets, wallets, and accessories) for measurement accuracy. Measurement protocols were given to participants one day before the tests. Measurement guidelines included (i) no meals and/or drinks three hours prior to measurements, (ii) no strenuous exercise 12 hours prior to measurements, (iii) no alcohol consumption 24 hours prior to measurements, (iv) absence of menstruation at the time of measurements, and (v) empty bladder before measurements. Participants were measured at approximately the same time of day. In case participants did not, or could not, observe the premeasurement guidelines, appointments were made to meet with them again when the guidelines were met. The body weight was measured in kilograms with one decimal point, waist circumference was determined to ±0.1 cm, and percent body fat was measured in percentage with ±0.1%. Asian criteria were used for BMI classifications [21] with the following cut-off points: BMI < 18.5 kg/m^2 indicating underweight, BMI of 18.5–22.9 kg/m^2 defined as a normal range, BMI of 23–24.9 kg/m^2 considered as overweight, BMI of 25–30 kg/m^2 defined as obesity class I, and BMI ≥ 30 kg/m^2 defined as obesity class II. The threshold for overweight/obesity in this study was, therefore, defined as a BMI ≥ 23 kg/m^2. Asian criteria for body fat percentage were also used to identify overweight/obesity with the following cut-off points: for males >20% for ages 20–29 and males >25%

for ages 30–50; for females >24% for ages 20–29 and 35% for ages 30–50 [22].

2.3. Leisure-Time Physical Activity (LPTA) Assessment. The LPTA was self-reported using a modified version of Bouchard's [23] physical activity questionnaire. The questionnaire consisted of 13 open-ended daily physical activities. Duration of each activity was divided into 15 minutes per session, with the frequency of that activity per week. The energy expenditure was calculated according to the intensity of physical activities (kcal/kg/15 min). For each 15-minute period, energy expenditure is qualified on a scale from 1 to 9. Approximate median energy expenditure for each of the nine categories in kcal/kg per 15 minutes is applied to compute daily energy expenditure for each individual [23]. Briefly, subjects were asked to report all leisure physical activities that were performed at least 3 times per week in 15-minute sessions (frequency/week < 3 times, 3–5 times, and >5 times and time/session < 15 min, 15–30 min, 31–45 min, 46–60 min, and >60 min). Detailed information was then collected about the type of LTPA (such as walking, cycling, swimming, jogging, and gardening), as well as the frequency and duration of each LTPA reported.

2.4. Dietary Assessment. The dietary assessment for energy and nutrients intake was determined by a single 24-hour dietary recall and a nonweighed 2-day food record. The 24-hour dietary recall was conducted by interviewers who were trained and experienced nutritionists. An information session was organized to give each study participant instructions on how to record the daily food intake. In the food diaries, days were divided into seven eating occasions, namely, breakfast, lunch, dinner, and snacks (before breakfast, mid-morning, afternoon, and late-evening snacks). Information on the type (including brand names) and amounts of food consumed was collected through an open entry format. For the verification and estimation of the size of individual food portions, the participants were instructed with the help of three-dimensional food models and household utensils to enhance the accuracy of portion size estimation. Participants were also given oral instructions on how to record their food intake and were shown how to use the household scale. Written instructions were incorporated in the food diaries as well, along with contact information in case any questions arose during the recording of food consumption. Participants were asked to record all food and beverages consumed for two consecutive days (Friday and Saturday) and to indicate dates and times when meals were taken. Since most of the meals reported as "brunch" were the subjects' first eating occasion of the day, breakfast and brunch were collapsed into one group referred to as breakfast. The importance of maintaining regular diets and recording all food and drink consumed was emphasized. The quality control of all food diaries was handled and reviewed by an experienced observer to avoid inconsistency and to maintain accurate data entries. After completion, the diaries were processed to record food quantities by experienced nutritionists using a standard manual on food portions and household measures [24]. The food codes were those used in the Thai Dietary Database (fourth edition), Institute of Nutrition, Mahidol University [25]. The calculation of nutrients was done by means of INMUCAL-Nutrient Software Version 2. While most macronutrients (carbohydrate, protein, and fat) could be found in that software's database, it was not possible to get all micronutrients for local food items. Consequently, micronutrients whose data were available in 80% or more of (except about 50% for zinc) food items in the INMUCAL database were included in the statistical analysis. The data on regional foods not covered by the software were sourced from the Thai Food Composition Table of the Nutrition Division, Department of Health, Ministry of Public Health of Thailand [26]. Adequacy of each nutrient intake was determined based on the 2003 Dietary Reference Intake for Thai People [17].

Incomplete food records and records for less than three days were excluded (76), leaving 224 complete records for data analysis (78 for men and 146 for women). The number of valid 24-hour dietary recalls and nonweighed 2-day food records were obtained for 78% and 73% of male and female participants, respectively. Nine nonplausible reporters, identified as subjects reporting energy intakes of <800 or >4,200 kcal/day for males or <600 and >3,500 kcal/day for females, and unrealistic values of other nutrients were excluded from the analysis [27]. The standardization of the exclusion procedure was guaranteed by the fact that the evaluation of the food diaries was done by the same nutritionist with a long-standing experience in the field of nutrition and epidemiological research. The valid data obtained for nutrients and energy intake was then compared with the Thai DRIs for adults aged 19–50 years. In this study, the Thai DRIs were based on the Recommended Dietary Allowances (RDA) for energy and most nutrients and on recommendations of Adequate Intake (AI) values for calcium and manganese. The Thai DRIs did not provide the tolerable upper intake level (UL), except for vitamin C, calcium, and sodium, as shown in Table 1 [17].

Another recommendation set by the Thai DRIs is the following daily macronutrient energy distribution for adults aged 19–50 years: 45%–65% from carbohydrate, 10%–15% from protein, and 20%–35% from fat.

Two variables—adherent and nonadherent—were used to measure participants' adherence to the Thai DRIs. Adherence to recommendations was defined as the level of energy and nutrient intake falling within 80–120% of the Thai DRIs. Nonadherence referred to insufficient or excessive intakes. Since the estimated average requirements (EAR) have not been established for the Thai DRIs, insufficient or inadequate energy and nutrients intake for each participant by age and sex was defined as that less than 80% of the Thai DRIs, estimated according to the formula RDA = 1.2 EAR of the US Institute of Medicine (IOM) [28]. Excessive intakes are those higher than the UL (for vitamin C, calcium, and sodium only) and those above 120% of the Thai DRIs.

2.5. Other Assessments. Smoking, socioeconomic factors, alcohol consumption, and medical history were also assessed by using a self-administered questionnaire. Based on the

Table 1: The Thai Dietary Reference Intakes (Recommended Dietary Allowances and Adequate Intakes of energy and selected nutrients) for adults aged 19–50 years.

Energy and nutrients	Sex				Tolerable upper intake levels
	Males		Females		
	Aged 19–30 yrs	Aged 31–50 yrs	Aged 19–30 yrs	Aged 31–50 yrs	
(1) Energy (kcal/d)	2,150	2,100	1,750	1,750	ND
(2) Protein (g/d)	57	57	52	52	ND
(3) Fiber (g/d)	25	25	25	25	ND
(4) Calcium (mg/d)	800	800	800	800	2,500
(5) Phosphorus (mg/d)	700	700	700	700	ND
(6) Sodium (mg/d)	500–1,475	475–1,450	400–1,200	400–1,200	2,400
(7) Potassium (mg/d)	2,450–4,100	2,450–4,100	2,050–3,400	2,050–3,400	ND
(8) Iron (mg/d)	10.4	10.4	24.7	24.7	ND
(9) Vitamin A (μg/d)	700	700	600	600	ND
(10) Vitamin E (mg/d)	15	15	15	15	ND
(11) Vitamin B_1 (mg/d)	1.2	1.2	1.1	1.1	ND
(12) Vitamin B_2 (mg/d)	1.3	1.3	1.1	1.1	ND
(13) Niacin (mg/d)	16	16	14	14	ND
(14) Vitamin C (mg/d)	90	90	75	75	2,000
(15) Magnesium (mg/d)	310	320	250	260	ND
(16) Selenium (μg/d)	55	55	55	55	ND
(17) Zinc (mg/d)	13	13	7	7	ND

ND = not determined due to lack of suitable data.
Source: Thai Dietary Reference Intakes for Energy and Selected Nutrients (2003). The report may be accessed via http://nutrition.anamai.moph.go.th/temp/main/view.php?group=2&id=132.

questionnaire, the participants were classified as those who never smoked in their life, former smokers (have not smoked in the past 12 months), and current smokers. For alcohol consumption, the participants were categorized into lifetime abstainers, former drinkers (have not drunk in the past 12 months), and current alcohol users. If participants reported that they were either current smokers or current alcohol users, they were asked to specify the frequency (i.e., daily or occasionally) of smoking or alcohol consumption.

2.6. Statistical Analysis. Predictive Analytics Software for Windows (PASW) Version 18 was used for data analysis. Descriptive statistics in this study are means and standard deviations for normally distributed continuous data and percentages of participant's adherence, or nonadherence, to dietary recommendations. Tests for normality were performed using the Kolmogorov-Smirnov test. Since not all the dietary intakes were distributed normally, median values and interquartile range (IQR) of the amounts estimated in a single 24-hour dietary recall and a nonweighed 2-day food record are also presented. A nonparametric test—Mann-Whitney *U* test—was used to identify median differences of macronutrient and energy intakes between genders and between nutritional status groups among males. The Kruskal Wallis test was used to identify differences of macronutrient and energy intakes between nutritional status groups among female participants. The Kruskal Wallis test was also used to identify the differences of BMI and energy and macronutrient intakes between four groups of eating occasions: 3 meals, 4 meals, 5 meals, and ≥6 meals. The Bonferroni corrections for multiple comparisons were then analyzed by the Mann-Whitney U test for group comparisons at the P value < 0.0125. The differences of the proportion of participants adhering to dietary recommendations among genders and nutritional status groups were also identified by using the chi-square test at P value < 0.05. Comparisons of the three-day average macronutrient and energy intakes between breakfast, lunch, and dinner were analyzed by using a nonparametric one-way repeated-measures ANOVA (Friedman's test) with the P value of <0.05 as a threshold for significance. That analysis was followed by comparing the average of macronutrient and energy intakes for breakfast, lunch, and dinner using nonparametric paired t-tests (Wilcoxon signed-rank tests). A Bonferroni correction was applied for the *post hoc* detection of significant pairwise differences, with the significance level set at P value < 0.0125.

3. Results

3.1. Sociodemographic and Anthropometric Profile. Table 2 presents the sociodemographic and anthropometric characteristics of study subjects consisting of 65.1% males and 34.9% females. The average age was 36.7 years (SD 8.1). Almost one-half of male participants were in a group older than 40 years. Among female participants, 38.6% were in the 30–39 age group and 39.3% were older than 40 years. More than a half of participants were single: 53.3% for males and 53.6% for females. The majority of study subjects (86.0%) had a

TABLE 2: Sociodemographic and anthropometric profile of participants.

Characteristics	All (n = 215) Number (percentage)	Men (n = 75) Number (percentage)	Women (n = 140) Number (percentage)
(1) Age in years			
20–29	49 (22.8)	18 (24.0)	31 (22.1)
30–39	75 (34.9)	21 (28.0)	54 (38.6)
≥40	91 (42.3)	36 (48.0)	55 (39.3)
(2) Marital status			
Single	114 (53.3)	40 (53.3)	75 (53.6)
Married	93 (43.4)	34 (45.3)	59 (42.1)
Divorced/widowed/separated	7 (3.3)	1 (1.4)	6 (4.3)
(3) Educational level			
Diploma	30 (14.0)	14 (18.7)	16 (11.4)
Bachelor degree	122 (56.7)	43 (57.3)	79 (56.4)
Master degree or higher	63 (29.3)	18 (24.0)	45 (32.2)
(4) Monthly income (baht)			
Less than 20,000	77 (35.8)	20 (26.7)	57 (40.7)
20,000–29,999	61 (28.4)	23 (30.7)	38 (27.1)
30,000–39,000	33 (15.3)	8 (10.6)	25 (17.9)
≥40,000	44 (20.5)	24 (32.0)	20 (14.3)
(5) Smoking			
Never smokers	194 (90.2)	55 (73.3)	139 (99.3)
Former smokers	7 (3.3)	6 (8.0)	1 (0.7)
Current smokers	14 (6.5)	14 (18.7)	0 (0.0)
Occasional smokers	14 (100.0)	14 (100.0)	0 (0.0)
(6) Alcohol users			
Lifetime abstainers	60 (27.9)	13 (17.3)	47 (33.6)
Former drinkers	40 (18.6)	11 (14.7)	29 (20.7)
Current alcohol users	115 (53.5)	51 (68.0)	64 (45.7)
Occasional drinkers	115 (100.0)	51 (100.0)	64 (100.0)
(7) Nutritional status classification			
Body mass index (BMI)			
Underweight	21 (9.8)	0 (0.0)	21 (15.0)
Normal weight	96 (44.6)	27 (36.0)	69 (49.3)
Overweight/obese[a]	98 (45.6)	48 (64.0)	50 (35.7)
Percent body fat (%BF)			
Normal range	104 (74.3)	37 (49.3)	104 (74.3)
High range[b]	36 (25.7)	38 (50.7)	36 (25.7)
Waist circumference (WC)			
Normal waist	140 (65.1)	48 (64.0)	92 (65.7)
Abdominal obesity[c]	75 (34.9)	27 (36.0)	48 (34.3)
(8) Sitting time at work (hour/day)			
Mean ± SD	6.0 ± 1.1	5.7 ± 1.1	6.1 ± 1.1
(9) Leisure time energy expenditure (kcal/day)			
Mean ± SD	124.9 ± 119.8	157.4 ± 138.6	107.5 ± 104.8
(10) Basal metabolic rate (BMR) (kcal)			
Mean ± SD	1,323.8 ± 207.2	1,552.1 ± 143.0	1,201.5 ± 111.5
(11) Total energy intake: basal metabolic rate (EI: BMR)			
Mean ± SD	1.1 ± 0.2	1.0 ± 0.2	1.2 ± 0.2

[a]Overweight/obese: BMI ≥ 23 kg/m^2 (Asian definition).

[b]High range: males of ages 20–29 > 20%, ages 30–50 > 25%; females of ages 20–29 > 24%, ages 30–50 > 35%.

[c]Abdominal obesity: males ≥ 90 cm; females ≥ 80 cm (Asian definition).

bachelor degree or higher. Monthly incomes of less than 20,000 Thai baht were reported by 40% of females and by 26% of males. All participants were nonhabitual smokers and drinkers. Only 18.7% ($n = 14$) of males reported occasional smoking, while more than two-thirds of males (68.0%) and nearly half of females (45.7%) reported occasional drinking.

Based on BMI measurements, the prevalence of overweight/obesity among males was almost two times higher than among females: 64.0% and 35.7%, respectively. The prevalence of abdominal obesity was almost the same for male and female participants: 36.0% and 34.3%, respectively. One-half of males (50.7%) and one-fourth (25.7%) of females were in a high range of percent body fat. The sitting time during the working hours of all participants was between 4 and 9 hours, with an average of 6.0 hours per day (SD 1.1). The average of the leisure time energy expenditure was 124.9 kcal/day (SD 119.8), and the average basal metabolic rates were 1,552 kcal (SD 143) for men and 1,206 kcal (SD 111) for women.

3.2. Daily Macronutrient and Energy Intakes. A preliminary analysis using nonparametric Kruskal Wallis test identified the differences of energy and macronutrient intakes between age groups and sociodemographic characteristics at a significance level of P value < 0.05. The results showed that there were no significant differences in daily energy and nutrient intakes among the three age groups (20–29, 30–39, and ≥40) and other sociodemographic characteristics (education level, marital status, and monthly income) of participants. Consequently, in further analysis, all age groups and sociodemographic characteristics were combined, and the results are presented in Table 3.

The median values of energy and macronutrient intakes (carbohydrate, protein, and fat) showed significant differences between genders, with intakes by men being higher than those by women (P value < 0.05). The median energy intake for men was 1,485 kcal (IQR 270) and was 1,428 kcal (IQR 280) for women. That accounted for 71.2% of the Thai DRIs for men and for 81.1% for women; the difference was statistically significant at the P value < 0.05. The median carbohydrate intake was 203 grams/day (IQR 50) for men and 194 grams/day (IQR 43) for women, representing 54.4% and 56.0% of their total energy intake, respectively. The intakes of protein for men and women were 61 grams/day (IQR 15) and 56 grams/day (IQR 13), respectively, accounting for 15.9% of daily energy intake for men and 16.2% for women. Both genders showed higher consumption of protein compared to the Thai DRIs: 106.7% for men and 108.6% for women. The median daily intake of fat was 52 grams/day (IQR 18) for men and 44 grams/day (IQR 16) for women, representing 29.6% and 28.6% of their daily energy intake, respectively.

A comparison of the energy and macronutrient intake between nutritional status groups, classified by BMI (Table 4 for men and Table 5 for women), found that there was no significant difference in median daily energy and macronutrient intakes between normal weight and overweight/obese males and between underweight, normal weight, and overweight/obese females. The only exception was the carbohydrate intake by males. The median carbohydrate intake of overweight/obese males was 217 grams/day (IQR 50) and was 197 grams/day (IQR 50) for those of normal weight, with a statistically significant difference at $P < 0.05$.

3.3. Micronutrients Intake and Dietary Fiber. The results reported in Table 3 show that intakes of most micronutrients—calcium, potassium, vitamin A, vitamin E, vitamin B_1, vitamin C_1, magnesium, selenium, and zinc—were less than 80% of the Thai DRIs. The intakes of phosphorus, vitamin B_2 and niacin were found adequate in the range of 84.8%–110.6% of the Thai DRIs for both genders. The iron intake by males was adequate at 94.6% of the Thai DRIs, but it was very low—39.2% of the Thai DRIs—in the case of female participants, indicating a statistically significant difference of iron consumption between genders (P value < 0.05). A significant difference between genders was also found with regard to the intake of niacin, magnesium, and zinc. The sodium intake exceeded by far the Thai DRIs: males' median intake of 1,967 mg/day (IQR 973) was 135.3% of DRIs, and the females' intake of 2,021 mg/day (IQR 1,091) was 168.4% of DRIs. The median intake of dietary fiber of 7.8 grams/day (IQR 4.8) for men and 8.0 grams/day (IQR 4.8) for women accounted for 31.2% and 32.0% of the Thai DRIs, respectively. Results reported in Tables 4 and 5 show that there were no statistically significant differences for the median intake of micronutrients and fiber between nutritional status groups of both genders.

3.4. Adherence to Dietary Recommendations. A proportion of study participants conforming to the Thai DRIs for energy and nutrient intakes are shown in Table 3. Less than a half of female study subjects (48.6%) and less than one-fourth of males (21.3%) adhered to energy recommendations, showing a statistically significant difference between genders (P value < 0.05). That showed that more than three-quarters of male participants (78.7%) and half of females (49.3%) had an insufficient energy intake. With regard to the protein intake, 76.0% of men and 65.7% of women adhered to the Thai DRIs.

In the area of micronutrients, 62.7% of men and 65.0% of women adhered to phosphorus intake recommendations. But the adherence to the Thai DRIs was much lower for calcium: only 24.0% of men and 18.6% of women in our study adhered to calcium intake recommendations. There was also a very low, 5.0%, adherence to iron intake by women. By contrast, 42.7% of men were found to be in compliance with iron intake recommendations. About three-quarters of males (72.0%) and two-thirds (67.9%) of females adhered to sodium intake recommendations. For vitamin C and niacin intakes, the rates of adherence ranged from 49.3% to 73.3% for both genders. However, the adherence to recommended vitamin A intakes was very low: 10.7% for males and 11.4% for females. A lack of adherence to the Thai DRIs was also noted in the case of both genders with respect to intakes of fiber and other micronutrients, such as potassium, vitamin E, magnesium, selenium, and zinc.

There were no significant differences in adherence to recommended energy and nutrient intakes between nutritional status groups. The only significant difference (P value < 0.05)

TABLE 3: Energy and nutrient intakes classified by genders.

Variables	Male (n = 75)					Female (n = 140)				
	Median (IQR)	%DRI	Adherent n (%)	Nonadherent; n (%) Insufficient n (%)	Nonadherent; n (%) Excessive n (%)	Median (IQR)	%DRI	Adherent n (%)	Nonadherent; n (%) Insufficient n (%)	Nonadherent; n (%) Excessive n (%)
Energy (kcal)	**1,485 (270)**[a]	**71.2**[b]	**16 (21.3)**[c]	59 (78.7)	0 (0.0)	**1,428 (280)**[a]	**80.1**[b]	**68 (48.6)**[c]	69 (49.3)	3 (2.1)
Carbohydrate (g)	**203 (50)**[a]	NA	NA	NA	NA	**194 (43)**[a]	NA	NA	NA	NA
Protein (g)	**61 (15)**[a]	106.7	57 (76.0)	0 (0.0)	18 (24.0)	**56 (13)**[a]	108.6	92 (65.7)	9 (6.4)	39 (27.9)
Animal protein (g)	30 (14)[a]	NA	NA	NA	NA	31 (11)[a]	NA	NA	NA	NA
Fat (g)	**52 (18)**[a]	NA	NA	NA	NA	**44 (16)**[a]	NA	NA	NA	NA
Carbohydrate (energy percentage)	**54.4 (7.9)**[a]	NA	67 (89.3)	5 (6.7)	3 (4.0)	**56.0 (9.1)**[a]	NA	133 (95.0)	4 (2.9)	3 (2.1)
Protein (energy percentage)	15.9 (3.2)	NA	57 (76.0)	0 (0.0)	18 (24.0)	16.2 (3.0)	NA	92 (65.7)	9 (6.4)	39 (27.9)
Fat (energy percentage)	29.6 (6.6)	NA	60 (80.0)	3 (4.0)	12 (16.0)	28.6 (7.6)	NA	118 (84.3)	5 (3.6)	17 (12.1)
Dietary fiber (g)	7.8 (4.8)	31.2	0 (0.0)	75 (100.0)	0 (0.0)	8 (4.8)	32.0	1 (0.7)	139 (99.3)	0 (0.0)
Calcium (mg)	492 (200)	62.5	18 (24.0)	57 (76.0)	0 (0.0)	444 (199)	55.2	26 (18.6)	114 (81.4)	0 (0.0)
Phosphorus (mg)	684 (200)	97.7	47 (62.7)	18 (24.0)	10 (13.3)	655 (147)	93.6	91 (65.0)	31 (21.1)	18 (12.9)
Sodium (mg)	1,967 (973)	**135.3**[b]	54 (72.0)	0 (0.0)	21 (28.0)	2,021 (1,091)	**168.4**[b]	95 (67.9)	0 (0.0)	45 (32.1)
Potassium (mg)	1,251 (549)	51.1	0 (0.0)	75 (100.0)	0 (0.0)	1,219 (501)	59.5	4 (2.9)	136 (97.1)	0 (0.0)
Iron (mg)	9.8 (5.1)	**94.6**[b]	**25 (33.3)**[c]	32 (42.7)	18 (24.0)	9.5 (3.3)	**39.2**[b]	**7 (5.0)**[c]	133 (95.0)	0 (0.0)
Vitamin A (μg)	346 (567)	49.5	8 (10.7)	50 (66.7)	17 (22.7)	316 (409)	52.6	16 (11.4)	95 (67.9)	29 (20.7)
Vitamin E (mg)	0.04 (0.4)	0.2	0 (0.0)	75 (100.0)	0 (0.0)	0.17 (0.4)	1.1	3 (2.1)	134 (95.8)	3 (2.1)
Vitamin B_1 (mg)	0.9 (0.5)	**74.3**[b]	24 (32.0)	45 (60.0)	6 (8.0)	0.9 (0.4)	**79.7**[b]	49 (35.0)	71 (50.7)	20 (14.3)
Vitamin B_2 (mg)	1.3 (.8)	**96.9**[b]	26 (34.7)	26 (34.7)	23 (30.6)	1.2 (0.6)	**110.6**[b]	55 (39.3)	28 (20.0)	57 (40.7)
Niacin (mg)	**13.8 (4.5)**[a]	86.3	55 (73.3)	16 (21.3)	4 (5.3)	**11.9 (4.1)**[a]	84.8	97 (69.3)	36 (25.7)	7 (5.0)
Vitamin C (mg)	64 (103)	71.4	41 (54.7)	34 (45.3)	0 (0.0)	68 (122)	90.5	69 (49.3)	71 (50.7)	0 (0.0)
Magnesium (mg)	**15.8 (19.7)**[a]	**6.7**[b]	0 (0.0)	75 (100.0)	0 (0.0)	**22.9 (26.4)**[a]	**10.3**[b]	0 (0.0)	140 (100.0)	0 (0.0)
Selenium (μg)	0.3 (1.1)	0.7	0 (0.0)	75 (100.0)	0 (0.0)	0.4 (0.7)	0.7	0 (0.0)	140 (100.0)	0 (0.0)
Zinc (mg)	**2.1 (1.0)**[a]	**16.1**[b]	0 (0.0)	75 (100.0)	0 (0.0)	**2.3 (1.2)**[a]	**33.5**[b]	0 (0.0)	140 (100.0)	0 (0.0)

[a] P values are for differences in median dietary intake between genders (Mann-Whitney U test); significant differences shown in bold at P value < 0.05.

[b] P values are for differences in %DRIs between genders (Mann-Whitney U test); significant differences shown in bold at P value < 0.05.

[c] P values are for differences in proportion of participants' adherence to recommendations between genders (chi-square test); significant differences shown in bold at P value < 0.05.

TABLE 4: Energy and nutrient intakes of male participants classified by nutritional status.

Variables	BMI < 23 kg/m² (*n* = 27)					BMI ≥ 23 kg/m² (*n* = 48)				
	Median (IQR)	%DRI	Adherent *n* (%)	Nonadherent; *n* (%) Insufficient *n* (%)	Nonadherent; *n* (%) Excessive *n* (%)	Median (IQR)	%DRI	Adherent *n* (%)	Nonadherent; n (%) Insufficient *n* (%)	Nonadherent; n (%) Excessive *n* (%)
Energy (kcal)	1,496 (306)	71.2	12 (44.4)	15 (55.6)	0 (0.0)	1,536 (306)	71.5	22 (45.8)	26 (54.2)	0 (0.0)
Carbohydrate (g)	**197 (50)**[a]	NA	NA	NA	NA	**217 (50)**[a]	NA	NA	NA	NA
Protein (g)	60 (17)	106.1	21 (77.8)	0 (0.0)	6 (22.2)	61 (17)	107.8	36 (75.0)	0 (0.0)	12 (25.0)
Animal protein (g)	29.6 (12.9)	NA	NA	NA	NA	29.4 (11.6)	NA	NA	NA	NA
Fat (g)	52 (15)	NA	NA	NA	NA	51 (15)	NA	NA	NA	NA
Carbohydrate (energy percentage)	52.3 (8.6)	NA	22 (81.5)	3 (11.1)	2 (7.4)	55.0 (7.9)	NA	45 (93.7)	1 (2.1)	2 (4.2)
Protein (energy percentage)	16.4 (4.1)	NA	11 (40.7)	0 (0.0)	16 (59.3)	15.8 (3.2)	NA	18 (37.5)	0 (0.0)	30 (62.5)
Fat (energy percentage)	30.3 (8.7)	NA	21 (77.8)	1 (3.7)	5 (18.5)	29.2 (6.3)	NA	39 (81.2)	2 (4.2)	7 (14.6)
Dietary fiber (g)	6.9 (3.3)	27.6	0 (0.0)	27 (100.0)	0 (0.0)	8.6 (4.4)	34.4	0 (0.0)	48 (100.0)	0 (0.0)
Calcium (mg)	498.6 (226.4)	62.3	6 (22.2)	21 (77.8)	0 (0.0)	492.3 (302.8)	60.9	12 (25.0)	36 (75.0)	0 (0.0)
Phosphorus (mg)	615.2 (224.2)	87.9	13 (48.1)	10 (37.0)	4 (14.8)	707.8 (179.5)	101.1	34 (70.8)	8 (16.7)	6 (12.5)
Sodium (mg)	2,159 (908)	141.4	19 (70.4)	0 (0.0)	8 (29.6)	1,842 (1,117)	129.5	35 (72.9)	0 (0.0)	13 (27.1)
Potassium (mg)	1,150 (543)	46.9	0 (0.0)	27 (100.0)	0 (0.0)	1,264 (521)	51.6	0 (0.0)	48 (100.0)	0 (0.0)
Iron (mg)	8.7 (5.5)	85.9	9 (33.3)	12 (44.4)	6 (22.2)	10.3 (4.6)	97.4	23 (49.9)	13 (27.1)	12 (25.0)
Vitamin A (μg)	346 (445.2)	50.1	4 (14.8)	18 (66.7)	5 (18.5)	316 (712.4)	47.9	4 (8.3)	32 (66.7)	12 (25.0)
Vitamin E (mg)	0.04 (0.3)	0.05	0 (0.0)	27 (100.0)	0 (0.0)	0.06 (0.4)	0.32	0 (0.0)	48 (100.0)	0 (0.0)
Vitamin B_1 (mg)	0.8 (0.5)	67.2	9 (33.3)	18 (66.7)	0 (0.0)	0.9 (0.6)	76.1	15 (31.2)	27 (56.3)	6 (12.5)
Vitamin B_2 (mg)	1.2 (0.9)	93.7	9 (33.3)	9 (33.3)	9 (33.3)	1.3 (0.8)	96.9	17 (35.4)	17 (35.4)	14 (29.2)
Niacin (mg)	12.9 (4.2)	81.3	12 (44.4)	13 (48.2)	2 (7.4)	14.2 (4.7)	87.7	29 (60.4)	17 (35.4)	2 (4.2)
Vitamin C (mg)	59.6 (111.9)	66.2	12 (44.4)	15 (55.6)	0 (0.0)	65.0 (98.8)	71.8	22 (45.8)	26 (54.2)	0 (0.0)
Magnesium (mg)	13.9 (12.1)	4.5	0 (0.0)	27 (100.0)	0 (0.0)	21.7 (23.5)	6.7	0 (0.0)	48 (100.0)	0 (0.0)
Selenium (μg)	0.2 (1.3)	0.5	0 (0.0)	27 (100.0)	0 (0.0)	0.4 (0.9)	0.7	0 (0.0)	48 (100.0)	0 (0.0)
Zinc (mg)	1.8 (0.9)	14.3	0 (0.0)	27 (100.0)	0 (0.0)	2.2 (1.1)	16.9	0 (0.0)	48 (100.0)	0 (0.0)

[a] *P* values are for differences in median dietary intakes between nutritional status groups (Mann-Whitney *U* test); significant differences shown in bold at *P* value < 0.05.

TABLE 5: Energy and nutrient intakes of female participants classified by BMI.

Variables	BMI < 18.5 kg/m^2 (n = 21)					BMI 18.5–22.9 kg/m^2 (n = 69)					BMI ≥ 23 kg/m^2 (n = 50)				
	Median (IQR)	%DRI	Adherent; n (%)	Nonadherent; n (%) Insufficient n (%)	Nonadherent; n (%) Excessive n (%)	Median (IQR)	%DRI	Adherent; n (%)	Nonadherent; n (%) Insufficient n (%)	Nonadherent; n (%) Excessive n (%)	Median (IQR)	%DRI	Adherent; n (%)	Nonadherent; n (%) Insufficient n (%)	Nonadherent; n (%) Excessive n (%)
Energy (kcal)	1,334 (263)	76.3	8 (38.1)	13 (61.9)	0 (0.0)	1,441 (300)	82.3	38 (55.1)	30 (43.5)	1 (1.4)	1,401 (289)	79.4	22 (44.0)	26 (52.0)	2 (4.0)
Carbohydrate (g)	178.5 (39.5)	NA	NA	NA	NA	202.8 (43.6)	NA	NA	NA	NA	192 (42)	NA	NA	NA	NA
Protein (g)	54.8 (8.6)	105.3	16 (76.2)	1 (4.8)	4 (19.0)	56.7 (13.4)	109.0	45 (65.2)	4 (5.8)	20 (29.0)	58 (15)	109.8	31 (62.0)	4 (8.0)	15 (30.0)
Animal protein (g)	30.9 (8.4)	NA	NA	NA	NA	32.5 (12.2)	NA	NA	NA	NA	29.1 (10.5)	NA	NA	NA	NA
Fat (g)	47.4 (15.7)	NA	NA	NA	NA	43.8 (15.8)	NA	NA	NA	NA	43.9 (15.6)	NA	NA	NA	NA
Carbohydrate (energy percentage)	52.2 (7.7)	NA	19 (90.4)	1 (4.8)	1 (4.8)	56.9 (7.8)	NA	64 (92.8)	3 (4.3)	2 (2.9)	53.5 (8.7)	NA	50 (100.0)	0 (0.0)	0 (0.0)
Protein (energy percentage)	16.7 (2.4)	NA	5 (23.8)	0 (0.0)	16 (76.2)	15.9 (2.6)	NA	24 (34.8)	2 (2.9)	43 (62.3)	16.4 (3.0)	NA	13 (26.0)	1 (2.0)	36 (72.0)
Fat (energy percentage)	30.9 (7.3)	NA	17 (81.0)	2 (9.5)	2 (9.5)	26.9 (7.4)	NA	59 (85.5)	2 (2.9)	8 (11.6)	29.3 (7.9)	NA	42 (84.0)	1 (2.0)	7 (14.0)
Dietary fiber (g)	8.1 (3.6)	32.4	0 (0.0)	21 (100.0)	0 (0.0)	9.3 (5.1)	37.2	1 (1.4)	68 (98.6)	0 (0.0)	8.2 (5.1)	32.8	0 (0.0)	50 (100.0)	0 (0.0)
Calcium (mg)	490.8 (234.7)	61.3	3 (14.3)	18 (85.7)	0 (0.0)	445.0 (210.8)	55.3	16 (23.2)	53 (76.8)	0 (0.0)	427.3 (180.3)	52.1	7 (14.0)	43 (86.0)	0 (0.0)
Phosphorus (mg)	657.4 (151.8)	93.9	13 (61.9)	6 (28.6)	2 (9.5)	660.3 (148.9)	94.3	45 (65.2)	14 (20.3)	10 (14.5)	647.9 (142.4)	92.6	33 (66.0)	11 (22.0)	6 (12.0)
Sodium (mg)	1,902 (926)	158.5	16 (76.2)	0 (0.0)	5 (23.8)	1,979 (1,068)	164.9	51 (73.9)	0 (0.0)	18 (26.1)	1,979 (1,068)	183.1	28 (56.0)	0 (0.0)	22 (44.0)
Potassium (mg)	1,236 (322)	60.3	1 (4.8)	20 (95.2)	0 (0.0)	1,229 (473)	59.9	2 (2.9)	67 (97.1)	0 (0.0)	1,196 (646)	58.3	1 (2.0)	49 (98.0)	0 (0.0)
Iron (mg)	9.2 (2.4)	37.4	0 (0.0)	21 (100.0)	0 (0.0)	9.8 (3.7)	40.3	3 (4.3)	66 (95.7)	0 (0.0)	9.3 (3.9)	38.9	4 (8.0)	46 (92.0)	0 (0.0)
Vitamin A (μg)	310.5 (350.9)	51.8	2 (9.5)	15 (71.4)	4 (19.0)	323.3 (402.5)	51.2	11 (15.9)	43 (62.3)	15 (21.7)	311.5 (355.8)	51.9	3 (6.0)	37 (74.0)	10 (20.0)
Vitamin E (mg)	0.2 (0.7)	6.6	1 (4.8)	20 (95.2)	0 (0.0)	0.2 (0.4)	8.8	1 (1.4)	66 (95.7)	2 (2.9)	0.1 (0.4)	9.6	1 (2.0)	48 (96.0)	1 (2.0)
Vitamin B_1 (mg)	0.8 (0.4)	72.3	5 (23.8)	13 (61.9)	3 (14.3)	0.9 (0.4)	79.8	27 (39.1)	35 (50.7)	7 (10.1)	0.9 (0.4)	82.2	17 (34.0)	23 (46.0)	10 (20.0)
Vitamin B_2 (mg)	1.0 (0.5)	92.3	10 (47.6)	6 (28.6)	5 (23.8)	1.3 (0.8)	117.0	23 (33.3)	13 (18.8)	33 (47.8)	1.2 (0.5)	111.5	22 (44.0)	9 (18.0)	19 (38.0)
Niacin (mg)	11.8 (3.4)	84.6	13 (61.9)	8 (38.1)	0 (0.0)	11.9 (3.9)	85.2	38 (55.1)	26 (37.7)	5 (7.2)	11.4 (4.1)	78.8	22 (44.0)	26 (52.0)	2 (4.0)
Vitamin C (mg)	86.1 (106.5)	114.8	**13 (61.9)**[a]	8 (38.1)	0 (0.0)	77.5 (138.5)	103.4	**38 (55.1)**[a]	31 (49.9)	0 (0.0)	50.8 (70.3)	67.8	**18 (36.0)**[a]	32 (64.00)	0 (0.0)
Magnesium (mg)	15.5 (28.5)	10.1	0 (0.0)	21 (100.0)	0 (0.0)	22.8 (20.8)	10.1	0 (0.0)	69 (100.0)	0 (0.0)	24.4 (31.1)	10.8	0 (0.0)	50 (100.0)	0 (0.0)
Selenium (μg)	0.4 (0.7)	0.7	0 (0.0)	21 (100.0)	0 (0.0)	0.4 (1.0)	0.7	0 (0.0)	69 (100.0)	0 (0.0)	0.4 (0.7)	0.7	0 (0.0)	50 (100.0)	0 (0.0)
Zinc (mg)	2.7 (1.4)	37.9	0 (0.0)	21 (100.0)	0 (0.0)	2.4 (1.2)	33.8	0 (0.0)	69 (100.0)	0 (0.0)	2.2 (1.2)	31.6	0 (0.0)	50 (100.0)	0 (0.0)

[a] P values are for differences in proportion of participants' adherence to recommendations between nutritional status groups (chi-square test); significant differences shown in bold at P value < 0.05.

TABLE 6: Differences of macronutrient and energy intakes for 3 main meals (breakfast, lunch, and dinner).

	Breakfast Median (IQR)	Lunch Median (IQR)	Dinner Median (IQR)	Snacks Median (IQR)
Male				
Energy (kcal)	392 (126)[a,b,c,d]	501 (129)[a,b,f]	517 (183)[a,c,g]	75 (195)[a,d,f,g]
Carbohydrate (g)	54 (20)[a,b,c,d]	71 (21)[a,b,f]	64 (27)[a,c,g]	26 (28)[a,d,f,g]
Protein (g)	16 (7)[a,b,c,d]	20 (6)[a,b,f]	21 (8)[a,c,g]	4 (5)[a,d,f,g]
Fat (g)	13 (7)[a,b,c,d]	17 (7)[a,b,f]	16 (10)[a,c,g]	4 (6)[a,d,f,g]
Female				
Energy (g)	372 (142)[a,b,c,d]	461 (138)[a,b,e,f]	432 (162)[a,c,e,g]	217 (179)[a,d,f,g]
Carbohydrate (g)	50 (18)[a,b,c,d]	62 (20)[a,b,e,f]	57 (22)[a,c,e,g]	25 (34)[a,d,f,g]
Protein (g)	16 (8)[a,b,c,d]	18 (7)[a,b,f]	19 (8)[a,c,g]	3 (4)[a,d,f,g]
Fat (g)	12 (8)[a,b,c,d]	14 (8)[a,b,f]	14 (9)[a,c,g]	3 (5)[a,d,f,g]

[a] P values are for differences among the three meals and snacks (Friedman's test); significant differences shown in bold at P value < 0.05.
[b,c,d,e,f,g] Significantly different from breakfast at P value < 0.0125 (Wilcoxon nonparametric t-test with Bonferroni correction).

TABLE 7: Comparison of BMI and macronutrient intakes for different eating occasions.

	Number of eating occasions/day			
	3	4	5	≥6
Male	(n = 30)	(n = 29)	(n = 11)	(n = 5)
BMI (kg/m^2)*	24.1 (3.8)	25.4 (3.3)	22.9 (2.2)	24.9 (4.6)
Energy (kcal/d)**	1,455 (22)[a]	1,578 (222)[a]	1,622 (344)[a]	1,684 (418)[a]
Carbohydrate (g/d)**	195 (49)[a,b,c]	218 (43)[a,b]	237 (60)[a,c]	215 (37)[a]
Protein (g/d)**	60 (13)	61 (14)	56 (13)	73 (24)
Fat (g/d)**	48 (13)	52 (17)	58 (23)	62 (25)
Female	(n = 32)	(n = 52)	(n = 44)	(n = 12)
BMI (kg/m^2)*	21.7 (3.4)	22.8 (3.4)	21.5 (2.8)	22.6 (5.8)
Energy (kcal/d)**	1,343 (265)[a,d]	1,380 (266)[a]	1,460 (299)[a]	1,488 (149)[a,d]
Carbohydrate (g/d)**	179 (48)[a,c,d]	197 (36)[a]	205 (44)[a,c]	209 (54)[a,d]
Protein (g/d)**	56 (12)	56 (13)	56 (52)	60 (14)
Fat (g/d)**	44 (19)	44 (15)	44 (16)	50 (16)

*Mean (SD), **median (IQR).
[a] Significant differences between 4 categories of eating occasions (Kruskal Wallis test), P value < 0.05.
[b,c,d] Significant differences between 2 categories of eating occasions (Mann-Whitney U test with Bonferroni correction), P value < 0.0125.

was found with regard to females' vitamin C intakes. The highest level—61.9%—of adherence to the recommended vitamin C intake was shown in the group of underweight females, followed by 55.1% in a normal weight group and by 36.0% in overweight/obese female participants.

3.5. Meal Patterns. Table 6 shows that significant differences (P value < 0.05) were found with respect to energy and macronutrient intakes for the three daily meals and snacks. Male participants showed the highest energy intake in kcal/day for dinner (517; IQR 183), followed by lunch (501; IQR 129), breakfast (392; IQR 126), and snacks (75; IQR 195). There were no statistically significant differences of males' energy intake (kcal/day) and the intakes of carbohydrate, protein, and fat (grams/day) between lunch and dinner. The highest energy intake for women was at lunch (461; IQR 138), followed by dinner (432; IQR 162), breakfast (372; IQR 142), and snacks (217; IQR 179). The highest carbohydrate intake (grams/day) by females was found at lunch, followed by dinner, breakfast, and snacks. The females' intakes of protein and fat were not statistically different between lunch and dinner.

The daily energy distribution derived from the three meals and snacks is presented in Figure 1.

Table 7 shows the comparison of BMI and macronutrient intakes for different eating occasions. That comparison showed no statistically significant difference between eating occasions for both genders. Participants of both genders with eating occasions ≥6 times/day also reported the highest level of energy and all macronutrient intakes. Female participants' energy intake of 1,488 kcal/day for eating occasions ≥6 times/day was significantly higher than 1,343 kcal/day for those who had eating occasions 3 times/day (P value < 0.0125). For male participants, the highest carbohydrate intake—237 grams/day—was reported by those who had eating occasions 5 times/day, and that was significantly higher than 195 grams/day for those who had only 3 eating occasions per day (P value < 0.0125). Among females, the

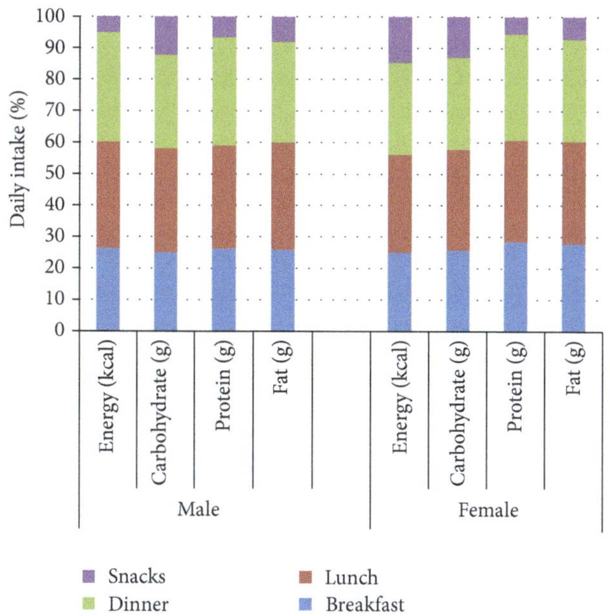

FIGURE 1: Daily percentage of energy distribution for breakfast, lunch, dinner, and snacks.

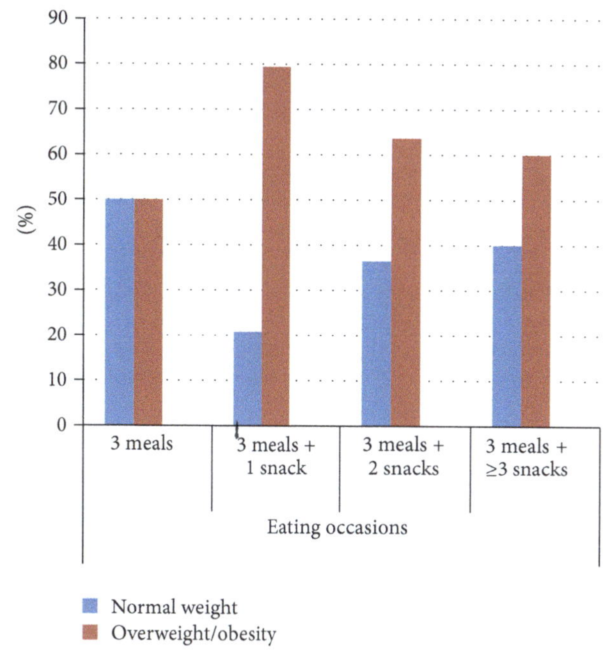

FIGURE 2: Comparison of eating occasions and nutritional status of male participants.

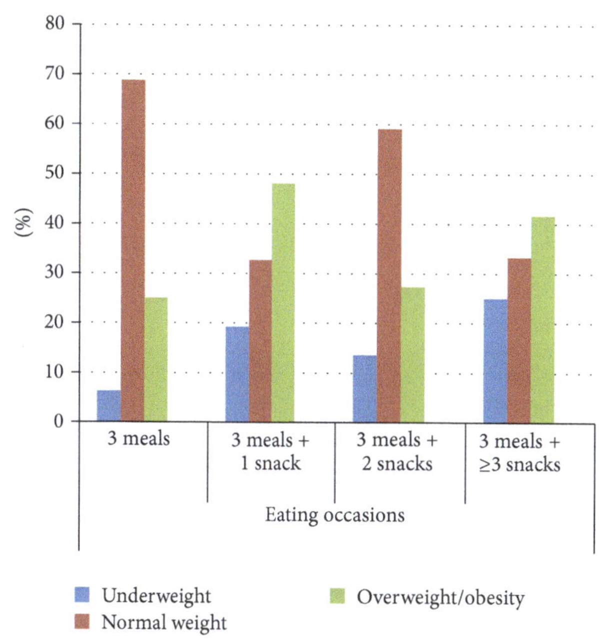

FIGURE 3: Comparison of eating occasions and nutritional status of female participants.

highest carbohydrate intake—209 grams/day—was recorded for those with eating occasions ≥6 times/day, followed by 205 grams/day for those with 5 eating occasions/day and 179 grams/day for those who had 3 eating occasions/day (P value < 0.05). Figures 2 and 3 further illustrate that subjects were more frequently classified as normal weight if they had 3 eating occasions/day relative to those who had 4, 5, or 6 eating occasions/day.

Normal weight and overweight/obese male participants showed no difference with regard to the energy intake from the three daily meals and snacks.

There were no statistically significant differences of protein and fat intake among eating occasions for both genders.

4. Discussion

Thailand is a developing country where rapid socioeconomic changes are affecting health and lifestyles of urban dwellers with sedentary occupations. By directly influencing the people's nutrition patterns and physical activity, these changes are creating serious health risks for the growing population subgroup of sedentary workers. While a large amount of research has been devoted to health risks facing people in sedentary occupations, a relatively limited attention was paid to problems of their nutrition patterns and dietary intakes.

This study shows the seriousness of diet-related health risks for Thai sedentary workers because it found that 64.0% of males and 35.7% of females in that population subgroup were found to be overweight/obese (BMI ≥ 23 kg/m^2). Also, 50.7% of males and 25.7% of females in sedentary occupations were in the high range of the percent body fat, and more than one-third (34.9%) of the entire study sample had abdominal obesity. These findings are similar to those reported for other countries where research focused on health hazards of occupational sitting. The prominent examples of such studies are Choi et al. [29] in the United States, Mummery et al. [30] in Australia, Larsson et al. [31] in Sweden, Ishizaki et al. [32] in Japan, and Shimokawa et al. [33] in China.

The study by Mummery et al. [30] about the relationship between sedentary work and overweight/obesity in Australia reports that the probability of having a BMI ≥ 25 kg/m^2 is greatly increased when the occupational sitting time is ≥5 hours/day, even in cases where physical activity levels reflect national guidelines. Our study found that the average sitting time in the Thai sample of sedentary workers was 6 hours/day, suggesting that this particular population subgroup in Thailand is exposed to a much higher risk of overweight/obesity and attendant diet-related pathologies.

That risk is also increased by our findings that the sedentary workers in our sample fell far short of adhering to the Thai DRIs for most nutrients, such as dietary fiber, potassium, calcium, iron, vitamin A, vitamin E, vitamin B_1, vitamin B_2, vitamin C, magnesium, selenium, and zinc. There were only few nutrients, mainly from the macronutrient category, where the study subjects conformed to the Thai DRIs.

The medians of energy intakes in our sample were also lower than those recommended in the Thai DRIs. The reported energy intakes of 1,485 kcal/day for males and 1,428 kcal/day for females accounted for only 71.2% and 80.1%, respectively, of the recommended Thai DRIs [17]. Relatively low energy intakes for both genders have also been reported in the Thai NHES IV whose estimates were based on a 24-hour dietary recall in a subsample of 2,969 people aged 15 years and over [34]. That study reported energy intakes of 1,600 kcal/day for males and 1,300 kcal/day for females. This may be partly due to the common problem that people tend to underreport their dietary intakes in 24-hour dietary recalls and in food records [35]. More generally, underreporting of energy intake is a well-known problem in self-reported dietary assessments and has been well documented in studies conducted in both developed [36, 37] and developing countries [38]. Using the Goldberg cut-off 1 test [39], we found that male and female participants in our sample underreported their energy intake by 26% and 11%, respectively. (The ratio of mean energy intake (EI) and the basal metabolic rate (BMR) < 1.35 is considered to represent underreporting. The average basal metabolic rate in this study was 1,552 kcal/day for men and 1.201 kcal/day for women. Thus, the energy requirement (BMR × 1.35) for male and female sedentary workers would be 2,000 kcal/day and 1,600 kcal/day, resp.) These results show that males' underreporting of energy intake is twice as high as the females' underreporting in a sample where 64% of males and 35.7% of females were classified as overweight/obese. Our findings are similar to other studies showing that overweight/obese individuals tend to underreport their dietary intake by as much as 20–50% [40, 41].

Median macronutrient intakes of carbohydrates and fats found in this study were all within their respective Thai DRIs [17]. The carbohydrate intake accounted for 54.4% of the total daily energy for men and for 56.0% for women. These results are almost identical to those reported in the latest national dietary survey by NHES IV of the general Thai adult population, showing that the percentage of energy distribution from carbohydrates was 52.7% for men and 55.5% for women [34, 42]. It is interesting to note that earlier Thai nutrition surveys found that the carbohydrate intake as a percentage of total energy declined from 78.9% in 1960 [43] to 62.1% in 2006 [44]. That was considered to result from a shift away from the traditional Thai staples that were high in carbohydrates to modern nutrition patterns mainly based on protein-rich food groups.

Protein intakes found in this study were higher than those recommended by the Thai DRIs. That higher protein intake was consistent with the research showing that, over the last five decades, the Thai protein consumption as part of the daily energy intake increased from 10% in 1960 [43] to 16% in 2009 [34]. It is interesting to note, however, that although the amount of protein intake in this study was higher than the Thai DRIs, it was still consistent with the Thai NHES IV, where recommended protein intakes were defined by acceptable macronutrient distribution ranges (AMDRs) developed by the US Institute of Medicine (IOM) [45].

With respect to the fat intake, this study found that the consumption of fats accounted for 29.6% of male and 28.6% of female daily energy. Our findings also conform to the data reported in Thai nutrition surveys, indicating that the fat consumption increased from 8.9% of total energy intake in 1960 to 28% in 2009 [34, 43, 46]. This is another confirmation of how much the traditional Thai diets, based on starchy staples, have been replaced by diets higher in animal protein [47]. These dietary changes are directly linked to the rising prevalence of dyslipidemia in the Thai population [48].

Micronutrients, required for virtually all metabolic and developmental processes, have been found with inadequate intakes in our study sample. Lower-than-recommended intakes were noted in the case of potassium, calcium, iron, vitamin A, vitamin E, vitamin B_1, vitamin B_2, magnesium, selenium, and zinc, indicating that, similar to other developing countries, micronutrient deficiencies are still a major public health problem in Thailand [49]. Inadequate micronutrient intakes are considered to reflect changing nutrition patterns associated with increasing risks of overweight, obesity, and other diet-related diseases [43].

Calcium intakes for most participants (80%) in our study were insufficient and amounted to 60% of the Thai DRIs. The NHES IV survey also reported very low calcium intakes. In that survey, the calcium intake for Thai adults (19–59 years age group) represented only 30% of the Thai DRIs, and only 24.1% of them consumed milk and dairy products [34]. Historically, milk has never been an important part of Thai nutrition because there were virtually no dairy cattle until the second half of the last century [50]. The inadequate calcium intake is an additional health risk for an already vulnerable sedentary population, because cross-sectional and longitudinal studies have widely reported that calcium deficiencies had adverse effects on blood pressure, metabolic balance, and weight management [51–53]. Female study participants showed an insufficient intake of iron. At 9.5 mg/day, the females' iron intake in our study is only 39.2% of the Thai DRIs. That is confirming the finding in the latest Thai dietary survey, which reported that, at 8.5 mg/day, the iron dietary intake by adult Thai women accounted for about one-third of the Thai DRIs [34]. Although it seems that the situation has improved somewhat in recent years, the iron deficiency had been noted as an important issue for the Thai female population since the mid-1970s [54].

Our finding of insufficient intakes of antioxidant vitamins (vitamins A, C, and E), which are present in many plant-based foods, is an indication of Thailand's low consumption of fruits and vegetables. That is confirmed by the Thai NHES III and IV nutrition surveys [55]. These surveys showed that only 36.5% of participants consumed fruit and 68.0% consumed vegetables on a daily basis in relatively small portions of 100 grams to 162.5 grams [34]. Antioxidant deficiencies compromise the oxidative state of the individual and could, therefore, increase the risk of comorbidities associated with overweight/obesity, such as heart disease, certain types of cancer, and insulin resistance [56, 57]. The low intake of selenium and zinc may be typical of our sample, drawn from the Bangkok-based sedentary workers, but the insufficient consumption of fish and shellfish is not typical of the diet followed by the general Thai population [44].

Our study indicates that the Thai's sodium intake remains excessive. Almost one-third of our study subjects reported a sodium consumption ≥2,400 mg/day, confirming similar results of the high intake of sodium in many studies, including the NHES IV [58, 59]. Typically, Thai dishes use condiments such as fish sauce, soya sauce, oyster sauce, table salt, and shrimp paste that are very high in sodium content. Excessive sodium intake is a worldwide problem [60]. Thailand, like many other countries, continues to conduct public health campaigns to lower the sodium consumption [61], because some evidence is showing that a high sodium intake may be contributing to problems of hypertension, cardiovascular diseases, and chronic kidney ailments in the Thai population [62].

We also found an insufficient potassium intake by both genders. Potassium is a very important mineral to regulate blood pressure and lower the risks of stroke and heart disease. Since potassium is mainly found in plant-based foods, its deficiency in the Thai diet is reflecting the already mentioned low consumption of fruits and vegetables.

Significantly inadequate fiber consumption is another important problem found in this study. Fiber intakes of 7.8 grams/day for men and 8.0 grams/day for women are far below the 25 grams/day recommended by the Thai DRIs. Our findings are consistent with those of the Thai NHES IV, reporting fiber intakes of 7.5 grams/day for males and 8.5 grams/day for females [55]. The problem of the low dietary fiber intake in this study is due to inadequate consumption of fruits, vegetables, and grains, partly because these foods are not readily available in the office area or in office canteens. Most of the study subjects reported having breakfast at the office, consisting mainly of coffee, pastries, sandwiches, congee (based on rice with small quantities of minced pork/chicken), or grilled pork and sticky rice. Only a few study subjects reported having a complete meal (composed of carbohydrate, protein, fruits, and vegetables) or a salad for breakfast. Our interviewers observed that lunch was typically taken in office canteens where not enough vegetables were served. Changes toward greater consumption of fiber-rich foods are an important public health issue in Thailand, because that would lower the risk of metabolic syndrome, diabetes, and cardiovascular diseases [63, 64].

Our study also shows that progress has to be made toward improving the quality of food in worksite canteens, as has been the case in some other countries [65, 66].

The study's findings about meal pattern characteristics (Figure 1) show that breakfast provides 26% of daily energy for both genders, which is only slightly above the 20%–25% general range for meals energy distribution. Lunch provided 34% of daily energy for men and 32% for women (well within the general range of 30%–35%). The dinner supplied 35% of daily energy for men and 30% for women. Snacks provided 5% of daily energy for men and 12% for women (the recommended range is 10%–20%). These results show that our study participants had a proper energy distribution between breakfast, lunch, dinner, and snacks and that most of their daily energy came from main meals.

Our results presented in Table 7 confirm those of previous studies that a higher number of eating occasions lead to higher energy intake and that higher carbohydrate consumption was associated with higher BMI measurements. A recent study reported that more daily meals were consumed by obese than by lean women [67]. Snacks, in particular, were identified to contain a higher proportion of carbohydrates and a smaller proportion of fats than regular meals [68]. The growing trend of grazing rather than following the traditional pattern of three proper meals is considered to be a major factor in the etiology of obesity. Forslund et al. showed that obese Swedish women snacked more often and had higher energy intake than their lean peers [67]. The French women with a habit of frequent snacking were also found to have a higher energy intake than those who did not snack [69].

4.1. Strengths and Limitations of This Study. The main limitation of this study resides in underreporting of energy intake. That is a well-known and well-documented problem of self-reported dietary assessment methods such as the 24-hour dietary recall and a nonweighed food record. We have calculated, for example, that male and female participants in this study underreported their respective energy intakes by 26% and 11%. Similar problems have also been noted in national nutrition surveys, including those conducted in Thailand [70, 71]. The 24-hour dietary recall relies on the respondents' memory, and the food record has a heavy record-keeping burden, showing, in some cases, that energy intakes decrease with increasing days of reporting the food consumption [72].

The other limitation of this study is due to the limitation of the Thai dietary database (INMUCAL-NV.2.0). That problem is discussed by Satheannoppakao et al. [59]. For example, some commonly used Thai food ingredients such as fermented fish ("Nam-Pla-Ra") are high in sodium, but the nutrient analysis shows no results for sodium content. Also, this software version cannot calculate the proportion of saturated fat in total fat. This nutrient analysis can provide up to 34 items of information on energy and nutrients for 2,054 food products with 50%–99.8% food biochemical references. But for vitamin B_6 and vitamin B_{12}, the program gives only 6.5% and 9.1% of biochemical references, which made it impossible for us to report these vitamins' intakes.

The important strength of this study is in methods and procedures used during the data collection and processing. Precise and comprehensive instructions were given to study subjects for the completion of a single 24-hour dietary recall and a nonweighed 2-day food record during orientation sessions. Food models were also used to enhance the accuracy of (a) portion estimates, (b) the record of food items consumed, and (c) the daily variation in dietary intakes.

In spite of a relatively small sample size ($n = 215$), this study's results on energy and nutrient intakes are quite similar and in some cases identical to those produced by studies, such as the Thai National Health Examination Surveys (NHES IV 2008-2009 [34]), based on much larger samples using the same dietary assessment methods. That is most probably because the relevant sample in that study was of a similar size to the one we used. Indeed, the NHES survey was based on a subsample of 2,969 people aged 1–60 years, but only 935 people were in the 15–59 age group. If, according to the physical activity assessment in that survey, these 935 subjects in that age group are further classified into three groups—nonsedentary, sedentary, and unemployed—the resulting sample size would very likely be similar to the sample size ($n = 215$) used in our study. Moreover, our study included only healthy sedentary workers having no diet-related diseases, but the NHES IV subsample did not exclude unhealthy subjects who might have had diet-related pathologies.

These study sample comparisons lead us to believe that our dietary assessment is suitable and that it can be representative of the nutrition patterns of Thailand's urban sedentary workers.

4.2. Implications of the Study. Our research shows that energy and nutrient intakes of Thai sedentary workers should be brought in greater compliance with the Thai DRIs. The way to do that is by improving public health messages about healthy dietary guidelines. That means that there is a need for public health intervention initiatives to inform and educate the public about healthy eating and healthy lifestyles. That is the new line of research highlighted by our study. Another possible topic for further study is the improvement of food quality in worksite eateries. A much broader study may also be warranted, based on a considerably larger sample size than the one we used, to evaluate diet-related health risks of the sedentary population.

5. Conclusions

This study examined the extent to which energy and nutrient intakes of a sample of sedentary workers in the Bangkok metropolitan area conform to the Thai DRIs. A declining trend noted for energy and carbohydrate intakes and an increasing consumption of protein and fat are reflecting shifts in the Thai diet away from traditional carbohydrate-rich staples to nutrition patterns based on proteins and fats. Intakes of dietary fiber and most micronutrients, including calcium, vitamins A, C, and E, magnesium, selenium, and zinc, did not meet the Thai DRIs. Higher intakes of foods rich in fat and lower intakes of fiber and antioxidants are putting sedentary workers at a greater risk of diet-related chronic diseases. Our study suggests that there is a need to improve dietary patterns for sedentary workers through worksite wellness and nutrition education initiatives. Based on our research, these nutrition education programs should emphasize that increasing consumption of certain nutrients found in this study (e.g., fat and sodium) and lower intakes of others (e.g., fiber, calcium, vitamin A, vitamin E, and vitamin C) are linked to a greater risk of cardiovascular diseases and certain forms of cancers. Worksites offer a large and captive audience to promote dietary behavior that would increase people's adherence to the recommended energy and nutrient intakes and physical activity programs that would reduce health risks inherent in sedentary lifestyles. Improving the quality of foods served in worksite canteens by providing healthier meal selections should be a priority in Thailand's efforts to promote public health.

This first and unique attempt to focus on nutrition problems of people in sedentary occupations is a positive contribution to Thailand's public health research. By offering a thorough evaluation of the extent to which sedentary workers' nutrition patterns conform to the Thai DRIs, this study's findings can be used in nutrition intervention programs to prevent diet-related diseases in this high-risk population subgroup.

Conflict of Interests

The authors declare no conflict of interests regarding the publication of this paper.

Acknowledgment

This study was partially supported by the Department of Health of the Thailand Ministry of Public Health.

References

[1] M. T. Hamilton, D. G. Hamilton, and T. W. Zderic, "Role of low energy expenditure and sitting in obesity, metabolic syndrome, type 2 diabetes, and cardiovascular disease," *Diabetes*, vol. 56, no. 11, pp. 2655–2667, 2007.

[2] R. R. Pate, J. R. O'Neill, and F. Lobelo, "The evolving definition of "sedentary"," *Exercise and Sport Sciences Reviews*, vol. 36, no. 4, pp. 173–178, 2008.

[3] S. Bertrais, J. P. Beyeme-Ondoua, S. Czernichow, P. Galan, S. Hercberg, and J. M. Oppert, "Sedentary behaviors, physical activity, and metabolic syndrome in middle-aged French subjects," *Obesity Research*, vol. 13, no. 5, pp. 936–944, 2005.

[4] P. T. Katzmarzyk, T. S. Church, C. L. Craig, and C. Bouchard, "Sitting time and mortality from all causes, cardiovascular disease, and cancer," *Medicine and Science in Sports and Exercise*, vol. 41, no. 5, pp. 998–1005, 2009.

[5] B. M. Popkin, "The nutrition transition and its health implications in lower-income countries," *Public Health Nutrition*, vol. 1, no. 1, pp. 5–21, 1998.

[6] B. M. Popkin, "Urbanization, lifestyle changes and the nutrition transition," *World Development*, vol. 27, no. 11, pp. 1905–1916, 1999.

[7] W. Aekplakorn, M. C. Hogan, V. Chongsuvivatwong et al., "Trends in obesity and associations with education and Urban

or rural residence in Thailand," *Obesity*, vol. 15, no. 12, pp. 3113–3121, 2007.

[8] W. Aekplakorn, *The Report of the 4th Thai National Health Examination Survey in 2008-2009*, National Health Examination Survey Office, Health System Research Institute, Nonthaburi, Thailand, 2009.

[9] W. Aekplakorn, R. Inthawong, P. Kessomboon et al., "Prevalence and trends of obesity and association with socioeconomic status in Thai adults: national health examination surveys, 1991–2009," *Journal of Obesity*, vol. 2014, Article ID 410259, 8 pages, 2014.

[10] WHO Expert Consultation, "Appropriate body-mass index for Asian populations and its implications for policy and intervention strategies," *The Lancet*, vol. 363, pp. 157–163, 1998.

[11] D. M. Freedman, E. Ron, R. Ballard-Barbash, M. M. Doody, and M. S. Linet, "Body mass index and all-cause mortality in a nationwide US cohort," *International Journal of Obesity*, vol. 30, no. 5, pp. 822–829, 2006.

[12] K. F. Adams, A. Schatzkin, T. B. Harris et al., "Overweight, obesity, and mortality in a large prospective cohort of persons 50 to 71 years old," *The New England Journal of Medicine*, vol. 355, no. 8, pp. 763–778, 2006.

[13] A. Must, J. Spadano, E. H. Coakley, A. E. Field, G. Colditz, and W. H. Dietz, "The disease burden associated with overweight and obesity," *The Journal of the American Medical Association*, vol. 282, no. 16, pp. 1523–1529, 1999.

[14] S. D. H. Malnick and H. Knobler, "The medical complications of obesity," *QJM*, vol. 99, no. 9, pp. 565–579, 2006.

[15] M. A. Raebel, D. C. Malone, D. A. Conner, S. Xu, J. A. Porter, and F. A. Lanty, "Health services use and health care costs of obese and nonobese individuals," *Archives of Internal Medicine*, vol. 164, no. 19, pp. 2135–2140, 2004.

[16] M. L. Daviglus, K. Liu, L. L. Yan et al., "Relation of body mass index in young adulthood and middle age to medicare expenditures in older age," *Journal of the American Medical Association*, vol. 292, no. 22, pp. 2743–2749, 2004.

[17] Nutrition Division, *Dietary Referrence Intake for Thais 2003*, Ministry of Public Health, Bangkok, Thailand, 4th edition, 2003.

[18] F. Faul, E. Erdfelder, A.-G. Lang, and A. Buchner, "G* Power 3: A flexible statistical power analysis program for the social, behavioral, and biomedical sciences," *Behavior Research Methods*, vol. 39, no. 2, pp. 175–191, 2007.

[19] J. Cohen, *Statistical Power Analysis for the Behavioral Sciences*, Lawrence Erlbaum Associates, Hillsdale, NJ, USA, 2nd edition, 1988.

[20] J. Wang, J. C. Thornton, S. Bari et al., "Comparisons of waist circumferences measured at 4 sites," *The American Journal of Clinical Nutrition*, vol. 77, no. 2, pp. 379–384, 2003.

[21] World Health Organization Western Pacific Region, International Association for the Study of Obesity, and International Obesity Task Force, *The Asia Pacific Perspective: Redefining Obesity and Its Treatment*, Health Communications, Sydney, Australia, 2000.

[22] L. Li, C. Wang, Y. Bao, L. Peng, H. Gu, and W. Jia, "Optimal body fat percentage cut-offs for obesity in Chinese adults," *Clinical and Experimental Pharmacology and Physiology*, vol. 39, no. 4, pp. 393–398, 2012.

[23] C. Bouchard, A. Tremblay, C. Leblanc, G. Lortie, R. Savard, and G. Theriault, "A method to assess energy expenditure in children and adults," *The American Journal of Clinical Nutrition*, vol. 37, no. 3, pp. 461–467, 1983.

[24] P. P. Sirichakwal, K. Sranacharoenpong, and K. Tontisirin, "Food based dietary guidelines (FBDGs) development and promotion in Thailand," *Asia Pacific Journal of Clinical Nutrition*, vol. 20, no. 3, pp. 477–483, 2011.

[25] Institute of Nutrition, *Manual of INMUCAL-Nutrients*, Mahidol University, Nakhon Pathom, Thailand, 4th edition, 2009.

[26] Nutrition Division, *Nutritive Values of Thai Foods*, Department of Health, Ministry of Public Health, Nonthaburi, Thailand, 2001.

[27] A. F. Subar, F. E. Thompson, V. Kipnis et al., "Comparative validation of the block, willett, and national cancer institute food frequency questionnaires: the eating at America's Table Study," *The American Journal of Epidemiology*, vol. 154, no. 12, pp. 1089–1099, 2001.

[28] Institute of Medicine (US) Food and Nutrition Board, *Dietary Reference Intakes: A Risk Assessment Model for Establishing Upper Intake Levels for Nutrients*, National Academies Press (US), Washington, DC, USA, 1998.

[29] B. Choi, P. L. Schnall, H. Yang et al., "Sedentary work, low physical job demand, and obesity in US workers," *American Journal of Industrial Medicine*, vol. 53, no. 11, pp. 1088–1101, 2010.

[30] W. K. Mummery, G. M. Schofield, R. Steele, E. G. Eakin, and W. J. Brown, "Occupational sitting time and overweight and obesity in Australian workers," *American Journal of Preventive Medicine*, vol. 29, no. 2, pp. 91–97, 2005.

[31] C. A. Larsson, L. Krøll, L. Bennet, B. Gullberg, L. Råstam, and U. Lindblad, "Leisure time and occupational physical activity in relation to obesity and insulin resistance: a population-based study from the Skaraborg Project in Sweden," *Metabolism—Clinical and Experimental*, vol. 61, no. 4, pp. 590–598, 2012.

[32] M. Ishizaki, Y. Morikawa, H. Nakagawa et al., "The influence of work characteristics on body mass index and waist to hip ratio in Japanese employees," *Industrial Health*, vol. 42, no. 1, pp. 41–49, 2004.

[33] S. Shimokawa, H.-H. Chang, and P. Pinstrup-Andersen, "Understanding the differences in obesity among working adults between Taiwan and China," *Asia Pacific Journal of Clinical Nutrition*, vol. 18, no. 1, pp. 88–95, 2009.

[34] W. Aekplakorn, P. Kessomboon, R. Sangthong et al., "Urban and rural variation in clustering of metabolic syndrome components in the Thai population: results from the fourth National Health Examination Survey 2009," *BMC Public Health*, vol. 11, article 854, 2011.

[35] L. Johansson, K. Solvoll, G.-E. A. Bjørneboe, and C. A. Drevon, "Under- and overreporting of energy intake related to weight status and lifestyle in a nationwide sample," *The American Journal of Clinical Nutrition*, vol. 68, no. 2, pp. 266–274, 1998.

[36] R. R. Briefel, C. T. Sempos, M. A. McDowell, S. Chien, and K. Alaimo, "Dietary methods research in the third National health and nutrition examination survey: underreporting of energy intake," *The American Journal of Clinical Nutrition*, vol. 65, no. 4, pp. 1203S–1209S, 1997.

[37] P. Ferrari, N. Slimani, A. Ciampi et al., "Evaluation of under- and overreporting of energy intake in the 24-hour diet recalls in the European Prospective Investigation into Cancer and Nutrition (EPIC)," *Public Health Nutrition*, vol. 5, no. 6B, pp. 1329–1345, 2002.

[38] F. B. Scagliusi, E. Ferriolli, and A. H. Lancha Jr., "Underreporting of energy intake in developing nations," *Nutrition Reviews*, vol. 64, no. 7, pp. 319–330, 2006.

[39] G. R. Goldberg, A. E. Black, S. A. Jebb et al., "Critical evaluation of energy intake data using fundamental principles of energy

physiology: 1. Derivation of cut-off limits to identify under-recording," *European Journal of Clinical Nutrition*, vol. 45, no. 12, pp. 569–581, 1991.

[40] S. W. Lichtman, K. Pisarska, E. R. Berman et al., "Discrepancy between self-reported and actual caloric intake and exercise in obese subjects," *The New England Journal of Medicine*, vol. 327, no. 27, pp. 1893–1898, 1992.

[41] A. H. Goris, M. S. Westerterp-Plantenga, and K. R. Westerterp, "Undereating and underrecording of habitual food intake in obese men: selective underreporting of fat intake," *American Journal of Clinical Nutrition*, vol. 71, no. 1, pp. 130–134, 2000.

[42] *The Fifth National Nutirtional Survey of Thailand 2003*, Department of Health, Ministry of Public Health, Bangkok, Thailand, 2003.

[43] V. Kosulwat, "The nutrition and health transition in Thailand," *Public Health Nutrition*, vol. 5, no. 1, pp. 183–189, 2002.

[44] National Bureau of Argricultural Commodity and Food Standard, *Food Consumption Data of Thailand*, National Bureau of Argricultural Commodity and Food Standard, Bangkok, Thailand, 2006.

[45] Institute of Medicine, *Dietary Reference Intakes for Energy, Carbohydrate, Fiber, Fat, Fatty Acids, Cholesterol, Protein, and Amino Acids 2005*, Institute of Medicine, 2011, http://www.nap.edu/catalog/10490.html.

[46] N. Jitnarin, V. Kosulwat, N. Rojroongwasinkul, A. Boonpraderm, C. K. Haddock, and W. S. C. Poston, "Risk factors for overweight and obesity among Thai adults: results of the national Thai food consumption survey," *Nutrients*, vol. 2, no. 1, pp. 60–74, 2010.

[47] P. Pingali, "Westernization of Asian diets and the transformation of food systems: implications for research and policy," *Food Policy*, vol. 32, no. 3, pp. 281–298, 2007.

[48] W. Aekplakorn, S. Taneepanichskul, P. Kessomboon et al., "Prevalence of dyslipidemia and management in the Thai population, National Health Examination Survey IV, 2009," *Journal of Lipids*, vol. 2014, Article ID 249584, 13 pages, 2014.

[49] J. R. Díaz, A. de las Cagigas, and R. Rodríguez, "Micronutrient deficiencies in developing and affluent countries," *European Journal of Clinical Nutrition*, vol. 57, no. 1, pp. S70–S72, 2003.

[50] T. Kehren and C. Tisdell, *The Thai Dairy Industry: Its Economic Evolution Raised by Land Rights and Cattle Diseases*, Department of Economics, University of Queensland, Brisbane, Australia, 1997, http://ageconsearch.umn.edu/bitstream/164582/2/WP31.pdf.

[51] M. R. S. G. Torres, T. D. S. Ferreira, D. C. Carvalho, and A. F. Sanjuliani, "Dietary calcium intake and its relationship with adiposity and metabolic profile in hypertensive patients," *Nutrition*, vol. 27, no. 6, pp. 666–671, 2011.

[52] M. A. Beydoun, T. L. Gary, B. H. Caballero, R. S. Lawrence, L. J. Cheskin, and Y. Wang, "Ethnic differences in dairy and related nutrient consumption among US adults and their association with obesity, central obesity, and the metabolic syndrome," *The American Journal of Clinical Nutrition*, vol. 87, no. 6, pp. 1914–1925, 2008.

[53] S. Eilat-Adar, J. Xu, C. Loria et al., "Dietary calcium is associated with body mass index and body fat in American Indians," *Journal of Nutrition*, vol. 137, no. 8, pp. 1955–1960, 2007.

[54] P. Winichagoon, "Prevention and control of anemia: thailand experiences," *The Journal of Nutrition*, vol. 132, no. 4, pp. 862S–866S, 2002.

[55] W. Satheannoppakao, W. Aekplakorn, and M. Pradipasen, "Fruit and vegetable consumption and its recommended intake associated with sociodemographic factors: Thailand National Health Examination Survey III," *Public Health Nutrition*, vol. 12, no. 11, pp. 2192–2198, 2009.

[56] D. Bonnefont-Rousselot, "The role of antioxidant micronutrients in the prevention of diabetic complications," *Treatments in Endocrinology*, vol. 3, no. 1, pp. 41–52, 2004.

[57] S. Shah, M. Iqbal, J. Karam, M. Salifu, and S. I. McFarlane, "Oxidative stress, glucose metabolism, and the prevention of type 2 diabetes: pathophysiological insights," *Antioxidants and Redox Signaling*, vol. 9, no. 7, pp. 911–929, 2007.

[58] P. Pavadhgul, S. Sunthonwaraluk, S. Srisorachatr, and P. Temcharoen, "Dietary sodium intake by semi-quantitative food frequency questionnaire among undergraduate students of Mahidol University," *Journal of the Medical Association of Thailand*, vol. 92, supplement 7, pp. S75–S82, 2009.

[59] W. Satheannoppakao, R. Kasemsup, R. Inthawong et al., "Sodium intake and socio-demographic determinants of the non-compliance with daily sodium intake recommendations: Thai NHES IV," *Journal of the Medical Association of Thailand*, vol. 96, pp. S161–S170, 2013.

[60] World Health Organization, *Report of a WHO Forum and Technical Meeting 5–7 October 2006, Paris*, WHO Press, Geneva, Switzerland, 2007.

[61] C. Supornsilaphachai, *Evolution of Salt Reduction Initiatives in Thailand: Lessons for Other Countries in the South-East Asia Region*, vol. 17, Regional Health Forum, 2013.

[62] F. J. He and G. A. MacGregor, "A comprehensive review on salt and health and current experience of worldwide salt reduction programmes," *Journal of Human Hypertension*, vol. 23, no. 6, pp. 363–384, 2009.

[63] M. Galisteo, J. Duarte, and A. Zarzuelo, "Effects of dietary fibers on disturbances clustered in the metabolic syndrome," *Journal of Nutritional Biochemistry*, vol. 19, no. 2, pp. 71–84, 2008.

[64] M. O. Weickert and A. F. H. Pfeiffer, "Metabolic effects of dietary fiber consumption and prevention of diabetes," *The Journal of Nutrition*, vol. 138, no. 3, pp. 439–442, 2008.

[65] A. Lassen, A. V. Thorsen, E. Trolle, M. Elsig, and L. Ovesen, "Successful strategies to increase the consumption of fruits and vegetables: results from the Danish "6 a day" work-site Canteen Model Study," *Public Health Nutrition*, vol. 7, no. 2, pp. 263–270, 2004.

[66] A. Lassen, K. S. Hansen, and E. Trolle, "Comparison of buffet and à la carte serving at worksite canteens on nutrient intake and fruit and vegetable consumption," *Public Health Nutrition*, vol. 10, no. 3, pp. 292–297, 2007.

[67] H. B. Forslund, A. K. Lindroos, L. Sjöström, and L. Lissner, "Meal patterns and obesity in Swedish women—a simple instrument describing usual meal types, frequency and temporal distribution," *European Journal of Clinical Nutrition*, vol. 56, no. 8, pp. 740–747, 2002.

[68] F. Bellisle, A. M. Dalix, L. Mennen et al., "Contribution of snacks and meals in the diet of French adults: a diet-diary study," *Physiology & Behavior*, vol. 79, no. 2, pp. 183–189, 2003.

[69] A. Basdevant, C. Craplet, and B. Guy-Grand, "Snacking patterns in obese French women," *Appetite*, vol. 21, no. 1, pp. 17–23, 1993.

[70] D. Garriguet, "Under-reporting of energy intake in the Canadian Community Health Survey," *Health Reports/Statistics Canada, Canadian Centre for Health Information*, vol. 19, no. 4, pp. 37–45, 2008.

[71] S. Kye, S.-O. Kwon, S.-Y. Lee et al., "Under-reporting of energy intake from 24-hour dietary recalls in the korean national health and nutrition examination survey," *Osong Public Health and Research Perspectives*, vol. 5, no. 2, pp. 85–91, 2014.

[72] S. C. Liberato, J. Bressan, and A. P. Hills, "Assessment of energy and macronutrient intake in young men: a comparison of 4-day food record and 24-hour dietary recall," *Revista de Nutrição*, vol. 22, no. 5, pp. 621–630, 2009.

17

Nutrigenomics: Definitions and Advances of This New Science

N. M. R. Sales,[1] P. B. Pelegrini,[2] and M. C. Goersch[3]

[1] *Universidade Anhembi Morumbi/Gerência de Nutrição Enteral e Parenteral, Rua Pedro Ivo 22, 01323-070 São Paulo, SP, Brazil*
[2] *BioLife Brasil Ltda, Distrito Agroindustrial de Formosa, Quadra 01, Lote 15, 73801-970 Formosa, GO, Brazil*
[3] *Gerência de Nutrição Enteral e Parenteral-GANEP, Rua Pedro Ivo 22, 01323-070 São Paulo, SP, Brazil*

Correspondence should be addressed to P. B. Pelegrini; pbpelegrini@gmail.com

Academic Editor: M. Meydani

The search for knowledge regarding healthy/adequate food has increased in the last decades among the world population, researchers, nutritionists, and health professionals. Since ancient times, humans have known that environment and food can interfere with an individual's health condition, and have used food and plants as medicines. With the advance of science, especially after the conclusion of the Human Genome Project (HGP), scientists started questioning if the interaction between genes and food bioactive compounds could positively or negatively influence an individual's health. In order to assess this interaction between genes and nutrients, the term "Nutrigenomics" was created. Hence, Nutrigenomics corresponds to the use of biochemistry, physiology, nutrition, genomics, proteomics, metabolomics, transcriptomics, and epigenomics to seek and explain the existing reciprocal interactions between genes and nutrients at a molecular level. The discovery of these interactions (gene-nutrient) will aid the prescription of customized diets according to each individual's genotype. Thus, it will be possible to mitigate the symptoms of existing diseases or to prevent future illnesses, especially in the area of Nontransmissible Chronic Diseases (NTCDs), which are currently considered an important world public health problem.

1. Introduction

Food intake and the environment are the two main factors that affect the health or illness of an individual [1]. Studies in nutritional area have increased the understanding of how to maintain healthy a group of individuals that live in different dietary conditions [2–4]. However, after the conclusion of the Human Genome Project (HGP), new insights about the influence of nutrients into people's diet were postulated, which included (i) will gene expression in response to metabolic process, at cellular level, influence the health of an individual? (ii) Are gene expression and metabolic response the result of the interaction between genotype and environment/nutrient? (iii) Understanding how this interaction process occurs between gene and nutrient could lead to the prescription of specific diets for each individual. Hence, in order to answer those questions, Nutrigenomics was introduced [2, 5–7]. The studies on Nutrigenomics are focused on the effects of the nutrients over the genome, proteome, and metabolome, as illustrated on Figure 1 [2, 7–10].

Therefore, Nutrigenomics is the area of nutrition that uses molecular tools to search, access, and understand the several responses obtained through a certain diet applied between individuals or population groups [2, 5, 11, 12]. It seeks to elucidate how the components of a particular diet (bioactive compound) may affect the expression of genes, which may have increased its potential or which can be suppressed [2, 5, 12]. This response will depend on how genes will show a changed activity or alter gene expression (Figure 1) [2]. Some examples of this gene-nutrient interaction are their capacity on binding to transcription factors. This binding enhances or interferes with the ability of transcription factors on interacting with elements that will lead to the binding control of RNA polymerase. Earlier studies performed with vitamins A, D and fatty acids have shown that they can trigger direct actions in activating nuclear receptors and induce gene transcription [5, 13, 14]. Compounds such as resveratrol present in wine and soy genistein may indirectly influence the molecular signaling pathways, such as the factor kappa B [8, 13, 14]. The involvement of these factors in the activation

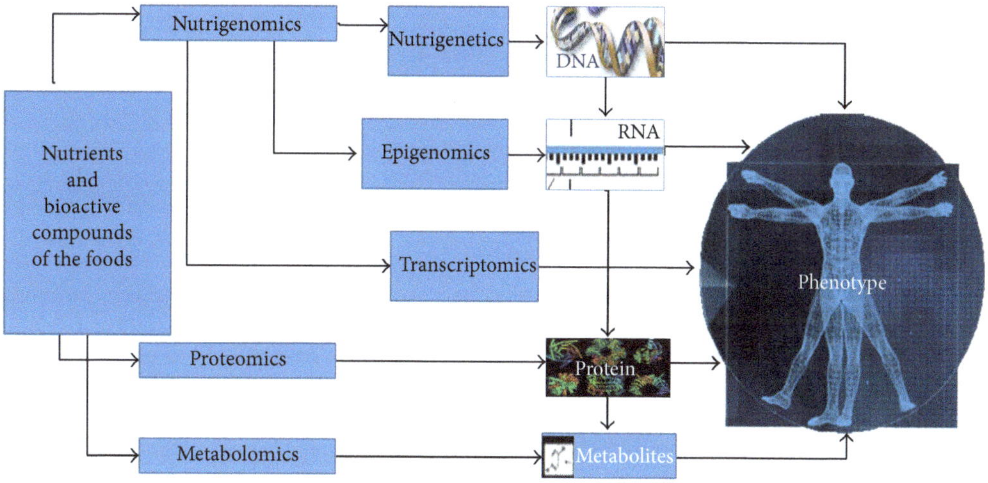

FIGURE 1: "Omics" sciences used in understanding the relationship between nutrition versus health versus disease (source: [4], with modifications; [9] with modifications).

and regulation of key molecules is associated with diseases ranging from inflammation to cancer [13, 14].

With information obtained from the HGP, it was found that humans have 99.9% identity between their genomes. A distinct difference between their weight, height, eye color/hair, and other features is only 0.1% of the gene sequence and this difference, among other factors, also determines the nutritional requirements and the risk of developing some of the NTCDs [2, 3, 16]. Single Nucleotide Polymorphisms (SNPs) are the main reason for this genetic variation, and it can often change the encoded protein [5, 17]. Studies have shown that certain genes and their variants can be regulated or are influenced by nutrients/food compounds from the diet and that these molecular variations may have beneficial actions to the health of an individual [2, 3, 17, 18].

1.1. From Nutrition to Nutrigenomics. Research on nutritional area can be seen since 400 BC, when Hippocrates speculated the hypothesis that the warm body temperature was innate. Around 1700 AD, the so-called "analytical Chemistry Era" began [12]. During this period, Lavoisier discovered how food was metabolized by the body, generating water, carbon dioxide, and energy [19]. In the 19th century, Liebig identified carbohydrates, proteins, lipids, and other macronutrients that released heat [12, 19]. During the next Era, called "Chemical and Analytical Era of Nutrition", which occurred between the 18th and 20th centuries, Antonie Lavoisier made important discoveries on food metabolism and their relation with energy production, including its relevance on breathing and oxidation [12, 19]. Later, during the "Biological Era", as the 19th century is known, studies on metabolism and chemistry were done, helping the science of nutrition on defining their role in the development and prevention of chronic diseases, such as cancer, cardiovascular, neurodegenerative, and bone metabolism disorders [12, 19]. Nowadays, the "Pos-Genomic Era" is being experienced. This era is characterized by the integration of three fields: biological, social, and environmental, where scientific discoveries on nutritional pathophysiology and metabolism are included [12, 19].

1.2. Nutrigenomics and Other Omics Sciences. After the HGP, discussions and actions began on a new biological era, the "Post-Genomic Era", where the evolution of bioinformatics provided advances in "omics" science research. These sciences use biotechnology to isolate and characterize a greater number of biomolecules from the same group, such as DNA, RNA, proteins, or metabolites. Hence, after Genomics, other Omic Sciences appeared as revolutionary tools, such as Proteomics, Metabolomics and Transcriptomics [8, 20].

Therefore, the first definition of Nutrigenomics referred only to studies on the effects of nutrients/bioactive food on gene expression of an individual. Nowadays, this definition expanded and, recently, Nutrigenomics also involves the studies on nutritional factors that act protecting the genome. Thus, this new science seeks to understand the influence of dietary components on the Genome, Transcriptome, Proteome, and Metabolome [1–3]. Nutrigenomics, in a single experiment, can generate multiple responses, so it is necessary to be innovative in the approaches of this area [2]. Moreover, Nutrigenomics is able to extract useful biological information from the data collected. The final answer can only be achieved after a series of investigations or surveys among different groups and teams. Therefore, there is a growing partnership between countries/teams/research groups, involving Nutrition field, Biology, Medicine, Genomics, and Bioinformation [1, 4, 5].

1.3. Nutritional Epigenomics. Epigenomics can be defined as the study of the complete set of epigenetic modifications in a cell or in a tissue at a given time [6, 21]. The epigenome consists of chemical compounds that modify or mark the genome in such way that it can indicate what a cell can do and where and when to do it. These marks are called epigenetic marks [6]. These epigenetic marks are passed from one cell to another when they divide themselves and thus will be passed from generation to generation. These signatures are influenced by genotype in the surrounding media (environment, diet, and drugs, e.g.) and will determine the phenotype (Figure 1) [6–9].

1.4. Epigenetics. The epigenetic studies the modification of DNA and proteins, linking the DNA and histones, which may cause changes in chromatin structure without changing the sequence of the nucleotides [7, 10]. Epigenetics is the transmitted information based on gene expression as its changes begin slowly but are progressive and potentially reversible [7, 11]. An example is the folate metabolism, where folic acid, which is a nutrient related to the genetic integrity, ensures a balanced amount of deoxyribonucleotides for DNA replication [2, 11]. It acts as a cofactor for enzymes associated with the biosynthesis of nucleotides and thymidylate (protein found in the RNA molecule), as universal donor of methyl and DNA methylation reactions [7, 11].

Modulations in gene expression may be caused by epigenetic mechanisms through changes in chromosome structure [4, 12]. Examples are DNA methylation and histone acetylation. Earlier studies have shown that DNA methylation is directly related to the remodeling of chromatin and this, in turn, is induced by the nutrient enzyme DNA methyltransferase (DNMT), which catalyzes the transfer of a methyl group from S-adenosylmethionine to specific sites on the DNA [7, 12]. The S-adenosylmethionine metabolizes compounds from food as folic acid, vitamins B6, B12, B2, choline, and methionine [2, 12]. A deficiency of these can lead to changes in carbon metabolism and thus impair DNA methylation, increasing the risk of development of NTCD. The DNA hypermethylation suppresses the gene responsible for transcription, once hypomethylation is associated with malignancies, such as prostate cancer and hepatocellular [4, 12].

1.5. Nutritional Transcriptomics. Transcriptomics studies the complete set of activated RNA transcripts [11]. The mRNAs are produced by a given moment and in a given tissue of a selected organism; therefore, gene expression varies according to the different circumstances and periods of time [8, 11]. Transcription factors, when activated, migrate to the nucleus and bind to a specific sequence of DNA in the promoter region of genes and, there, act by inhibiting or facilitating transcription [4, 8]. These transcription factors can be stimulated by (i) physiological signals, such as those triggered by nutrients/bioactive food compounds or the metabolites resulting from them; (ii) hormones, pharmacological treatments, and diseases, among others [8, 13]. They act as sensors regulating/modulating transcription of the cells as needed [13]. In nutrition research, transcriptomics can assist in providing information about the mechanisms or underlying effects of a particular nutrient or diet. It can also help identifying genes, proteins, or metabolites that change in the state of prediseases, as well as assisting on recognizing and characterizing the pathways regulated by nutrients or bioactive compounds in foods [8, 9, 11, 13].

1.6. Proteomics. Proteomics is the science that studies the complete set of proteins involved in the biological processes of a certain species [11, 21]. These proteins act in the cell, tissue, or organ in its normal state, but in different physiological or pathological situations, they may change their expression level, or even their activity, likewise in transcriptomics [8, 11, 21].

Proteins are an important class of molecules that are found in all living cells. They play a variety of roles in the cell such as structural, mechanical, biochemical, cell signaling, transport, and storage [2, 6]. They are also an essential part of the human diet [2, 6]. The number of proteins produced by an organism is much larger than the number of genes that it possesses. This happens due to the numerous posttranscriptional/translational modifications. Proteomics uses a set of technologies designed to study the expression of proteins. For this, it utilizes strategies, such as chromatographic techniques associated with electrophoresis, prefractionation of samples by extraction sequences, and organellar proteome analysis, among others [3, 4, 8, 14]. Hence, Proteomics is a primordial resource for Nutrigenomics, once that it is the gap between genome sequences and cell behavior, becoming the biological tool used to understand the process of genetic function determination, and of how genome is activated in response to certain diet [3, 4, 8, 14]. An example is the activity of butyrate, which can alter the expression of several proteins from the ubiquitin-proteasome system. This alteration suggests that proteolysis can be the mechanism by which butyrate can regulate key-proteins on the control of cell cycle, apoptosis, and cell differentiation [4].

1.7. Metabolomics. Metabolome consists of a set of small primary/secondary metabolites and body fluids of an organism or species. Metabolomics is the area of functional genomics that studies the changes in metabolites, whose goal is to isolate and characterize them [1, 3, 22]. Research advances in this area may facilitate the understanding of how the genotype is related to the phenotype of an individual. Nutritionally, metabolomics has many applications, once that it allows knowing the arrangements and metabolic disorders caused due to a person's diet and how these changes may affect the one's health or disease [1, 3, 8, 16, 22]. Hence, metabolomics also studies the metabolism under environmental and genetic perturbations [11], which can be analyzed and interpreted with the help of bioinformatics and statistical tools [11, 17].

The metabolites are dissolved in the cell cytosol and are small organic molecules that interact directly with the proteins and other macromolecules. They are divided into primary and secondary metabolites [1, 11, 22]. Primery metabolites are directly involved with the routes of synthesis and degradation of macromolecules, while secondary metabolites are more common in plants and fungi, and act as structural components and their defense [3, 22]. The metabolites in living beings can act as substrates, such as inhibitors or activators of an enzyme, molecular precursors, wastes of synthesis, or degradation of macromolecules, among others [3].

In the area of nutrition, metabolomics allows the understanding of metabolic arrangements and instabilities that are the cause or suffer interference from the diet [1, 11]. This contributes to the knowledge of how the excess or lack of some nutrients or compounds (secondary metabolites) present in food can affect the health/illness of an individual.

These compounds (nutrients or not) interact in several ways within the body, changing the metabolome pathways. As an example, perilla alcohol (monoterpene extracted from strawberry) can act as an anticancer molecule under certain organic stimulations [1, 3].

1.8. Nutrigenomics and Nontransmissible Chromic Diseases (NTCDs). Nutrition is the process that offers different substances to an organism that can work as energy supplier (carbohydrate and fat), as cell structure sources (proteins), and on metabolism control (vitamins and minerals), thus maintaining its homeostasis [16]. The nutritional state of an individual is the result of the interaction between various factors, such as genetic background, physical body, and emotional and social state [4]. Diet is a key role factor, once those nutrients and other bioactive compounds present in food can either be beneficial or initiate several diseases [4, 8, 16]. Among the illnesses related to food consumption, there are the celiac disease, phenylketonuria, and NTCDs, such as cancer, diabetes, and dislipidemies [8]. In this way, the health state of a person will depend on the interaction between their genes and their food diet [11]. Therefore, Nutrigenomics, along with other omic sciences, aim to clarify the interaction between genes and bioactive compounds from food sources [4, 8, 11, 12, 16, 17].

1.9. Obesity. Researches have indicated that not only environmental factors, but also genetics aspects are related to health problems, including DCNTs and Metabolic Syndromes (MS) [8, 23]. Recent studies showed that 80% of the difference observed in the body mass index (BMI) of twins are related to genetic factors [23]. As obesity causes a chronic process of inflammation, the use of Nutrigenomics to modulate this manner is highly promising [8]. Other reports demonstrated that some food contain anti-inflammatory bioactives, such as the caffeic acid (found in Yerba mate), tyrosol (found in olive oil), quercetin (present in fruits and greeneries), and lycopene (present in tomatoes, guavas, and watermelon). These molecules act inhibiting the expression of COX2 and iNOS genes through the reducing the translocation of the Kappa-B nuclear factor from the cytoplasm to the nucleus [8, 16].

There are several other ways where bioactive compounds from food can interfere on genes. One of the primary mechanisms for gene expression modulation is during transcription [2, 4], where the synthesis of inflammatory mediators occurs and has a fundamental role on numerous chronic illnesses, including obesity [2, 4, 24]. Hence, interleukin-1 is one of these mediators, which, after activation, stimulates the production of many other molecules during the inflammation cascade. The bioactive compound α-tocoferol, found in green tea, acts by decreasing the level of this chronic inflammatory process that occurs in obese individuals. Therefore, studies indicate that this component can assist on the treatment of obesity [2, 24].

1.10. Cancer. The need for certain micronutrientes by the organism depends on the person's age, genetic background, and physical state [4, 25]. Earlier studies showed that the deficiency of micronutrients, such as folic acid, vitamins B12, B6, C, and E, selenium, niacin, and zinc can cause changes into the DNA similar to what is seen after radiation exposure [2, 4, 25]. These alterations can lead to the rupture of the DNA double strand, oxidative lesions, or both. Furthermore, they demonstrated to be narrowly related to the development of cancer [25].

Moreover, molecules present in contaminated food can produce toxic metabolites that may interact with DNA, modifying its structure and inducing mutations. It is the case of aflatoxin B1, which forms an adding compound able to bind to the N-7 position of guanine residue, generating a new product. This new molecule cleaves, then, the interaction between one sugar and one nitrogenous base of a nucleotide, leading to the formation of an apurinic site. The mutation can, thus, cause severe damages on the liver, including necrosis, cirrhosis, and carcinoma [7, 14].

During the metabolism of folate, the folic acid found in food sources, is absorbed by the intestine and, through many chemical processes of catabolism and synthesis, it is transformed into 5-methyltetrahydrofolate. This chemical component is necessary for the synthesis of methionine, which in turn is used during the process of DNA mutilation. Thus, a diet poor in folic acid can alter this process and interfere on DNA replication, leading also to an increased risk of cancer development [2–4, 25].

Nevertheless, various minerals work as protectors against cancer development [25]. Among them, there are (i) selenium, which stimulates the production of glutathione peroxidase enzyme that acts on the reduction of hydrogen peroxide and maintain the integrity of cell membranes; (ii) prostacyclins, which decrease the oxidative damage of important molecules, such as DNA, lipids, and lipoproteins; (iii) zinc, which act on processes for the maintenance of genomic stability, genetic expression, and apoptosis modulation [3, 25].

1.11. Type II Diabetes. Diabetes counts for more than 90% of all diseases of the world [2]. Type II Diabetes is a multifactorial pathogenesis that involves the interaction between genetics and environmental factors [23]. Genomics studies showed that there are 65 SNPs associated with the risk of developing type II Diabetes [26].

With the advances of genome sequencing and the decoding of the human genome, tests for the detection of SNPs related to Type II Diabetes became available to the public. In these exams, the patient is able to know it there is a genetic predisposition to develop the disease [26]. However, caution must be taken on applying this test for clinical practice, once that studies have shown that patients that had negative results on this exam for the presence of Type II Diabetes SNPs felt so secure about it that they stopped taking care of their food diets. Consequently, some later developed Diabetes due to food income and insulin tolerance [26]. Nevertheless, patients who received a positive result for the presence of type II Diabetes change their life style, especially in their food intake, which later decreased the development of the disease by this group [26].

Table 1: Classification of Ronteltap and colleagues [15], showing the new marketing strategies into nine business models.

Business model	Marketing strategy
Orientation according to the lifestyle of the individual	Shared responsibility between employer and employee, where a healthy lifestyle can contribute to a better productivity.
We are togheter and strong	Improve the lifestyle with the help of social groups.
Gyms	Based on a set of changes into the lifestyle, including physical activities, diet therapies, and use of food supplement, among others.
Do a healthy diet yourself	This model offers a diagnostic tool where, based on the person's food ingestion, it can create a diet plan to improve a healthy diet.
In, out	Similar to the 4th plan reported above. In this, the person describes the food intake but also some phenotype parameters in order to receive diet counseling.
Test and execute until the end	Offers an interactive feedback about a person's health progress, which allows toadjust the diet counseling as needed.
Orientation about the lifestyle	Inclusion of genetic information, data about food intake, and the person's phenotype, in order to obtain a personalized counseling on lifestyle, diet terapy, control of stress activity, and time management.
Face to face	In person treatment, where anthropometric data and food consumption of the cliente are obtained by a face-to-face interview.
We tell you	Use of mass media as a healthy diet education method.

2. Conclusion and Perspectives

Nutrigenomics shows a new way of working with nutrition and now, the knowledge of how food interferes with the genetic code and how the organism responds to these interferences and with the phenotype can be clarified.

Furthermore, the development inside Nutrition Sciences, together with communication and marketing fields, provided the emergence of a personalized nutritional counseling based in Nutrigenomics. After analysis, it was observed that nine different models can be built (Table 1).

Moreover, other sophisticated programs are available, where it is even possible to obtain a nutritional counseling based on the client's DNA. In this way, Gollust et al. [27] published a report where he studied the intention of patients on using online hypothetic programs for personalized nutritional counseling. During the research, the patients described that the presence of a physician during the nutritional evaluation and the compilation of data into the program was essential to bring a trusting diet plan and nutritional counseling through the online system [27].

Furthermore, in order to evaluate the impact of personal genomics to the population, a study placed in The United States of America intended to analyze the motivation, knowledge, risks, and benefits of sharing personal data to build a personalized field of every individual's health condition [27]. Hence, it was observed that 92% of the interviewed ones were willing to share their personal information to professional of health area, while 32% considered themselves unaware of what a genomic evaluation could do to their personal data [27].

Henceforward, the recent advances on Nutrition Sciences allow the application of a personal counselling, in contrary to the general use of standard nutritional diets that was being applied for many years. The personalized nutritional counseling can be used not only to change diet habits and improve life style, but also mainly will permit a better diagnostic of certain diseases, retard the evolution of chronic illnesses, and assist on the treatment of others. Although Nutrigenomics is already available for clinical use, there are still few places applying this tool in the heatlh field [15].

Since the introduction of Nutrigenomics into the "omics" group, there was an upgrade of how physicians and other professionals on evaluating and treating different diseases, especially DCTNs. Nevertheless, there is still a long way to go on Nutrigenomics, as further research needs to be done in order to connect a patient's symptons/diseases with their genetic profile, food diet, and environmental habits.

Moreover, with the knowledge integration of technologies into health sciences, novel business models for personalized nutritional counseling were highlighted, based on the person's DNA. In this way, entities connected to the health system must act in order to regulate these business models in order to preserve the integrity of patients, increasing the performance of the system for nutritional orientation.

Therefore, scientists and professionals of human health sciences can constantly contribute to Nutrigenomics through research and on the development of new tools that can assist on a better quality of life and healthy diet to the population.

Conflict of Interests

The authors declare that there is no conflict of interests regarding the publication of this paper.

References

[1] A. Ronteltap, J. C. M. van Trijp, and R. J. Renes, "Consumer acceptance of nutrigenomics-based personalised nutrition," *British Journal of Nutrition*, vol. 101, no. 1, pp. 132–144, 2009.

[2] S. M. F. Cozzolino and C. Cominetti, *Biochemical and Physiological Bases of Nutrition in Different Stages of Life in Health and Disease*, Monole, São Paulo, Brazil, 1st edition, 2013.

[3] S. M. F. Cozzolino, *The Bioavailability of Nutrients*, Monole, São Paulo, Brazil, 4th edition, 2012.

[4] N. M. B. Costa and C. O. B. Rosa, *Functional Foods: Bioactive Components and Physiological Effects. 1 Reprint*, Rúbio, Rio de Janeiro, Brazil, 2011.

[5] M. J. Dauncey, "Recent advances in nutrition, genes and brain health," *Proceedings of the Nutrition Society*, vol. 71, no. 4, pp. 581–591, 2012.

[6] J. Sanhueza and A. Valenzuela, "Nutrigenomics: revealing the molecular aspects of a personalized nutrition," *Chilean Journal of Nutrition*, vol. 39, no. 1, pp. 1–12, 2012.

[7] A. A. C. Moraes and F. E. L. Pereira, "Nutrigênomics of the soy," in *Parenteral Nutrition in Clinical Practice*, D. Waitzberg and L. Enteral, Eds., pp. 2039–2047, Atheneu, São Paulo, Brazil, 4th edition, 2009.

[8] L. Dalmiel, T. Vargas, and A. R. Molina, "Nutritional genomics for the characterization of the effect of bioactive molecules in lipid metabolism and related pathways," *Electrophoresis Journal*, vol. 33, no. 15, pp. 2266–2289, 2012.

[9] D. De Lorenzo, "Nutrigenomics: for a personalized nutrition," *Journal Digitalis*, no. 22, pp. 1–3, 2008.

[10] J. M. Ordovás and R. Carmena, Nutrigenetics, 2012, http://www.fundacionmhm.org/pdf/Mono9/Articulos/articulo1.pdf.

[11] B. Liu and S. B. Qian, "Translational regulation in nutrigenomics," *American Society for Nutrition*, vol. 2, pp. 511–519, 2011.

[12] I. B. M. Cruz, M. Taufer, and C. H. A. Schwanke, "Genomics in the era of aging and its potential application in gerontology and geriatrics," in *Institute of Geriatrics and Gerontology PUCRS: The Cradle of Academic Geriatrics In Brazil*, A. C. A. Souza, Ed., pp. 83–84, Pontifícia Universidade Católica do Rio Grande do Sul (EDIPUCRS), Porto Alegre, Brazil, 1st edition, 2003.

[13] E. Fialho, F. S. Moreno, and T. P. P. Ong, "Nutrition in the post-genomics: fundamentals and applications of omics tools," *Journal of Nutrition*, vol. 21, no. 6, pp. 757–766, 2008.

[14] L. K. Mahan and S. Scott Stump, *Food, Nutrition & Diet Therapy*, Roca, São Paulo, Brazil, 6th edition, 2005.

[15] A. Ronteltap, H. V. Trijp, A. Berezowska, and J. Goossens, "Nutrigenomics-based personalised nutritional advice: in search of a business model," *Genes & Nutrition*, vol. 8, no. 2, pp. 153–163, 2013.

[16] T. P. Ong and M. M. Rogero, "Nutrigenomics: importance of nutrient-gene interaction for health promotion," *Journal of the ABESO*, no. 40, 2009.

[17] F. Norheim, I. M. F. Gjelstad, M. Hjorth et al., "Molecular nutrition research-the modern way of performing nutritional science. Switzerland," *Journal Nutrients*, no. 4, pp. 1898–1944, 2012.

[18] T. M. M. Fujii, R. Medeiros, and R. yamada, "Nutrigenomics and nutrigenetics: important concepts for the science of nutrition," *Nutrire: Journal of the Brazilian Society of Food and Nutrition*, vol. 35, no. 1, pp. 149–166, 2010.

[19] F. A. G. Vasconcelos, "The science of nutrition in transit: from nutrition and dietetics to nutrigenomics," *Revista de Nutricao*, vol. 23, no. 6, pp. 935–945, 2010.

[20] A. Koutsos, A. Manaia, and J. Willingale-Theune, *Introdução aos Microarrays de ADN*, Versão 2.3, European Molecular Biology Laboratory (EMBL), Cambridge, UK, 2009–2011.

[21] National Human Genome Research Institute, "Talking Glossary of Genetic Terms," 2013, http://www.genome.gov/Research-Funding/.

[22] T. L. R. Rocha, R. G. S. Evaristo, L. P. Silva et al., *Metabolomics: Applications and Perspectives*, EMBRAPA, Brasilia, Brazil, 1st edition, 2006.

[23] C. M. Phillips, "Nutrigenetics and metabolic disease: current status and implications for personalised nutrition. Switzerland," *Journal Nutrients*, vol. 5, no. 1, pp. 32–57, 2013.

[24] E. P. Cruz, G. M. Mier, and A. Z. Rivera, "Nutritional genomics: prospects for the future," *Journal of Endocrinogy and Nutrition. Mexico*, vol. 13, no. 4, pp. 190–196, 2005.

[25] V. Almendro and P. Gascan, "Nutrigenomics and Cancer," 2012, http://www.fundacionmhm.org/pdf/Mono9/Articulos/articulo9.pdf.

[26] A. A. Vorderstrasse, A. Cho, C. Eu Voils, L. A. Orlando, and G. S. Ginsburg, "Personalized medicine clinical utility of genetic risk testing in primary care the example of type 2 diabetes," *Personalized Medicine*, vol. 10, no. 6, pp. 549–563, 2013.

[27] S. E. Gollust, E. S. Gordon, C. Zayac et al., "Motivations and perceptions of early adopters of personalized genomics: perspectives from research participants," *Public Health Genomics*, vol. 15, no. 1, pp. 22–30, 2011.

18

Leucine Modulates Mitochondrial Biogenesis and SIRT1-AMPK Signaling in C2C12 Myotubes

Chunzi Liang,[1] Benjamin J. Curry,[2] Patricia L. Brown,[1] and Michael B. Zemel[3]

[1] *Department of Nutrition, University of Tennessee, Knoxville, 1215 W. Cumberland Avenue, 229 Jessie Harris Building, Knoxville, TN 37996-1920, USA*
[2] *Ension, Inc., 11020 Solway School Road, Suite 108, Knoxville, TN 37931, USA*
[3] *NuSirt Biopharma, 11020 Solway School Road, Suite 109, Knoxville, TN 37931, USA*

Correspondence should be addressed to Chunzi Liang; cliang3@vols.utk.edu

Academic Editor: Robert B. Rucker

Previous studies from this laboratory demonstrate that dietary leucine protects against high fat diet-induced mitochondrial impairments and stimulates mitochondrial biogenesis and energy partitioning from adipocytes to muscle cells through SIRT1-mediated mechanisms. Moreover, β-hydroxy-β-methyl butyrate (HMB), a metabolite of leucine, has been reported to activate AMPK synergistically with resveratrol in C2C12 myotubes. Therefore, we hypothesize that leucine-induced activation of SIRT1 and AMPK is the central event that links the upregulated mitochondrial biogenesis and fatty acid oxidation in skeletal muscle. Thus, C2C12 myotubes were treated with leucine (0.5 mM), alanine (0.5 mM), valine (0.5 mM), EX527 (SIRT1 inhibitor, 25 μM), and Compound C (AMPK inhibitor, 25 μM) alone or in combination to determine the roles of AMPK and SIRT1 in leucine-modulation of energy metabolism. Leucine significantly increased mitochondrial content, mitochondrial biogenesis-related genes expression, fatty acid oxidation, SIRT1 activity and gene expression, and AMPK phosphorylation in C2C12 myotubes compared to the controls, while EX527 and Compound C markedly attenuated these effects. Furthermore, leucine treatment for 24 hours resulted in time-dependent increases in cellular NAD^+, SIRT1 activity, and p-AMPK level, with SIRT1 activation preceding that of AMPK, indicating that leucine activation of SIRT1, rather than AMPK, is the primary event.

1. Introduction

Impaired mitochondrial function in skeletal muscle is one of the major predisposing factors to metabolic diseases, such as insulin resistance, type 2 diabetes, and cardiovascular diseases [1]. Indeed, lower mitochondrial content and decreased expression of oxidative enzymes are observed in patients with type 2 diabetes [2]. SIRT1 and AMP-activated protein kinase (AMPK) are known to promote mitochondrial biogenesis and oxidative capacity and prevent the mitochondrial dysfunction in skeletal muscle [3, 4].

SIRT1, a nicotinamide adenine dinucleotide- (NAD^+-) dependent deacetylase, is the key enzyme that mediates caloric restriction- (CR-) induced longevity in mammals [5]. By sensing intracellular NAD^+/NADH ratio, SIRT1 regulates target gene expression via changing acetylation status of histones of transcriptional factors, such as peroxisome proliferator-activated receptor gamma coactivator 1-alpha (PGC-1α), tumor suppressor p53 (p53), nuclear factor kappa-light-chain-enhancer of activated B cells (NF-κB), and forkhead box O3 (FOXO3) [6], resulting in modulation of wide range of cellular fundamental processes, including DNA repairing, energy metabolism, and cell apoptosis [7, 8]. Overexpression and activation of SIRT1 protect against high fat diet- (HFD-) induced metabolic abnormalities in mice, such as insulin resistance, glucose intolerance, and liver steatosis, without extending their lifespan [9, 10]. Therefore, small molecules that could activate SIRT1 and mimic the CR impacts have drawn considerable attention.

Leucine, a branched-chain amino acid (BCAA), plays a distinct role in energy metabolism in addition to its pivotal function in protein synthesis [11, 12]. For example, leucine promotes energy partitioning from adipocytes to muscle cells, leading to decreased lipid storage in adipocytes and increased fat utilization in muscle cells [13]. Leucine administration increases insulin sensitivity and glucose tolerance by promoting glucose uptake and fatty acid oxidation in skeletal muscle in HFD-fed mice [14–17]. In fact these effects are mediated partially through SIRT1-dependent pathway, as *Sirt1* knockout significantly attenuates these effects [18, 19]. Further, recent data indicate that leucine can directly activate SIRT1 by promoting the enzyme affinity for its substrates and NAD^+ [18], resulting in elevated mitochondrial biogenesis and fatty acid oxidation in both adipocytes and myotubes [20, 21].

HMB, a minor metabolite of leucine, has been reported to stimulate AMPK phosphorylation synergistically with metformin, resulting in significant increases in insulin sensitivity and glucose tolerance in mice [22]. Similar to SIRT1, AMPK is an evolutionary conserved enzyme and acts as an energy status sensor via intracellular AMP or AMP/ATP ratio in eukaryotes [3]. In response to nutrient restriction, activated AMPK promotes a cell catabolic shift with increased ATP production to rescue the cellular fuel crisis [23]. Furthermore, phosphorylated AMPK is highly associated with SIRT1 activation in both *in vivo* and *in vitro* studies [5, 24], and part of these two enzymes signaling pathways are overlapped [25].

These findings provide a mechanistic framework for leucine-modulation of mitochondrial biogenesis [21, 26]. We hypothesize that leucine activation of SIRT1 and AMPK is the major event that regulates fatty acid oxidation and mitochondrial biogenesis in skeletal muscle. Accordingly, we examined the effects of leucine, valine (branched-chain amino acid control), and alanine (nonbranched chain amino acid control) on mitochondrial content, mitochondrial biogenesis-related gene expression, SIRT1 activity, and AMPK phosphorylation in C2C12 myotubes. In addition, we used EX-527 (SIRT1 selective inhibitor) and Compound C (specific AMPK inhibitor) to probe the roles of each enzyme in leucine-modulation of energy metabolism in C2C12 myotubes.

2. Materials and Methods

2.1. Cell Culture. C2C12 myoblast cells were seeded at a density of 1.2×10^6 cells per well in 6-well plates and incubated in Dulbecco's modified eagle medium (DMEM) containing 4.5 g/L D-glucose, 10% fetal bovine serum (FBS), and 1% penicillin-streptomycin at 37°C and 5% CO_2. After the cells reach 90% confluence, the medium was switched to a standard differentiation medium (DMEM supplemented with 2% horse serum and 1% penicillin-streptomycin) for 2 to 4 days. The differentiation medium was changed every other day to allow 90% of the cells to fully form myotubes (3–5 days later) before additional treatments began.

The dosages of reagents were 0.5 mM for leucine, 0.5 mM for alanine, 0.5 mM for valine, 100 nM for resveratrol, 25 μM for EX527, 25 μM for Compound C, and 50 μM for AICAR. The incubation lengths were from 1 to 48 hours as indicated in the figure legend.

2.2. RNA Extraction and Quantitative Real-Time PCR (RT-PCR) Analyses. Total RNA was extracted from C2C12 myotubes using Ambion Totally RNA Isolation Kit (Ambion, Inc., Austin, TX, USA) according to the manufacturers' instructions. The RNA content was determined using NanoDrop ND-1000 Spectrophotometer (NanoDrop Technologies Inc., Wilmington, DE, USA). RNA quality was assessed by the 260 nm/280 nm ratio (1.8–2.0) and 260 nm/230 nm ratio (2.0). The mRNA expression of selected genes related to mitochondrial biogenesis, including *Sirt1*, *Sirt3*, *PGC-1α*, cytochrome c oxidase subunit 5b (*Cox5b*), heat shock cognate protein 1 (*Hspd 1*), and *Cox2*, was analyzed using a TaqMan Universal PCR Master Mix kit (Applied Biosystem) according to the manufacturers' instructions. The primers and probes sets were obtained from Applied Biosystems TaqMan Gene Expression Assays primers and probe set collection. The quantitative RT-PCR reactions were carried out in 96-well format using ABI 7300HT instrument (Applied Biosystem) according to the instructions. Mouse 18S ribosomal RNA was used as the housekeeping gene. Data for each gene was normalized to *18S* and presented as a ratio to the transcript of interest to *18S*.

2.3. SIRT1 Activity Measurement. SIRT1 Fluorometric Drug Discovery Kit (BML-AK555, ENZO Life Science International, Inc., PA, USA) was used to measure SIRT1 activity in C2C12 myotubes, following the manufacturer instruction. In this assay, SIRT1 activity is determined by the degree of deacetylation of a standardized substrate that contains an acetylated lysine residue. This Fluor de Lys substrate is a peptide containing amino acids 379–382 of human p53 (Arg-His-Lys-Lys [Ac]) and serves as a direct target for SIRT1. SIRT1 activity is proportionally related to the degree of deacetylation of Lys-382 and the corresponding fluorescence signal changes.

Cell lysates were harvested by homogenizing cells in RIPA buffer (Sigma-Aldrich, MO, USA), which contains protease inhibitor cocktail (MP Biomedicals LLC, Solon, OH, USA) (100 : 1 v/v). After 5 seconds of ultrasonication on ice, the cell lysates were centrifuged at 12,000 ×g for 5 minutes. The supernatant was used for SIRT1 activity assessment and other experiments. According to the protocol, 5 μL of cell lysate was used for the endogenous SIRT1 activity detection. Samples were incubated in a phosphate-buffered saline solution with peptide substrate (25 μM) and NAD^+ (500 μM) at 37°C on a horizontal shaker for 45 minutes. The deacetylation reaction was stopped with the addition of the stop solution (2 mM nicotinamide) and developer that binds to the deacetylated lysine to form a fluorophore. Following 10 minutes of incubation at 37°C, fluorescence intensity was measured using Glomax Multi Detection System (Promega, WI, USA), with excitation and emission wavelengths of 360 nm and 450 nm, respectively. Resveratrol (100 mM) and suramin sodium (25 mM) were used as positive and negative controls, respectively. To normalize the data of SIRT1 activity, concentrations of the sample cellular protein were measured via BCA-assay (Thermo Scientific Inc, Waltham, MA, USA). Data for each sample SIRT1 activity is presented as a ratio to the protein content.

2.4. Cellular NAD$^+$. NAD$^+$ was measured in C2C12 myotubes using a colorimetric assay (Cayman Chemical Company, Ann Arbor, MI, USA) that uses an alcohol dehydrogenase reaction to reduce NAD$^+$ in cell lysates to NADH and the NADH is used to reduce a tretrazolium salt substrate (WST-1) to formazan. Formazan absorbance, measured at 450 nm, is proportional to the NAD$^+$ in the cell lysate.

2.5. Fatty Acid Oxidation. Cellular fatty acid oxidation was measured using [^{3}H]-palmitate, as described in our previous studies [13]. C2C12 cells in 12-well plates were washed with 2 mL of cold PBS solution twice and incubated in 1 mL of Hank's basic salt solution containing 0.5 mg/mL BSA, 22 μM-unlabeled palmitate plus 5 μM [^{3}H]-palmitate (32.4 mCi/μm) for 2 hours. All of the reaction solutions were collected from each well, and then 200 μL of 10% trichloroacetic acid and 70 μL 6 N NaOH were added in the solution. Mixtures were then removed from each well and placed in corresponding poly-prep chromatography columns with 1.5 mL Dowex-1 overnight. The ^3H_2O that passed through the column was collected into a scintillation vial, and radioactivity was measured with a liquid scintillation counter. The protein level of each well was measured using BCA-assay (Thermo Scientific Inc., Waltham, MA, USA) and was used to normalize the palmitate oxidation data.

2.6. Western Blotting. Primary antibodies for total AMPK, phospho-AMPKα (Thr172), total ACC (Acetyl-CoA Carboxylase), and phospho-ACC (Ser79) were obtained from Cell Signaling Technology Inc. (Danvers, MA, USA). Horseradish peroxidase- (HRP-) conjugated goat anti-rabbit secondary antibody was obtained from Thermo Scientific Inc. (Waltham, MA, USA).

Following the indicated treatments, C2C12 myotubes were washed twice with ice cold PBS, and the total cell lysates were prepared using RIPA buffer plus protease/phosphatase inhibitor cocktails with 100 : 1 : 1 (v/v/v, ratio) (Sigma-Aldrich). Following a 10-minute centrifugation at 14,000 ×g, the supernatants were collected for the determination of protein content using BCA assay kit (Thermo Scientific Inc., Waltham, MA, USA) and western blotting. Equal amounts of total cell lysates (20 μg) were loaded to 10% SDS-PAGE (10 cm × 10 cm, Criterion precast gel, Bio-Rad Laboratories, Hercules, CA, USA) and transferred to PVDF membrane (polyvinylidene difluoride membrane) (Bio-Rad, Hercules, CA). The membrane was incubated in 25 mL blocking buffer (1× TBS, 0.1% Tween-20 with 5% w/v nonfat dry milk) for 1 hour at room temperature. Then the membrane was incubated in TBST containing 5% dry milk with primary antibody (1 : 1000) with gentle agitation at 4°C overnight, washed three times with TBST, and incubated with TBST containing rabbit HRP-conjugated secondary antibody (1 : 5000) for 1 hour at RT. Bound antibodies were visualized by chemiluminescence (ECL Western Blotting Substrate, Thermo Scientific) and membranes were exposed to X-ray films (Phenix Research Product, Candler, NC) for protein band detection. The films were scanned using an HP Scanjet 39070 (Palo Alto, CA 94304) and stored as tagged image file format (TIFF) at 300 dpi. The protein bands were quantified by densitometry using BioRad ChemiDoc instrumentation and software of Image Lab 4.0 (Bio-Rad Laboratories).

2.7. Measurement of Mitochondrial Contents. Mitochondrial abundance in C2C12 myotubes was assessed by 10-N-nonyl acridine orange (NAO) dye (Life Sciences, PA, USA) according to manufacturer's instruction. After desired treatment, cells in 96-well plates were treated with 10 μM NAO dye, following 2-hour incubation at 37°C in the dark. NAO is not fluorescent, but it can be oxidized into the fluorescent-NAO by oxidative species and accumulated in mitochondrial membrane. The absorbance in each well was measured at 570 nm wavelengths (Promega, WI, USA) and normalized with cellular protein level. The image of mitochondria was taken using a Nikon Eclipse Ti-E Ti-E *Fluorescence* microscope (Nikon Metrology, Inc., US) equipped with an automated stage and a 20x objective. A 3 × 3 large image scan was taken in each of 5 random fields by multichannel capture (channel 1: excitation/emission = 488/517, channel 2: excitation/emission = 550/567 nm).

2.8. Statistical Analysis. Data is presented as means ± standard deviation (SD). Levene's test was used to determine homogeneity of variance among groups using SPSS 21.0 statistical software (IBM, Armonk, NY) and where necessary natural log transformation was performed before analysis. Multiple comparisons were analyzed by one-way analysis of variance (ANOVA) using least significant difference when equal variance was assumed, and Games-Howell test was used when equal variance was not assumed. The independent sample t-test was used to compare two conditions. Differences were considered statistically significant at $P < 0.05$.

3. Results

3.1. Leucine Treatment Induced the Mitochondrial Biogenesis in C2C12 Myotubes. Leucine significantly increased mitochondria content in C2C12s compared to alanine and valine ($P = 0.007$) (Figure 1(a)). These effects were accompanied by increases in mRNA levels of *PGC-1α* (198%, $P = 0.01$) and *SIRT3* (167%, $P = 0.03$) (Figure 1(b)). SIRT1 activity ($P = 0.017$) and fatty acid oxidation ($P = 0.03$) in the C2C12 myotubes were significantly elevated by leucine compared to the control groups (Figures 1(c) and 1(d), resp.).

3.2. SIRT1 Is Required for Leucine-Induced Mitochondria Biogenesis in C2C12 Myotubes. We used a selective SIRT1 inhibitor (EX527) to determine the role of SIRT1 in leucine-induced mitochondrial biogenesis. Leucine increased mitochondrial biogenesis as demonstrated by significant increases in mitochondrial content ($P = 0.001$), palmitate oxidation ($P = 0.038$) and expression of mitochondrial biogenesis-related gene markers *PGC-1α* ($P = 0.003$), *SIRT3* ($P = 0.031$), and *COX5b* ($P = 0.015$) (Figures 2(a), 2(c) and 2(d), dark panel), and these effects were markedly attenuated by EX527 administration (Figures 2(a), 2(c) and 2(d), grey panel). Comparing the relative *Sirt1* expression, leucine and resveratrol

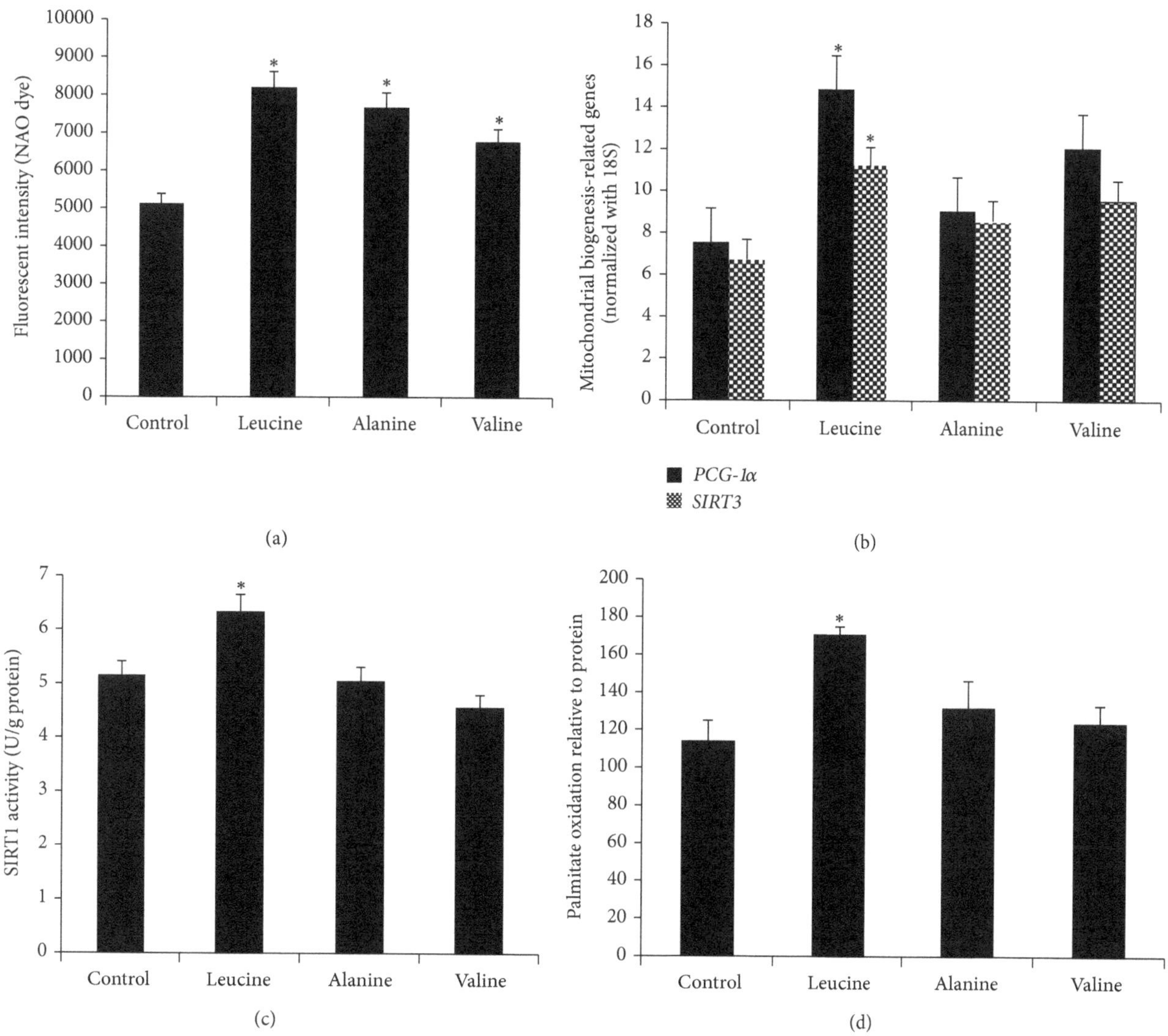

FIGURE 1: Leucine treatment induces mitochondrial biogenesis and SIRT1 enzymatic activity in C2C12 myotubes. (a) Mitochondrial content was quantitated with NAO dye (10 μM) 48 hours after treatment with leucine (0.5 mM), alanine (0.5 mM), and valine (0.5 mM); (b) mRNA expression levels of *PGC-1α* and *Sirt3* with the same treatments were evaluated by quantitative RT-PCR. The relative mRNA expression was normalized to 18S and expressed as dark bars for *PGC-1α* and grey bars for *Sirt3*. (c) Cellular SIRT1 activity and (d) palmitate oxidation were measured after treatment for 48 hours. The results were normalized to cellular protein level for each sample. Data are mean ± SE ($n = 4$). *Significantly different from controls with $P < 0.05$.

(positive control) markedly increased *Sirt1* mRNA level ($P = 0.020$); the SIRT1 inhibitor (EX527) plus leucine treatment ($P = 0.015$) revealed the same pattern (Figure 2(b)).

3.3. Leucine Stimulates Phosphorylation of AMPK in a SIRT1-Dependent Manner. Six hours of leucine treatment resulted in a 3-fold increase in AMPK phosphorylation in the C2C12 myotubes, which was significantly different from baseline, valine, and alanine. Consistent with this observation, phosphorylation of ACC, a downstream target enzyme of AMPK, was also elevated by leucine compared to the controls ($P = 0.014$) (Figure 3(a)), while EX527 treatment resulted in corresponding suppression of AMPK phosphorylation ($P = 0.012$) (Figure 3(b)), indicating the necessity of SIRT1 for leucine-induced AMPK activation.

3.4. Leucine Stimulates SIRT1 Activity, Phosphorylation of AMPK, and Cellular NAD^+ in a Time-Dependent Manner. To determine the interplay between SIRT1 and AMPK, we measured the cellular NAD^+ level, SIRT1 activity, and phosphorylation of AMPK at time points: 0, 1, 4, 6, 12, and 24 hours by leucine treatment in C2C12 myotubes. Leucine increased SIRT1 activity at 1 ($P = 0.028$), 12 ($P = 0.042$) and 24 hours ($P = 0.010$) compared to the baseline (Figure 4(a)). However, no change was observed for the NAD^+ content and p-AMPK level during the first 4 hours. NAD^+ level was elevated almost twofold higher at time points 4 ($P = 0.025$)

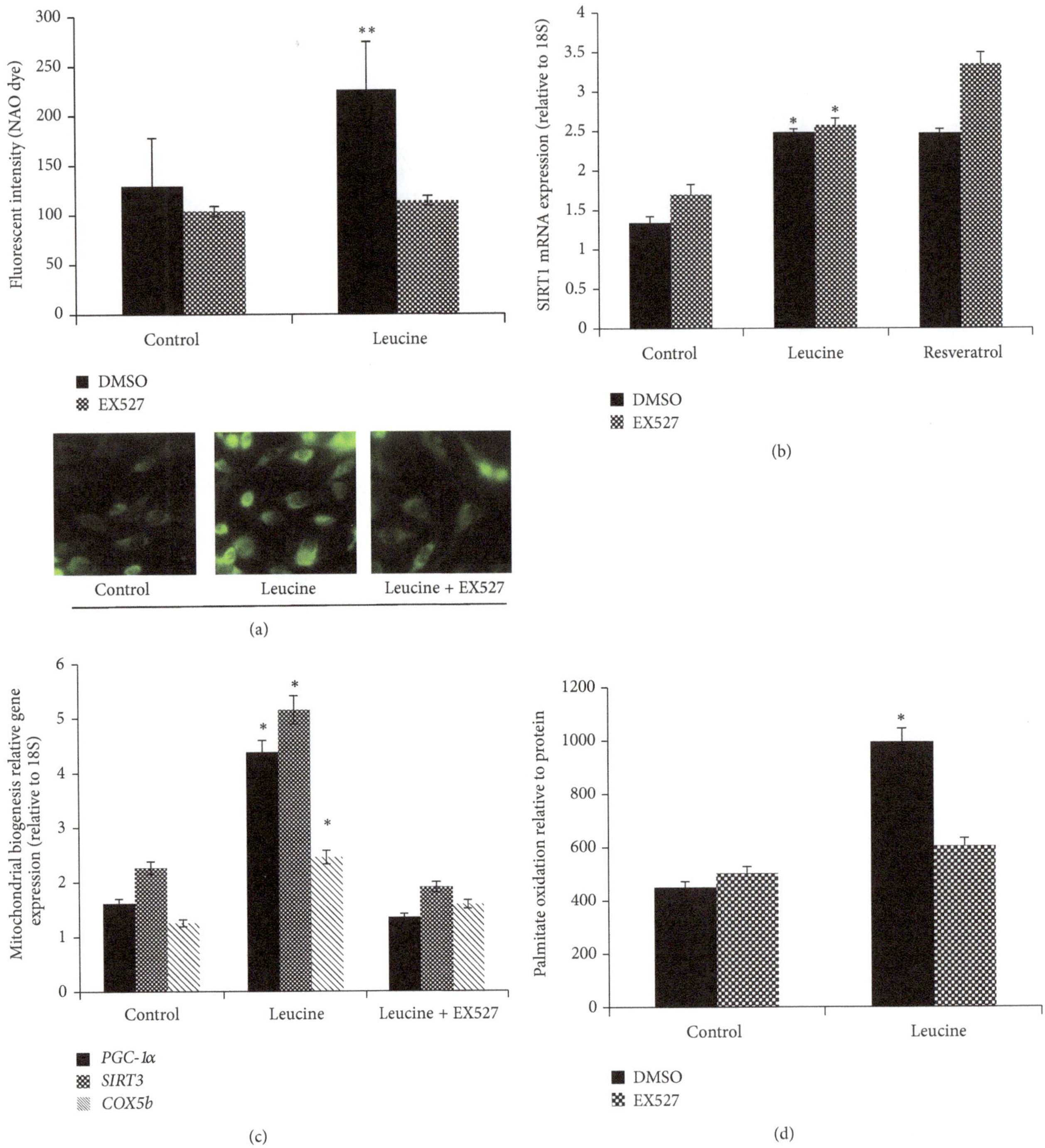

FIGURE 2: Leucine improves mitochondrial biogenesis in C2C12 myotubes in a SIRT1-dependent manner. (a) Mitochondrial content was measured using NAO (10 μM) dye after 48-hour leucine (dark bars), leucine plus SIRT1 inhibitor (EX527 25 μM; grey bars) for 48 hours in C2C12 myotubes. (b, c) SIRT1 activity and mitochondrial biogenesis- related genes (*PGC-1α, Sirt3, and COX5b*) mRNA levels were measured after the same treatments. The relative SIRT1 activity was normalized to cellular protein level, and mRNA level was normalized to housekeeper gene *18S*. (b) Dark bars are DMSO control, grey bars are EX527. (c) Dark bars are *PGC-1α*, grey bars are *Sirt3*; striped bars are *COX5b*. (d) Palmitate oxidation level was detected after the same treatment, and the results were normalized to cellular protein for each sample. Data are mean ± SE ($n = 4$). Different letters indicate significant differences within a given variable. Dark bars are DMSO control and grey bars are EX527. *Significantly different from controls, and **significantly different from control and EX527 groups with $P < 0.05$.

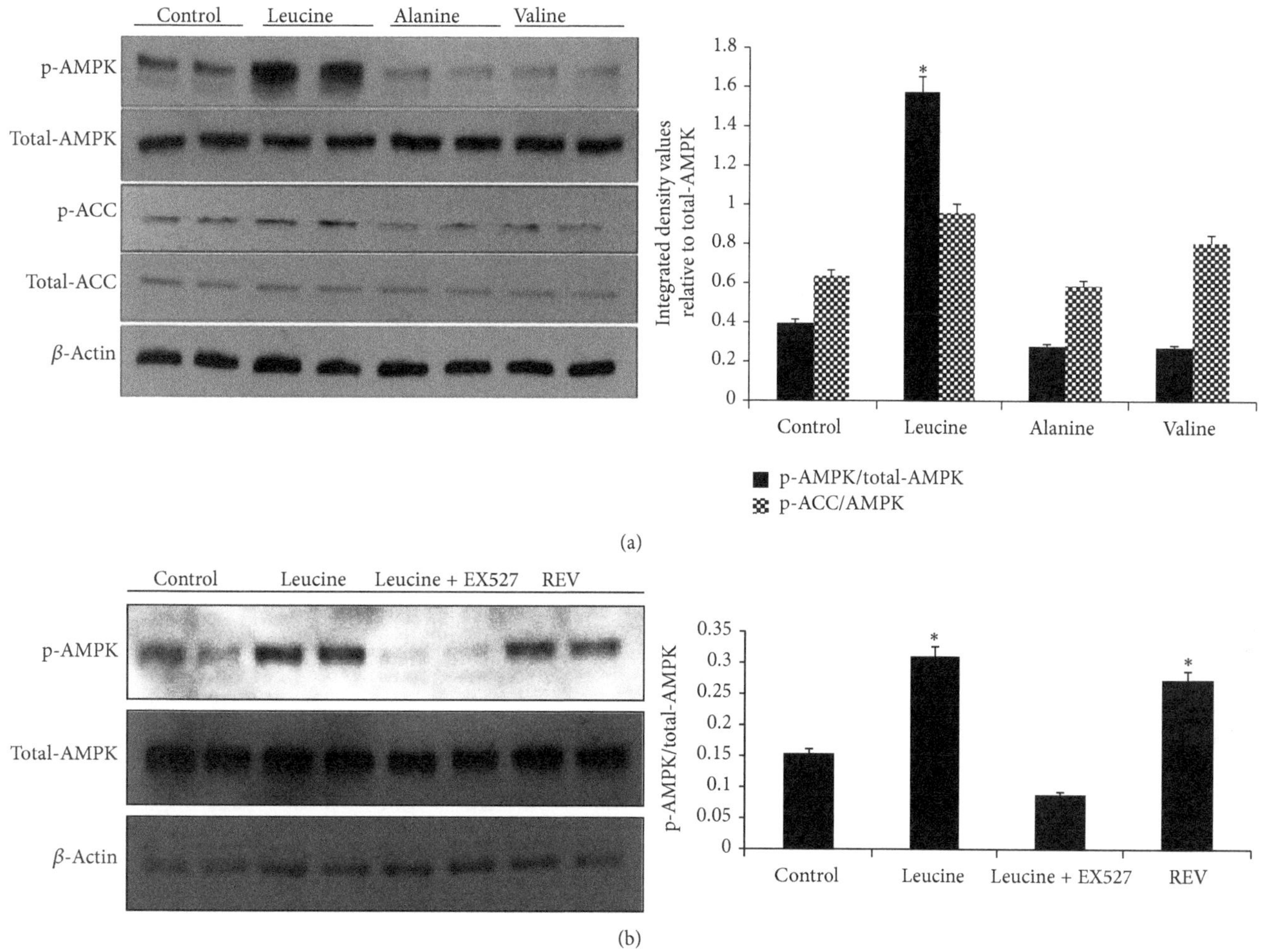

FIGURE 3: Leucine-induced phosphorylation of AMPK and ACC requires SIRT1 in C2C12 myotubes. (a) C2C12 myotubes were serum starved overnight and treated with leucine (0.5 mM), alanine (0.5 mM), valine (0.5 mM), and DMSO for 6 hours. The cell lysates were assessed by western blotting analysis with specific antibodies against phosphor-AMPKα (Thr 172), phosphor-ACC (Ser 79), total AMPKα (Thr 172), and beta-actin. Integrated density values for the p-AMPK and p-ACC were normalized to total-AMPK band density and represented as dark or gray bars. (b) C2C12 myotubes were treated with 0.2% FBS medium overnight and then treated with leucine (0.5 mM), resveratrol (100 nM), and leucine plus EX527 (25 μM) for 6 hours. Whole cell lysates were prepared and detected by western blotting with specific antibodies against phosphor-AMPKα, AMPKα, and beta-actin. Integrated density value for phosphor-AMPK was normalized to total-AMPK. *Significantly different from controls with $P < 0.05$.

and 24 hours ($P = 0.010$) compared to baseline level and otherwise remained low (Figure 4(b)); the levels of p-AMPK were markedly increased and stayed high from 4 to 24 hours ($P = 0.000$) (Figure 4(c)).

3.5. Leucine-Induced Mitochondrial Biogenesis in C2C12 Myotubes Requires AMPK. We next examined whether AMPK also mediates leucine's impacts on mitochondrial biogenesis in C2C12 myotubes. As shown in Figure 5, leucine treatment markedly increased the mitochondrial component genes expression (Figure 5(a), dark columns), *Hspd1* ($P = 0.003$) and *COX2* ($P = 0.003$). Similar effects were found for genes encoding mitochondrial biogenesis regulatory proteins, [*PGC-1α* ($P = 0.001$), *Sirt1* ($P = 0.022$)] and component proteins [*Cox5b* ($P = 0.04$)], (Figure 5(b), dark columns), while Compound C treatment markedly impaired all these inductions (Figure 5, grey panels).

4. Discussion

These data indicate that leucine stimulates significant muscular metabolic changes, including SIRT1 activation, AMPK phosphorylation, and mitochondrial biogenesis in C2C12 myotubes. These changes may contribute to leucine's beneficial effects on energy metabolism and insulin sensitivity in both animal and human models [17, 19, 27, 28].

A previous clinical trial has shown that high dairy intake (rich in BCAAs) induces significant suppression of reactive oxygen species (ROS) and inflammatory stress, indicated by decreased plasma tumor necrosis factor alpha (TNF-α),

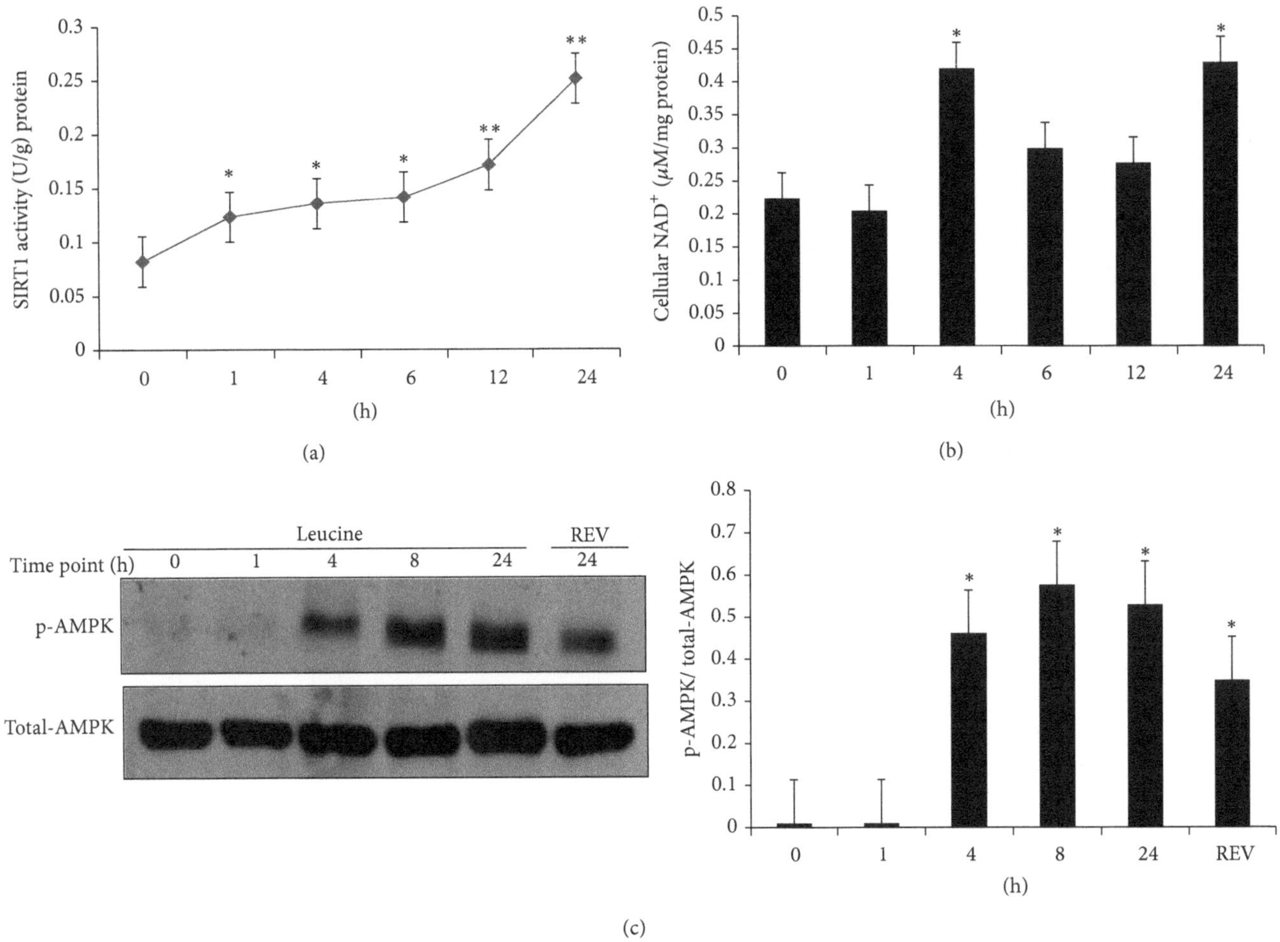

FIGURE 4: Leucine stimulates SIRT1 activity, AMPK phosphorylation, and cellular NAD^+ in a time-dependent manner. C2C12 myotubes were serum starved overnight and treated with leucine (0.5 mM). Cell lysate was collected and analyzed for cellular SIRT1 activity, western blotting of p-AMPK and cellular NAD^+ levels at indicated certain time points. (a) SIRT1 activity. (b) Cellular NAD^+. Both SIRT1 activity and NAD^+ level were normalized to cellular protein for each sample. (c) Phosphorylation level of AMPK was detected using western blotting following the same time course in C2C12 cells, with resveratrol serving as positive control. Data are mean ± SE ($n = 3$). Different letters indicate significant differences between dark or gray bars. *Significantly different from point 0, and **significantly different from time point 1.

interleukin 6 (IL-6), and monocyte chemoattractant protein-1 (MCP-1) levels [19]. Doubling leucine intake in mice has been found to reverse multiple HFD-induced metabolic abnormalities, including glucose intolerance, hepatic steatosis, and inflammation [11]. These effects are accompanied by corresponding increases in mitochondrial oxidative capacity and mitochondrial content. Since mitochondrial dysfunction and mitochondrial content loss are directly linked to the development of metabolic disorders [1, 29], increased mitochondrial biogenesis appears to rescue part of these obesity-related abnormalities [30].

Consistent with our previous studies [13], here we show that 0.5 mM leucine treatment, which is comparable to the plasma leucine concentration achieved by a leucine-rich diet [31], can markedly increase mitochondrial content, mitochondrial biogenesis-related gene expression, and fatty acid oxidation in C2C12 myotubes, compared to valine and alanine.

The data herein demonstrate that the improvement of fatty acid oxidation and mitochondrial content by leucine is accompanied by increased SIRT1 activity in C2C12 cells. SIRT1 has been demonstrated to play significant roles in leucine's effects on energy metabolism. In Macotela's study [11], leucine restores HFD-reduced hepatic NAD^+ and SIRT1 expression back to normal levels. Similarly, Li et al. demonstrate that leucine increases SIRT1 expression and decreases acetylation level of PGC-1α, resulting in attenuation of HFD-induced mitochondrial dysfunction, insulin resistance, and obesity in mice [32]. Furthermore, Sun and Zemel found that leucine induces mitochondrial biogenesis in muscle cells by stimulating the expression of PGC-1α and NRF-1 via a SIRT1-dependent pathway [26]. These findings, along with the observations reported here, are in agreement with our recent work that leucine could activate SIRT1 enzyme through allosteric interaction in adipocytes and myotubes [25].

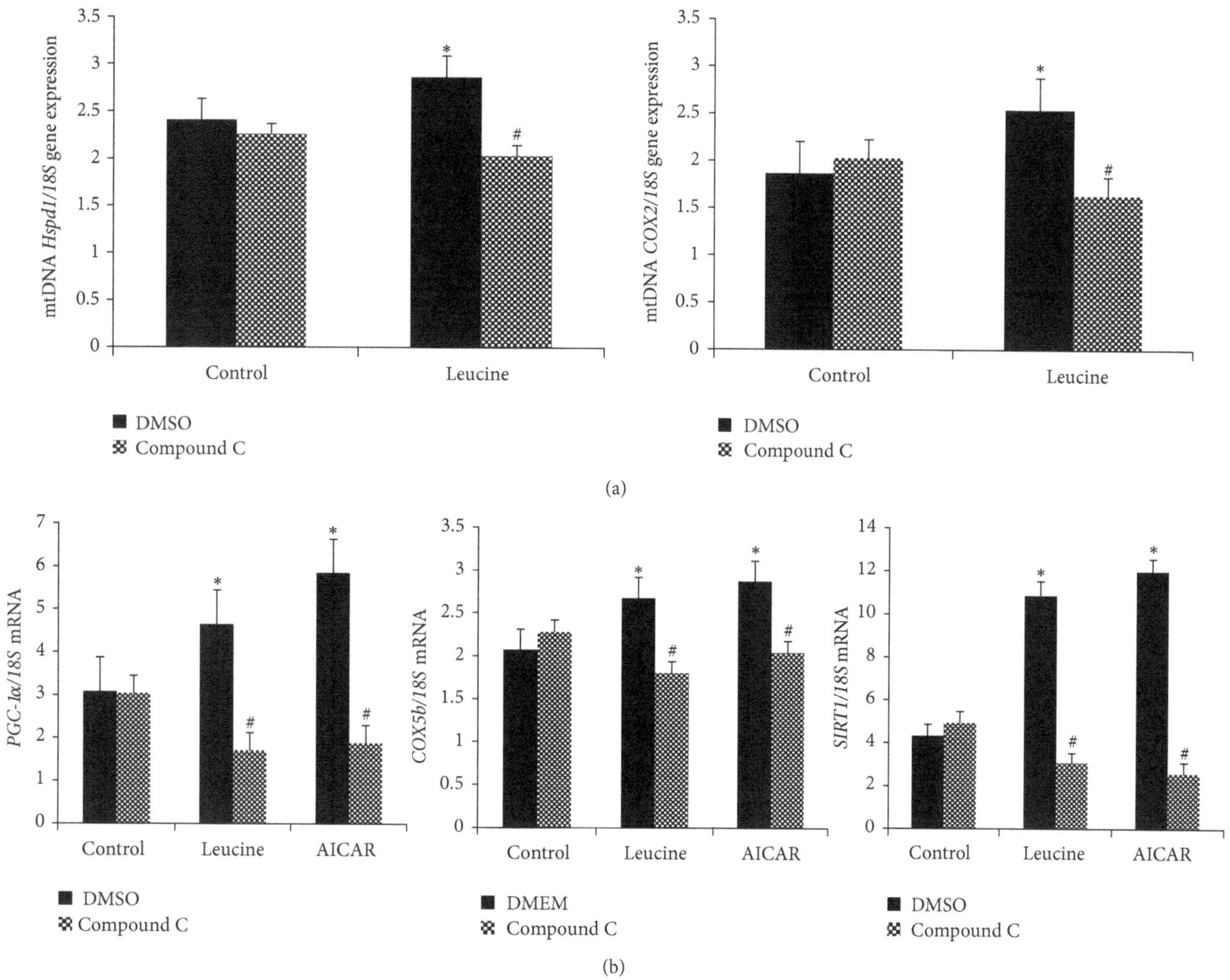

FIGURE 5: Leucine-induced mitochondrial biogenesis in C2C12 myotubes requires AMPK. (a) C2C12 myotubes were treated with leucine (0.5 mM), AICAR (20 μM), and Compound C (25 μM) for 24 hours. mtDNA levels of the cells were analyzed by the mitochondrial markers gene expression, *Hspdl* and *COX2*, using real-time PCR. (b) *Sirtl* and mitochondrial biogenesis related mRNA level of *PGC-1α* and *COX5b* were evaluated also by RT-PCR after treating with leucine and Compound C for 24 hours was measured; all the mRNA levels were normalized to 18S housekeeping gene. Data are mean ± SE ($n = 4$). Dark bars are vehicle control; grey bars are Compound C. *Significantly different from controls. #Significant Compound C effects.

To establish whether or not SIRT1 is required for leucine-induced mitochondrial biogenesis, EX527, a selective SIRT1 enzyme inhibitor, was used to treat the cells in combination with leucine. EX527 significantly attenuated leucine-induced mitochondrial content, mitochondrial biogenesis-related genes expression, and fatty acid oxidation in C2C12 myotubes. The observations reported here are consistent with Price's work,in which SIRT1 knockout completely blocked resveratrol-induced mitochondrial biogenesis and β-oxidation in skeletal muscle [33], further supporting the essential roles of SIRT1. However, the leucine-induced *Sirt1* gene expression was not affected by EX527, possibly due to the unique inhibition mechanism of EX527 on SIRT1 catalytic activity [34].

We also found that AMPK phosphorylation, which is elevated in response to metabolic stress [35], was also increased by leucine in C2C12 myotubes. This change might help to explain the increased fatty acid oxidation in the cells. Similarly, in mice, leucine supplementation has been reported to activate AMPK synergistically with resveratrol and metformin, resulting in increased insulin sensitivity and glucose tolerance [22]. On the other hand, Compound C, an inhibitor of AMPK, markedly blocked leucine's effects on mitochondrial biogenesis, indicating that like SIRT1 elevated mitochondrial biogenesis and fatty acid oxidation by leucine requires AMPK in C2C12 myotubes.

Notably, we found that leucine-induced AMPK phosphorylation was markedly blocked by EX527, suggesting that AMPK might serve as a downstream target of SIRT1. In support of this concept, Price et al. reported that SIRT1 activation is required for AMPK phosphorylation and improvement of mitochondrial function via deacetylation and activation of LKB1, a primary upstream kinase of AMPK [33, 36], while Park et al. found that resveratrol activates SIRT1 via an indirect pathway involving calmodulin-dependent protein kinase kinase β (Camkkβ) and AMPK activation [36]. Currently available evidences suggest that AMPK and SIRT1 display mutual interactions with each other; AMPK could activate SIRT1 by increasing cellular NAD^+ level via promoting expression of nicotinamide phosphoribosyltransferase (Nampt), a rate-limiting enzyme in NAD^+ biosynthesis; however, SIRT1 can also directly deacetylate and activate LKB1, resulting in the activation and phosphorylation of AMPK [37].

Our time-course data suggest that SIRT1 may be the initial target of leucine. SIRT1 activity was increased within the first hour of leucine treatment, while cellular NAD^+ and p-AMPK levels remained unchanged. Considering that the increased *Sirt1* mRNA and SIRT1 activity level occurred at some time after the leucine treatment for 24 hours, it is possible that SIRT1 activity is elevated by leucine first, and then activation of AMPK is a subsequent event, which may be responsible for the further SIRT1 activation at the later time points.

Our data may also reflect dose-dependent effects of leucine treatment. For example, high-dose leucine infusion and supplementation have been shown to induce insulin resistance and glucose intolerance in both human and animal models [38, 39], possibly via activation of mammalian target of rapamycin- (mTOR-) insulin receptor substrate 1 (IRS-1) signaling pathways [40]. In contrast, modest increases in leucine intake, sufficient to induce plasma leucine elevations to ~0.5 mM, significantly reduced obesity-related oxidative and inflammatory stress, resulting in improvement of insulin sensitivity in humans [19]. Similarly, Vaughan et al. found that leucine in the 0.1–0.5 mM range induces a dose-dependent increases of PGC-1α expression, leading to significant elevated mitochondrial density and oxidative capacity in skeletal muscle cells [17]. Consistent with these evidences, we found comparable levels of leucine promoted mitochondrial biogenesis and fatty acid oxidation in C2C12 myotubes.

There are several limitations to this study. One of them is the use of the Fleur de Lys assay to measure SIRT1 activity. Studies have challenged the validity of the assay, as some of them have found that sirtuin-activating compounds (STACs) only increased SIRT1 activity by using fluorophore-tagged substrates but not the matching nontagged peptides, which also might explain why the activation can be found exclusively *in vitro* but not *in vivo* [41, 42]. According to Gertz et al., the fluorophore can act synergistically with STACs to promote binding between substrates and SIRT1 enzyme [43]. Furthermore, evidence suggests that resveratrol-induced SIRT1 activation is actually mediated through an indirect signaling pathway involved in cAMP phosphodiesterases (PDE) and AMPK *in vivo* [36]. However, Hubbard et al. recently provided more evidences to support the allosteric binding and activation theory between STACs and SIRT1. They found that specific hydrophobic motifs in SIRT1 substrates and a single amino acid (Glu^{230}) in SIRT1 enzyme mediate the structure change during the

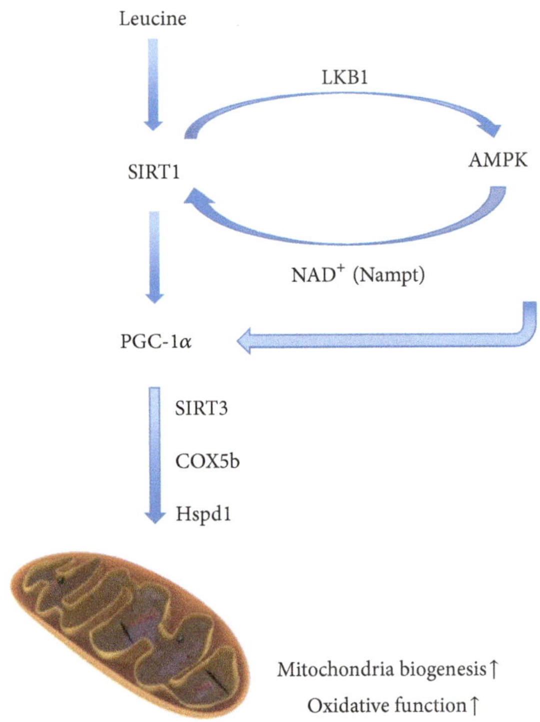

FIGURE 6: Proposed mechanism of leucine-induced mitochondrial biogenesis. In C2C12 myotubes, leucine treatment leads to activation of SIRT1. SIRT1 then deacetylates and activates LKB1, which subsequently induces AMPK phosphorylation and activation. In turn, activated AMPK could promote SIRT1 activation via intracellular NAD^+ level by changing expression and activity of Nampt. Activated AMPK and SIRT1 further activate PGC-1α via phosphorylation and deacetylation, resulting in elevated mitochondrial biogenesis and oxidative function.

deacetylation [44]. As a highly hydrophobic amino acid, leucine might directly activate SIRT1 through conformation change. Indeed, recent evidence demonstrates that leucine exerts direct effects on SIRT1 kinetics by decreasing 50% km for NAD^+ and substrates. With the presence of leucine and HMB, lower concentration of resveratrol is required for the activation of SIRT1 [45]. Therefore, further experiments using fluorophore-free substrates to measure the SIRT1 activity are needed to elucidate the exact pathways of leucine-activated SIRT1. A second limitation is lack of data assessing the cellular acetylation status of LKB1 and PGC-1α, as well as Nampt phosphorylation and expression.

In summary, with the present work, we demonstrate that leucine improves mitochondrial biogenesis and fatty acid oxidation in C2C12 myotubes through SIRT1 and AMPK-dependent pathway, with secondary activation of AMPK mediated by SIRT1 (Figure 6).

Conflict of Interests

The authors declare that there is no conflict of interests regarding the publication of this paper.

Acknowledgments

The authors thank Drs. Ling Zhao and Antje Bruckbauer for technical support in cell culture and SIRT1 activity measurement.

References

[1] P. Schrauwen, V. Schrauwen-Hinderling, J. Hoeks, and M. K. C. Hesselink, "Mitochondrial dysfunction and lipotoxicity," *Biochimica et Biophysica Acta*, vol. 1801, no. 3, pp. 266–271, 2010.

[2] D. L. Johannsen and E. Ravussin, "The role of mitochondria in health and disease," *Current Opinion in Pharmacology*, vol. 9, no. 6, pp. 780–786, 2009.

[3] C. Cantó, L. Q. Jiang, A. S. Deshmukh et al., "Interdependence of AMPK and SIRT1 for metabolic adaptation to fasting and exercise in skeletal muscle," *Cell Metabolism*, vol. 11, no. 3, pp. 213–219, 2010.

[4] C. Cantó, Z. Gerhart-Hines, J. N. Feige et al., "AMPK regulates energy expenditure by modulating NAD^+ metabolism and SIRT1 activity," *Nature*, vol. 458, no. 7241, pp. 1056–1060, 2009.

[5] K. T. Howitz, K. J. Bitterman, H. Y. Cohen et al., "Small molecule activators of sirtuins extend Saccharomyces cerevisiae lifespan," *Nature*, vol. 425, no. 6954, pp. 191–196, 2003.

[6] B. Schwer and E. Verdin, "Conserved metabolic regulatory functions of sirtuins," *Cell Metabolism*, vol. 7, no. 2, pp. 104–112, 2008.

[7] L. Guarente, "Sirtuins, aging, and metabolism," *Cold Spring Harbor Symposia on Quantitative Biology*, vol. 76, pp. 81–90, 2011.

[8] S.-I. Imai and L. Guarente, "Ten years of NAD-dependent SIR2 family deacetylases: Implications for metabolic diseases," *Trends in Pharmacological Sciences*, vol. 31, no. 5, pp. 212–220, 2010.

[9] D. Herranz and M. Serrano, "SIRT1: recent lessons from mouse models," *Nature Reviews Cancer*, vol. 10, no. 12, pp. 819–823, 2010.

[10] P. Puigserver and B. M. Spiegelman, "Peroxisome proliferator-activated receptor-γ coactivator 1α (PGC-1α): transcriptional coactivator and metabolic regulator," *Endocrine Reviews*, vol. 24, no. 1, pp. 78–90, 2003.

[11] Y. Macotela, B. Emanuelli, A. M. Bång et al., "Dietary leucine—an environmental modifier of insulin resistance acting on multiple levels of metabolism," *PLoS ONE*, vol. 6, no. 6, Article ID e21187, 2011.

[12] M. H. Stipanuk, "Leucine and protein synthesis: mTOR and beyond," *Nutrition Reviews*, vol. 65, no. 3, pp. 122–129, 2007.

[13] X. Sun and M. B. Zemel, "Leucine and calcium regulate fat metabolism and energy partitioning in murine adipocytes and muscle cells," *Lipids*, vol. 42, no. 4, pp. 297–305, 2007.

[14] J. Donato Jr., R. G. Pedrosa, V. F. Cruzat, I. S. D. O. Pires, and J. Tirapegui, "Effects of leucine supplementation on the body composition and protein status of rats submitted to food restriction," *Nutrition*, vol. 22, no. 5, pp. 520–527, 2006.

[15] L. Rachdi, V. Aïello, B. Duvillié, and R. Scharfmann, "L-Leucine alters pancreatic β-cell differentiation and function via the mTor signaling pathway," *Diabetes*, vol. 61, no. 2, pp. 409–417, 2012.

[16] N. E. Zanchi, H. Nicastro, and A. H. Lancha Jr., "Potential antiproteolytic effects of L-leucine: observations of in vitro and in vivo studies," *Nutrition and Metabolism*, vol. 5, no. 1, article 20, 2008.

[17] R. A. Vaughan, R. Garcia-Smith, N. P. Gannon, M. Bisoffi, K. A. Trujillo, and C. A. Conn, "Leucine treatment enhances oxidative capacity through complete carbohydrate oxidation and increased mitochondrial density in skeletal muscle cells," *Amino Acids*, vol. 45, no. 4, pp. 901–911, 2013.

[18] M. B. Zemel and A. Bruckbauer, "Effects of a leucine and pyridoxine-containing nutraceutical on fat oxidation, and oxidative and inflammatory stress in overweight and obese subjects," *Nutrients*, vol. 4, no. 6, pp. 529–541, 2012.

[19] M. B. Zemel, X. Sun, T. Sobhani, and B. Wilson, "Effects of dairy compared with soy on oxidative and inflammatory stress in overweight and obese subjects," *The American Journal of Clinical Nutrition*, vol. 91, no. 1, pp. 16–22, 2010.

[20] M. Lagouge, C. Argmann, Z. Gerhart-Hines et al., " Resveratrol improves mitochondrial function and protects against metabolic disease by activating SIRT1 and PGC-1α," *Cell*, vol. 127, no. 6, pp. 1109–1122, 2006.

[21] A. Bruckbauer, M. B. Zemel, T. Thorpe et al., "Synergistic effects of leucine and resveratrol on insulin sensitivity and fat metabolism in adipocytes and mice," *Nutrition & Metabolism*, vol. 9, no. 1, article 77, 2012.

[22] A. Bruckbauer and M. B. Zemel, "Synergistic effects of metformin, resveratrol, and hydroxymethylbutyrate on insulin sensitivity," *Diabetes, Metabolic Syndrome and Obesity: Targets and Therapy*, vol. 6, pp. 93–102, 2013.

[23] S. A. Hawley, D. A. Pan, K. J. Mustard et al., "Calmodulin-dependent protein kinase kinase-β is an alternative upstream kinase for AMP-activated protein kinase," *Cell Metabolism*, vol. 2, no. 1, pp. 9–19, 2005.

[24] J. M. Harig, J. A. Barry, V. M. Rajendran, K. H. Soergel, and K. Ramaswamy, "D-glucose and L-leucine transport by human intestinal brush-border membrane vesicles," *The American Journal of Physiology—Gastrointestinal and Liver Physiology*, vol. 256, no. 3, part 1, pp. G618–G623, 1989.

[25] A. Bruckbauer and M. B. Zemel, "Effects of dairy consumption on SIRT1 and mitochondrial biogenesis in adipocytes and muscle cells," *Nutrition and Metabolism*, vol. 8, article 91, 2011.

[26] X. Sun and M. B. Zemel, "Leucine modulation of mitochondrial mass and oxygen consumption in skeletal muscle cells and adipocytes," *Nutrition & Metabolism*, vol. 6, article 26, 2009.

[27] H. Nicastro, G. G. Artioli, A. Dos Santos Costa et al., "An overview of the therapeutic effects of leucine supplementation on skeletal muscle under atrophic conditions," *Amino Acids*, vol. 40, no. 2, pp. 287–300, 2011.

[28] P. L. Campos-Ferraz, T. Bozza, H. Nicastro, and A. H. Lancha, "Distinct effects of leucine or a mixture of the branched-chain amino acids (leucine, isoleucine, and valine) supplementation on resistance to fatigue, and muscle and liver-glycogen degradation, in trained rats," *Nutrition*, vol. 29, no. 11-12, pp. 1388–1394, 2013.

[29] M. D. Brand, "Uncoupling to survive? The role of mitochondrial inefficiency in ageing," *Experimental Gerontology*, vol. 35, no. 6-7, pp. 811–820, 2000.

[30] J. A. Baur, K. J. Pearson, N. L. Price et al., "Resveratrol improves health and survival of mice on a high-calorie diet," *Nature*, vol. 444, no. 7117, pp. 337–342, 2006.

[31] J. E. Tang, D. R. Moore, G. W. Kujbida, M. A. Tarnopolsky, and S. M. Phillips, "Ingestion of whey hydrolysate, casein, or soy protein isolate: effects on mixed muscle protein synthesis at rest and following resistance exercise in young men," *Journal of Applied Physiology*, vol. 107, no. 3, pp. 987–992, 2009.

[32] H. Li, M. Xu, J. Lee, C. He, and Z. Xie, "Leucine supplementation increases SIRT1 expression and prevents mitochondrial dysfunction and metabolic disorders in high-fat diet-induced obese mice," *American Journal of Physiology: Endocrinology and Metabolism*, vol. 303, no. 10, pp. E1234–E1244, 2012.

[33] N. L. Price, A. P. Gomes, A. J. Y. Ling et al., "SIRT1 is required for AMPK activation and the beneficial effects of resveratrol on mitochondrial function," *Cell Metabolism*, vol. 15, no. 5, pp. 675–690, 2012.

[34] M. Gertz, F. Fischer, G. T. T. Nguyen et al., "Ex-527 inhibits Sirtuins by exploiting their unique NAD^+-dependent deacetylation mechanism," *Proceedings of the National Academy of Sciences of the United States of America*, vol. 110, no. 30, pp. E2772–E2781, 2013.

[35] Y. Wang, Y. Liang, and P. M. Vanhoutte, "SIRT1 and AMPK in regulating mammalian senescence: a critical review and a working model," *FEBS Letters*, vol. 585, no. 7, pp. 986–994, 2011.

[36] S.-J. Park, F. Ahmad, A. Philp et al., "Resveratrol ameliorates aging-related metabolic phenotypes by inhibiting cAMP phosphodiesterases," *Cell*, vol. 148, no. 3, pp. 421–433, 2012.

[37] N. B. Ruderman, D. Carling, M. Prentki, and J. M. Cacicedo, "AMPK, insulin resistance, and the metabolic syndrome," *Journal of Clinical Investigation*, vol. 123, no. 7, pp. 2764–2772, 2013.

[38] F. Tremblay, M. Krebs, L. Dombrowski et al., "Overactivation of S6 kinase 1 as a cause of human insulin resistance during increased amino acid availability," *Diabetes*, vol. 54, no. 9, pp. 2674–2684, 2005.

[39] C. B. Newgard, J. An, J. R. Bain et al., "A branched-chain amino acid-related metabolic signature that differentiates obese and lean humans and contributes to insulin resistance," *Cell Metabolism*, vol. 9, no. 4, pp. 311–326, 2009.

[40] T. Haruta, T. Uno, J. Kawahara et al., "A rapamycin-sensitive pathway down-regulates insulin signaling via phosphorylation and proteasomal degradation of insulin receptor substrate-1," *Molecular Endocrinology*, vol. 14, no. 6, pp. 783–794, 2000.

[41] H. Dai, L. Kustigian, D. Carney et al., "SIRT1 activation by small molecules: kinetic and biophysical evidence for direct interaction of enzyme and activator," *Journal of Biological Chemistry*, vol. 285, no. 43, pp. 32695–32703, 2010.

[42] D. Beher, J. Wu, S. Cumine et al., "Resveratrol is not a direct activator of sirt1 enzyme activity," *Chemical Biology and Drug Design*, vol. 74, no. 6, pp. 619–624, 2009.

[43] M. Gertz, G. T. T. Nguyen, F. Fischer et al., "A molecular mechanism for direct sirtuin activation by resveratrol," *PLoS ONE*, vol. 7, no. 11, Article ID e49761, 2012.

[44] B. P. Hubbard, A. P. Gomes, H. Dai et al., "Evidence for a common mechanism of SIRT1 regulation by allosteric activators," *Science*, vol. 339, no. 6124, pp. 1216–1219, 2013.

[45] A. Bruckbauer and M. B. Zemel, "Synergistic effects of polyphenols and methylxanthines with leucine on AMPK/sirtuin-mediated metabolism in muscle cells and adipocytes," *PLoS ONE*, vol. 9, no. 2, Article ID e89166, 2014.

19

Study of Body Composition and Metabolic Parameters in HIV-1 Male Patients

Gurudath Gundurao Sreekantamurthy, N Biplab Singh, Th Bhimo Singh, Th Suraj Singh, and Karam Romeo Singh

Department of General Medicine, Regional Institute of Medical Sciences, Imphal 795004, India

Correspondence should be addressed to Gurudath Gundurao Sreekantamurthy; gurudathsdumc@gmail.com

Academic Editor: Phillip B. Hylemon

Background. HIV patients on highly active antiretroviral therapy (HAART) containing protease inhibitors (PIs) had been often associated with lipodystrophy. However, there are only few studies on association of nucleoside and nonnucleoside reverse transcriptase inhibitors (NRTI and NNRTI) with lipodystrophy. *Study Design.* One hundred and one HIV male patients were categorised into ART naïve ($n = 22$), zidovudine ($n = 22$), stavudine ($n = 18$), tenofovir ($n = 15$), and PIs ($n = 24$) based HAART. Their clinicoepidemiological data had been entered in preformed pro forma. The body composition, using TANITA machine and metabolic parameters like lipid profile, blood sugars was analysed. *Results.* Clinically, lipoatrophy of face was most prevalent in HIV patients on stavudine (15 patients, 83.3%) and PIs (20 patients, 83.3%) based HAART. The mean BMI among study population was in normal range. Excess visceral fat was most prevalent among patients on PIs, 4 patients (16.7%). The waist-hip ratio was significantly higher in PIs ($P = 0.01$) based HAART. There was no significant difference among different study populations in terms of BMI ($P = 0.917$), body water ($P = 0.318$), body fat ($P = 0.172$), bone mass ($P = 0.200$), and muscle mass ($P = 0.070$). Hypertriglyceridiemia was found in stavudine, tenofovir, and protease inhibitors regimens. Low levels of high density lipoprotein (HDL) was found zidovudine, stavudine, and PIs regimens. Fasting and postprandial hyperglycaemia was found PIs and impaired glucose tolerance in stavudine regimen. *Conclusion.* Patients on PIs were associated with truncal obesity and lipoatrophy of face, along with dyslipidemia and hyperglycaemia. Stavudine based regimen is associated with hypertriglyceridiemia and low HDL along with lipoatrophy of face.

1. Introduction

Lipodystrophy had been well described and has frequent disturbing side effect with protease inhibitors. The main manifestations of the lipodystrophy (LD) syndrome in patients with normal serum cortisol levels are fat accumulation in the dorsocervical spine (also known as buffalo hump), in the breasts, and in the abdomen or fat reduction in the face and in the limbs. Lipoatrophy may occur isolated or together with fat accumulation. These body changes are accompanied by dyslipidemia and hyperglycemia [1]. However, there are only few reports of hyperlipidemia and lipodystrophy associated with HAART without PI [2, 3]. A recent study by Smith et al. [4] demonstrated an excess of cardiovascular risk factors in HIV patients receiving HAART, and lipodystrophy syndrome has been considered a worrisome factor. This study was performed to analyze the body composition and metabolic variations in HIV-1 infected individuals during treatment with HAART of different ART regimen and without HAART in the north-eastern part of India.

2. Subjects and Methods

A hospital based cross-sectional study conducted from October 2011 to September 2013 included random 101 seropositive HIV-1 male patients *aged more than 15 years* who are attending Medicine Out-Patient Department (OPD) and ART centre, Centre of Excellence, Regional Institute of Medical Sciences, for routine check-up on OPD basis. These patients were categorised into the following groups: ART

Table 1: Duration of HIV infection and duration of ART in different study population.

Parameters	ART naïve	Zidovudine based regimen	Stavudine based regimen	Tenofovir based regimen	Protease inhibitors regimen
Duration of ART in months (mean)	0	45	64	51	111
Duration of HIV in months (mean)	5.1	51.7	75.6	66.3	126

Table 2: Prevalence of lipoatrophy of face in different study population.

Lipoatrophy of face	ART naïve *n* (%)	Zidovudine based regimen *n* (%)	Stavudine based regimen *n* (%)	Tenofovir based regimen *n* (%)	Protease inhibitors regimen *n* (%)
Present	0 (0%)	3 (13.6%)	15 (83.3%)	3 (20.0%)	20 (83.3%)
Absent	22 (100%)	19 (86.4%)	3 (16.7%)	12 (80%)	4 (16.7%)
Total	22 (100%)	22 (100%)	18 (100%)	15 (100%)	24 (100%)

naïve patients ($n = 22$), HIV patients on zidovudine based regimen (AZT and 3TC and NVP or EFV, $n = 22$), stavudine based regimen (d4T and 3TC and NVP or EFV, $n = 18$), tenofovir based regimen (TDF and 3TC and NVP or EFV, $n = 15$), and protease inhibitors based regimen (*atazanavir 300 mg boosted with low dose ritonavir 100 mg, n = 24). The minimum duration of each ART regimen was for at least 6 months*. The ART was given to patients according to National AIDS Control Organization (NACO) guidelines, India. Patients who were known case of diabetes mellitus, chronic kidney disease, congestive cardiac failure, already on hypolipidemic drugs before starting ART, and suffering from tuberculosis and other active opportunistic infections were excluded from the study.

2.1. Data Collection. Detailed history regarding clinicoepidemiological data including, facial dysmorphism (lipoatrophy of face), exposure, HIV infection duration, duration of the present regimen, alcoholism (>80 gms/day), and cigarette smoking (>5 cigarettes/day) were recorded in a preformed pro forma. Anthropometric measurements, like body mass index (BMI) and waist-hip ratio, were measured. Waist circumference was measured at midpoint between lowermost palpable rib and topmost point of iliac crest along midaxillary line, and hip circumference was measured at widest portion of buttocks. Waist-hip ratio >0.9 indicates truncal obesity and high risk for cardiovascular complications [5]. BMI of study population was analysed in accordance with Asian populations [6]. Total body water (in percentage), bone mass (in kgs), visceral fat (in kgs), body fat (in percentage), and muscle mass (in kgs) were analysed using TANITA machine, a body composition guide for inner scan manufactured by Tanita India Private Limited, Mumbai. Visceral fat ranging from 1 to 12 kgs was considered as healthy range and more than 12 as excess [7]. Body fat percentage was analyzed according to weight [8]. Blood biochemistry including lipid profile, kidney function test, and blood sugar levels (random, fasting and postprandial) was performed.

2.2. Statistical Analysis Used. SPSS 11 was used to analyse the data. ANOVA and post hoc Dunnet's T3 test were used to determine significant difference among study population.

3. Results

3.1. Baseline Parameters. The age range of study population was ART naïve patients, 24 to 51 years; zidovudine based patients, 26 to 53 years; stavudine based patients, 32 to 52 years; tenofovir based patients, 37 to 60 years; and protease inhibitors patients, 32 to 60 years. The mean duration of HIV infection and the present ART regimen (in months) is shown in Table 1. Lipoatrophy of face was most prevalent in HIV-positive population who are on stavudine based regimen (15 out of 18 patients, 83.3%) and protease inhibitors based regimen (20 out of 24 patients, 83.3%). None of the ART naïve patients had lipoatrophy of face. Three patients (13.6%) on zidovudine based regimen and 3 patients (20.0%) on tenofovir based regimen had lipoatrophy of face (refer Table 2).

3.2. Body Composition in Study Population (Refer to Tables 3 and 4). Mean BMI of different categories of study population were in normal range, and there was no significant difference in mean BMI among different categories of study population ($P = 0.917$). Among study population, excess visceral fat was most prevalent in protease inhibitors group (4 out of 24 patients, 16.7%). The mean total visceral fat among study population was in healthy range, and there was statistically insignificant difference among study population. Though statistically insignificant, mean difference of total visceral fat between PIs (6.21 ± 4.79) and ART naïve (4.09 ± 3.66) patients was about 2.11. Mean body fat percentage was in normal range among study population and did not vary significantly among different study groups ($P = 0.172$). The mean total body water percentage was relatively higher in zidovudine based regimen (61.98 ± 7.86) and stavudine based regimen (61.4 ± 6.63) than other groups (ART naïve regimen was 59.80 ± 9.73, tenofovir based regimen was 58.90 ± 4.9, and

TABLE 3: Prevalence of excess and healthy visceral fat among study population.

Visceral fat	ART naive	Zidovudine based regimen	Stavudine based regimen	Tenofovir based regimen	PIs based regimen	Total
Healthy	21 (13.1%)	20 (12.5%)	18 (11.2%)	14 (8.8%)	20 (12.5%)	160 (100%)
Excess	1 (4.2%)	2 (8.3%)	0 (0%)	1 (4.2%)	**4 (16.7%)**	24 (100%)
Total	22 (12.0%)	22 (12.0%)	18 (9.8%)	15 (8.2%)	24 (13.0%)	184 (100%)

TABLE 4: Body composition parameters in study population.

Parameters (mean ± SD)	ART naive	Zidovudine based regimen	Stavudine based regimen	Tenofovir based regimen	Protease inhibitors	*P* value
BMI	20.85 ± 2.35	21.15 ± 2.39	21.48 ± 2.28	20.63 ± 3.39	21.22 ± 3.22	$P = 0.917$
Total visceral fat (in Kgs)	4.09 ± 3.66	5.18 ± 4.46	5.11 ± 3.92	6.20 ± 4.16	6.21 ± 4.79	$P = 0.467$
Total body fat (in %)	13.60 ± 4.71	12.25 ± 5.81	13.82 ± 5.64	14.96 ± 5.96	16.63 ± 7.71	$P = 0.172$
Total body water (in %)	59.80 ± 9.73	61.98 ± 7.86	61.4 ± 6.63	58.90 ± 4.91	57.64 ± 7.69	$P = 0.318$
Total muscle mass (in Kgs)	45.05 ± 4.10	47.69 ± 3.53	47.5 ± 5.14	44.46 ± 6.63	44.56 ± 5.05	$P = 0.070$
Total bone mass (in Kgs)	2.52 ± 0.23	2.69 ± 0.34	2.73 ± 0.83	2.45 ± 0.32	2.50 ± 0.24	$P = 0.200$
Waist-hip ratio	0.87 ± 0.02	0.85 ± 0.04	0.87 ± 0.03	0.86 ± 0.04	0.91 ± 0.04	$P = 0.001$

protease inhibitors regimen was 57.64 ± 6.99), and there was no significant difference among study population ($P = 0.318$). There was no statistically significant difference among study population in terms of mean bone mass ($P = 0.200$) and mean muscle mass ($P = 0.070$). Waist-hip ratio (mean) in protease inhibitors group was 0.91±0.04, indicating central obesity. The mean waist-hip ratio significantly varied among study population, being higher in HIV patients receiving protease inhibitors ($P = 0.001$).

3.3. Metabolic Parameters (Refer to Table 5). Hypertriglyceridemia was found in stavudine (173.06 ± 48.11), tenofovir (164.07 ± 77.40), and protease inhibitors (249.83 ± 125.3) based regimens. Low HDL was found in zidovudine (32.75 ± 7.19), stavudine (26.72 ± 7.12), and protease inhibitors (32.79 ± 11.0) based regimens. High VLDL was found only in protease inhibitors regimen (33.9 ± 14.10). Patients with protease inhibitors were found to have both fasting (118.04 ± 46.03) and postprandial (205.38 ± 105.26) hyperglycemia. Patients on stavudine based regimen had only postprandial hyperglycemia (145.22 ± 30.11). There was a significant difference in total cholesterol ($P = 0.001$), triglycerides ($P = 0.000$), HDL ($P = 0.003$), LDL ($P = 0.009$), VLDL ($P = 0.001$), RBS ($P = 0.000$), FBS ($P = 0.001$), and PPBS ($P = 0.000$) among study population.

4. Discussion

The present study is a cross-sectional comparative data regarding body composition and metabolic parameters derangements among HIV patients receiving protease inhibitors and NRTIs. 83.3% patients receiving protease inhibitors had facial dysmorphism in form of lipoatrophy and 16.7% patients had excess visceral fat with normal total body fat percentage. In addition to this, there was increased waist-hip ratio, more than 0.90, indicating truncal obesity. There were also metabolic abnormalities in the form of hypertriglyceridemia, low HDL, high VLDL, and fasting and postprandial hyperglycemia. These features indicate regional fat redistribution from periphery to visceral resulting in insulin resistance rather than mere fat loss or fat gain. Similar finding was also described by previous study by Dinges et al. [9] and Carr et al. [10]. Mulligan et al. [11] analysed the paired data in HIV infected patients before and after initiating PIs ($n = 20$). They found increase lean body mass and fat mass in protease inhibitor group. They also found significant increase in glucose (+9 ± 3 mg/dL, $P = 0.0136$), insulin (+12 ± 4.9 mg/dL, $P = 0.023$), triglycerides (+53 ± 17 mg/dL, $P = 0.0069$), and total and LDL cholesterol (+32 ± 11 mg/dL and +18 ± 5 mg/dL, $P = 0.0082$ and 0.0026, resp.). In present study, there was no significant difference between the study groups in terms of BMI, total body water, bone mass and muscle mass, and visceral and body fat. Samaras et al. [12] in their international cross-sectional data involving 788 cohorts of HIV patients on HAART noticed 14% prevalence of metabolic syndrome according to International Diabetic Federation criteria and 18% prevalence of the same, according to US national cholesterol education program adult treatment panel III (ATP III) criteria, in patients receiving PIs, the highest being in indinavir and the least in amprenavir. 49% of patients had at least 2 features of metabolic syndrome but did not meet the diagnostic criteria. In present study also, it is emphasising the same fact of insulin resistance in the form of truncal adiposity and hyperglycemia in patients receiving PIs. Mulligan et al. [11] studied 148 HIV-1 patients (116 patients were on protease inhibitors and 32 had never taken protease inhibitors) and in 47 healthy men found that development of syndromic features like central adiposity along with hypertriglyceridemia, hypercholesterolemia, and

Table 5: Metabolic parameters in study population.

Metabolic parameters	ART naïve	Zidovudine based regimen	Stavudine based regimen	Tenofovir based regimen	Protease inhibitors based regimen	P value
Total cholesterol	120 ± 34.6	134.77 ± 31.29	154.50 ± 41.4	147.87 ± 37.9	173.46 ± 52	0.001
Triglycerides	117.81 ± 44.1	117.32 ± 38.3	173.06 ± 48.1	164.07 ± 77.4	249.83 ± 125.3	0.000
HDL	38.85 ± 9.1	32.75 ± 7.1	26.72 ± 7.12	37.47 ± 14.6	32.79 ± 11.0	0.003
LDL	76.8 ± 21.8	83.85 ± 14.5	94.11 ± 12.4	87.7 ± 20.8	98.3 ± 27.8	0.009
VLDL	20.86 ± 11.9	21.57 ± 6.2	25.88 ± 11.4	25.53 ± 6.1	33.9 ± 14.1	0.001
RBS	104.05 ± 22.3	97.74 ± 16.2	123.67 ± 35.9	117.53 ± 43.5	188.13 ± 81.8	0.000
FBS	88.86 ± 8.0	82.64 ± 7.6	96.67 ± 13.0	95.60 ± 39.0	118.04 ± 46.0	0.001
PPBS	129.19 ± 26.7	116.86 ± 21.4	145.22 ± 30.1	133.90 ± 57.6	205.38 ± 105.2	0.000

insulin resistance with type 2 diabetes mellitus occurred mainly with the use of all potent protease inhibitors and hence concluded it to be a class effect.

Current data suggests that abnormal fat distribution occurs in HIV patients receiving non-PIs based regimen. Peripheral fat wasting was noticed in clinically stable asymptomatic HIV patients and may be due to long term use of NRTIs especially stavudine. Lipoatrophy of face (buccal pad fat atrophy) was observed in 83.3% of patients receiving stavudine based HAART, 13.6% of patients in zidovudine group, 20.0% of patients on tenofovir group, and none in ART naïve group. Along with this, stavudine subgroup also had hypertriglyceridemia, low HDL, and postprandial hyperglycemia, which were almost absent in other NRTIs groups. On the contrary, hypertriglyceridemia was also found in the tenofovir group. This may be due to their previous treatment with stavudine based therapy. Saint-Marc et al. [13] in their cross-sectional data on 43 HIV patients on NRTI found peripheral lipodystrophy (Bichat atrophy) in 17 patients (63%) receiving stavudine, 3 patients (18.75%) receiving zidovudine and none in ART naïve. Along with lipodystrophy, patients on stavudine had hypertriglyceridiemia which is similar to result of the present study. Domingo et al. [14], in his study involving 150 HIV patients receiving HAART, 75 stavudine based patients, and 75 zidovudine based patients noticed significant lower body fat in patients taking stavudine. Ninety-four patients had regional fat redistribution in terms of lipoatrophy. Mallal et al. [15] analysed regional fat measured by dual energy X-ray absorptiometry (DEXA) in 161 HIV patients and found progressive subcutaneous fat wasting occurring in stavudine-containing regimens compared with zidovudine-containing regimens (relative risk was 1.085 per month; $P < 0.0001$) and concluded that NRTIs especially stavudine do have an independent contribution to fat wasting. Other body composition parameters like total body water, muscle mass, and bone mass are not significantly affected among study population. HDL was significantly low in ART naïve and all NRTI groups except for tenofovir group. Though LDL was in normal range, it differed significantly ($P = 0.009$) among study population, being high in stavudine based regimen (94.11 ± 12.4) and protease inhibitors (98.3 ± 27.8).

Conflict of Interests

The authors declare that there is no conflict of interests regarding the publication of this paper.

References

[1] A. Carr, K. Samaras, S. Burton et al., "A syndrome of peripheral lipodystrophy, hyperlipidaemia and insulin resistance in patients receiving HIV protease inhibitors," *AIDS*, vol. 12, no. 7, pp. F51–F58, 1998.

[2] C. Grunfeld, "Disturbances in lipid metabolism due to HIV infection and its therapy," in *Proceedings of the Conference on Retroviruses and Opportunistic Infections*, vol. 6, Chicago, Ill, USA, 1999.

[3] R. Thiébaut, F. Dabis, D. Malvy, H. Jacqmin-Gadda, P. Mercié, and V. Daucourt Valentin, "Serum triglycerides, HIV infection, and highly active antiretroviral therapy, Aquitaine Cohort, France, 1996 to 1998," *Journal of Acquired Immune Deficiency Syndromes*, vol. 23, no. 3, pp. 261–265, 2000.

[4] C. J. Smith, I. Levy, C. A. Sabin, E. Kaya, M. A. Johnson, and M. C. I. Lipman, "Cardiovascular disease risk factors and antiretroviral therapy in an HIV-positive UK population," *HIV Medicine*, vol. 5, no. 2, pp. 88–92, 2004.

[5] *Waist Circumference and Waist-Hip Ratio. A Report of WHO Expert Consultation*, Geneva, Switzerland, 2008.

[6] M. Kanazawa, N. Yoshiike, T. Osaka, Y. Numba, P. Zimmet, and S. Inoue, "Criteria and classi—cation of obesity in Japan and Asia-Oceania," *Asia Pacific Journal of Clinical Nutrition*, vol. 11, pp. 732–737, 2002.

[7] Z. Wang et al., "Japanese-American differences in visceral adiposity and a simplified estimation method for visceral adipose tissue," in *North American Association for the Study of Obesity. Annual Meeting*, 2004, Abstract 518-P.

[8] D. Gallagher, S. B. Heymsfield, M. Heo, S. A. Jebb, P. R. Murgatroyd, and Y. Sakamoto, "Healthy percentage body fat ranges: an approach for developing guidelines based on body mass index," *American Journal of Clinical Nutrition*, vol. 72, no. 3, pp. 694–701, 2000.

[9] W. L. Dinges, D. Chen, P. G. Snell, P. T. Weatherall, D. M. Peterson, and A. Garg, "Regional body fat distribution in HIV-infected patients with lipodystrophy," *Journal of Investigative Medicine*, vol. 53, no. 1, pp. 15–25, 2005.

[10] A. Carr, K. Samaras, D. J. Chisholm, and D. A. Cooper, "Abnormal fat distribution and use of protease inhibitors," *The Lancet*, vol. 351, no. 9117, p. 1736, 1998.

[11] K. Mulligan, C. Grunfeld, V. W. Tai et al., "Hyperlipidemia and insulin resistance are induced by protease inhibitors independent of changes in body composition in patients with HIV infection," *Journal of Acquired Immune Deficiency Syndromes and Human Retrovirology*, vol. 23, no. 1, pp. 35–43, 2000.

[12] K. Samaras, H. Wand, L. Matthew, S. Emery, D. Cooper, and A. Carr, "Prevalence of metabolic syndrome in HIV-infected patients receiving highly active antiretroviral therapy using International Diabetes Foundation and Adult Treatment Panel III criteria: associations with insulin resistance, disturbed body fat compartmentalization, elevated C-reactive protein, and hypoadiponectinemia," *Diabetes Care*, vol. 30, no. 1, pp. 113–119, 2007.

[13] T. Saint-Marc, M. Partisani, I. Poizot-Martin et al., "A syndrome of peripheral fat wasting (lipodystrophy) in patients receiving long-term nucleoside analogue therapy," *AIDS*, vol. 13, no. 13, pp. 1659–1667, 1999.

[14] P. Domingo, M. A. Sambeat, A. Pérez, J. Ordoñez, J. Rodriguez, and O. Vázquez, "Fat distribution and metabolic abnormalities in HIV-infected patients on first combination antiretroviral therapy including stavudine or zidovudine: role of physical activity as a protective factor," *Antiviral Therapy*, vol. 8, no. 3, pp. 223–231, 2003.

[15] S. A. Mallal, M. John, C. B. Moore, I. R. James, and E. J. McKinnon, "Contribution of nucleoside analogue reverse transcriptase inhibitors to subcutaneous fat wasting in patients with HIV infection," *AIDS*, vol. 14, no. 10, pp. 1309–1316, 2000.

20

Low-Salt Intake during Mating or Gestation in Rats Is Associated with Low Birth and Survival Rates of Babies

Ranna Chou, Anna Hara, DongDong Du, Namiko Shimizu, Hiroe Sakuyama, and Yoshio Uehara

Division of Clinical Nutrition, Faculty of Home Economics, Kyoritsu Women's University, 2-2-1 Hitotsubashi, Chiyoda, Tokyo 101-8437, Japan

Correspondence should be addressed to Yoshio Uehara; yuehara@kyoritsu-wu.ac.jp

Academic Editor: Christel Lamberg-Allardt

We investigated the influence of maternal salt restriction during mating or gestation on birth rate and offspring growth in Dahl salt-sensitive rats (DS). DS were divided into 5 groups: DS fed a low-salt (0.3% NaCl, w/w) (DS-low) or high-salt (4% NaCl, w/w) diet (DS-high) during mating and DS-high or DS-low during gestation, and DS fed regular chow (0.75% NaCl, w/w) (DS-regular) throughout mating and gestation. During the unspecified periods, the rats were given regular chow. DS-low during mating delivered fewer infants than high-salt mothers ($P < 0.05$). The birth rate on regular chow was 87%. Six out of 11 DS-low rats during pregnancy produced pups while the rats fed a high-salt diet all delivered pups ($P < 0.025$). The pup survival rate was 67% for high-salt mothers during mating and 54% for mothers on a low-salt diet. The pup survival rate was 95% for mothers on a high-salt diet during pregnancy and 64% for mothers on a low-salt diet ($P < 0.0001$). Seven out of 8 DS-regular rats during mating delivered 59 neonates. However, 66% of the neonates survived. A low-salt diet during mating or pregnancy lowers birth rate and the neonates from low-salt mothers during pregnancy were more likely to die than those from high-salt mothers.

1. Introduction

For the last several decades, increased salt intake has been blamed for hypertension and related diseases, and a world-wide campaign on salt reduction commenced. Since the 1980s many investigators have reported serious problems in adult physiological and psychological health that are related to excessive salt intake. Despite such trends towards salt reduction and to our surprise, a recent meta-analysis including 25 previous studies on salt intake and cardiovascular events demonstrated that excessively low-salt intake is associated with an increase in cardiovascular events rather than benefits [1].

Some researchers reported that, particularly during pregnancy, low-salt intake is associated with adverse effects, that is, low birth weight and insulin resistance in offspring in adulthood [2–4]. Growth inhibition of astrocytes in the neonate brain, which occurs due to low-salt intake during pregnancy, is possibly involved in the dysfunction of the autonomic system in adulthood, thereby affecting hemodynamic or metabolic changes in adulthood [5]. Based on these data, there is a need to more carefully determine what the necessary salt requirements are and consider factors such as age or the stage of growth.

Accordingly, an increasing amount of evidence has suggested that fetal programming during gestation may play an important role in the development of metabolism-related disorders during adulthood [2, 6, 7]. Conceivably, genes are subjected to the influence of maternal food intake and lifestyle during gestation. During our investigation of the association between fetal programming and a postweaning salt appetite, we recognized that pregnant mothers on a low-salt diet appeared to deliver fewer babies than mothers on a high-salt diet (unpublished data).

To our knowledge, however, few studies have investigated the influence that maternal low-salt intake has on the process

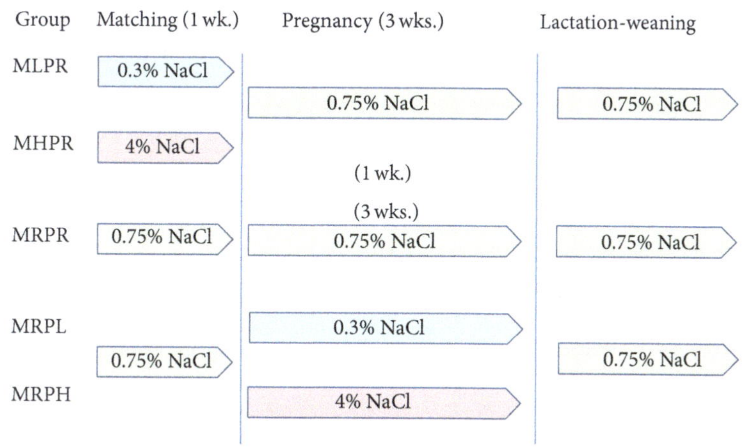

FIGURE 1: Protocol of the study. This graph shows each type of salt diet and the time period in which it was fed to the Dahl S rats. 0.3% NaCl, a low-salt (0.3% NaCl, w/w) diet; 0.75% NaCl, regular (0.75% NaCl, w/w) diet; and 4% NaCl, a high-salt (4% NaCl, w/w) diet.

of fetal growth. There is a necessity to separately examine the influence of low-salt diet on fertilization and on fetal growth. Moreover, it is of interest to disclose the influence of low-salt intake in mothers on the growth of babies after weaning.

Thus, in the present study, we hypothesized that low-salt intake during mating (early pregnancy period) or pregnancy period influences fertilization and fetal growth and causes a decrease in the number of newborn pups. To test this hypothesis, we investigated the effects of a low-salt diet during mating or pregnancy on birth and growth rates of offspring in Dahl salt-sensitive (Dahl S) rats. Moreover, we examined the influence of low-salt intake on the growth of babies after weaning.

2. Materials and Methods

2.1. Study Design. We obtained 4-week-old male and female Dahl salt-sensitive (Dahl S) rats from Sankyo Laboratories, Inc., Tokyo, Japan. The rats were fed regular chow (0.75% NaCl, w/w) (CEL Rodent Diet CE-7, CLEA Japan, Inc., Tokyo, Japan). At 10 weeks of age, the rats were allowed to mate for 1 week (mating period (early pregnancy period)). Next, the mated female rats were placed in separate cages for the pregnancy period. Three weeks after mating, the mothers began delivering pups. We monitored the number of mothers that delivered pups, the number of pups delivered, and the growth of pups until weaning.

Dahl S rats were divided into 5 groups (Figure 1): (1) Dahl S rats that were fed a low-salt (0.3% NaCl, w/w) diet (F2Dahl-0.3%, Oriental Yeast Co., Ltd., Tokyo, Japan) for the mating period, (2) Dahl S rats that were fed a high-salt (4% NaCl, w/w) diet (F2Dahl-4%, Oriental Yeast) for the mating period, (3) Dahl S rats that were fed a low-salt (0.3% NaCl diet, w/w) diet for the pregnancy period, (4) Dahl S rats that were fed a high-salt (4% NaCl, w/w) diet (F2Dahl-4%, Oriental Yeast) for the pregnancy period, and (5) Dahl S rats fed regular chow (0.75% NaCl, w/w) throughout the mating and pregnancy periods.

Dahl S rats were fed the regular (0.75% NaCl, w/w) chow during the periods unspecified. Water was available *ad libitum*.

2.2. Statistical Analysis. All statistical analyses were performed using STATISTICA software (StatSoft, Tulsa, OK). Values were expressed as means ± SD. Differences were assessed by χ^2 and one-way factorial analysis of variance (ANOVA) followed by post hoc least significant difference (LSD) test, Mann-Whitney U test, or Kaplan-Meier survival curve. P values less than 0.05 were considered statistically significant.

2.3. Guidelines for Handling Rats. We followed the guidelines for experimental animal handling, and our study was approved by the Animal Care Committee of the Kyoritsu Women's University. The experiment was conducted in accordance with the National Institutes of Health (NIH) guidelines.

3. Results

3.1. The Effects of a Low-Salt Diet on Neonate Birth Rates. Dahl S rats fed a low-salt diet during the mating (early pregnancy) period delivered less pups compared with those fed a high-salt diet (Table 1). In Dahl S rats that were fed regular chow, 6 out of 7 mothers delivered pups, and there were no differences in birth rates between the high-salt rats and the rats fed regular chow.

Similarly, Dahl S rats fed the low-salt diet during the pregnancy period delivered less pups compared with those fed the high-salt diet. There were no differences in the birth rates between Dahl S rats that were fed the high-salt rats and those on regular chow.

The number of babies delivered from one mother was equal among the 5 mother groups (Table 1).

Table 1: Number of mothers that delivered pups.

Group	Diet: Mating (NaCl, w/w)	Diet: Pregnant (NaCl, w/w)	Mother: Total (rats)	Mother: Delivered (rats)	Mother: Babies/mother* (rats)	Statistics: χ^2	Statistics: P	Statistics: Versus
MHPR	4.0%	0.75%	7	7	7.4 ± 2.3			
MLPR	0.3%	0.75%	7	4	8.7 ± 1.7	3.82	0.050	MHPR
MRPR	0.75%	0.75%	8	7	8.2 ± 1.4	0.94	0.332	MHPR
						1.76	0.184	MLPR
						1.45	0.228	MRPH
						2.33	0.127	MRPL
MRPH	0.75%	4.0%	11	11	10.2 ± 1.3			
MRPL	0.75%	0.3%	11	6	8.8 ± 3.0	6.87	0.011	MRPH

MHPR: mothers fed the high-salt diet during mating period and the regular chow during pregnancy period; MLPR: mothers fed the low-salt diet during mating period and the regular chow during pregnancy period; MRPR: mothers fed the regular chow during mating period and during pregnancy period; MRPH: mothers fed the regular chow during mating period and the high-salt diet pregnant period; MRPL: mothers fed the regular chow during mating period and the low-salt diet during pregnant period. Total: the total number of mothers; delivered: the number of mothers that gave birth to pups. The differences in delivered mothers were assessed by χ^2 test (df = 1). 0.75% NaCl (w/w) chow is usually used for maintaining rat colonies.
*: number of rats delivered per mother.

Table 2: Mortality rates of the newborn pups until weaning.

Group	Diet (NaCl, w/w): Matching	Diet (NaCl, w/w): Pregnant	Rats born (mother*)	Survival rate during lactation: Alive (%)	Survival rate during lactation: Dead (%)	**Statistics: χ^2	**Statistics: P	**Statistics: Versus
MHPR	4%	0.75%	52 (7)	35 (67)	17 (33)			
MLPR	0.3%	0.75%	35 (4)	19 (54)	16 (46)	1.51	0.21	MHPR
MRPR	0.75%	0.75%	59 (7)	39 (66)	20 (34)	0.02	0.89	MHPR
						1.30	0.25	MLPR
						26.82	0.00	MRPH
						0.05	0.82	MRPL
MRPH	0.75%	4.0%	113 (11)	107 (95)	5 (5)			
MRPL	0.75%	0.3%	53 (6)	34 (64)	19 (36)	19.88	0.00	MRPH

MHPR: mothers fed the high-salt diet during mating period and the regular chow during pregnancy period; MLPR: mothers fed the low-salt diet during mating period and the regular chow during pregnancy period; MRPR: mothers fed the regular chow during mating period and during pregnant period; MRPH: mothers fed the regular chow during mating period and the high-salt diet pregnancy period; MRPL: mothers fed the regular chow during mating period and the low-salt diet during pregnancy period. Rats born: the total number of newborn pups; alive: the number of rats that were alive until weaning; dead: the number of rats that died before weaning. *: mothers that delivered babies. **The differences in alive pups were assessed by χ^2 test (df = 1). 0.75% NaCl (w/w) chow (regular chow) is usually used for maintaining rat colonies.

3.2. The Effects of a Low-Salt Diet on the Growth of the Neonates. The pups that were born in each group and the offspring that survived during the lactation period are shown in Table 2. There were no differences in the number of pups that survived the lactation period between Dahl S rats fed the low- and high-salt diets during the mating (early pregnancy) period.

In contrast, only 64% pups from mothers that were fed the low-salt diet during the pregnancy period survived the lactation period, while the survival rate was 95% for the pups whose mothers were fed the high-salt diet ($\chi^2 = 19.88$; $P < 0.0001$).

Using the data from the pups of mothers that were fed the low- and high-salt diets during the pregnancy period, we determined the Kaplan-Meier survival curve in order to increase our understanding of the time course before death (Figure 2). Some of the pups died immediately after birth and a substantial number of pups died during the lactation period.

We assessed pup growth during the lactation period (Table 3). Female pups born to mothers that were fed the low-salt diet tended to have lower body weights than pups born to mothers that were fed the high-salt diet. However, the difference was not significant. On the other hand, the body weights of the male pups from mothers that were fed the low-salt diet were equal to those of pups from mothers that were fed the high-salt diet.

4. Discussion

In the present study, we attempted to examine the role that salt intake has on the health of mothers and their children. We demonstrated that mothers on a low-salt (0.3% NaCl) diet during the mating pregnancy period produced lower number of pups than those on a high-salt (4% NaCl) diet. Since 7 out of 8 mothers that were fed regular chow throughout

TABLE 3: Body weights of pups during lactation.

	Day 2	Day 9	Day 16	Day 23	Day 30
LS-male	7.0 ± 0.2	$12.6 \pm 1.3^{*}$	$24.6 \pm 1.7^{**}$	40.0 ± 2.0	69.5 ± 3.3
HS-male	7.1 ± 1.1	14.4 ± 2.0	26.9 ± 1.8	41.2 ± 2.2	73.4 ± 4.4
LS-female	6.8 ± 0.3	$13.4 \pm 0.9^{**}$	25.0 ± 1.4	42.2 ± 2.9	$71.0 \pm 3.0^{**}$
HS-female	7.3 ± 1.1	$14.9 \pm 1.9^{**}$	26.3 ± 2.8	43.0 ± 2.6	$75.3 \pm 6.0^{**}$

LS: offspring of Dahl S rats fed a low-salt diet during the pregnancy period; HS: offspring of Dahl S rats fed a high-salt diet during the pregnancy period. The differences were assessed by ANOVA and followed by post hoc analysis. $^{*}P < 0.1$ and $^{**}P < 0.05$ versus offspring of mothers fed the high-salt diet.

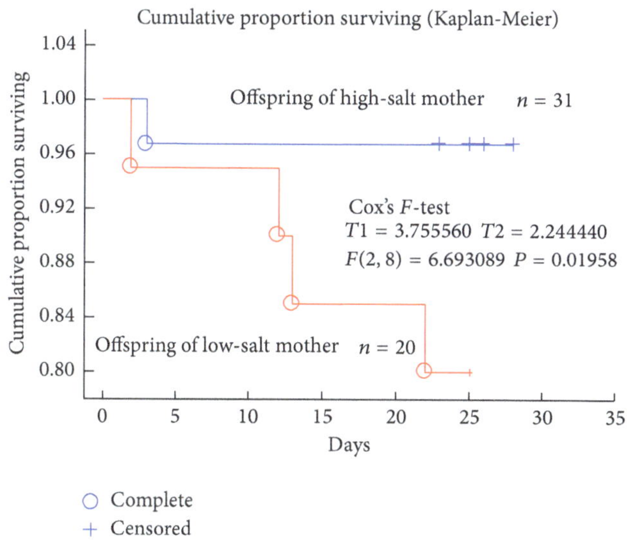

FIGURE 2: Cumulative proportion of offspring survival (Kaplan-Meier analysis). Survival prognosis was evaluated from their birth of the rat pups to weaning. Blue circles refer to the death (complete) of pups from high-salt intake mothers; red circles refer to the death (complete) of pups from a low-salt intake mothers. Plus (+), censor. The difference was analyzed by Cox's F-test ($T1 = 3.755$, $T2 = 2.244$, $F(2, 8) = 6.6930$, and $P = 0.01958$).

the mating and pregnancy periods produced babies, the low-birth rate of pups from mothers that were fed the low-salt diet during the mating period may be due to the failure to achieve fertilization. However, the mating season usually lasts 6-7 days in rats, and this means that time point of fertilization varied in mothers tested. The separation on the influence on fertilization and pregnancy process is technically difficult in the present experimental setting.

Similarly, the mothers that were fed the low-salt diet during the pregnancy period produced less pups than those that were fed the high-salt diet. As shown in mothers that were fed 0.75% NaCl throughout the mating and pregnancy periods, 7 out of 8 mothers produced pups, suggesting that the low-birth rate of pups from mothers that were fed the low-salt diet during the pregnancy period was due to a failure to maintain the pregnancy. In the present study, we were not able to conclude that fertilization is affected by the low-salt diet; however, it may be concluded that low-salt diet during the mating period and following pregnancy decreases birth rate.

In addition, we demonstrated that there were no differences in the number of babies born per pregnant mother among the experimental groups. These findings suggested that ovulation may not be affected by a low-salt diet.

The mechanism underlying why fewer pregnant mothers on a low-salt diet delivered pups is unclear. We speculate that low-salt intake enhanced the activity of the renin angiotensin system (RAS), and this may affect the intrauterine growth of pups. Moreover, some studies suggest that the overexpression of RAS transgenic mice results in a placenta-fetus imbalance and leads to preeclampsia [8–10]. In the present study, it is feasible to conclude that enhanced RAS activity following salt restriction led to a placenta and fetal imbalance, thereby decreasing the number of mothers that produced pups. Furthermore, recent studies have described the epigenetic mechanism for growth of the babies during pregnancy period [2, 4]. Events that are triggered after a low-salt challenge, that is, upregulation of the renin-angiotensin system, are believed to be involved in the epigenetic mechanism.

We demonstrated in the present study that rat pups born to mothers that were on a low-salt diet during a pregnancy period were more likely to die before weaning than those born to mothers that were fed a high-salt diet. However, it was also demonstrated that there were no differences in survival rate among the pups from mothers fed 0.75% NaCl or a low-salt diets during a pregnancy period. The 0.75% NaCl chow did not influence the birth rate of pups, but it lowered the survival rate of pups from mothers in comparison with pups from mothers fed a high-salt diet during a pregnancy period. This was demonstrated in Kaplan-Meier survival analysis. Since 95% of the pups from mothers on a high-salt diet during a pregnancy period survived until weaning, it was suggested that the regular (0.75% NaCl) diet was less likely to prevent the death of pups during a lactation period. The sensitivity of salt intake to benefits for mothers and pups differed.

In the present study we investigated Dahl S rats for the implications of mating or gestational salt intake in the birth and growth of the offspring. It is unclear whether salt sensitivity contributes to the relationship of mothers that were on a low-salt diet and their offspring. The future direction of research regarding salt intake will entail evaluating whether these findings are also true in other genetic rat strains of spontaneous hypertension.

We investigated the influence of salt intake on gestation and baby growth using Dahl S rats. Dahl S rats are a genetic model for salt-induced hypertension in humans [11, 12]. This strain develops hypertension with insulin resistance, depending on their salt intake. In addition, salt-induced hypertension is associated with target organ damage, that is, cerebral stroke, heart failure, and renal failure. These properties are similar to the conditions that develop in humans who have metabolic syndrome. There have been few studies on a public basis about salt sensitivity in Japan. However, it is believed that not less than 50% of Japanese people may be classified as salt sensitive. This figure is higher than that observed in Western countries [13, 14]. Based on the data, Dahl S rats are more appropriate for investigating the potential benefits or adverse effects of salt intake than the rat model for spontaneous hypertension (SHR).

The role of salt intake in the development of hypertension in humans has been the main theme over several decades, and this led to a world-wide campaign of salt reduction. However, the extent of salt reduction is controversial. Recently, a meta-analysis involving 25 studies reported that excessive salt intake is associated with an increased incidence of cardiovascular diseases [1]. Moreover, low-salt intake during gestation affects the development of fetuses and may contribute to hypertension or insulin resistance in their adulthood. In the present study, we provided evidence that the amount of salt intake for health benefits varies depending on each life stage. We may not extrapolate from the present results to humans. Investigation on humans is required.

In conclusion, low-salt intake during mating and pregnancy is associated with low-birth rates and high death rates among the surviving offspring. We urge additional investigations into the effects of paragestational salt reduction in humans as the level of salt intake that is required for optimal health varies depending on the subjects' age or stage of growth. Furthermore, the current finding provides evidence that supports the recommendation that the amount of salt consumed during pregnancy should be considered carefully based on data regarding the relationship between mothers and their offspring.

Abbreviations

Dahl S rats:	Dahl salt-sensitive rats
SHR:	Spontaneously hypertensive rats
RAS:	Renin angiotensin system
SD:	Standard deviation
ANOVA:	One-way factorial analysis of variance.

Conflict of Interests

The authors report no conflict of interests regarding the publication of this paper.

Acknowledgment

The authors acknowledge Grants-in-Aid from the Center for Interdisciplinary Studies of Science and Culture, Kyoritsu Women's University & Junior College. The authors alone are responsible for the content and writing of the paper.

References

[1] N. Graudal, G. Jurgens, B. Baslund, and M. H. Alderman, "Compared with usual sodium intake, low- and excessive-sodium diets are associated with increased mortality: a meta-analysis," *American Journal of Hypertension*, 2014.

[2] A. F. Vidonho Jr., A. A. da Silva, S. Catanozi et al., "Perinatal salt restriction: a new pathway to programming insulin resistance and dyslipidemia in adult Wistar rats," *Pediatric Research*, vol. 56, no. 6, pp. 842–848, 2004.

[3] S. M. Leandro, L. N. S. Furukawa, M. H. M. Shimizu et al., "Low birth weight in response to salt restriction during pregnancy is not due to alterations in uterine-placental blood flow or the placental and peripheral renin-angiotensin system," *Physiology and Behavior*, vol. 95, no. 1-2, pp. 145–151, 2008.

[4] K. L. Lopes, L. N. S. Furukawa, I. B. de Oliveira, M. S. Dolnikoff, and J. C. Heimann, "Perinatal salt restriction: a new pathway to programming adiposity indices in adult female Wistar rats," *Life Sciences*, vol. 82, no. 13-14, pp. 728–732, 2008.

[5] A. Shimo, "Have we been wrong about salt?" in CHATELAINE, September 2011, CA. Macleans, http://www.chatelaine.com/health/have-we-been-wrong-about-salt/.

[6] G. Piecha, N. Koleganova, E. Ritz et al., "High salt intake causes adverse fetal programmingvascular effects beyond blood pressure," *Nephrology Dialysis Transplantation*, vol. 27, no. 9, pp. 3464–3476, 2012.

[7] E. N. Alves-Rodrigues, M. M. Veras, K. T. Rosa et al., "Salt intake during pregnancy alters offspring's myocardial structure," *Nutrition, Metabolism and Cardiovascular Diseases*, vol. 23, no. 5, pp. 481–486, 2013.

[8] K. B. Brosnihan, L. Hering, R. Dechend, M. C. Chappell, and F. Herse, "Increased angiotensin II in the mesometrial triangle of a transgenic rat model of preeclampsia," *Hypertension*, vol. 55, no. 2, pp. 562–566, 2010.

[9] L. Anton, D. C. Merrill, L. A. A. Neves et al., "The uterine placental bed renin-angiotensin system in normal and preeclamptic pregnancy," *Endocrinology*, vol. 150, no. 9, pp. 4316–4325, 2009.

[10] D. M. Shah, "The role of RAS in the pathogenesis of preeclampsia," *Current Hypertension Reports*, vol. 8, no. 2, pp. 144–152, 2006.

[11] L. K. Dahl, "Effects of chronic excess salt feedings. Induction of self sustaining hypertension in rats," *Journal of Experimental Medicine*, vol. 114, pp. 231–236, 1961.

[12] J. P. Rapp, "Dahl salt-susceptible and salt-resistant rats. A review," *Hypertension*, vol. 4, no. 6, pp. 753–763, 1982.

[13] N. Sasaki, "High blood pressure and salt intake of the Japanese," *Japanese Heart Journal*, vol. 3, pp. 313–324, 1962.

[14] I. J. Brwon, I. Tzoulaki, V. Candeias, and P. Elliot, "Salt intakes around the world: implications for public health," *International Journal of Epidemiology*, vol. 38, pp. 791–813, 2009.

21

Effects of Canned Pineapple Consumption on Nutritional Status, Immunomodulation, and Physical Health of Selected School Children

Mavil May C. Cervo,[1] Luisito O. Llido,[1,2] Erniel B. Barrios,[3] and Leonora N. Panlasigui[1]

[1] *School of Nutrition, Philippine Women's University, 1004 Manila, Philippines*
[2] *Clinical Nutrition Services, St. Luke's Medical Center, 1102 Quezon City, Philippines*
[3] *School of Statistics, University of the Philippines-Diliman, 1101 Quezon City, Philippines*

Correspondence should be addressed to Mavil May C. Cervo; mavil_may.cervo@yahoo.com

Academic Editor: Phillip B. Hylemon

This randomized, controlled trial examined the effects of canned pineapple consumption on immunomodulation, nutritional status, and physical health of ninety-eight (98) school children with mean age of 8.44 ± 0.20. The study participants were divided into three groups: Group A (33) includes subjects who were not given canned pineapple, Group B (33) includes those who were given 140 g, and Group C (32) includes those given 280 g of canned pineapple for nine weeks. Each major group was further divided into two groups: normal (N) and underweight (U) based on 2007 WHO Growth Reference Standards. Sociodemographic, anthropometric, physical examination, dietary intake, hemoglobin level, and immunological data were analyzed. Results showed a decrease in incidence of viral and bacterial infections for both Group B and Group C (normal and underweight) after canned pineapple consumption. Granulocyte production increased by 0.77–26.61% for normal weight subjects and 14.95–34.55% for underweight. CD16+56 count augmented by 20.44–22.13% for normal weight and 3.57–15.89% for underweight subjects. Thus, intake of both one can (140 g) and two cans (280 g) of canned pineapple may shorten the duration and incidence of infection and may increase the production of granulocytes and CD16+56, but intake of two cans (280 g) demonstrated higher granulocyte and CD16+56 production. This trial is registered with Philippine Health Research Registry: PHRR140826-000225.

1. Introduction

Despite the various improvements in the treatment of perpetually evolving viral and bacterial infections, they are still considered as significant problems in health care today, especially for developing countries like the Philippines. According to UNICEF (2012), most child deaths result from one of the following five causes or combination of acute respiratory infections, diarrhea, measles, malaria, and malnutrition. Pneumonia continues to be one of the leading causes of death among children not only in the Philippines, but also around the world. Of the estimated 6.9 M child deaths each year, 18% or 1.2 M are due to pneumonia and 11% or 750,000 are due to diarrhea, highly concentrated in the poorest regions and countries [1]. Mortality due to childhood pneumonia and diarrhea is strongly linked to poverty-related factors like malnutrition as well as viral and bacterial infections.

Based on the report submitted by the National Statistical Coordination Board (NSCB) in 2010, 5.97% or 5970 of 5–9-year-old children died due to pneumonia in 2009 in the Philippines. It is also the second leading cause of mortality in 5–14-year-old children and the leading cause of death in 1–4-year-old children. Among the top 10 leading causes of death in children in the country, five of these are caused by viral and bacterial infections [2]. The incidence of these infections is influenced by the efficiency of both the innate and the adaptive immune system of individuals and their interactions. Thus, it is of great importance to examine and utilize factors that may strengthen the immune system in

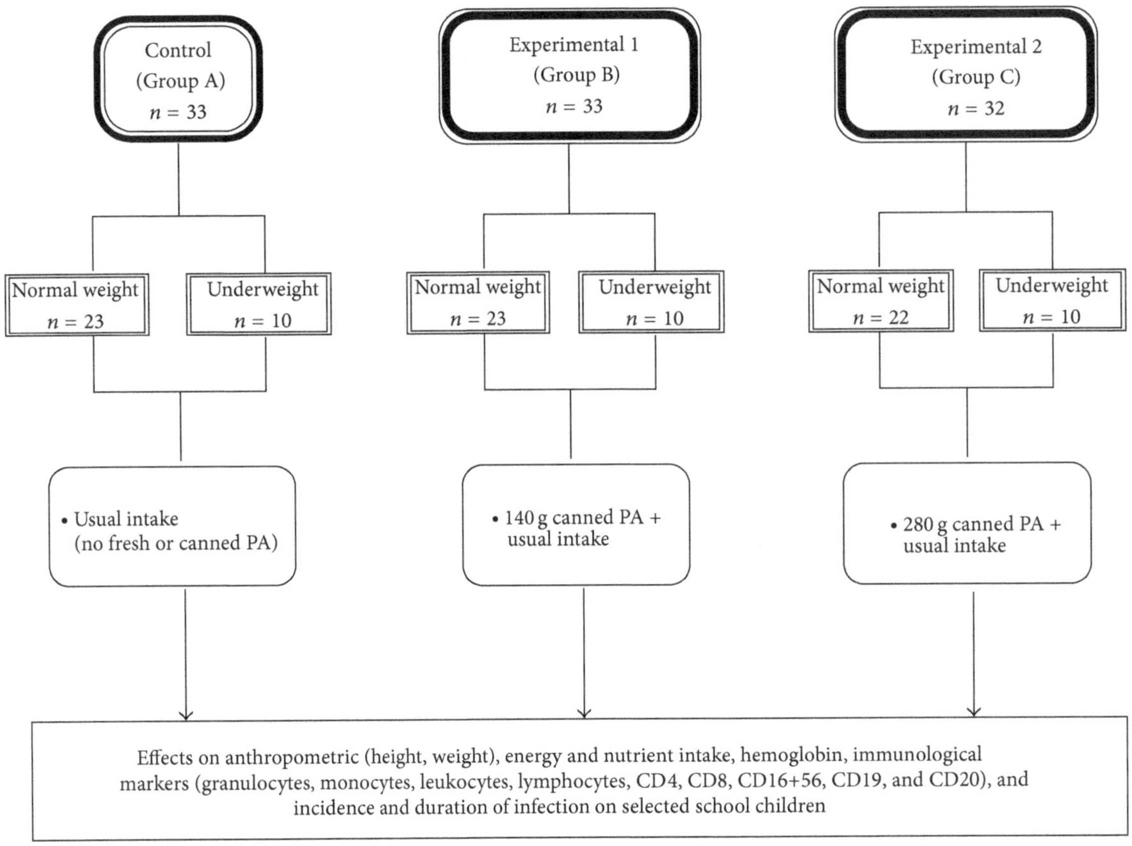

FIGURE 1: Experimental design ($n = 98$).

order to reduce the incidence of viral and bacterial infections and hence the overall mortality rates.

Among the known factors that influence the strength of the immune system is malnutrition. The relationship between macro- and micronutrient deficiencies and immunity has been clearly established. It has been observed that malnutrition appears to lead to increased infection in some instances and increased resistance to infection in others [3]. Growing evidence suggests that eating the recommended amount of fruits and vegetables protects against a number of diseases. The potential benefits of increased fruit and vegetable intake may have been contributed by the antioxidant and phytochemical content of these food groups.

Although several studies have been done on various fruits and immunity, the pineapple (*Ananas comosus*), the country's top produce, has not been examined in children. Most of the studies conducted on pineapple and immunity used bromelain, which is often extracted in pineapple's stem. Also, most of these studies were done in vitro or in mice. Pineapple contains several phytochemicals like coumaric acid, ferulic acid, chlorogenic acid, and ellagic acid, as well as micronutrients that have established influences such as vitamin C, manganese, thiamin, riboflavin, pyridoxine, copper, and dietary fiber. Regular intake of pineapple may increase consumption of these phytochemicals and micronutrients and potentially influence some immunological markers and help improve the child's physical health. Hence, consumption of pineapple may have potential benefits in the alleviation of viral and bacterial infections in the country which may lead to decrease prevalence of death among children. In lieu, this research was done to determine the effect of nine-week canned pineapple consumption on various immune markers, nutritional status, and physical health of selected school children.

2. Materials and Methods

2.1. Experimental Design (See Figure 1). This randomized, controlled trial involved 98 subjects which were divided into three major groups. Group A served as the control group or those subjects who were not given canned pineapple and maintained their usual intake. Group B served as the first treatment or those subjects who were given 140 g of canned pineapple, while Group C is the second treatment or those who were given 280 g of canned pineapple for nine weeks. Each major group was further divided into two subgroups: normal (N) and underweight (U) based on 2007 WHO Growth Reference Standards. Anthropometric parameters (weight and height), energy and nutrient intake, immunological markers (granulocytes, monocytes, leukocytes, lymphocytes, CD4, CD8, CD16+56, CD19, and CD20), hemoglobin, and incidence and duration of infection or illness were analyzed in the study.

2.2. Study Site Selection. The study was conducted in an elementary school (Fernando Ma. Guerrero Elementary School) in Paco, Manila, Philippines, with the highest number of students and the most number of students who are underweight based on the report submitted to the Department of Education-Division of Manila.

2.3. Subjects. A total of 98 students with mean age of 8.44 ± 0.20, divided randomly into three groups (A, B, and C), were included in the study. Each major group was further divided into two groups: normal (N) and underweight (U) based on 2007 WHO Growth Reference Standards.

The inclusion criteria were normal and underweight school children at the ages of 6–12 years old, having parents' or guardian's consent, having no known allergy to pineapple, being not included in the school's feeding program, and having no history of pineapple consumption for the past eight weeks. Subjects who fit the following criteria were excluded in the study: severely underweight, overweight, and obese children; menstruating; with previous history of any vitamin or mineral supplements or any pharmacotherapy within the past 2 months; having an acute or chronic infection or wound either at present or up to 2 weeks before the start of the study; having vaccination within 4 weeks before the start of the study or having any chronic or inflammatory disease; having undergone surgery; and any subject who will require hospitalization and aggressive treatment or medications.

2.3.1. Selection Process. Subjects underwent extensive selection process to ensure that they satisfy the set inclusion criteria. All students aged 6–12 from the identified elementary school were given a pretested and validated survey form to assess whether they fit the set inclusion criteria. The survey form was verified using a one-on-one interview with the subjects and their parents. Those who meet the criteria (256) were then subjected to anthropometric assessment (weight and height). Physical examination was also done to eliminate subjects who have infection or any form of disease. Out of 256, only 99 students satisfy the set inclusion criteria. Potential participants were oriented with their parents or guardians regarding the details of the study. Parents or guardians of amenable subjects signed a volunteer consent form that was patterned after the ethical research guidelines by the Institutional Ethics Review Committee. Parents were also asked to fill up the 3-day food record (two nonconsecutive weekdays and one weekend). Ninety-nine students were then divided into three groups. Each group has 33 students at first. However, one student from Group CN dropped out of the study because she stopped going to school. These three groups were further divided into two groups: those who are normal and those who are underweight. For each major group, twenty-three are classified as normal and 10 of which are classified as underweight.

2.4. Test Food. Group B was given 140 g (1 can) of canned pineapple (pineapple tidbits), while Group C was given 280 g (2 cans) of canned pineapple every day for nine weeks. The test food was produced by Del Monte plantation and cannery in Mindanao, Philippines.

2.5. Feeding Intervention. The intervention period (feeding) lasted for nine weeks. The control (Group A) was not given any intervention thus maintaining their usual dietary intake. Experimental (Groups B and C) was instructed to consume 140 g of canned pineapple or 280 g of canned pineapple, respectively, a day for nine weeks while maintaining their usual dietary intake. The test food was prepared in the school's canteen and delivered to all participants under the experimental group every morning. In order to ensure that the participants really consumed the test food, the subjects' respective teachers were asked to monitor their students. The researcher and helpers collect the containers and weigh if there are any leftovers. During weekends, all subjects under the treatment group were required to go to school to consume their test foods.

2.6. Data Collection and Analysis. Subjects under the experimental group were fed with canned pineapple every day for nine weeks. The participants were monitored and assessed using different parameters.

Medical history and family history of the subjects were taken before the start of the study to determine allergies or other diseases. Presence of blood coagulation problems is ruled out. Sociodemographic profile of the subjects was taken before the start of the study.

Anthropometric data include weight (measured using Platform balance, Detecto Eye Level Physician Scale, MO, USA) and height (using Microtoise, Tanita Corporation, Tokyo, Japan). Anthropometric data were gathered before and at the 4th and 9th weeks of study intervention.

Usual dietary intake was assessed using three-day food record (FR). Food record was used to quantify the actual food intake of the subjects during the study. This was taken before (baseline), 4th week (test 1), and 9th week (final) of study intervention. Each record includes two nonconsecutive weekdays and one weekend. FR data was validated through interview. The nutrient intake of the subjects was computed using the Philippine Food Composition Tables (FNRI-DOST, 1997) and Food Exchange Lists for Meal Planning (FNRI-DOST, 1994).

To determine the incidence of viral and bacterial infections, physical examination was done every week. Laboratory tests measure levels of metabolites, particularly in blood, to evaluate the body's state of health or its response to various treatments. In this study, hemoglobin and immunological tests were done before and after the study. Immunological test was done to determine the effect of canned pineapple consumption on immune system. It provides data on immunological markers such as CD4, CD8, CD19, CD20, and CD16+56. It also includes data on leukocyte, lymphocyte, granulocyte, and monocyte count. Since the immunological test is quite expensive and due to the limited budget, only five subjects per group (total of 30 subjects) were included in the immunological analysis.

Table 1: Baseline characteristics of subjects.

Parameters	AN		BN		CN		AU		BU		CU	
	Mean	SEM	Mean	SEM	Mean	SEM	Mean	SEM	Mean	SEM	Mean	SEM
Height (cm)	119.92	2.08	126.36	2.87	128.33	2.72	114.55	3.13	115.40	2.42	118.28	2.73
Weight (kg)	22.78	1.21	27.75	2.33	27.81	1.64	18.36	1.41	18.80	0.96	19.35	1.04
Immunological markers (cells/mL)												
WBC	6684.00	717.79	7508.00	716.45	7532.50	641.70	8015.00	728.13	7487.50	816.92	7577.50	873.11
Granulocytes	3347.00	510.13	3837.40	452.64	4121.75	285.08	3874.50	199.33	3952.00	657.58	4265.25	667.15
Lymphocytes	2748.00	340.59	3209.00	293.66	2884.25	344.97	3533.75	657.82	2926.50	154.88	2820.25	286.25
Monocytes	589.00	99.39	461.60	43.62	526.50	72.30	606.75	78.55	609.00	106.87	492.00	66.27
CD4	942.80	108.21	884.40	242.88	954.75	114.31	1189.25	255.82	956.25	62.67	989.00	121.30
CD8	716.00	60.86	1176.60	164.23	1036.25	180.35	1030.00	179.79	1055.75	67.78	865.25	168.02
CD16+56	444.46	66.65	386.60	128.67	385.00	54.27	531.00	63.29	462.00	135.16	440.00	52.45
CD20	639.80	75.00	647.20	60.97	490.75	67.58	697.50	138.03	597.25	60.92	602.75	139.94
CD19	641.40	75.60	635.80	66.05	481.50	74.01	769.00	199.55	595.50	65.69	601.75	132.05

Physical examination and blood extraction were done in the school auditorium. Ten milliliters of blood was extracted from the antecubital vein for immunological analysis. Immune panel data from the subjects' blood samples was obtained using flow cytometry analysis performed on a Becton-Dickinson FACScan (Becton-Dickinson, Mountain view, CA, USA). Hemoglobin and immunological analyses were done at St. Luke's Medical Center, Quezon City.

Each subject was given a notebook that serves as their diary to record possible discomfort related to gastrointestinal effects such as diarrhea, abdominal distension, flatulence, nausea, or any allergic reaction. Entries were checked every day after delivery of test foods to individual classroom of subjects.

All members of the research team have the necessary background and skills needed for the study. They were oriented on the data collection techniques to ensure the uniformity of the methods that were employed. The medical technologists as well as medical doctors were blinded to the group assignments of the study participants to prevent bias. All forms were pretested and validated prior to the actual study. Laboratory equipment was calibrated and underwent usual quality control methods. Collected data was reviewed for completeness and accuracy by the principal investigator.

2.7. Statistical Analysis. Changes in different parameters were determined from baseline until the end of intervention. Anthropometric, dietary (from food record), and biochemical results were presented as mean ± standard error (SE). Determination of the significant differences was analyzed using paired sample t-test and analysis of covariance (ANCOVA) in repeated measures and using Duncan multiple-range tests (DMRT). Multiple-linear regression was used to determine the relationship of nutrients on different immune markers. The SAS version 9.1 data analysis software (SAS Institute Inc., NC, USA) was used in testing the significance among the data. Microsoft Office Excel 2010 (Microsoft Corporation, WA, USA) was used in processing the other numerical data gathered. The level of significance is set at $P < 0.05$. Results of physical examination and sociodemographic data were analyzed using descriptive statistics and were presented in frequencies and percentages.

Table 2: Mean change in levels of innate immunological markers among groups.

Group	Immunological markers (cells/mL)			
	Granulocytes*	WBC	Lymphocytes	Monocytes
Normal				
AN	−507.97	259.64	559.72	18.20
BN	34.50	373.92	191.98	52.41
CN	1315.34	1294.63	323.97	2.07
Underweight				
AU	−203.04	297.50	240.00	16.25
BU	669.34	890.00	234.75	28.50
CU	1707.97	1777.50	541.00	131.00

*Significant at $P < 0.05$.

3. Results

3.1. Baseline Characteristics of Subjects (See Table 1). Table 1 shows the baseline characteristics of subjects based on anthropometric measurements and immunological markers.

3.2. Immunological Markers. Table 2 shows the average increase in levels of innate immunological markers. It is evident that granulocyte count is significantly affected by baseline level ($P < 0.0002$) as well as the treatment ($P < 0.0378$). Granulocyte count increases as canned pineapple consumption increases. Among the normal weight subjects, Group AN (control) exhibited a reduction in granulocyte count by 14.19% and those who have taken one can of pineapple (Group BN) yielded an increase of 0.77%. Group C or those who consumed two cans of pineapple produced

TABLE 3: Mean change in levels of adaptive immunological markers among groups.

Group	Immunological markers (cells/mL)				
	CD4	CD8	CD16+56*	CD20	CD19
Normal					
AN	84.65	187.20	−76.30	−45.80	−53.80
BN	−110.87	66.91	99.48	−106.02	−97.92
CN	63.88	38.25	110.64	−119.06	−119.06
Underweight					
AU	32.00	127.25	−8.32	−65.50	102.37
BU	93.37	115.25	17.37	103.36	−99.50
CU	42.75	44.00	79.46	163.76	−75.00

*$P < 0.05$.

26.61% more granulocytes during the study period, while underweight subjects produce more granulocytes with intake of canned pineapple. Group AU's granulocyte count significantly decreased by 5.67%, while Groups BU and CU escalated by 14.95% and 34.55%, respectively.

After nine weeks of study, white blood cells (WBCs) are affected by baseline level ($P < 0.0068$) but do not vary significantly across different treatments ($P < 0.3282$). Among the normal weight subjects, Group CN has the highest average increase in WBC count followed by Groups BN and AN, respectively. For those who are underweight, Group CU has the highest increase in WBC count followed by Groups BU and AU. Lymphocyte count is not affected by baseline level ($P < 0.6734$) and is not significantly affected by the treatment ($P < 0.4622$) as well. Group AN has the highest increase in level of lymphocytes followed by Group CN and Group BN among the normal subjects. Among underweight subjects, Group CU has the highest raise in lymphocyte count followed by Groups AU and BU, respectively.

Just like WBC, monocyte count is found to be affected by baseline level ($P < 0.0011$), but it was not significantly affected by the treatment ($P < 0.8274$). Among normal weight subjects, Group BN yielded the highest average increase followed by Groups AN and CN. For the underweight subjects, Group CU has the highest increase followed by Groups BU and AU.

Table 3 illustrates the average change in adaptive immunological markers. CD4 count is not affected by the baseline and treatment level at $P < 0.3834$ and $P < 0.4241$, respectively. Among normal subjects, Group AN has the highest average increase followed by Group CN, while CD4 count of BN decreased. Among underweight subjects, Group BU has the highest increase in CD4 count followed by Groups CU and AU.

Among subjects with normal weight, Group AN has the highest average increase in level of CD8 count followed by Group BN and Group CN, while, for those who are underweight, Group AU has the highest increase in CD8 count followed by Groups BU and CU. However, using the analysis of covariance, CD8 count is found to be not significantly affected by baseline level ($P < 0.4202$) and by treatment ($P < 0.5082$).

TABLE 4: Interaction of selected nutrients and immune markers.

Dependent variable	Independent variable	Coefficient	Probability*
Granulocyte	Vitamin C	86.75	0.0269
WBC	Vitamin C	103.45	0.0264
CD16+56	Iron	68.61	0.0080

*$P < 0.05$.

CD20 count for the normal weight groups diminished during the course of the study. Group CN has the highest average decrease followed by Groups BN and AN. Among underweight subjects, Group CU has the highest average increase followed by Group BU, while Group AU's CD20 count declined after nine weeks of study intervention. Though ANCOVA, the results were not affected by baseline or treatment at $P < 0.1516$ and $P < 0.7786$.

Though CD19 count was affected by baseline level ($P < 0.0426$) it was not significantly affected by the treatment ($P < 0.8332$). Like CD20, CD19 count of each group with normal weight decreased. Group CN has the highest average decrease in level of CD20 count followed by Groups BN and AN, respectively. Among underweight subjects, only Group AU's CD19 count increased, while those of Groups BU and CU diminished.

Among the CD markers, only CD16+56 count is significantly affected by both baseline and treatment at $P < 0.0238$ and $P < 0.0386$. Among the subjects with normal weight, Group CN's CD16+56 count augmented with an average of 22.13% followed by Group BN with 20.44%, while Group AN decreased with an average of 19.93%. For those who are underweight, Group CU's CD16+56 count escalated by 15.89% followed by BU with 3.57%. Just like with the normal weight subjects, Group AU declined by 2.17%. These data show that intake of canned pineapple can increase CD16+56 count by 20.44 to 22.13% for those who have normal weight and 3.57 to 15.89% for those who are underweight.

3.3. Nutrients and Immunological Markers (See Table 4). Based on the results of the linear regression analysis, as shown in Table 4, among the nutrients studied, only vitamin C was found to be associated with granulocyte count and white blood cell count. Results indicate that every 1 mg increase in vitamin C intake will result in an increase of 86.75 cells/mL increase in granulocyte count and an increase of 103.45 cells/mL in WBCs. In terms of CD16+56, only iron was found to be associated. In every 1 mg increase in iron intake, there will be 68.61 cells/mL increase in CD16+56.

3.4. Hemoglobin. Level of hemoglobin had increased for both normal and underweight subjects. Among normal subjects, Group CN has the highest increase by 0.60 g/dL followed by Groups BN (0.59 g/dL) and AN (0.30 g/dL), respectively, while for those underweight subjects, Group CU has the highest increase by 0.57 g/dL followed by Groups BU and AU with 0.40 g/dL and 0.19 g/dL, respectively. All data are found to be statistically significant except for Groups AU and BU (Table 5).

Table 5: Mean change in hemoglobin level among groups.

Group	Hg (g/dL) Baseline Mean	Baseline SEM	Posttest Mean	Posttest SEM	P value*
Normal					
AN	12.80	0.23	13.10	0.19	$P = 0.023^*$
BN	12.76	0.20	13.34	0.19	$P = 0.002^*$
CN	12.68	0.23	13.28	0.15	$P = 0.008^*$
Underweight					
AU	12.95	0.52	13.14	0.35	$P = 0.301$
BU	11.77	0.20	12.17	1.05	$P = 0.371$
CU	12.76	0.33	13.33	0.29	$P = 0.039^*$

$^*P < 0.05$.

3.5. Incidence and Duration of Infection (see Table 6). Physical examination of normal weight subjects shows that compared to control (Group AN), Groups BN and CN had zero incidence of infection at week 1 and had much lower incidences during the succeeding weeks which even became zero during the last week of the study intervention. The same thing goes for those subjects who are underweight. Group AU has the highest incidence of infection starting from week 1 till the final week of study intervention compared with Groups BU and CU. It is also noteworthy that the incidence of infection is high during the second, third, and fifth weeks of study intervention. This is probably due to occurrence of typhoons and climate change.

Aside from the incidence of infection, the duration or extent of an infection was analyzed for those who had an infection. For normal weight subjects, Group AN's average length or duration of infection lasted for 4–8 days with an average of 5.28 or 5 days. Groups BN and CN lasted for 3-4 days with an average of 3.26 and 3.30 days (3 days), respectively. For those who are underweight, Group AU's average duration of infection lasted for 4–8 days with an average of 5.62 or 5-6 days, while both Groups BU and CU lasted for 3–5 days with an average of 4.1 (4 days) and 3.5 (3-4 days), respectively. These data only reveal that the average duration of infection is much shorter for subjects who consumed canned pineapple compared with the control.

4. Discussion

Pineapple is a multiple fruit that is truly abundant in the Philippines. It is a good source of many nutrients such as vitamins B2 (riboflavin), C (ascorbic acid), manganese, and phytochemicals [4]. Consumption of pineapple and its components has been linked with immunity [5–18]. Some researchers also associated pineapple with cancer [19–23], wound healing [24–26], diabetes mellitus and liver disorders [27–31], diarrhea or other disorders of the gastrointestinal tract [32, 33], and even tuberculosis [34].

Microbial agents are encountered throughout life. Young children have an average of six to seven colds per year, but 10–15% of children have at least 12 infections per year [35]. In this study, incidence of viral and bacterial infections on both normal and undernourished children increased tremendously during the second, third, and fifth weeks of study intervention which may be associated with the occurrence of typhoons and climate change. Groups B and C, both normal and underweight, had zero incidence of infection at week 1 and have much lower incidences during the succeeding weeks which even became zero during the last week of the study intervention. Aside from the incidence of infection, the duration or extent of an infection was analyzed for those who had an infection. Data reveal that the average duration of infection is much shorter for subjects who consumed canned pineapple compared with the control. In addition, recovery also took much longer duration for underweight subjects than those who are normal in weight. Subjects mostly suffered from common colds, cough, fever, and upper respiratory tract infection during the study.

The human immune system has a central role in protecting against various infectious agents such as bacteria and viruses. The essential features of the immune system are an innate component that functions as a first-line of defense and an adaptive component that takes longer to mobilize but confers specificity and exhibits memory. Unlike adaptive immune system, which may take days to mobilize, innate system is extremely quickly mobilized [36].

Among subjects with normal weight, Group CN has the highest average increase in levels of immunological markers such as granulocytes and CD16+56. Group CN has four times higher on the level of granulocytes than Group AN (control). For the normal weight subjects, Group AN exhibited reduction in granulocyte count by 14.19%. Group BN yielded an increase of 0.77% and Group CN produced 26.6% more granulocytes during the study period. For the underweight subjects, Group AU's granulocyte count decreased by 5.67%, while Groups BU and CU increased by 14.95% and 34.55%, respectively. Findings suggest that canned pineapple can possibly be a booster in the production of granulocytes in both normal and underweight subjects. By consuming canned pineapple, normal weight subjects can boost their granulocyte count by 0.77–26.61% and 14.95–34.55% for those who are underweight. It also demonstrates that canned pineapple has a more pronounced effect on granulocyte production of underweight subjects compared with those who are normal in weight. Granulocytes play an important role in the innate immune system. It has a phagocytic capacity in destroying or killing viruses and bacteria and is also important in intracellular communication [37]. Granulocytes include neutrophils, eosinophils, and basophils.

In terms of CD16+56, Group AN showed a decline by 19.93%, while Groups BN and CN produced more CD16+56 by 20.44% and 22.13% correspondingly. For the underweight subjects, Group AU's CD16+56 count also declined by 2.17%, while Group BU and CU augmented by 3.57% and 15.89% respectively. Based on the results, it is also apparent that increasing consumption of canned pineapple can increase CD16+56 count for both normal and underweight subjects. CD16+56 mediates antibody-dependent cytotoxicity of foreign cells, phagocytosis, and other antibody-dependent responses and is also involved in the activation pathways of natural killer cells [36]. Though level of CD16+56 escalated

Table 6: Incidence of infection among groups.

Group	Students with infection (%)								
	Week 1	Week 2	Week 3	Week 4	Week 5	Week 6	Week 7	Week 8	Week 9
Normal									
AN	8.70	30.43	34.78	13.04	21.74	13.04	13.04	13.04	4.35
BN	0.00	13.04	4.35	0.00	13.04	4.35	4.35	4.35	0.00
CN	0.00	9.09	4.54	4.54	9.09	9.09	0.00	0.00	0.00
Underweight									
AU	30.00	30.00	30.00	10.00	30.00	30.00	10.00	10.00	20.00
BU	10.00	20.00	10.00	0.00	10.00	10.00	0.00	0.00	0.00
CU	10.00	10.00	0.00	10.00	0.00	0.00	0.00	0.00	0.00

on both Groups B and C (both normal and underweight subjects), it is evident that it is still within the normal range (212–612 cells/mL); thus, the increase in amount may not lead to autoimmunity.

Among the immunological parameters, only granulocytes and CD16+56 count are found to be statistically significant. This suggests that canned pineapple consumption has distinct effect on both innate and adaptive immune system. Though there is no significant effect on CD4 and CD8, which are often used in many studies, the significant increase in CD16+56 demonstrates the effect of canned pineapple consumption on the adaptive immune system. One important feature of both granulocytes and CD16+56 is their capacity to express natural killer cells which are important in killing certain tumor cells and are cytotoxic for both viruses and bacteria. Natural killer cells have receptor and inhibitory receptor and thus can selectively kill virus-infected and transformed cells while sparing normal cells [37].

In addition, it is best to remember that the innate and adaptive immune systems are interconnected and overlapping cells. The adaptive immune system is usually triggered by the innate immune system and only comes into play if the innate immune system fails to overwhelm the microorganism or the microorganism has found a way to avoid interaction with the innate system [38]. This means that, without the innate immune system, the adaptive immune system will not be able to do its job efficiently for the adaptive immune system depend on prior activation and participation of the innate immune system. As shown in this study, canned pineapple consumption was able to significantly increase the production of granulocytes which is an important marker of the innate immune system as well as CD16+56 which, on the other hand, is a marker of the adaptive immune system.

Hemoglobin level increased for both normal and underweight subjects especially to those who consumed canned pineapple. Before the study commenced, 20% of the subjects from Group BU had a hemoglobin level of less than 10 g/dL and therefore suffered from anemia. After nine weeks of canned pineapple consumption, their hemoglobin level became normal. The vitamin C content of the canned pineapple may be somehow responsible for this positive effect since vitamin C enhances iron absorption. Moreover, the increase in CD16+56 can also be a factor since this marker is essential in regulation of hematopoiesis.

The problem of disease and infection is multifaceted. Malnutrition, including both calorie and micronutrient deprivation, contributes to reduced immunity and increases the prevalence of infection. Infection and disease develop when the host's immune system is insufficient to entirely protect against potential pathogens. Once an infection is established, an array of immune responses is set into action to control the disease and avert it from reappearing. Based on the anthropometric data, there is a slight increase in height and weight among groups, which is expected since canned pineapple can only provide 76–152 kcal for one and two cans.

Food record data show that intake of subjects did not vary much all throughout the study period which also indicates that the improvement in the different parameters particularly on granulocyte and CD16+56 count can be attributed to the canned pineapple. Also, intakes of canned pineapple have contributed a marked increase in almost all nutrients especially riboflavin, iron, and vitamin C intakes among normal weight and underweight subjects when compared with baseline data. Forty-three percent of the subjects' mean vitamin C intake comes from canned pineapple (140 g) alone. However, most of the groups have inadequate consumption of calcium, iron, vitamin A, vitamin B1, and vitamin C throughout the study. This data reflects micronutrient deficiencies that may have direct effects on immunity and susceptibility to disease. These nutrients, with the exception of vitamin C, iron, and calcium, are not abundantly found in canned pineapples and thus inadequate intakes are not affected by the study. Moreover, the subjects' intake was already inadequate in calcium, iron, and vitamins A and C prior to canned pineapple supplementation. These micronutrients play an important role in immune system modulation. Vitamin A, for instance, is essential in the integrity of the immune system. All lymphocytes require vitamin A to develop and function properly. Consequently, vitamin A deficiency can alter the response of some antibodies to antigens and may also exert effects on the network of cytokines secreted during an immune response. Vitamin C is important in the biosynthesis of collagen and therefore promotes wound healing and also protects the skin. Furthermore, just like vitamin A, it is also an important antioxidant which protects the body from free radicals. Iron is essential for the normal function of the immune system. Since calcium is required in cellular movement, it is believed to have a vital role in increasing the body's resistance to infections. This inadequacy in micronutrients may somehow affect the results of the study particularly on the immune markers.

Among the nutrients studied, only vitamin C was found to be associated with granulocyte count and white blood cell count. Every 1 mg increase in vitamin C intake may lead to 86.75 cells/mL increase in granulocyte count and 103.45 cells/mL increase in WBCs. This result coincides with the findings of de la Fuente et al. [39] when administration of 1 g vitamin C to 30 subjects for 16 weeks increased the lymphoproliferative capacity and phagocytic function of neutrophils, which is a type of granulocytes. In terms of CD16+56, only iron was found to be associated ($P < 0.0080$). In every 1 mg increase in iron intake, there will be 68.61 cells/mL increase in CD16+56. This can also be due to the role of iron in normal functioning of the immune system.

Aside from the dietary intake and nutritional status of children, social and environmental influences have a major impact on the health and well-being of children. Family income is central to the health and well-being of children. Based on the sociodemographic form filled up by the subjects' parents, the mean monthly income is Php 6,434.00 with 43.3% of them having Php 2,500.00. With a minimum average of six members per household, Php 29.47 is spent for food per person daily. This means that some of these children have less access to nutritious foods which leads to poor health. This may also be the reason why subjects have inadequate intake of many micronutrients. Environmental factors such as living conditions as well as weather/climate may have affected the results of the study. Colds are common during rainy days because of the seasonal increase in viral infections particularly rhinovirus. The highest incidence of rhinovirus infection or common colds occurs from August to October. This can also be a factor why there is a remarkable increase in incidence of infection during the 2nd, 3rd, and 5th physical examinations.

Based on the results of the study, it can be concluded that canned pineapple consumption can lower the incidence of viral and bacterial infections and further increase the production of granulocytes by 0.77–26.61% for normal weight subjects and 14.95–34.55% for those who are underweight. It can also boost the production of CD16+56 by 20.44–22.13% for normal weight subjects and 3.57–15.89% for underweight subjects. Both one can (140 g) and two cans (280 g) of canned pineapple may increase the production of granulocytes and CD16+56, but two cans of (280 g) pineapple consumption demonstrated higher granulocyte and CD16+56 production.

The time of the study might not be ideal due to occurrences of typhoons. Furthermore, colds are common during rainy days because of seasonal increase in viral infection. For future studies, the researchers recommend to conduct the study during both rainy and dry seasons. More subjects should be recruited in the study and the intervention period should be much longer to see more significant results. Research using both fresh and canned pineapple fruit is recommended. Further studies on the effect and mechanism of pineapple and its component on nutritional status and immunology are also warranted.

Conflict of Interests

The authors declare that there is no conflict of interests. Del Monte Philippines Inc. provided the canned pineapple but had no role in study design, execution, data analysis, or interpretation.

Acknowledgments

The authors would like to thank the following because without their invaluable contribution this research would not have been possible: (i) the ninety-nine subjects who participated in the study together with their parents and family for their conscientious participation and cooperation, (ii) Fernando Ma. Guerrero Elementary School headed by its principal, Dr. Elizabeth R. Arrobang; school nurse, Mrs. Marife Silang; and all the teachers and the staff for their overwhelming support and use of facilities, (iii) Del Monte Philippines Inc. for supplying the test food and financial support, (iv) St. Luke's Medical Center and its four medical doctors and the medical technologists for diligently examining and extracting blood from the subjects, and (v) Mrs. Joy Noree Maramba-Turqueza, Mrs. Ailyn Kuan-Del Rio, and Mrs. Lorelie Teves-Maagma for their contributions during the collation and processing of data as well as the two assistants who are responsible for preparing the test food.

References

[1] UNICEF-The UN Inter-Agency Group for Child Mortality Estimation (IGME), *Levels and Trends in Child Mortality: Report 2012*, UNICEF-The UN Inter-Agency Group for Child Mortality Estimation (IGME), New York, NY, USA, 2012.

[2] National Statistical Coordination Board, "Ten Leading Causes of Child Mortality," January 2011, http://www.nscb.gov.ph.

[3] B. W. Ritz and E. M. Gardner, "Malnutrition and energy restriction differentially affect viral immunity," *Journal of Nutrition*, vol. 136, no. 5, pp. 1141–1144, 2006.

[4] A. Haripyaree, K. Guneshwor, and M. Damayanti, "Evaluation of antioxidant properties of phenolics extracted from *Ananas comosus* L," *Notulae Scientia Biologicae*, vol. 2, no. 2, pp. 68–71, 2010.

[5] H. Barth, A. Guseo, and R. Klein, "In vitro study on the immunological effect of bromelain and trypsin on mononuclear cells from humans," *European Journal of Medical Research*, vol. 10, no. 8, pp. 325–331, 2005.

[6] C. R. Engwerda, D. Andrew, A. Ladhams, and T. L. Mynott, "Bromelain modulates T cell and B cell immune responses in vitro and in vivo," *Cellular Immunology*, vol. 210, no. 1, pp. 66–75, 2001.

[7] D. J. Fitzhugh, S. Shan, M. W. Dewhirst, and L. P. Hale, "Bromelain treatment decreases neutrophil migration to sites of inflammation," *Clinical Immunology*, vol. 128, no. 1, pp. 66–74, 2008.

[8] L. P. Hale, P. K. Greer, and G. D. Sempowski, "Bromelain treatment alters leukocyte expression of cell surface molecules involved in cellular adhesion and activation," *Clinical Immunology*, vol. 104, no. 2, pp. 183–190, 2002.

[9] L. P. Hale, D. J. Fitzhugh, and H. F. Staats, "Oral immunogenicity of the plant proteinase bromelain," *International Immunopharmacology*, vol. 6, no. 13-14, pp. 2038–2046, 2006.

[10] L. P. Hale, M. Chichlowski, C. T. Trinh, and P. K. Greer, "Dietary supplementation with fresh pineapple juice decreases inflammation and colonic neoplasia in IL-10-deficient mice with colitis," *Inflammatory Bowel Diseases*, vol. 16, no. 12, pp. 2012–2021, 2010.

[11] J.-R. Huang, C.-C. Wu, R. C.-W. Hou, and K.-C. Jeng, "Bromelain inhibits lipopolysaccharide-induced cytokine production in human THP-1 monocytes via the removal of CD14," *Immunological Investigations*, vol. 37, no. 4, pp. 263–277, 2008.

[12] S. Mondal, S. Bhattacharya, and J. N. Pandey, "Evaluation of acute anti-inflammatory effect of Ananas comosus leaf extracts in rats," *Pharmacologyonline*, vol. 3, pp. 1312–1315, 2011.

[13] S. Müller, R. März, M. Schmolz, B. Drewelow, K. Eschmann, and P. Meiser, "Placebo-controlled randomized clinical trial on the immunomodulating activities of low- and high-dose bromelain after oral administration—new evidence on the antiinflammatory mode of action of bromelain," *Phytotherapy Research*, vol. 27, no. 2, pp. 199–204, 2013.

[14] M. P. Nantz, C. A. Rowe, C. Nieves Jr., and S. S. Percival, "Immunity and antioxidant capacity in humans is enhanced by consumption of a dried, encapsulated fruit and vegetable juice concentrate," *Journal of Nutrition*, vol. 136, no. 10, pp. 2606–2610, 2006.

[15] J. E. Onken, P. K. Greer, B. Calingaert, and L. P. Hale, "Bromelain treatment decreases secretion of pro-inflammatory cytokines and chemokines by colon biopsies in vitro," *Clinical Immunology*, vol. 126, no. 3, pp. 345–352, 2008.

[16] R. Pavan, S. Jain, Shraddha, and A. Kumar, "Properties and therapeutic application of Bromelain: a review," *Biotechnology Research International*, vol. 2012, Article ID 976203, 6 pages, 2012.

[17] E. R. Secor Jr., W. F. Carson IV, A. Singh et al., "Oral bromelain attenuates inflammation in an ovalbumin-induced murine model of asthma," *Evidence-based Complementary and Alternative Medicine*, vol. 5, no. 1, pp. 61–69, 2008.

[18] E. R. Secor Jr., S. J. Shah, L. A. Guernsey, C. M. Schramm, and R. S. Thrall, "Bromelain limits airway inflammation in an ovalbumin-induced murine model of established asthma," *Alternative Therapies in Health and Medicine*, vol. 18, no. 5, pp. 9–17, 2012.

[19] A. Amini, A. Ehteda, S. Masoumi, J. Akhter, K. Pillai, and D. L. Morris, "Cytotoxic effects of bromelain in human gastrointestinal carcinoma cell lines (MKN45, KATO-III, HT29-5F12, and HT29-5M21)," *OncoTargets and Therapy*, vol. 6, pp. 403–409, 2013.

[20] A. B. Awad and C. S. Fink, "Phytosterols as anticancer dietary components: evidence and mechanism of action," *Journal of Nutrition*, vol. 130, no. 9, pp. 2127–2130, 2000.

[21] R. Béez, M. T. P. Lopes, C. E. Salas, and M. Hernández, "*In vivo* antitumoral activity of stem pineapple (*Ananas comosus*) bromelain," *Planta Medica*, vol. 73, no. 13, pp. 1377–1383, 2007.

[22] S. Dhandayuthapani, H. D. Perez, A. Paroulek et al., "Bromelain-induced apoptosis in GI-101A breast cancer cells," *Journal of Medicinal Food*, vol. 15, no. 4, pp. 344–349, 2012.

[23] A. E. Millen, A. F. Subar, B. I. Graubard et al., "Fruit and vegetable intake and prevalence of colorectal adenoma in a cancer screening trial," *The American Journal of Clinical Nutrition*, vol. 86, no. 6, pp. 1754–1764, 2007.

[24] K. Aichele, M. Bubel, G. Deubel, T. Pohlemann, and M. Oberringer, "Bromelain down-regulates myofibroblast differentiation in an in vitro wound healing assay," *Naunyn-Schmiedeberg's Archives of Pharmacology*, vol. 386, no. 10, pp. 853–863, 2013.

[25] L. Rosenberg, Y. Krieger, E. Silberstein et al., "Selectivity of a bromelain based enzymatic debridement agent: a porcine study," *Burns*, vol. 38, no. 7, pp. 1035–1040, 2012.

[26] S.-Y. Wu, W. Hu, B. Zhang, S. Liu, J.-M. Wang, and A.-M. Wang, "Bromelain ameliorates the wound microenvironment and improves the healing of firearm wounds," *Journal of Surgical Research*, vol. 176, no. 2, pp. 503–509, 2012.

[27] W. Xie, D. Xing, H. Sun, W. Wang, Y. Ding, and L. Du, "The effects of *Ananas comosus* L. leaves on diabetic-dyslipidemic rats induced by alloxan and a high-fat/high-cholesterol diet," *The American Journal of Chinese Medicine*, vol. 33, no. 1, pp. 95–105, 2005.

[28] W. Xie, W. Wang, H. Su, D. Xing, Y. Pan, and L. Du, "Effect of ethanolic extracts of *Ananas comosus* L. leaves on insulin sensitivity in rats and HepG2," *Comparative Biochemistry and Physiology C: Toxicology & Pharmacology*, vol. 143, no. 4, pp. 429–435, 2006.

[29] W. Xie, W. Wang, H. Su, D. Xing, G. Cai, and L. Du, "Hypolipidemic mechanisms of *Ananas comosus* L. leaves in mice: different from fibrates but similar to statins," *Journal of Pharmacological Sciences*, vol. 103, no. 3, pp. 267–274, 2007.

[30] T. Dougnon, "Biochemical effectiveness in liver detoxication of fresh pineapple (Ananas comosus) with the wistar rats, previously intoxicated by Doliprane," *Journal of Cell and Animal Biology*, vol. 2, no. 2, pp. 31–35, 2008.

[31] M. P. Riya, K. A. Antu, T. Vinu, K. C. Chandrakanth, K. S. Anilkumar, and K. G. Raghu, "An in vitro study reveals nutraceutical properties of *Ananas comosus* (L.) Merr. var. *Mauritius* fruit residue beneficial to diabetes," *Journal of the Science of Food and Agriculture*, vol. 94, no. 5, pp. 943–950, 2014.

[32] F. Borrelli, R. Capasso, B. Severino et al., "Inhibitory effects of bromelain, a cysteine protease derived from pineapple stem (*Ananas comosus*), on intestinal motility in mice," *Neurogastroenterology and Motility*, vol. 23, no. 8, pp. 745–e331, 2011.

[33] D. S. Chandler and T. L. Mynott, "Bromelain protects piglets from diarrhoea caused by oral challenge with K88 positive enterotoxigenic *Escherichia coli*," *Gut*, vol. 43, no. 2, pp. 196–202, 1998.

[34] S. Mahajan, V. Chandra, S. Dave, R. Nanduri, and P. Gupta, "Stem bromelain-induced macrophage apoptosis and activation curtail mycobacterium tuberculosis persistence," *The Journal of Infectious Diseases*, vol. 206, no. 3, pp. 366–376, 2012.

[35] R. M. Kliegman, R. E. Behrman, H. B. Jenson, and B. Stanton, *Nelson Textbook of Pediatrics*, W.B. Saunders Company, 17th edition, 2003.

[36] R. Nairn and M. Helbert, *Immunology for Medical Students*, Mosby, Elsevier, 2002.

[37] I. M. Roitt and P. J. Delves, *Roitts Essential Immunology*, Blackwell Publishing, 10th edition, 2001.

[38] C. A. Janeway, P. Travers, M. Walport, and M. J. Shlomchik, *Immunology*, Garland Publishing, New York, NY, USA, 5th edition, 2001.

[39] M. de la Fuente, M. D. Ferrández, M. S. Burgos, A. Soler, A. Prieto, and J. Miquel, "Immune function in aged women is improved by ingestion of vitamins C and E," *Canadian Journal of Physiology and Pharmacology*, vol. 76, no. 4, pp. 373–380, 1998.

22

Accuracy of Self-Reported Physical Activity Levels in Obese Adolescents

Sarah A. Elliott,[1,2] Kimberley A. Baxter,[1] Peter S. W. Davies,[1] and Helen Truby[2]

[1] *Children's Nutrition Research Centre, Queensland Children's Medical Research Institute, Royal Children's Hospital, The University of Queensland, Herston, QLD 4029, Australia*

[2] *Department of Nutrition and Dietetics, Faculty of Medicine, Nursing and Health Sciences, Monash University, Notting Hill, VIC 3168, Australia*

Correspondence should be addressed to Sarah A. Elliott; sarah.alexandra.elliott@gmail.com

Academic Editor: Johannes B. van Goudoever

Introduction. Self-reported measures of habitual physical activity rely completely on the respondent's ability to provide accurate information on their own physical activity behaviours. Our aim was to investigate if obese adolescents could accurately report their physical activity levels (PAL) using self-reported diaries. *Methods.* Total energy expenditure (TEE) was measured using doubly labelled water (DLW) and resting energy expenditure (REE) was measured via indirect calorimetry. Activity energy expenditure (AEE) and PAL values were derived from measured TEE and REE. Self-reported, four-day activity diaries were used to calculate daily MET values and averaged to give an estimated PAL value (ePAL). *Results.* Twenty-two obese adolescents, mean age 13.2 ± 1.8 years, mean BMI 31.3 ± 4.6 kg/m^2, completed the study. No significant differences between mean measured and estimated PAL values were observed (1.37 ± 0.13 versus 1.40 ± 0.34, $P = 0.74$). Bland Altman analysis illustrated a significant relationship ($r = -0.76$, $P < 0.05$) between the two methods; thus the bias was not consistent across a range of physical activity levels, with the more inactive overreporting their physical activity. *Conclusion.* At an individual level, obese adolescents are unlikely to be able to provide an accurate estimation of their own activity.

1. Introduction

The accurate and reliable assessment of physical activity among children and adolescents is important for many reasons. Physical activity is often assessed in large population based studies to detect group changes in physical activity levels, estimate population prevalence and trends, and evaluate the efficacy of interventions which aim to alter physical activity levels. The assessment of physical activity can also be used to guide dietetic advice, by assisting with the calculation of individual energy requirements.

Numerous methods exist for the assessment of physical activity. Broadly, the various techniques can be grouped as self-report, observation, heart rate monitoring (HRM), motion sensors, and doubly labelled water (DLW), which measures total energy expenditure (TEE) over time [1]. While the DLW technique is considered the most accurate method available for the assessment of daily TEE, which includes energy spent in physical activity in free-living subjects, however, its high cost and demanding methodology render the DLW method unavailable for routine clinical use. Additionally, other techniques, such as HRM and motion sensors, also have methodological drawbacks [1]. Consequently, often pragmatic considerations such as cost and participant burden lead to self-report as the tool of choice. Diaries represent a low cost and time efficient way to record physical activity patterns, timing, and relative intensity of physical activity, compared with criterion measures [2].

Self-reported questionnaires are useful in large-scale studies, but they rely on the child (or parent) to recall physical activity behaviour information accurately. Numerous self-reported instruments are available, with several being used in children. The four major types are (i) self-administered recall, (ii) interview-administered recall, (iii) diaries, and (iiii) proxy reports [3].

While some studies have reported a 73.4% to 86.3% agreement between these instruments and direct observation, these techniques should be used cautiously in pediatric populations in which recalling such information may be difficult. More recent evidence of the reliability and validity of physical activity assessments in children has been examined by Kohl III et al. [4], who reported a low to moderate validity for self-reported measures of physical activity.

Evidence on the use of self-reported measures of physical activity in health children is contradictory. Sallis et al. [5] in 1993, suggested that physical activity recalls of children as young as the fifth grade are of adequate reliability and validity to use in research on physical activity in children. However, Arvidsson et al. [6] have documented that, in comparison to DLW measures of physical activity, the physical activity questionnaire for adolescents was not able to accurately predict physical activity in a group of Swedish adolescents. As a result of this conflicting evidence, Corder et al. [7], just recently, assessed the validity and reliability of four self-reports to assess physical activity and time spent at moderate and vigorous intensity physical activity in British young people between 4 and 17 years of age, against measures of DLW. The strength of association between questionnaire and criterion methods varied ($r = 0.09$ to $r = 0.46$). While some questionnaires were able to accurately assess group-level physical activity, the error was large for individual-level estimates.

While there have been a variety of studies investigating the use of self-reported measure of physical activity in healthy children and adolescents, little evidence of the use of such tools is available in obese children and adolescents. Dated studies by Heitmann and Lissner [8] and Waxman and Stunkard [9] reported that obese boys were less active than nonobese boys at home, but there were no differences in their activities when they were in a playground. It was also suggested that self-recorded activity records suffered from the same potential compliance problems as self-reported dietary intake and thus should be compared with more direct measures of activity [4, 10].

The accuracy of physical activity diaries, as stated by Sallis et al. [5], is highly dependent on the cooperation of the subject. In healthy children and adults, diary based self-reported instruments can provide, with good participation compliance, accurate assessments of habitual physical activity level (PAL) values [11]. However, self-reported measures of activity are fraught with difficulties in obese adult populations, who tend to overreport activity and underreport dietary energy intake [12]. It is unknown if these reporting difficulties are also found in obese children and adolescents. Thus, the aim of this study was to assess how accurately obese children and adolescents self-report their PAL values and to quantify the energy expended during physical activity.

2. Methods

Participants were aged between 10.00 and 17.99 years with a body mass index (BMI) greater than the 90th percentile for age and sex [13]. Subjects were obese but otherwise healthy, free from psychiatric morbidity, and not taking any medications known to alter metabolism or body composition. The experimental protocol was approved by the Royal Children's Hospital, Brisbane (2006/096), and the University of Queensland Human Ethics Committee (2007000797). Written informed assent and consent were obtained by all adolescents and their parents prior to the commencement of the study.

2.1. Anthropometry. Height was measured to the last completed millimetre using a wall-mounted stadiometer (Holtain Instruments Limited Crymych, UK). Weight was measured to the nearest 0.05 kg using calibrated electronic scales (Tanita BWB-600 Wedderburn Scales, Australia). Body mass index (BMI) was calculated (weight (kg)/height (m^2)) and Z scores (SD scores), for BMI, were calculated using the reference values for children from the Centers for Disease Control and Prevention [13]. Pubertal (Tanner [14]) stage was determined by a pediatrician.

2.2. Total Energy Expenditure. Total energy expenditure was measured using the DLW method as described in detail elsewhere [15]. A baseline urine sample was collected for the determination of the background isotope enrichment level. Subjects were then given an oral dose of DLW (^2H_2O and $H_2{}^{18}O$), approximately 0.083 g $^2H_2$ (99.8 atom% excess; Sigma Aldrich, Milwaukee, WI) and 2.083 g ^{18}O (10 atom% excess; Taiyo Nippon Sanso, Yokogawa, Japan) per kg of body water. Spot urine samples were collected after the dose at 5 hr and on ten consecutive days. Analysis of the isotopic enrichment was determined in duplicate with an Isoprime Dual Inlet Stable Isotope Ratio Mass Spectrometer (Isoprime Dual Inlet SIRM, MassLynx 4.0i Software, Isoprime, Manchester, UK) coupled in-line with a Multiprep-Gilson autosampler. Hydrogen analyses were done by a 3 hr equilibration with hydrogen gas at 40°C using Hokko coils. Oxygen analyses were done by a 10 hr equilibration with CO_2 at 40°C. All samples were analyzed in duplicate and laboratory standards were calibrated using the international suite of waters SMOW, SLAP, and GISP. Results were reported in ‰ (delta per mil units) relative to SMOW. Total energy expenditure was calculated using the multipoint method. The $^2H_2$ and ^{18}O zero-time intercepts and elimination rates (kd and ko) were calculated using linear regression of the log of isotopic enrichments, relative to predose enrichment using postdose urine samples over the 10-day sample period. The zero-time intercepts were used to determine the dilution spaces at the time of the dose. Total body water (TBW) was then calculated using the $^2H_2$ and ^{18}O dilutions spaces (^2H dilution space/1.04 and ^{18}O dilution space/1.01, resp.). The production rates of carbon dioxide (rCO_2) and water (rH_2O) from the isotope elimination rates (kd and ko) and TBW were determined by the method of Wells et al. [16]. Oxygen consumption was then determined using the following equation:

$$\text{Oxygen consumption (l/min)} = \frac{\text{carbon dioxide production (l/ min)}}{0.85}. \quad (1)$$

Table 1: Physical characteristics of all subjects.

	All (n 22) mean ± SD	Boys (n 11) mean ± SD	Girls (n 11) mean ± SD
Age (y.)	13.2 ± 1.8	13.4 ± 2.4	13.00 ± 1.5
Height (cm)	161.9 ± 12.1	162.5 ± 13.5	161.3 ± 11.1
Tanner stage (% 1, 2, 3, 4, and 5)	18, 9, 36, 9, 28	36, 0, 37, 18, 9	0, 18, 36, 0, 46
Weight (kg)	82.9 ± 20.5	82.8 ± 21.6	83.0 ± 20.5
BMI (kg/m^2)	31.3 ± 4.6	31.0 ± 4.22	31.7 ± 5.1
BMI Z score	2.16 ± 0.33	2.18 ± 0.36	2.13 ± 0.30
REE (kcal/24 hr)	2042 ± 442	2078 ± 452	2004 ± 452
TEE (kcal/24 hr)	2846 ± 995	3018 ± 1313	2657 ± 531
AEE (kcal/24 hr)	519 ± 452	638 ± 729	387 ± 26
PAL	1.40 ± 0.34	1.43 ± 0.40	1.36 ± 0.27
ePAL	1.37 ± 0.13	1.39 ± 0.10	1.35 ± 0.45
Active behaviour (hrs)	3.0 ± 0.1	3.1 ± 0.1	3.0 ± 0.3
Sedentary behaviour (hrs)	10.7 ± 2.0	11.4 ± 2.3	10.1 ± 2.1

ePAL, PAL estimated by diary.
Active behaviour was defined as any activity in which the MET value ranged from 3.5 to 5.0, such as playing outside and sports. Sedentary behaviour was defined as any activity in which the MET values ranged from 1.0 to 1.5, such as going to school, transportation, and screen time.

A respiratory quotient of 0.85 was used for each subject, as this value is indicative of the catabolism of a mixed diet over a period of days [17]. Using the production rate of carbon dioxide and the consumption rate of oxygen, TEE was then calculated using Weir's equation [17]. Consider

$$\begin{aligned} &\text{TEE (kcal/24 hrs)} \\ &\quad = 3.941 \times O_2 \text{ consumption } (\text{l/day}) \qquad (2) \\ &\qquad + 1.106 \times CO_2 \text{ production } (\text{l/day}). \end{aligned}$$

2.3. Resting Energy Expenditure. Resting energy expenditure (REE) was measured via indirect calorimetry using a ventilated hood system with the subject resting supine after an overnight fast (Deltratrac II Metabolic Monitor, MBM-200, Datex-Deltatrac, Datex-Engstrom Division, Helsinki, Finland). The calorimeter was calibrated with a reference gas mixture of 95.00% O_2 and 5.00% CO_2. O_2 consumption and CO_2 production were measured at one minute intervals for 30 minutes and averaged over the whole measurement period. The subjects were required to rest in a supine position for 20 minutes prior to the commencement of the test. The first five to ten minutes were excluded from the analysis to account for environmental adjustment by the subjects and gas adaptation in the hood. REE was calculated from the measured oxygen consumption and carbon dioxide production similar to TEE according to the formula by Weir [17].

By measuring both TEE and REE, we therefore derived a measured activity energy expenditure (TEE − REE = AEE) or PAL values (PAL = TEE/REE).

2.4. Physical Activity. Physical activity was measured via a four-day self-reported diary, using a simplified version of activity dairies as described by Bouchard et al. [18], in conjunction with the measurement period of TEE using DLW. Diaries were completed on three week days and one weekend day [18]. Subjects were given verbal and written instructions with an example of how to complete the diary. Each day was divided into 96 × 15 minute intervals and the subjects were asked to record their activities on each day. On completion, these activities were categorised into nine levels according to their average energy costs, representing multiples of their respective metabolic equivalents (METs) assigned as per Ainsworth et al. [19, 20]. Total daily METs values were calculated and averaged to give a PAL value for each subject. At the end of the recording period, the diary was checked with a researcher for completeness and any clarification of recorded activities was sought from parents of the child participating. Compliance with completing the diary was 100% as subjects were required to return the diary at a scheduled appointment.

2.5. Statistical Analysis. Statistical computation was performed using the SPSS for Windows statistical package (Version 18.0; SPPS Inc, Chicago, IL). In order to evaluate the agreement between the two measurement instruments, Bland-Altman analysis was used [21]. Level of significance was set at 5% ($P < 0.05$) for all comparisons. Paired t tests were used to compare measured and estimated TEE and PAL values. Results are expressed as means ± SD.

3. Results

Twenty-two subjects, aged between 10.03 and 16.00 years, completed the study prior to entry into the Eat Smart program. All subjects were overweight with BMI Z scores ranging from 1.43 to 2.61. Summary characteristics are presented in Table 1. There were no statistically significant differences in any of the physical characteristics, TEE, or REE between boys and girls. Similarly, when TEE and REE were

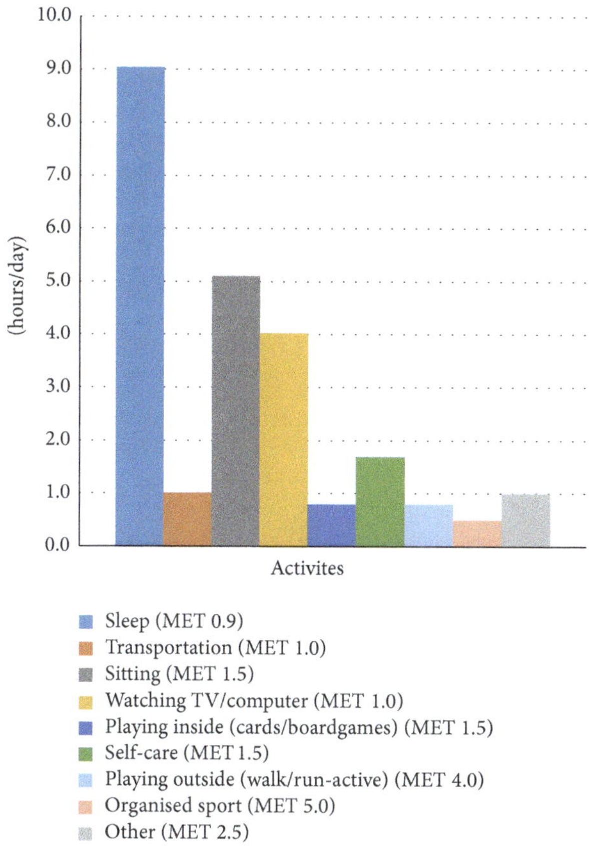

FIGURE 1: Time spent in physical activity estimated from physical activity diaries.

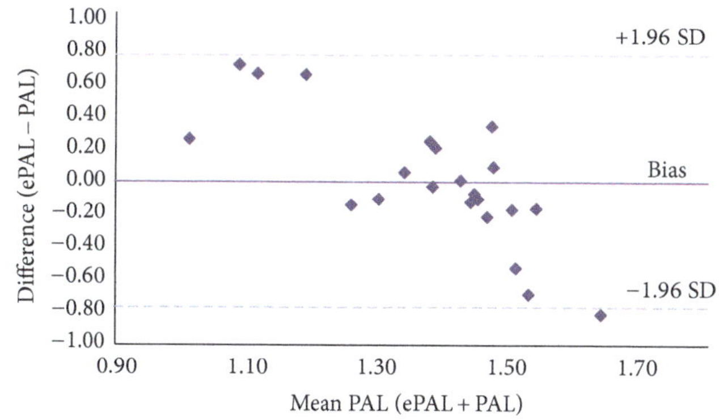

FIGURE 2: *Bland-Altman ePAL*. The Bland-Altman representation of the mean difference between estimated PAL values self-reported from physical activity diaries (ePAL) and measured PAL values obtained from measurements of TEE and REE (PAL = TEE/REE), in 22 obese children and adolescents.

adjusted for FFM, using log-log regression, no significant gender differences were apparent.

The proportion of energy used for physical activity (AEE) was, on average, 20% of their daily TEE. Time spent in sedentary activity (METs 1–1.5) occupied 45% of the day, while time spent in activity rated between 3.5 and 5.5 METs (i.e., playing outside and organised sport) occupied only 13% of the day. Figure 1 shows the hours spent in each physical activity diary category per day.

The group as a whole were sedentary, with girls reporting on average 4.7 hours per day of "screen time" and boys 3 hours per day. At a group level, there was no statistically significant difference between mean measured and mean estimated PAL values (1.37 ± 0.13 versus 1.40 ± 0.34, resp., $P = 0.74$). A Bland-Altman analysis illustrated a significant relationship ($r = -0.76$, $P \leq 0.05$) between the difference and the mean and between the two methods, demonstrating that the bias (−0.01 ± 0.39) was not consistent across the range of PAL values (Figure 2). Subjects with a low measured PAL tended to overreport their activity, while those with a higher measured PAL underreported their activity using the diary.

4. Discussion

Self-reported questionnaires are considered a useful tool to obtain qualitative data relating to habitual physical activity in children and adolescents. Physical activity diaries are often more economical and can provide information on the types and perceived intensity of activity not recorded from more objective measurement methods, such as accelerometers. While many physical activity diaries for the estimation of PAL values have been validated in healthy adults and children [22, 23], the use of these diaries to estimate PAL values in obese children and adolescents has not been previously reported.

In this present study, we compared estimated PAL values via self-reported diaries against energy expenditure measured via the reference DLW technique. Our findings show that obese adolescents tend to overestimate their PAL values when self-reporting their physical activity behaviours. At a group level, no statistically significant difference between measured and self-reported PAL values was observed. However, when determining the accuracy of these diaries to estimate PAL values using a Bland-Altman plot, correlation of significant magnitude was illustrated. Subjects with a low measured PAL tended to overreport their physical activity, following similar reporting patterns to obese adults [12].

Although in the presence of obesity the usefulness of self-reported measures of activity has been questioned, with obese adults shown to overreport their activity [24], many practitioners still utilise self-reported measures of activity to guide their choice of PAL value for the estimation of energy requirements.

The physical activity compendium developed by Ainsworth et al. [20], in 2002, was designed to facilitate the coding of physical activity diaries for the estimation of PAL values and provides activity codes and MET intensities for use in adults. Ridley and Olds [24] reported that using adult MET values was the most accurate assignment technique of energy costs to children's activities [25, 26], yet the energy cost of activity in obese children and adolescents may well be different. As mentioned previously, REE depends mainly on FFM, and as a result, in populations where FFM is altered, MET energy equivalents may vary significantly from the original kcal/kg/hour point of reference, as suggested by Ainsworth et al. [20].

A study by Ekelund et al. [27], investigating whether the intensity and duration of physical activity differed between obese and normal weight adolescents, has suggested that in obese children and adolescents the energy cost of any given physical activity is greater than in nonobese individuals. Moreover, Goran et al. [28] infer that perhaps the time spent in physical activity may be a more significant factor than energy expenditure attributable to physical activity, in the maintenance of energy balance.

The results from the present study, using a direct measure of physical activity, disagree with the findings of Goran et al. [28] and suggest that not only the amount of time spent participating in physical activity but also AEE might be associated with obesity and its development in children and adolescents. The mean PAL values for both boys (1.43) and girls (1.36) in the present study were much lower than the values previously reported in studies of children of similar ages [29–32]. In the current study, the mean values for AEE as a percentage of TEE (23.5% in males and 16.2% in females) were lower than those reported by Ekelund et al. [27] (38.7% in males and 36.3% in girls). The proportion of energy used for physical activity provided some insight as to whether physical activity, per se, could push the energy balance equation into deficit. The sample in this study used 20% of their daily energy expenditure for activity. In this sedentary population, a PAL value sufficient to push them into weight loss would require a substantial change in their activity. Schoeller [33] speculated that a relatively high PAL value of 1.75 is permissive for weight gain in a "society with unrestricted possibilities of excessive food consumption" [27, 33]. A recent study by Metcalf et al. [34] also implied that fatness leads to inactivity not the other way around. Currently, there is no consensus on whether energy intake or lack of energy expenditure is the driver for childhood obesity [35]. There is a need to develop a low cost, practical, and accurate measure of physical activity in obese children and adolescents. The accurate measurement of physical activity is critical for determining energy requirements, current levels of physical activity, monitoring compliance with physical activity guidelines, and understanding the dose-response relationship between physical activity and health.

One of the main limitations of this study was that data were captured at a single time point, when activity was objectively measured using DLW; hence, no data on reproducibility are available. Furthermore, the inference that subjects with a PAL value underreported their level of physical activity as demonstrated in Figure 2 needs to be interpreted with caution. While a significant trend was apparent as suggestive of the correlation between the mean and difference of the two methods, the group as a whole were very sedentary. Approximately 90% of subjects had a PAL value of less than 1.6. A "higher" observed PAL value may be because TEE was higher, due to the energy cost of activity being higher in obese, so whilst they do not think they are very active, the activity they do perform, cost more energy.

Conversely, given the poor performance of self-report, one would question the need of such investigation. The present study was also limited by its cross-sectional design. Thus, we cannot draw any conclusions as to whether an inactive lifestyle causes obesity or whether obesity leads to a physically inactive lifestyle. Additionally, the small sample size due mainly in part to the high cost of the 18-oxygen isotope when dosing large individuals meant that this was not a true validation study. However, one of the strengths of this study was that the major components of energy expenditure (TEE and REE) were measured using reference techniques; thus valid assessments of energy expended in physical activity could be accurately assessed.

5. Conclusion

Sedentary, obese adolescents tend to overestimate their activity levels or perhaps underreport the time spent in sedentary activity, such as screen time. In clinical practice, the misreporting of PAL values could have deleterious effects on any weight management programme, as energy requirements would be significantly overestimated or underestimated. It is therefore recommended that the use of a four-day self-reported physical activity diary for the estimation of a PAL value in obese children and adolescents is unreliable for this purpose.

Estimating the true energy cost of physical activity remains a challenge. It is therefore important that physical activity records are accurate for use in individuals, to assist in the prediction of energy requirements. Physical activity records have the unique advantage of providing additional information on the types of activity and time devoted by individuals to specific activities. However, as sedentary obese children and adolescents tend to overreport their PAL values, objective measurement instruments or techniques need to be developed, to be used either in place of, or in combination, with diaries.

Conflict of Interests

The authors declare that there is no conflict of interests regarding the publication of this paper.

Acknowledgments

The authors are grateful for funding from University of Queensland and the ANZ Trustees funding.

References

[1] J. Dollman, A. D. Okely, L. Hardy, A. Timperio, J. Salmon, and A. P. Hills, "A hitchhiker's guide to assessing young people's physical activity: deciding what method to use," *Journal of Science and Medicine in Sport*, vol. 12, no. 5, pp. 518–525, 2009.

[2] E. C. Rush, M. E. Valencia, and L. D. Plank, "Validation of a 7-day physical activity diary against doubly-labelled water," *Annals of Human Biology*, vol. 35, no. 4, pp. 416–421, 2008.

[3] J. F. Sallis, "Self-report measures of children's physical activity," *The Journal of School Health*, vol. 61, no. 5, pp. 215–219, 1991.

[4] H. W. Kohl III, J. E. Fulton, and C. J. Caspersen, "Assessment of physical activity among children and adolescents: a review and synthesis," *Preventive Medicine*, vol. 31, no. 2, pp. S54–S76, 2000.

[5] J. F. Sallis, M. J. Buono, J. J. Roby, F. G. Micale, and J. A. Nelson, "Seven-day recall and other physical activity self-reports in children and adolescents," *Medicine and Science in Sports and Exercise*, vol. 25, no. 1, pp. 99–108, 1993.

[6] D. Arvidsson, F. Slinde, and L. Hulthèn, "Physical activity questionnaire for adolescents validated against doubly labelled water," *European Journal of Clinical Nutrition*, vol. 59, no. 3, pp. 376–383, 2005.

[7] K. Corder, E. M. F. van Sluijs, A. Wright, P. Whincup, N. J. Wareham, and U. Ekelund, "Is it possible to assess free-living physical activity and energy expenditure in young people by self-report?" *The American Journal of Clinical Nutrition*, vol. 89, no. 3, pp. 862–870, 2009.

[8] B. L. Heitmann and L. Lissner, "Dietary underreporting by obese individuals—is it specific or non-specific?" *British Medical Journal*, vol. 311, no. 7011, pp. 986–989, 1995.

[9] M. Waxman and A. J. Stunkard, "Caloric intake and expenditure of obese boys," *The Journal of Pediatrics*, vol. 96, no. 2, pp. 187–193, 1980.

[10] C. W. Yu, R. Y. Sung, R. So et al., "Energy expenditure and physical activity of obese children: cross-sectional study.," *Hong Kong Medical Journal*, vol. 8, no. 5, pp. 313–317, 2002.

[11] C. Matthews, "Use of self-report instruments to assess physical activity," in *Physical Activity Assessments for Health-Related Research*, G. J. Welk, Ed., Human Kinetics Publishers, Champaign, Ill, USA, 2002.

[12] S. W. Lichtman, K. Pisarska, E. R. Berman et al., "Discrepancy between self-reported and actual caloric intake and exercise in obese subjects," *The New England Journal of Medicine*, vol. 327, no. 27I, pp. 1893–1898, 1992.

[13] R. Kuczmarski, C. Ogden, and S. Guo, "2000 CDC growth charts for the United States: methods and development. National Centre for Health Statistics," *Vital Health Statistics*, vol. 11, no. 246, pp. 1–190, 2000.

[14] J. Tanner, *Growth at Adolescence*, Blackwell Scientific, Oxford, UK, 2nd edition, 1962.

[15] P. Davies, "Measurement of energy expenditure and body composition using stable isotopes," *Developmental Physiopathology and Clinics*, vol. 2, no. 2, pp. 95–110, 1991.

[16] J. C. K. Wells, D. Haroun, J. E. Williams et al., "Evaluation of lean tissue density for use in air displacement plethysmography in obese children and adolescents," *European Journal of Clinical Nutrition*, vol. 65, no. 10, pp. 1094–1101, 2011.

[17] J. B. Weir, "New methods for calculating metabolic rate with special reference to protein metabolism," *Journal of Physiology*, vol. 109, no. 1-2, pp. 1–9, 1949.

[18] C. Bouchard, A. Tremblay, C. Leblanc, G. Lortie, R. Savard, and G. Thériault, "A method to assess energy expenditure in children and adults," *The American Journal of Clinical Nutrition*, vol. 37, no. 3, pp. 461–467, 1983.

[19] B. E. Ainsworth, W. L. Haskell, A. S. Leon et al., "Compendium of physical activities: classification of energy costs of human physical activities," *Medicine and Science in Sports & Exercise*, vol. 25, no. 1, pp. 71–80, 1993.

[20] B. Ainsworth, W. L. Haskell, M. C. Whitt et al., "Compendium of physical activities: an update of activity codes and MET intensities," *Medicine and Science in Sports and Exercise*, vol. 32, supplement, pp. S498–S505, 2000.

[21] J. M. Bland and D. G. Altman, "Statistical methods for assessing agreement between two methods of clinical measurement," *The Lancet*, vol. 1, no. 8476, pp. 307–310, 1986.

[22] L.-. Bratteby, B. Sandhagen, H. Fan, and G. Samuelson, "A 7-day activity diary for assessment of daily energy expenditure validated by the doubly labelled water method in adolescents," *European Journal of Clinical Nutrition*, vol. 51, no. 9, pp. 585–591, 1997.

[23] J. M. Conway, J. L. Seale, D. R. Jacobs Jr., M. L. Irwin, and B. E. Ainsworth, "Comparison of energy expenditure estimates from doubly labeled water, a physical activity questionnaire, and physical activity records," *The American Journal of Clinical Nutrition*, vol. 75, no. 3, pp. 519–525, 2002.

[24] K. Ridley and T. S. Olds, "Assigning energy costs to activities in children: a review and synthesis," *Medicine and Science in Sports and Exercise*, vol. 40, no. 8, pp. 1439–1446, 2008.

[25] B. Torun, "Energy requirements of children and adolescents," *Public Health Nutrition*, vol. 8, no. 7A, pp. 968–993, 2005.

[26] World Health Organisation, *Energy and Protein Requirements. Report of a Joint FAO/WHO/UNU Expert Consultation*, WHO Technical Report Series No. 724, World Health Organisation, Geneva, Switzerland, 1985.

[27] U. Ekelund, J. Åman, A. Yngve, C. Renman, K. Westerterp, and M. Sjöström, "Physical activity but not energy expenditure is reduced in obese adolescents: a case-control study," *The American Journal of Clinical Nutrition*, vol. 76, no. 5, pp. 935–941, 2002.

[28] M. Goran, W. H. Carpenter, A. McGloin et al., "Energy expenditure in children of lean and obese parents," *American Physiology Society*, vol. 268, no. 5, part 1, pp. e917–e924, 1995.

[29] L. G. Bandini, D. A. Schoeller, and W. H. Dietz, "Energy expenditure in obese and nonobese adolescents," *Pediatric Research*, vol. 27, no. 2, pp. 198–203, 1990.

[30] U. Ekelund, M. Sjöströ, A. Yngve et al., "Physical activity assessed by activity monitor and doubly labeled water in children," *Medicine and Science in Sports and Exercise*, vol. 33, no. 2, pp. 275–281, 2001.

[31] D. J. Hoffman, A. L. Sawaya, W. A. Coward et al., "Energy expenditure of stunted and nonstunted boys and girls living in the shantytowns of Sao Paulo, Brazil," *American Journal of Clinical Nutrition*, vol. 72, no. 4, pp. 1025–1031, 2000.

[32] A. E. Black, "Physical activity levels from a meta-analysis of doubly labeled water studies for validating energy intake as measured by dietary assessment," *Nutrition Reviews*, vol. 54, no. 6, pp. 170–174, 1996.

[33] D. A. Schoeller, "Balancing energy expenditure and body weight," *The American Journal of Clinical Nutrition*, vol. 68, no. 4, pp. 956S–961S, 1998.

[34] B. S. Metcalf, J. Hosking, A. N. Jeffery, L. D. Voss, W. Henley, and T. J. Wilkin, "Fatness leads to inactivity, but inactivity does not lead to fatness: a longitudinal study in children (EarlyBird 45)," *Archives of Disease in Childhood*, vol. 96, no. 10, pp. 942–947, 2011.

[35] S. N. Bleich, R. Ku, and Y. C. Wang, "Relative contribution of energy intake and energy expenditure to childhood obesity: a review of the literature and directions for future research," *International Journal of Obesity*, vol. 35, no. 1, pp. 1–15, 2011.

23

Acute Effects of Exogenous Hormone Administration on Postprandial Acylation Stimulating Protein Levels in Ovariectomized Rats after a Fat Load

Bashair Al Riyami,[1] Marah El-Tahir,[2] Sultan Al Maskari,[1] Eugene H. Johnson,[3] and Jumana Saleh[1]

[1] *Department of Biochemistry, College of Medicine, Sultan Qaboos University, Al-Khod, 123 Muscat, Oman*
[2] *Department of Microbiology and Immunology, College of Medicine, Sultan Qaboos University, Al-Khod, 123 Muscat, Oman*
[3] *Department of Animal and Veterinary Sciences, College of Agricultural and Marine Sciences, Sultan Qaboos University, Al-Khod, 123 Muscat, Oman*

Correspondence should be addressed to Jumana Saleh; jumanasaleh@hotmail.com

Academic Editor: Duo Li

Background. ASP, a potent lipogenic factor, was linked to female fat metabolism in association studies. *Aim.* To investigate acute effects of sex hormone treatment on postprandial ASP levels *in vivo*. *Methods.* 24 female rats were randomly divided into 4 groups including controls. The rats were ovariectomized and injected with progesterone, estrogen, or testosterone. An hour later, olive oil was administered orally. Plasma ASP and triglycerides were measured at several postprandial time points. Area under the curve (TG-AUC) represented TG clearance. *Results.* Only the progesterone treated group had a significant postprandial ASP increase at two hours compared to basal levels (439.8 ± 62.4 versus $253.4 \pm 59.03\,\mu$g/mL, $P = 0.04$). Interestingly, increased ASP levels coordinated negatively with corresponding TG levels and TG-AUC postprandially, mostly evident in the opposite effects in the progesterone and testosterone treated groups. ASP levels increased 3-fold in the progesterone versus testosterone treated groups, whereas TG-AUC was significantly lower. *Conclusion.* These findings suggest that progesterone enhances ASP production and TG clearance simultaneously, supporting the notion of a stimulatory role for progesterone on ASP-mediated TG clearance. This is the first functional study demonstrating a cause-effect relationship between hormone treatment and ASP levels *in vivo* and may contribute to understanding the mechanism of progesterone function as a female lipogenic hormone.

1. Introduction

Acylation stimulating protein (ASP) is a small basic protein that consists of 76-amino acids produced from adipocytes. ASP is produced by the interaction of complement C3, factor B, and adipsin produced in adipose tissue [1]. Despite numerous identified adipose tissue derived factors, ASP uniquely demonstrated a great effect in lipogenesis stimulation. ASP actively stimulates TG synthesis in human fibroblasts and to a much greater extent in adipocytes [2]. This occurs through the enhanced esterification of fatty acids in adipose tissue by increasing the activity of diacylglycerol acyltransferase (DGAT), the rate limiting enzyme in TG synthesis, and indirectly enhances the activity of lipoprotein lipase (LPL) in adipocytes by increasing fatty acid trapping within adipocytes [3]. ASP was also found to inhibit hormone-sensitive lipase (HSL) [4]. Interestingly, ASP was shown to be as potent as insulin in its fat storage effects. However, ASP effects were independent, and additive, to insulin. Dietary triglyceride rich lipoproteins, chylomicrons (CHYLO) were the most potent stimulators of ASP production *in vitro* [5]. In animal studies, using a C3 knockout (ASP deficient) mice showed delayed postprandial TG clearance and reduced TG storage capacity after a fat load [2, 6], which highlighted the important role of ASP in the postprandial TG clearance. Such an effect was diminished by the administration of exogenous ASP accelerating the clearance of free fatty acids and TG from the circulation after oral fat administration [2]. ASP treatment stimulated weight gain and decreased energy expenditure in the mice [7]. In humans, after a fat load, the production of ASP increased postprandially in correlation with plasma TG clearance [8]. Adipose metabolism is mediated through a complex network of hormonal signals including sex hormones [9]. The relationship between sex hormones and ASP production and function is not yet understood. Extensive studies exploring the role of sex hormones in rodents and humans have demonstrated the involvement of sex hormones in the modulation of adipose tissue metabolism through lipolytic and lipogenic pathways. Estrogen has been described as a lipolytic hormone [10], while progesterone was shown to enhance fat storage and exhibit lipogenic effects [11]. The role of sex hormones in fat distribution remains poorly understood and it is mainly attributed to their effects on lipogenic hormones, transcription factors, and enzymes that act together to determine the outcome. The effects of sex hormones on fat storage factors, such as insulin, have been investigated in several studies; however, no clear connection has been established due to conflicting evidence [12]. Recent studies have shown novel findings linking ASP to fat metabolism and hormonal changes in females. ASP was found to strongly associate with fluctuating progesterone levels in females across the menstrual cycle whereas no association was found with insulin [13]. At late gestation, ASP levels increased significantly along with natural increases in female hormones, particularly progesterone which increases late during pregnancy, whereas maternal triglyceride levels also strongly correlated with fetal birth weight [14]. Interestingly, also cord blood ASP levels correlated positively with fetal birth weight [15]. In addition, elevations in ASP levels were seen in women with metabolic diseases that included hormonal disturbances such as polycystic ovary syndrome (PCOS) [16, 17], suggesting that ASP metabolism may be altered in association with sex hormones disturbances. The connection of ASP to female fat metabolism and hormonal changes is collated in a recent review [18]. Although associations of ASP levels with female hormone alterations have been reported, a cause-effect relationship of sex hormone and ASP production and function has not been established. Therefore, animal studies investigating the direct role of sex hormones on the function of ASP in postprandial fat clearance may be of significant importance in this regard. In this study, we investigated the effect of exogenous hormone administration (of different sex hormones) on postprandial ASP and TG after a fat load in ovariectomized rats.

2. Materials and Methods

2.1. Animals. All animal experiments carried out in this study were handled and conducted in the small animals' house and approved by the Animal Ethics Committee at Sultan Qaboos University. The animals were housed in a temperature controlled room (23°C) with a 12 : 12 light : dark cycle and free access to food and water.

2.2. Experimental Design. 24 female Wister rats (8 weeks of age) were randomly divided into 4 groups ($n = 6$). All rats were ovariectomized (OVX). On day 14 after the OVX procedure, a postprandial experiment was conducted under different hormone treatments. The concentrations of ovarian steroids were expected to reach nadir within 10 to 23 days after the OVX procedure [19], which was confirmed before running the main experiment on a different set of rats.

2.3. Surgical Procedure. At age of 8 weeks, rats were ovariectomized under general anesthesia using ketamine/xylazine combination, dosed at 75 mg/kg body weight and 6 mg/kg body weight. The ovariectomy procedure was conducted as described by Waynforth and Flecnell [20]; briefly, a small midline dorsal skin incision (1-2 cm) was made halfway between the caudal edge of the ribcage and the base of the tail. A small cut then was made into the muscle wall and the ovary was gently pulled and a ligature was placed below the ovary, and the ovary was then removed. The remaining tissue was replaced into the peritoneal cavity. Muscle layer and skin were sutured. The ovary on the other side is removed in a similar manner.

2.4. Hormone Treatments. After surgery, ovariectomized rats were allowed to recover from surgery stress for 14 days. All hormones were purchased from Sigma Aldrich. All hormones were delivered by sesame oil subcutaneously. The progesterone and estrogen doses used in this experiment were modified to maintain acceptable high physiological levels that mimic late gestation pregnancy in rats (Table 1) [21–24].

2.5. Postprandial Fat Clearance. Fourteen days after being ovariectomized, fasting rats were injected subcutaneously with single dose of different hormone treatments delivered in 0.2 mL sesame oil. A control group received sesame oil injection only. One hour later, a fat load consisting of 1.5 mL olive oil was administered by oral gavage [25, 26]. Plasma blood samples were collected at basal levels (time 0) followed by 2, 4, 6, and 8 hours after oil administration. Blood was collected into EDTA containing tubes, centrifuged at 2000 rpm for 10 minutes at 4°C. Plasma was isolated and stored at −80°C until analysis.

2.6. Analysis. Plasma ASP was measured in plasma using rat C3a ELISA (Novatein Biosciences Inc., USA). TG levels were measured by enzymatic colorimetric method using commercial assay kits (Roche) and automated *cobas 111c analyzer* (Roche). All protocols were conducted according to the kits manufacturer's instruction.

2.7. Statistical Analysis. Differences in ASP and TG baseline levels and the weights of rats (controls and hormone treated groups) were compared by one-way ANOVA followed by Bonferroni post analysis based on the assumption of homogeneity of variances for the tested groups. Tamhane's T2 post hoc analysis was used when the assumption of equal variances was not met. To evaluate the effect of hormonal intervention on ASP level changes over the postprandial period, two-way

TABLE 1: Hormones concentrations delivered in 0.2 mL sesame oil.

Treatment	n	Dose
Control-OVX	6	0.2 mL oil
Progesterone-OVX	6	4 mg/0.2 mL oil
17-β estradiol-OVX	6	5 μg/0.2 mL oil
Progesterone and 17-β estradiol-OVX	6	5 μg and 4 mg/0.2 mL oil
Testosterone propionate	6	20 μg/0.2 mL oil

TABLE 2: Mean weight and age of rats ($n = 6$) and basal ASP (μg/mL) and TG (mmol·mL) levels after hormone treatment.

	Age (weeks)	Weight (g)	ASP (μg/mL)	TG (mmol/L)
Control	10	214.6 ± 4.3	33.7 ± 4.9	0.74 ± .048
Progesterone	10	211.7 ± 4.8	32.3 ± 2.6	0.63 ± .067
Estrogen	10	195.1 ± 3.9	82.9 ± 22.2	0.55 ± .028
Testosterone	10	205.8 ± 6.4	85.2 ± 25.1	0.59 ± .039

One-way ANOVA, followed by post hoc analysis, showed no significant differences in rat weights at the time of OVX and at the day of the experiment. In addition, no significant differences for basal ASP or TG levels were detected before the experiment ($P < 0.05$).

repeated measures analysis of covariance (2-way ANCOVA) was performed for comparisons in ASP levels between the hormone treated groups after adjusting for weight as a covariate. The required assumption of sphericity was ensured by Mauchly's test. Greenhouse-Geisser correction was applied when sphericity was not met. Analysis included hormone treatments (controls, progesterone, estrogen, and testosterone) as the "between subjects factors" and time as the "within subjects factor" including basal, 2, 4, 6, and 8 hours. A significant (hormone treatment × time) interaction was followed by further breakdown of the analysis by performing one-way RM ANOVA to investigate postprandial ASP changes over time for each hormone treatment independently. Significant differences in ASP levels from basal were reported based on Bonferroni adjusted pairwise comparisons. ASP levels in the analysis were presented as mean differences from basal. For postprandial triglyceride (TG) changes (differences from basal), the area under the curve (TG-AUC) was measured to represent triglyceride clearance. TG-AUC was determined with the linear trapezoidal method calculated by Microsoft Excel Software. Differences in TG clearance between the different hormone treated groups were compared by one-way ANOVA. The correlation of mean ASP levels at specific time points with corresponding TG-AUC for each hormone treated group was presented in a simple scatter graph. Data in this study was analyzed using SPSS 17 statistical software package with significance set at $P < 0.05$.

3. Results

3.1. Postprandial ASP Level Variations. The rat's weights, basal ASP, and TG levels are presented in Table 2. One-way ANOVA showed no significant differences in rat weights at the time of ovariectomies or at the day of fat load administration. In addition, no significant differences in basal ASP or TG levels were detected. Two-way RM-ANCOVA demonstrated a main significant interaction between all hormone treatments and time (hormone treatment × time); $F_{12,72} = 2.04$, $P = 0.033$. This indicated a significant main hormone effect on ASP levels over the postprandial time course. Postprandial ASP level changes are presented in Figure 1. Further breakdown of the analysis by one-way RM ANOVA showed that ASP postprandial changes in the control group, which appeared as a biphasic increase across the postprandial time course, did not reach statistical significance ($P = 0.10$). Similarly, ASP levels in the estrogen and testosterone treated groups across the postprandial course were not significant, and therefore no further analysis was warranted in the repeated measures design. Only the progesterone treated group showed a significant ASP postprandial increase over basal levels at 2 hours ($P = 0.012$), and when compared separately to the controls by 2-way RM ANCOVA, ASP levels in the progesterone treated group were markedly higher than the controls at 2 hours (439.8 ± 62.4 versus 253.45 ± 59.03), $F_{4,32} = 2.8$, $P = 0.04$, *partial eta squared* = 0.26. ASP levels gradually decreased at 4 and 6 hours and were back to basal levels at 8 hours. Furthermore, ASP level differences at each postprandial time point were compared for the different hormone treated groups by one-way ANOVA. Post hoc comparisons showed that the greatest difference in ASP levels was between the progesterone and testosterone treated groups at 2 hours, where ASP levels in the progesterone treated group was about 3-fold ASP levels in the testosterone treated group (Figure 3(a)).

3.2. Postprandial TG Level Variations. On the other hand, postprandial TG levels over the time course are shown in Figure 2. Two-way RM-ANOVA showed significant (hormone × time) interactions across the postprandial course between the different groups reflecting a main effect of hormone treatment on postprandial TG levels over time. However, in this study, the main focus was on "TG clearance" which was represented by measuring TG-AUC. One-way ANOVA analysis showed a main significant difference in TG-AUC values between the groups, $F = 3.17$, $P = 0.047$ (Figure 3(b)). Similar to findings on ASP levels at 2 hours, post hoc comparisons of TG-AUC between the different hormone treatments showed that the greatest TG-AUC difference was between the progesterone and testosterone treated groups. However, the trend was opposite to the results seen for ASP levels at 2 hours, as TG-AUC in the progesterone treated group was significantly lower compared to the testosterone treated group representing increased TG clearance in the progesterone treated group compared to the testosterone treated groups that displayed delayed clearance.

3.3. Relation of ASP Levels in the Hormone Treated Groups with Corresponding TG Clearance. The opposite trend seen for the ASP levels at 2 hours and corresponding TG-AUC for the different hormone treatments (Figures 3(a) and 3(b)) are further displayed in a scatter graph demonstrating a strong negative association between the means of ASP levels

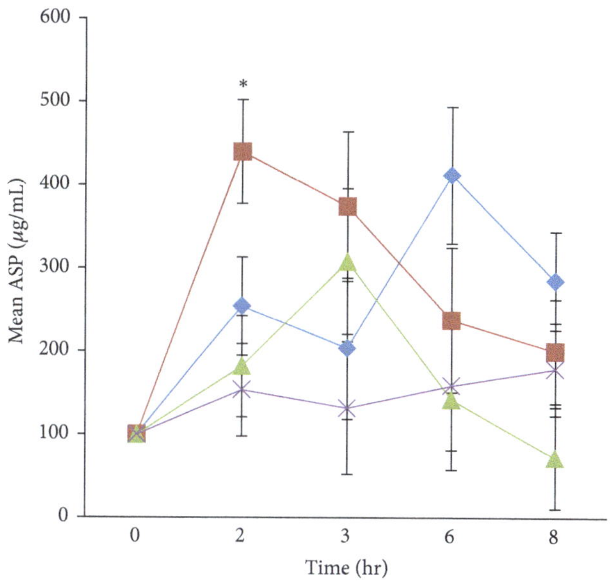

FIGURE 1: Changes in plasma ASP concentrations after an oral fat load for different hormone treated groups and control group (each group, $n = 6$). Results at each time point during postprandial period are expressed as means ± SEM (μg/mL) of difference in ASP levels from basal (by subtracting basal value at time zero (as shown in Table 2)) from each time point, and percentage increase was calculated. Two-way RM-ANCOVA demonstrated a significant main interaction between hormone treatment and time (hormone treatment × time); $F = 2.04$, $P = 0.033$, and *partial eta squared* = 0.25. Significant changes were set at $^*P < 0.05$, basal ASP levels.

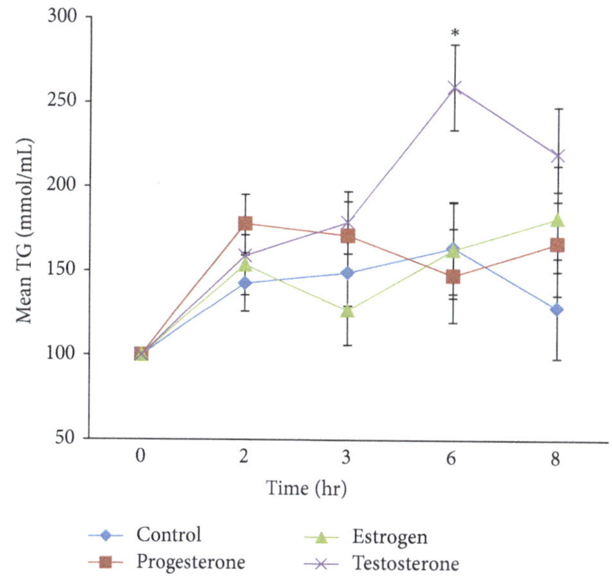

FIGURE 2: Changes in plasma triglyceride (TG) concentrations for different hormone treated groups and control group ($n = 6$). Results at each time point during postprandial period are expressed as means ± SEM (mmol/L) of difference in TG levels from basal (by subtracting basal values) from each time point and percentage increase was calculated. Two-way RM-ANCOVA adjusting for weight as a covariate demonstrated a significant interaction between hormone treatment and time (hormone treatment × time); $F = 2.31$, $P = 0.015$, and *partial eta squared* = 0.28. Significant changes were set at $^*P < 0.05$.

at 2 hours and corresponding TG-AUC for the different hormone treatments (Figure 4).

4. Discussion

Postprandial plasma ASP variations after a fat load were shown in several studies [1, 2, 8, 27–29]. One study on humans showed direct postprandial ASP changes *in vivo* by measuring venoarterial gradients across the subcutaneous adipose tissue bed after a fat load. In that study, local ASP production from human adipose tissue was accentuated postprandially after the fat load, and the maximal generation of ASP coordinated with maximal removal of TG [27]. In studies on mice, exogenous ASP treatment was associated with enhanced TG clearance [29]. On the other hand, ASP knock-out mice demonstrated delayed TG clearance, which was more pronounced in males, suggesting gender differences in ASP mediated postprandial TG clearance [2]. Evidence that links sex hormones to ASP variations is limited and mainly consists of results that show significant associations between plasma ASP and circulating female hormone levels in humans [13, 15, 30]. Direct effects of female hormones on ASP production and function were shown *in vitro* [31]. However, *in vivo* evidence identifying a direct role for sex hormones in plasma ASP variations is lacking. In this study, initially, acute hormone treatments after one hour had no significant effect on baseline ASP levels. Interestingly however, after the fat load, there was a significant main hormone effect on postprandial ASP levels over time. Postanalysis showed that the main hormone effect was due to progesterone treatment, as there was no significant increase in postprandial ASP levels in the controls or with estrogen and testosterone treatments. The key finding of this study was the effect of progesterone treatment on postprandial ASP levels. The progesterone treated group showed a marked surge in ASP levels at 2 hours which significantly contributed to around 26% variance in postprandial ASP levels (determined by partial eta squared). These findings are interesting especially in connection to a recent study showing a strong association of endogenous ASP fluctuations during the menstrual cycle with progesterone levels in females of the reproductive age. The study showed that ASP levels followed the same pattern of increasing progesterone levels from the follicular towards the luteal phase where ASP levels became significantly elevated [13]. Furthermore, a recent study showed that circulating progesterone levels correlated positively with complement C3 (ASP precursor) gene expression in adipose tissue [30]. Another human study showed that hormone replacement treatment in postmenopausal women increased circulating C3 levels; however, differential effects of female hormones were not determined [32]. In this study, we provide evidence

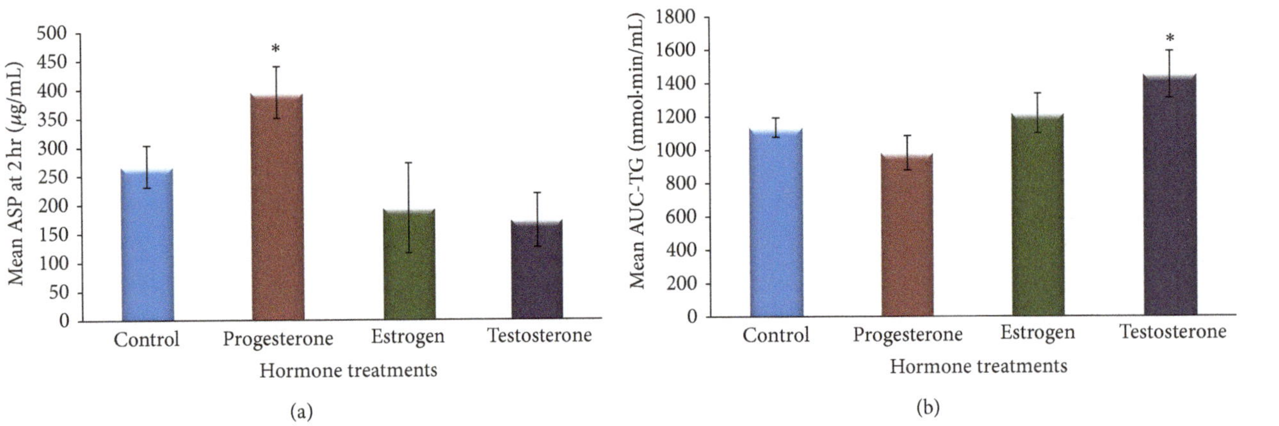

FIGURE 3: (a) ASP levels at 2 hrs showed that the greatest difference in ASP levels was between the progesterone treated group and the testosterone treated group ($P = 0.049$). (b) Areas under the curves (AUCs) for postprandial changes in plasma triglyceride (TG) concentrations for the different hormone treated groups and control group were measured to represent postprandial TG clearance after a fat load. The greatest difference in TG-AUC was greatest between the progesterone treated group and the testosterone treated group (0.041). Results are expressed as mean ± SEM. Significant differences were set at $^{*}P < 0.05$.

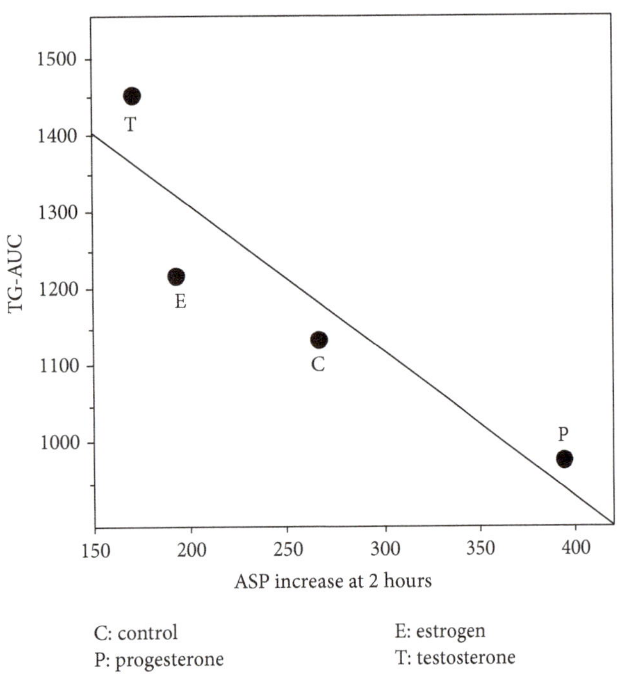

FIGURE 4: A simple scatter graph representing a negative association between mean postprandial ASP level increase at 2 hours and corresponding mean TG-AUC of the different hormone treated groups.

consistent with a direct role for progesterone in enhancing endogenous ASP production *in vivo*.

It is already established that ASP production from adipocytes is largely enhanced by TG-rich chylomicron treatment *in vitro* [33]. After a fat load, *in vivo* TG-rich chylomicrons are highly abundant and in combination with progesterone treatment may contribute to even more ASP production leading to increased systemic ASP levels. Furthermore, the opposite trends shown considering increased postprandial ASP levels and the corresponding change in TG levels in the different hormone treated groups (Figures 1 and 2) and the negative association of ASP levels with TG-AUC (Figures 3 and 4) are indicative of enhanced TG clearance, concurrent with increased ASP levels. This was mostly evident in the progesterone treated group. Conversely, lower ASP levels in the testosterone treated group were associated with delayed TG clearance (presented as increased AUC) which is consistent with evidence in the literature presenting testosterone as an antilipogenic factor [34]. The findings altogether suggest that ASP mediated TG clearance is subject to hormonal influence.

In conclusion, although progesterone is recognized as a lipogenic hormone [35], the mechanisms by which progesterone exerts its function in adipose tissue are poorly understood. Studies have suggested that progesterone stimulates fat clearance and storage by enhancing lipoprotein lipase activity [36] or increasing the expression of transcription factors that control synthesis of fatty acid synthase [37]. Several studies reported that progesterone has antiglucocorticoid effects leading to enhanced subcutaneous fat storage and reduced abdominal fat storage [36, 38]. The findings of this study suggest a causal association between hormone treatment and endogenous ASP production. Enhanced TG clearance in association with increased postprandial ASP levels in the progesterone treated group support the notion of a stimulatory role for progesterone in ASP mediated TG clearance.

The young age of the rats provided further advantage to this study as endocrine and tissue changes that occur with aging, coupled with estrogen treatment, were suggested to place ovariectomized rats at risk for adverse metabolic consequences [39].

To our knowledge, this is the first functional study *in vivo* that demonstrates a cause-effect relationship between acute hormone treatment and endogenous ASP levels *in vivo*.

In light of the potent lipogenic role of ASP, this study may contribute to further understanding of mechanisms by which progesterone exerts its lipogenic effects in females.

Conflict of Interests

The authors declare that there is no conflict of interests regarding the publication of this paper.

Acknowledgments

This work was supported by a student project grant from Sultan Qaboos University. The authors sincerely thank the technical staff in the Department of Biochemistry at College of Medicine and Health Sciences due to their support during the analysis. The authors want to express deep appreciation to Dr. Fahad Al-Zedjali, Mrs. Khalsa Al Malki, and Mrs. Zainab Al Hashmi. Special thanks are due to Dr. Anandarajah Srikandakumar and the staff in the small animal unit for their technical support in animal handling and welfare.

References

[1] K. Cianflone, Z. Xia, and L. Y. Chen, "Critical review of acylation-stimulating protein physiology in humans and rodents," *Biochimica et Biophysica Acta: Biomembranes*, vol. 1609, no. 2, pp. 127–143, 2003.

[2] I. Murray, A. D. Sniderman, P. J. Havel, and K. Cianflone, "Acylation stimulating protein (ASP) deficiency alters postprandial and adipose tissue metabolism in male mice," *The Journal of Biological Chemistry*, vol. 274, no. 51, pp. 36219–36225, 1999.

[3] M. Faraj, A. D. Sniderman, and K. Cianflone, "ASP enhances in situ lipoprotein lipase activity by increasing fatty acid trapping in adipocytes," *Journal of Lipid Research*, vol. 45, no. 4, pp. 657–666, 2004.

[4] V. van Harmelen, S. Reynisdottir, K. Cianflone et al., "Mechanisms involved in the regulation of free fatty acid release from isolated human fat cells by acylation-stimulating protein and insulin," *The Journal of Biological Chemistry*, vol. 274, no. 26, pp. 18243–18251, 1999.

[5] M. Maslowska, T. Scantlebury, R. Germinario, and K. Cianflone, "Acute in vitro production of ASP in differentiated adipocytes," *Journal of Lipid Research*, vol. 38, no. 1, pp. 21–31, 1997.

[6] S. Paglialunga, A. Fisette, Y. Yan et al., "Acylation-stimulating protein deficiency and altered adipose tissue in alternative complement pathway knockout mice," *American Journal of Physiology: Endocrinology and Metabolism*, vol. 294, no. 3, pp. E521–E529, 2008.

[7] S. Paglialunga, A. Fisette, M. Munkonda, Y. Gao, D. Richard, and K. Cianflone, "The effects of acylation stimulating protein supplementation VS antibody neutralization on energy expenditure in wildtype mice," *BMC Physiology*, vol. 10, no. 1, article 4, 2010.

[8] K. Cianflone, R. Zakarian, C. Couillard, B. Delplanque, J.-P. Despres, and A. Sniderman, "Fasting acylation-stimulating protein is predictive of postprandial triglyceride clearance," *Journal of Lipid Research*, vol. 45, no. 1, pp. 124–131, 2004.

[9] V. Mohamed-Ali, J. H. Pinkney, and S. W. Coppack, "Adipose tissue as an endocrine and paracrine organ," *International Journal of Obesity*, vol. 22, no. 12, pp. 1145–1158, 1998.

[10] P. S. Cooke and A. Naaz, "Role of estrogens in adipocyte development and function," *Experimental Biology and Medicine*, vol. 229, no. 11, pp. 1127–1135, 2004.

[11] M. J. Lobo, X. Remesar, and M. Alemany, "Effect of chronic intravenous injection of steroid hormones on body weight and composition of female rats," *Biochemistry and Molecular Biology International*, vol. 29, no. 2, pp. 349–358, 1993.

[12] R. Ahmadi and S. H. Oryan, "Effects of ovariectomy and estradiol valerate or progestetone on serum insulin level in rats," *International Journal of Medicine and Medical Sciences*, vol. 1, no. 6, pp. 263–266, 2009.

[13] J. Saleh, M. Al-Khanbashi, M. Al-Maarof, M. Al-Lawati, S. G. Rizvi, and K. Cianflone, "Acylation-stimulating protein increases and correlates with increased progesterone levels during the luteal phase of the menstrual cycle," *European Journal of Endocrinology*, vol. 160, no. 2, pp. 301–307, 2009.

[14] J. Saleh, K. Cianflone, T. Chaudhary, H. Al-Riyami, A.-R. Al-Abri, and R. Bayoumi, "Increased plasma acylation-stimulating protein correlates with hyperlipidemia at late gestation," *Obesity*, vol. 15, no. 3, pp. 646–652, 2007.

[15] J. Saleh, H. D. S. Al-Riyami, T. A. Chaudhary, and K. Cianflone, "Cord blood ASP is predicted by maternal lipids and correlates with fetal birth weight," *Obesity*, vol. 16, no. 6, pp. 1193–1198, 2008.

[16] Y. Wu, J. Zhang, Y. Wen, H. Wang, M. Zhang, and K. Cianflone, "Increased acylation-stimulating protein, C-reactive protein, and lipid levels in young women with polycystic ovary syndrome," *Fertility and Sterility*, vol. 91, no. 1, pp. 213–219, 2009.

[17] C. Oktenli, T. Ozgurtas, M. Dede et al., "Metformin decreases circulating acylation-stimulating protein levels in polycystic ovary syndrome," *Gynecological Endocrinology*, vol. 23, no. 12, pp. 710–715, 2007.

[18] J. Saleh, N. Al-Wardy, H. Farhan, M. Al-Khanbashi, and K. Cianflone, "Acylation stimulating protein: a female lipogenic factor?" *Obesity Reviews*, vol. 12, no. 6, pp. 440–448, 2011.

[19] Z. C. Murphy, P. Pezuk, M. Menaker, and M. T. Sellix, "Effects of ovarian hormones on internal circadian organization in rats 2013," in *Proceedings of the 45th Annual Meeting of the Society for the Study of Reproduction*, 2013.

[20] H. B. Waynforth and P. A. Flecnell, *Experimental and Surgical Technique in the Rat*, Academic Press, London, UK, 2nd edition, 1992.

[21] A. Liu, L. Gao, S. Kang et al., "Testosterone enhances estradiol's cardioprotection in ovariectomized rats," *Journal of Endocrinology*, vol. 212, no. 1, pp. 61–69, 2012.

[22] C.-L. Lu, J.-C. Hsieh, M.-L. Tsaur et al., "Estrogen rapidly modulates mustard oil-induced visceral hypersensitivity in conscious female rats: a role of CREB phosphorylation in spinal dorsal horn neurons," *American Journal of Physiology: Gastrointestinal and Liver Physiology*, vol. 292, no. 1, pp. G438–G446, 2007.

[23] P. R. R. Gangula, S. J. Wimalawansa, and C. Yallampalli, "Pregnancy and sex steroid hormones enhance circulating calcitonin gene-related peptide concentrations in rats," *Human Reproduction*, vol. 15, no. 4, pp. 949–953, 2000.

[24] W. Arkaravichien and K. E. Kendle, "Critical progesterone requirement for maintenance of pregnancy in ovariectomized rats," *Journal of Reproduction and Fertility*, vol. 90, no. 1, pp. 63–70, 1990.

[25] K.-H. Diehl, R. Hull, D. Morton et al., "A good practice guide to the administration of substances and removal of blood, including routes and volumes," *Journal of Applied Toxicology*, vol. 21, no. 1, pp. 15–23, 2001.

[26] S. Damsch, G. Eichenbaum, A. Tonelli et al., "Gavage-related reflux in rats: identification, pathogenesis, and toxicological implications (Review)," *Toxicologic Pathology*, vol. 39, no. 2, pp. 348–360, 2011.

[27] J. Saleh, L. K. M. Summers, K. Cianflone, B. A. Fielding, A. D. Sniderman, and K. N. Frayn, "Coordinated release of acylation stimulating protein (ASP) and triacylglycerol clearance by human adipose tissue in vivo in the postprandial period," *The Journal of Lipid Research*, vol. 39, no. 4, pp. 884–891, 1998.

[28] S. Meijssen, H. van Dijk, C. Verseyden, D. W. Erkelens, and M. C. Cabezas, "Delayed and exaggerated postprandial complement component 3 response in familial combined hyperlipidemia," *Arteriosclerosis, Thrombosis, and Vascular Biology*, vol. 22, no. 5, pp. 811–816, 2002.

[29] J. Saleh, J. E. Blevins, P. J. Havel, J. A. Barrett, D. W. Gietzen, and K. Cianflone, "Acylation stimulating protein (ASP) acute effects on postprandial lipemia and food intake in rodents," *International Journal of Obesity*, vol. 25, no. 5, pp. 705–713, 2001.

[30] R. Rezvani, A. Gupta, J. Smith et al., "Cross-sectional associations of acylation stimulating protein (ASP) and adipose tissue gene expression with estradiol and progesterone in pre- and postmenopausal women," *Clinical Endocrinology*, vol. 81, no. 5, pp. 736–745, 2014.

[31] Y. Wen, H. Wang, R. MacLaren, H. Lu, X.-F. Hu, and K. Cianflone, "Sex steroid hormones induce acylation stimulating protein resistance in 3T3-L1 adipocytes," *Journal of Cellular Biochemistry*, vol. 105, no. 2, pp. 404–413, 2008.

[32] Y. Liu and L. Lv, "Effect of hormone replacement therapy on serum complement (C3, C4) and immunoglobulin (IgG, IgM) levels in post-menopausal women," *Journal of Huazhong University of Science and Technology: Medical Sciences*, vol. 28, no. 1, pp. 102–103, 2008.

[33] T. Scantlebury, M. Maslowska, and K. Cianflone, "Chylomicron-specific enhancement of acylation stimulating protein and precursor protein C3 production in differentiated human adipocytes," *The Journal of Biological Chemistry*, vol. 273, no. 33, pp. 20903–20909, 1998.

[34] H. Zhang, Y. Liu, L. Wang et al., "Differential effects of estrogen/androgen on the prevention of nonalcoholic fatty liver disease in the male rat," *Journal of Lipid Research*, vol. 54, no. 2, pp. 345–357, 2013.

[35] R. K. Kalkhoff, "Metabolic effects of progesterone," *American Journal of Obstetrics and Gynecology*, vol. 142, no. 6, pp. 735–738, 1982.

[36] S. B. Pedersen, K. Kristensen, and B. Richelsen, "Antiglucocorticoid effects of progesterone in vivo on rat adipose tissue metabolism," *Steroids*, vol. 68, no. 6, pp. 543–550, 2003.

[37] D. Lacasa, E. Garcia, B. Agli, and Y. Giudicelli, "Control of rat preadipocyte adipose conversion by ovarian status: regional specificity and possible involvement of the mitogen-activated protein kinase-dependent and c-fos signaling pathways," *Endocrinology*, vol. 138, no. 7, pp. 2729–2734, 1997.

[38] P. Björntorp, "Hormonal control of regional fat distribution," *Human Reproduction*, vol. 12, no. 1, pp. 21–25, 1997.

[39] A. Selvamani and F. Sohrabji, "The neurotoxic effects of estrogen on ischemic stroke in older female rats is associated with age-dependent loss of insulin-like growth factor-1," *The Journal of Neuroscience*, vol. 30, no. 20, pp. 6852–6861, 2010.

Nutritional Knowledge and Attitudes of Adolescent Swimmers in Trinidad and Tobago

Marquitta C. Webb and Safiya E. Beckford

Department of Agricultural Economics and Extension, Faculty of Food and Agriculture, The University of the West Indies, St. Augustine, Trinidad and Tobago

Correspondence should be addressed to Marquitta C. Webb; marquitta3@gmail.com

Academic Editor: C. S. Johnston

Purpose. To investigate the level of nutrition knowledge and attitude of adolescent male and female swimmers training competitively in Trinidad and Tobago. *Methodology*. A self-administered questionnaire, which consisted of 21 nutrition knowledge and 11 attitude statements, was utilized to assess the level of nutrition knowledge and attitude of adolescent swimmers. For the assessment of nutrition knowledge, correct answers were given a score of "1" and incorrect answers were given a score of "0." For the evaluation of attitude towards nutrition, a score ranging from 1 to 5 was assigned to each response; "5" was given to the most positive response, and "1" was given to the most negative. Data were analyzed using SPSS version 21.0. *Results*. Two hundred and twenty swimmers with a mean age of 14.56 ± 2.544 completed the questionnaire. The mean nutrition knowledge score was 10.97 ± 2.897 and mean attitude score was 41.69 ± 6.215. Nutrition knowledge was positively and significantly related to the number of reported nutrition sources ($r = 0.172$, $P = 0.005$). Nutrition knowledge was positively and significantly related to the attitude ($r = 0.130$, $P = 0.027$). *Conclusions and Implication*. Athletes lack nutrition knowledge but have a positive attitude towards nutrition, which may indicate receptiveness to future nutrition education.

1. Introduction

Adolescence is the stage of life between ages of 11 and 21 years [1]. During this time, significant changes occur, which prepares a child for adulthood. The rapid growth and development during this stage increases energy and nutrient requirements [2–4]. Additionally, when an adolescent is an athlete, there are several important factors to consider in relation to nutrition for both growth and sports performance. Factors, such as lifestyle and dietary habits, have a tremendous impact on health status and sports performance [3]. As such, to sustain proper growth, development, and health while improving performance, adolescents engaged in sports could benefit from knowledge about the importance of good nutrition.

Nutrition is an important factor in the performance and health of athletes. High levels of nutrition knowledge and positive attitudes can result in increased performance and health of an athlete. Therefore, nutrition is an essential component of any athletic or physical activity program. Almost every process in the body involves some aspect of nutrition. Hence, knowledge of how the body utilizes nutrients and the relation of nutrients with metabolism in health and disease, energy production, and recovery from exercise are all important aspects of sports nutrition.

Existing studies [5, 6] are constant in their findings that athletes lack the nutrition knowledge they need in order to gain the benefits of increased performance and health. This may be due to misconceptions about nutrition and its importance to health and performance among athletes [7–13]. Additionally, it may be due to deficiencies in nutrition knowledge and misguided or incorrect information given by one or more unreliable sources [7–13]. Torres-McGehee and colleagues [14] examined nutrition knowledge in 4 domains of sports nutrition among athletes in various sports as well as coaches, athletic trainers, and conditioning specialists and found that athletes had inadequate nutrition knowledge scores of less than 75%. Hoogenboom and colleagues [15] investigated the nutritional knowledge and eating behaviors of female collegiate swimmers in Divisions I, II, and III and reported that on average these swimmers had a mean nutrition knowledge score of 54.53 ± 4.34, which was considered fair. These researchers concluded that these swimmers exhibited a lack of application of the knowledge they possessed. Areas in which adolescents typically lack nutrition knowledge include the dietary roles of protein and carbohydrates, role of vitamin and minerals, carbohydrate loading, glycemic index, muscle energy sources, energy expenditure and nutritional requirements, nutrient sources, and nutrients and performance [5, 8, 16]. Ozdoğan and Ozcelik [6] evaluated the nutrition knowledge of students in the sports teaching and coaching department at three universities using 30 true/false questions and found that the mean nutrition knowledge score was low (12.247 ± 3.525) and showed insufficiency of knowledge. Similarly, in a study of 185 Mid-American Conference College female softball players, which measured nutrition knowledge, practices, attitudes, and information sources using a questionnaire, Torres-McGehee and colleagues [14] found that the mean nutrition knowledge score was low (45.7 ± 4.7). However, when a standard 60% cut-off-point was set as the failing mark, two thirds (n = 120, 62%) of the softball players were judged as having failed the questionnaire, indicating a low level of nutrition knowledge. Likewise, Davar [16] reported that the total mean score for female hockey players was 22.85 (38.8%) and concluded that athletes lack nutrition knowledge. In contrast, Azizi and colleagues [9] utilized a questionnaire to investigate nutritional knowledge and attitudes of elite college athletes and reported that males and females had moderate nutrition knowledge, 52.36 ± 6.7 and 54.31 ± 6.3, respectively. However, although the scores were classified as moderate, the researchers concluded that nutrition knowledge needs to be improved in this population. In another study, Zawila and colleagues [8] indicated that female collegiate cross-country runners had a reasonable amount of knowledge regarding some aspects of nutrition. However, there were parts in which the lack of knowledge was evident. Runners scored higher in sections related to sport nutrition than sections related to general nutrition. Additionally, those exposed to nutrition courses in college showed a greater score for nutrition knowledge than those who were not exposed to nutrition class.

Besides knowledge, attitude towards nutrition is also an important factor in the performance of an athlete. There is a growing body of evidence that suggests athletes hold a positive attitude towards nutrition [12, 17, 18]. Azizi and colleagues [9] noted that nutrition attitude scores for males and females were 50.61 ± 5.1 and 52.03 ± 5.8, respectively, which demonstrated positive attitudes toward nutrition. Females were found to have a significant higher attitude score than males, which corresponded to a better attitude towards nutrition. Zawila and colleagues [8] conveyed that female collegiate cross-country runners had a mean total positive response for attitudes of 90.6% representing an overall positive attitude towards nutrition. The mean attitude towards nutrition score in the investigation of college softball players by Hornstrom and colleagues [11] was 1.9 ± 0.4 indicating a positive attitude (range 1–6, 1 signifying a more positive attitude and 6 signifying a more negative attitude). Davar [16] reported that the mean of female hockey players' total positive attitude response was 90.6% and 93.3% of the players agreed that providing nutrition education would have a positive impact on food selection. Thus, several authors have shown that athletes have a positive attitude towards nutrition and therefore may be open to acquiring nutrition information [8, 16].

There is a relationship between nutrition knowledge and attitude. This statement was supported by studies [9, 11] which indicated that nutrition knowledge and attitudes are significant and positively correlated; that is, the more the knowledge about nutrition the greater the positive attitude towards it. However, nutrition knowledge and attitudes had no impact on dietary practices [11].

A review of literature suggests that few studies have been conducted to explore the nutrition knowledge and attitude towards nutrition of athletes in developing countries, such as Trinidad and Tobago. Most of the literature focuses on elite and college level athletes with few articles on their adolescent counterparts. Additionally, there is no current information about the level of nutrition knowledge among adolescent swimmers in Trinidad and Tobago. Hence, the purpose of this study was to assess the level of nutrition knowledge and attitude of adolescent swimmers training competitively in Trinidad and Tobago. Study aims were to enable participants to examine their nutrition knowledge and attitude towards nutrition in a quest to enhance performance. Study hypotheses were that (1) adolescent competitive swimmers will lack nutrition knowledge, (2) adolescent competitive swimmers will have a positive attitude toward nutrition, and (3) there will be a positive correlation between nutrition knowledge and attitude toward nutrition.

2. Materials and Methods

2.1. Subjects. The questionnaire to evaluate the level of the nutrition knowledge and attitudes and nutrition of adolescent swimmers was carried out from September 2012 to November 2012. The participants involved were registered and trained competitively with private swimming clubs in Trinidad and Tobago. All of the clubs were registered with the Amateur Swimming Association of Trinidad and Tobago (ASATT). A total of 220 adolescent male and female swimmers, aged 11–21 years, were recruited for this study. To guarantee anonymity of responses and easy identification of the questionnaires by participants, identification numbers were randomly assigned to each questionnaire. The same instructions were read to each participant to ensure the questionnaire was completed correctly. Each questionnaire took approximately 15 minutes to complete. Ethical approval was obtained from the ethics committee. Additionally, signed informed consent was obtained from participants and/or their parent/guardian.

2.2. Procedure. In this cross-sectional descriptive study, a self-administered structured questionnaire was utilized to evaluate the nutrition knowledge and attitudes of adolescent swimmers training in competitive clubs. The survey instrument used in this study was developed from a combination of

TABLE 1: Profile of the study population (n = 220).

Variable	Classification	Frequency (n = 220)	Percentage (%)
Sex	Female	98	44.5
	Male	122	55.5
Ethnicity	African	70	31.8
	Indian	11	5.0
	Caucasian	7	3.2
	Asian	3	1.4
	Mixed	129	58.6
Age range	11-12	61	27.7
	13-14	59	26.8
	15–17	70	31.8
	18–21	30	13.6
Highest level of education	Primary	16	7.3
	Secondary	175	79.5
	Tertiary	29	13.2
Number of nutrition classes/courses/seminars attended	None	125	56.8
	1–3	76	35.0
	4 or more	18	8.2
Most recent class/course/seminar attended	Never	117	53.2
	2011-2012	74	33.6
	More than 2 years ago	29	13.2

previously administered questionnaires [6, 8]. The modified questionnaire had a reliability of 0.71 and a review by faculty members was utilized to establish its face validity. The questionnaire items included demographic characteristics of participants, 21 nutrition knowledge statements, to which participants replied "true," "false," or "not sure," and 11 statements to evaluate the attitudes toward nutrition, to which participants responded using a Likert scale with options ranging from "strongly agree" to "strongly disagree." For the assessment of nutrition knowledge, correct answers were given a score of "1" and incorrect answers were given a score of "0." For the evaluation of attitude towards nutrition a score ranging from 1 to 5 was assigned to each response; "5" was given to the most positive response and "1" was given to the most negative.

2.3. Statistical Analysis. All data were analyzed using the Statistical Package for Social Sciences (SPSS version 21.0, SPSS Inc., Chicago, IL, 2012). Descriptive statistics were used to compile the data collected. Analysis of variance (ANOVA) (confidence interval 95%) was used to compare nutrition knowledge and attitudes between sex, age range, highest level of education, and number of nutrition classes attended. Pearson correlation was used to evaluate the relationship between nutrition knowledge and sources of information and between nutrition knowledge and attitudes.

3. Results

3.1. Demographic Profile. Table 1 presents the demographics of the study population. Of the 220 swimmers who participated in the study, 122 (55.5%) were males and 98 (44.5%) females with a mean age of 14.56 ± 2.544. The majority (58.6%) of the participants were of mixed descent. Most of the respondents attended secondary school and were in the "15–17" age group. Over half (56.8%) of the participants had never had a nutrition class/course/seminar since beginning to swim. Additionally, more than half (53.2%) of the participants indicated never attending a nutrition class/course/seminar over the last year.

3.2. Nutrition Knowledge. The highest potential score for the nutrition knowledge was 21 and the lowest score was 0. The mean nutrition knowledge score for the entire sample was 10.97 ± 2.90 (Table 2). The highest score was 20 (95.24%) and the lowest was 2 (9.52%). Male participants had a mean score of 11.05 ± 2.793, which was higher than the female participants' mean score of 10.88 ± 3.033. Participants in the age group of 18–21 scored the highest out of the 4 specified age groups with a mean score of 12.97 ± 2.735. The 11-12 age group scored the lowest with 9.95 ± 2.362. Table 3 presents survey questions that were answered correctly less than 50% by the respondents. Participants scored below 50% for 10 of the 21 nutrition knowledge questions. There were no significant differences in the knowledge scores when compared between sex, age range, and highest level of education, however, there was a significant difference in knowledge scores between the categories (i.e., none, 1–3, 4 or more) for of number of nutrition classes attended. Hence, the more nutrition classes attended, the higher the nutrition knowledge scores. Pearson correlation revealed that nutrition knowledge was positively and significantly related to the number of reported nutrition sources (r = 0.172, P = 0.005). Additionally, nutrition knowledge was positively and significantly related to the

Table 2: Mean (SD) nutrition knowledge score according to the variables ($n = 220$).

Category	Group	Means ± standard deviation	P^{**}
Sample	Total	10.97 ± 2.897	
Sex	Male	11.05 ± 2.793	0.880
	Female	10.88 ± 3.033	
Age range	11-12	9.95 ± 2.362	0.158
	13-14	10.78 ± 3.074	
	15–17	11.17 ± 2.823	
	18–21	12.97 ± 2.735	

$^{**}P < 0.05$.

attitude of the respondents held towards nutrition ($r = 0.130$, $P = 0.027$).

3.3. Attitude towards Nutrition. The highest possible score for the attitude section was 55 and the lowest was 11. The mean positive attitude towards nutrition score, standard deviation, and ANOVA results are shown in Table 4 according to sex and age groups. The mean score for attitude towards nutrition for the entire sample was 41.69 ± 6.215. The highest score was 55 (100%) and the lowest 19 (34.55%). Male respondents had a slightly higher mean positive attitude towards nutrition score than female respondents, 41.96 ± 5.931 and 41.35 ± 6.566, respectively. As the age range increased so did the attitude towards nutrition scores. The age range of "18–21" had the highest score of 43.83 ± 6.438, followed by the "15–17" age range scoring 42.47 ± 6.236. The "11-12" and "13-14" age ranges had similar scores of 40.48 ± 6.239 and 40.47 ± 5.618, respectively. There were no significant differences in the attitude towards nutrition score between the means for sex, age range, highest level of education, number of nutrition classes attended since beginning to swim, and club.

4. Discussion

This current study investigated the nutrition knowledge and attitudes of adolescent swimmers training competitively in Trinidad and Tobago. Out of the 21 nutrition knowledge statements set, there were 5 statements that dealt specifically with macronutrients, 8 with micronutrients, and 2 with hydration, while the remaining 6 statements focused on diet practices. From the analysis of results, the nutrition knowledge of adolescent swimmers in this study was found to be insufficient. The mean knowledge score for the general sample was 10.97 ± 2.897, which was lower than the mean score of 12.247 ± 3.525 found in Ozdoğan and Ozcelik [6] for the same knowledge statements. In another study [5], the mean score for male respondents was higher than that for female participants. However, after comparison of means, there were no significant differences between genders. The mean knowledge score increased as the age range increased, but no significant difference was found between means. The mean scores for participants with or without nutrition classes were also compared (means not shown in results) and were found to be statistically significant ($P = 0.039$). This indicated that the more nutrition classes taken, the higher the score, which suggested that nutrition education may be beneficial to these groups of athletes. The areas of nutrition knowledge that seemed to be deficient in this sample, as demonstrated by less than 50% of participants responding correctly, were the role of protein in the body, the characteristics of carbohydrates, the role of micronutrients in the body (especially iron), types of fats, and proper energy sources from food.

Carbohydrates, in the form of glucose and glycogen, are the main sources of energy for muscle; proteins are only used for energy when the body is deficient in carbohydrates [8], which is important for athletes to know. For the statement "protein is the main energy source for the muscle," only 20% of the respondents answered correctly. This was similar to the 31.7% of female cross-country runners who correctly answered this statement [8]. In contrast, Ozdoğan and Ozcelik [6] found that 77.8% of the athletes answered this statement correctly. However, a diet rich in carbohydrates increases endurance performance because of the extra store of glycogen in the muscles. For the statement "carbohydrates are stored in muscles in the form of glycogen," 46.4% of participants answered correctly. The amount of carbohydrates consumed affects the amount of glycogen stored; if glycogen stores are low, performance can be negatively affected [19].

Although participants were fairly knowledgeable about the importance of fat in the diet they were not very well familiar with the types of fat, which was also reported by Davar [16] and Dunn and colleagues [20]. It is recommended that 30% of an athlete's caloric intake be from fat sources [21]. Fats have many roles in the body; however, saturated fats have been linked to heart disease due to their potential role in increasing LDL cholesterol. Hence, it is recommended that saturated fats be replaced with unsaturated fats (mono- and polyunsaturated fats) in the diet, which decreases the risk of heart disease [19]. For the statement "saturated and unsaturated oils both have an equal effect on the health," only 31.4% of the participants answered correctly.

Only 30% of the swimmers chose "false" for the statement, "table salt is an essential part of a healthy diet," which was similar to the 37.6% in Ozdoğan and Ozcelik [6]. Sodium is the major extracellular cation and is needed for nerve impulses and muscle contractions, maintenance of cellular volume, and acid base balance [19]. Sodium is naturally found in vegetables and cereals and consuming these items can provide the recommended amount of sodium making it unnecessary to have additional salt added to food. Additionally, salt (sodium chloride) has been linked to high blood pressure and edema, so excess amounts should be avoided [6].

The statement "iron-deficiency anemia results in a decrease in the amount of oxygen that can be carried in the blood" was not answered well by the participants with only 38.2% correctly choosing "true." In contrast, a similar question, "iron deficiency anemia results in decreased activity and performance," was answered well by participants in Davar's study [16]. Less than 40% of swimmers (37.7%) correctly chose "false" for the statement "iron in meat is absorbed at the same rate as iron in a plant food." This showed a lack of knowledge about sources of iron, which is supported

TABLE 3: Questions answered correctly by less than 50% of the participants ($n = 220$).

Statement	Frequency ($n = 220$)	Mean ± standard deviation
Protein is the main energy source for the muscle.	44 (20.0%)	0.20 ± 0.401
Iron-deficiency anemia results in a decrease in the amount of oxygen that can be carried in the blood.	84 (38.2%)	0.38 ± 0.487
Iron in meat is absorbed at the same rate as iron in a plant food.	83 (37.7%)	0.38 ± 0.486
Vitamin supplements are recommended for all physically active people.	23 (10.5%)	0.10 ± 0.307
During physical activity, feeling thirsty is enough to indicate the need for liquid.	45 (20.5%)	0.20 ± 0.404
Vitamins are good sources of energy.	29 (13.2%)	0.13 ± 0.339
Saturated and unsaturated oils both have an equal effect on the health.	69 (31.4%)	0.31 ± 0.465
Table salt is an essential part of a healthy diet.	66 (30.0%)	0.30 ± 0.495
Basic sugars like brown or granulated sugar, jam, and honey are the most suitable energy sources for sportsmen.	95 (43.2%)	0.43 ± 0.496
Carbohydrates are stored in muscles in the form of glycogen.	102 (46.4%)	0.46 ± 0.500

TABLE 4: Mean (SD) positive attitude towards nutrition scores according to the variables ($n = 220$).

Category	Group	Means ± standard deviation	P^{**}
Sample	Total	41.69 ± 6.215	
Sex	Male	41.96 ± 5.931	0.707
	Female	41.35 ± 6.566	
Age range	11-12	40.48 ± 6.239	0.074
	13-14	40.47 ± 5.618	
	15–17	42.47 ± 6.236	
	18–21	43.83 ± 6.438	

$^{**}P < 0.05$.

by the results in Davar's study [16]. Iron is needed for creation of hemoglobin and myoglobin, which accepts transports and releases oxygen around the body; it also serves as a cofactor for many enzymes needed for metabolism [19]. Absorption of iron is partially dependent on its source; animal sources of iron contain heme, which is easily absorbed, and plant sources contain nonheme iron, which is not absorbed as well (animal sources also contain nonheme iron) [19]. A deficiency in iron is one of the most common nutrient deficiencies and is observed in many athletes and nonathletes, especially in females [6].

Only 10.5% of the participants correctly chose "false" for the statement "vitamin supplements are recommended for all physically active people," which was the case in Zawila and colleagues [8] where only 10.0% of the participants correctly responded to the statement. In contrast, 67.9% of the participants in the study by Ozdoğan and Ozcelik [6] correctly answered the same question. For the statement "vitamins are good sources of energy" only 13.2% of the participants correctly identified "false" was the answer. This was lower than the 31.6% and 64.1% found in other studies [6, 8]. Nutrient supplementation does not benefit the performance of well-nourished athletes, so it can be said that if a balanced diet is achieved which meets caloric needs of the athlete, additional vitamin supplements are not required [19]. Vitamins also do not produce energy; however, they are needed in the processes required for energy production [20].

Swimmers who correctly selected "false" to the statement "during physical activity, feeling thirsty is enough to indicate the need for liquid" were a mere 20.5% of the sample. In the study performed by Ozdoğan and Ozcelik [6], the rate of people being knowledgeable about this was more than half the rate seen in the present study. However 85.9% of swimmers correctly indicated "true" as the response to "dehydration decreases performance," which was also the case in the Davar study [16] where it was reported that female hockey players were seemingly knowledgeable about this. In normal circumstances, thirst may be enough to stimulate fluid intake; however, in vigorous training situations and demanding environments thirst may be an inadequate indicator [22]. When an athlete is well hydrated before, during, and after exercise performance is optimal. However, if dehydrated performance decreases, there is an increased risk of potentially life-threatening heat related injury [23]. The swimmers in this study need to be educated about the importance of staying hydrated before, during, and after exercise.

The statement "basic sugars like brown or granulated sugar, jam, and honey are the most suitable energy sources for sportsmen" was correctly answered by 43.2% of the participants which was lower than the 72.3% of athletes that correctly answered in Ozdoğan and Ozcelik [6]. Foods like basic sugars, jam, and honey are high glycemic index foods, which enter the blood stream quickly and are carried away by insulin to the muscles; they cannot be used to maintain blood glucose levels over a long period of time and may cause fatigue [24]. Athletes should consume foods of moderate to low glycemic index, which are broken down over time and are able to provide a constant supply of glucose in the blood over a period of time [24]. Athletes in this study should also be educated about pre-, during, and postexercise nutritional practices, including foods with better glycemic indices.

In the results for the attitude section there were overall positive responses for 7 out of the 11 questions. The mean score of 41.69 ± 6.215 reflected an overall positive attitude towards nutrition. There was no significant difference found between any of the means when compared between sex, age groups, level of education, and previous exposure to nutrition classes. However, in another study it was reported that female athletes had a significantly higher attitude score than male athletes and also those enrolled in physical activity majors had a higher score than other majors [9].

For the statement "learning facts about nutrition is the best way to achieve favorable changes in food habits" 79.5% of participants either strongly agreed or agreed with the statement which was slightly less than in Davar [16] where 91.7% and 93.3% of athletes strongly agreed or agreed, respectively. Similarly, Zawila and colleagues [8] reported higher results for the same statement. There was a low acceptance (16.3%) of the statement "learning about nutrition is not important for athletes because they eat so much food and they always get the nutrients their bodies need" and a high acceptance of the statement "the relationship between good eating habits and good health should be stressed to the athlete" by the swimmers. The results to these statements showed that the athletes may be receptive to nutrition education. The same conclusion was made by Davar [16] and Zawila and colleagues [8] who found that there was a significantly positive correlation between nutrition knowledge and attitudes ($P = 0.027$). Many studies [9, 11, 25] have found the same relationship. Therefore, it may be said that nutrition education may not only increase nutrition knowledge but also have a positive effect on attitudes.

The current study is not without limitations. Studies have shown no relationship between nutrition knowledge and food behavior but a few do [26, 27]. However, in this study, we did not assess the relationship between knowledge, attitudes, and behaviors, or, more basically, food choices. There are several factors that influence food selection and nutrition, including flavor, religion, culture, and cognitive and social factors. Therefore, knowledge does not always translate into good food choices since athletes with a wealth of nutrition knowledge still eat poorly because of other factors. Another limitation is that we did not study the application of attitudes. There is no assumption that if knowledge is increased and attitude is good, the athlete will go on and make excellent food choices.

5. Conclusion

The findings serve as an indication of the need for nutrition education among swimmers and possibly other athletes. Athletes face enormous pressure to perform at their optimal level. However, in order to reach that optimal performance level the athletes must be equipped with excellent support which may be influenced by knowledge and attitude. Swimmers in Trinidad and Tobago were found to have inadequate nutrition knowledge, which could potentially contribute to their improved performance and health. Nutrition plays a critical role in the performance of an athlete; therefore by improving their knowledge about nutrition and good food choices, athletes could enhance their chances of reaching optimal level. Based on the poor nutrition knowledge but positive attitude towards nutrition, the swimmers may benefit from a nutrition education intervention program, which is geared towards improving eating habits for optimal performance and healthy lifestyle changes.

6. Implications for Research and Practice

Findings suggest that nutrition education should be a component of an athlete's training program. Data from this study provided some useful information which can be used to develop areas of an intervention program for swimmers. Future studies need to determine whether athletes in other sports possess the nutrition knowledge and attitudes toward nutrition that will enable them to perform at their optimal level.

Conflict of Interests

The authors declare that there is no conflict of interests regarding the publication of this paper.

References

[1] J. Stang, "Adolescent nutrition," in *Nutrition Through the Life Cycle*, J. E. Brown, Ed., pp. 356–384, Cengage Learning, Wadsworth, Ohio, USA, 4th edition, 2011.

[2] M. N. G. Oliveira, E. H. Almada Cezar, and E. D. A. Soares, "Comparison of the anthropometric assessment of adolescents of privileged and unprivileged social and economic levels," *Nutrition Research*, vol. 20, no. 10, pp. 1423–1436, 2000.

[3] J. A. Tur, M. S. Puig, E. Benito, and A. Pons, "Associations between sociodemographic and lifestyle factors and dietary quality among adolescents in Palma de Mallorca," *Nutrition*, vol. 20, no. 6, pp. 502–508, 2004.

[4] J. Przysławski, M. Stelmach, B. Grygiel-Górniak, M. Mardas, and J. Walkowiak, "Dietary habits and nutritional status of female adolescents from the great Poland region," *Polish Journal of Food and Nutrition Sciences*, vol. 61, no. 1, pp. 73–78, 2011.

[5] K. A. Beals, "Nutrition concerns of adolescent athletes," in *Nutritional Applications in Exercise and Sport*, I. Wolinsky and J. A. Driskell, Eds., pp. 59–80, CRC Press, Boca Raton, Fla, USA, 2001.

[6] Y. Ozdoğan and A. O. Ozcelik, "Evaluation of the nutrition knowledge of sports department students of universities," *Journal of the International Society of Sports Nutrition*, vol. 8, article 11, 2011.

[7] T. Abreu de Almeida and E. Abreu Soares, "Nutritional and anthropometric profile of adolescent volleyball athletes," *Revista Brasileira de Medicina do Esporte*, vol. 9, no. 4, pp. 198–203, 2003.

[8] L. G. Zawila, C.-S. M. Steib, and B. Hoogenboom, "The female collegiate cross-country runner: nutritional knowledge and attitudes," *Journal of Athletic Training*, vol. 38, no. 1, pp. 67–74, 2003.

[9] M. Azizi, F. Rahmani-Nia, M. Malaee, M. Malaee, and N. A. Khosravi, "A study of nutritional knowledge and attitudes of elite college athletes in Iran," *Brazilian Journal of Biomotricity*, vol. 4, pp. 105–112, 2010.

[10] A. Cupisti, C. D'Alessandro, S. Castrogiovanni, A. Barale, and E. Morelli, "Nutrition knowledge and dietary composition in Italian adolescent female athletes and non-athletes," *International Journal of Sport Nutrition*, vol. 12, no. 2, pp. 207–219, 2002.

[11] G. R. Hornstrom, C. A. Friesen, J. E. Ellery, and K. Pike, "Nutrition knowledge, practices, attitudes, and information sources of mid-american conference college softball players," *Food and Nutrition Sciences*, vol. 2, pp. 109–117, 2011.

[12] R. Sakamaki, K. Toyama, R. Amamoto, C.-J. Liu, and N. Shinfuku, "Nutritional knowledge, food habits and health attitude of Chinese university students—a cross sectional study," *Nutrition Journal*, vol. 4, pp. 1475–1480, 2005.

[13] N. Cotunga, C. E. Vickery, and S. McBee, "Sports nutrition for young athletes," *The Journal of School Nursing*, vol. 21, no. 6, pp. 323–328, 2005.

[14] T. M. Torres-McGehee, K. L. Pritchett, D. Zippel, D. M. Minton, A. Cellamare, and M. Sibilia, "Sports nutrition knowledge among collegiate athletes, coaches, athletic trainers, and strength and conditioning specialists," *Journal of Athletic Training*, vol. 47, no. 2, pp. 205–211, 2012.

[15] B. J. Hoogenboom, J. Morris, C. Morris, and K. Schaefer, "Nutritional knowledge and eating behaviors of female, collegiate swimmers," *North Amercian Journal of Sports Physical Therap*, vol. 4, pp. 139–148, 2009.

[16] V. Davar, "Nutrition knowledge and attitudes towards healthy eating of college-going women hockey players," *Journal of Human Ecology*, vol. 37, pp. 119–124, 2012.

[17] B. H. Jacobson, C. Sobonya, and J. Ransone, "Nutrition practices and knowledge of college varsity athletes: a follow-up," *The Journal of Strength & Conditioning Research*, vol. 15, pp. 63–68, 2001.

[18] Y. Wong, Y.-C. Huang, S.-L. Chen, and S. Yamamoto, "Is the college environment adequate for accessing to nutrition education: a study in Taiwan," *Nutrition Research*, vol. 19, no. 9, pp. 1327–1337, 1999.

[19] E. Whitney and S. R. Rolfes, Eds., *Understanding Nutrition*, Wadsworth Cengage Learning, Independence, Ky, USA, 12th edition, 2011.

[20] D. Dunn, L. W. Turner, and G. Denny, "Nutrition knowledge and attitudes of college athletes," *The Sport Journal*, vol. 10, pp. 144–151, 2007.

[21] B. Pramuková, V. Szabadosová, and A. Šoltésová, "Current knowledge about sports nutrition," *Australasian Medical Journal*, vol. 4, pp. 107–110, 2011.

[22] C. V. Gisolfi, "Fluid balance for optimal performance," *Nutrition Reviews*, vol. 54, no. 4, pp. S159–S168, 1996.

[23] American College of Sports Medicine, American Dietetic Association, Dietitians of Canada, "Joint position statement: nutrition and athletic performance," *Medicine & Science in Sports & Exercise*, vol. 49, pp. 709–731, 2009.

[24] P. M. Siu and S. H. S. Wong, "Use of the glycemic index: effects on feeding patterns and exercise performance," *Journal of Physiological Anthropology and Applied Human Science*, vol. 23, no. 1, pp. 1–6, 2004.

[25] M. Perron and J. Endres, "Knowledge, attitudes, and dietary practices of female athletes," *Journal of the American Dietetic Association*, vol. 85, no. 5, pp. 573–576, 1985.

[26] J. Wardle, K. Parmenter, and J. Waller, "Nutrition knowledge and food intake," *Appetite*, vol. 34, no. 3, pp. 269–275, 2000.

[27] M. Marmot and R. Wilkinson, *Social Determinants of Health*, Oxford University Press, Oxford, UK, 1999.

25

A Body Shape Index and Heart Rate Variability in Healthy Indians with Low Body Mass Index

Sharma Sowmya,[1] Tinku Thomas,[2] Ankalmadagu Venkatsubbareddy Bharathi,[3] and Sambashivaiah Sucharita[1,3]

[1] *Department of Physiology, St. John's Medical College, Bangalore 560034, India*
[2] *Division of Epidemiology and Statistics, St. John's Research Institute, Bangalore 560034, India*
[3] *Clinical Physiology Unit, Department of Physiology, St. John's Medical College, Bangalore 560034, India*

Correspondence should be addressed to Sambashivaiah Sucharita; sucharita.dr@gmail.com

Academic Editor: C. S. Johnston

Background. One third of Indian population is said to be suffering from chronic energy deficiency (CED), with increased risk of developing chronic diseases. A new anthropometric measure called A Body Shape Index (ABSI) is said to be a better index in predicting risks for premature mortality. ABSI is also in part said to be a surrogate of visceral fat. *Objective.* The present study aimed to explore the association between indices of HRV (heart rate variability), BMI, WC, and ABSI in healthy Indian males with low BMI (BMI < 18.5 kg/m^2) and to compare with normal BMI group (BMI 18.5 to 24.9 kg/m^2). *Methodology.* ABSI and BMI were derived from anthropometric parameters, namely, height, weight, and waist circumference in 178 males aged 18 to 78 years. Subjects were categorized into two groups based on their BMI. *Results and Conclusions.* Power spectral analysis of HRV demonstrated a significant negative correlation between Log HF (high frequency) and ABSI in both low BMI [−24.2 (9.4), $P < 0.05$] and normal BMI group [−23.41 (10.1), $P < 0.05$] even after controlling for age. Thus even with slight increase in BMI among low BMI individuals, there could be a greater risk of cardiovascular morbidity and mortality.

1. Introduction

Obesity has been the center of focus of majority of body composition related studies. This is partly due to a large body of data available from western population where obesity is considered a major public health concern [1–3]. Epidemiological studies with longitudinal follow-up have demonstrated association between simple anthropometric measures like height, weight (body mass index: BMI), and waist circumference (WC) with cardiovascular risks [3, 4]. Within a given BMI range, association with increased cardiovascular risks becomes even stronger as WC increases [5]. Studies have linked alterations in autonomic nervous activity with increased cardiovascular risk [6]. Heart rate variability (HRV) is a simple noninvasive, sensitive measure to evaluate autonomic system activity. Changes in HRV have been demonstrated to detect even subclinical states [7, 8]. Decreased heart rate variability (both low and high frequency) has been observed in obese, compared to individuals with normal BMI [9–11]. Pockets of data available from Indian population are not different from those of western data [12, 13]. However, India has an additional burden of chronic energy deficiency with one-third of population being of low BMI [14]. Recent evidence based on HRV measures from our laboratory demonstrated decreased HRV in undernourished subjects when compared to well-nourished/normal BMI group [15]. It has been speculated that undernourished/low BMI individuals might have more of visceral adiposity or ectopic fat than subcutaneous fat [16]. Although WC has been suggested as being superior to BMI in predicting cardiovascular risks in obese [17], its ability to predict increased cardiovascular risk in individuals with low BMI needs to be explored. Further, WC is highly sensitive to body size and fat distribution and is correlated with BMI. Thus, differentiating

BMI and WC as two independent epidemiological risk factors becomes difficult. Recently a new anthropometric measure, called A Body Shape Index (ABSI), has been derived from waist circumference which is independent of BMI and is said to be a better index than using either WC or BMI independently. Thus being independent of BMI, ABSI helps in understanding whether abdominal obesity has predictive ability that cannot be explained by BMI alone. Current study aimed to explore the association between indices of HRV and the anthropometric measures including BMI, WC, and ABSI in healthy Indians with low BMI and compare it with normal.

2. Materials and Methods

One hundred and seventy eight males aged 18 to 78 years were recruited from in and around the medical college. Based on their BMI, subjects were categorized into low BMI group ($<18.5\,kg/m^2$) and normal BMI group (between 18.5 and $24.9\,kg/m^2$). Subjects with any chronic ailments including diabetes, hypertension, and renal and endocrinal disorders were excluded. None of the subjects reported any symptoms suggestive of peripheral or autonomic neuropathy like giddiness on standing, urinary urgency, tingling sensation of limbs, or limb weakness or gastrointestinal symptoms like burning sensation in epigastric region, diarrhea, and constipation. The institutional ethical review board approved the research protocol.

All subjects underwent anthropometric assessment including weight recorded in minimal clothing to the nearest 0.1 kg, using a digital scale (Alfoset, Model HW-100KA1, Mumbai, India) and height being recorded to the nearest 0.1 centimeter, using a vertically mobile scale (Holtain, Crymych, UK). Body mass index was calculated as weight/height^2. WC was measured using a standard nonstretchable tape measure, at the narrowest point between the iliac crest and ribcage. The measured WC from our population was compared with the predicted WC derived from the equation $\log(WC) = (-2.589 \pm 0.020) + (0.6807 \pm 0.0052)\log \text{weight} - (0.814 \pm 0.020)\log \text{height}$ [18]. The measured WC was used for subsequent analysis. ABSI was defined as $WC/(BMI)^{2/3} \times (Height)^{1/2}$. In the above equation, weight was expressed in kilograms and height and WC in meters [18].

2.1. Heart Rate Variability Assessment

2.1.1. Preparation of the Subject. After explaining the process of ECG recording, subjects were asked to rest and relax in supine posture for 30 minutes. ECG recording was performed in a quiet room with soothing lighting and with comfortable temperature being maintained. Any electronic gadgets or metal or magnetic objects which could interfere with the ECG recording were removed. Skin was kept dry and rubbed with alcohol pad before application of electrodes.

2.1.2. Electrode Placement. Disposable electrodes were placed firmly onto the skin and we made sure that a good contact between the skin and the electrode was maintained. Electrodes were placed on the right arm, left arm, and left leg (lead II).

2.1.3. ECG Module. ECG module contains the electronics that detect electrical impulses from the heart and convert them into digital data that is transmitted to the computer. One end of the ECG module was connected to the subject through leads and another end to the USB port of the computer.

2.2. Processing of the ECG Signals. Analog lead II ECG signal obtained was digitized using analog to digital converter (CIO-AD16Jr A/D card). Sampling frequency of ECG signals was 1000 Hz using an IBM compatible PC and a data acquisition package (CVMS; World Precision Instruments Inc., Sarasota, FL, USA.) The data acquisition system included a threshold peak detection system, from which RR intervals were determined. The RR intervals were plotted as a tachogram (RR values versus interval number). Data segments of 128 sec duration were sampled at 2 Hz to create 256 point data sets. For each 10 min recording, an average of eight data sets of 256 points overlapping by half were processed. Heart rate variability analysis was performed using the frequency domain method. Power spectral density (PSD) was calculated using mathematical algorithm fast Fourier transform (FFT). The PSD was analyzed by calculating powers and peak frequencies for different frequency bands. No more than 3 data points were edited in a given segment [19].

2.2.1. Spectrum Estimation. FFT algorithm uses the signals in the time domain (RR interval) and transforms them into discrete frequency domain representation (power spectra). To prevent digital leakage during FFT Hanning digital filtering was used. The spectra obtained for the different data sets were averaged to reduce variance and to sharpen reproducible central peaks. Power was calculated in three bands. The 0.04–0.15 Hz band of RR power (referred to as the low-frequency band) reflects, at least in part, sympathetic nerve activity to the heart and partly parasympathetic activity, while the 0.15–0.4 Hz band (high-frequency band) reflects parasympathetic nerve activity to the heart. Absolute total power (TP) represents the total variance of the power spectral density, that is, 0–0.4 Hz band (msec^2). In addition to the absolute power, data for heart rate variability are also presented as normalized units, as recommended [20], where the power in the low- and high-frequency bands are expressed as percentage of the total power minus the power of the very-low-frequency band (0.0–0.04 Hz).

2.3. Statistical Analysis. The data was checked for normality using the Kolmogorov-Smirnov test. All normally distributed data are expressed as mean ± SD and median (Quartile1 and Quartile3), when not normally distributed. The differences between low and normal BMI study group were compared using independent t test. Bland-Altman plots [21] were constructed to examine limits of agreement between measured WC and the predicted WC. Heart rate variability indices were represented as log transformed (absolute units)

Table 1: Descriptive characteristics of study population.

Subject characteristics	Entire group (n = 178)	Normal BMI group (n = 105)	Low BMI group (n = 73)
Age (Yr)	40 ± 22	41 ± 23	35 ± 22
Weight (Kg)	54.3 ± 9.1	59.3 ± 7.9	47.1 ± 4.9
Height (m)	1.66 ± 0.07	1.66 ± 0.08	1.65 ± 0.07
Waist circumference (m)	0.74 ± 0.08	0.79 ± 0.01	0.68 ± 0.05*
BMI (Kg/m^2)	19.6 ± 2.6	21.4 ± 1.75	17.08 ± 0.9*
ABSI	0.079 ± 0.006	0.079 ± 0.006	0.078 ± 0.006
Heart rate (bpm)	63.1 ± 9.9	63.8 ± 9.7	62.1 ± 10.3
Low frequency power (LF) (ms^2)	568.6 (192.7–1155.9)	569.1 (196.1–1405.5)	562.6 (178.3–925.5)
High frequency power (HF) (ms^2)	538.4 (175.7–1309.4)	572.8 (146.3–1396.1)	522.8 (220.7–872.1)
Total power (TP) (ms^2)	1843.9 (718.6–3426.9)	2055.9 (685.1–4418.8)	1617.6 (90.3–2935)
Low frequency (nu)	55.2 (38.7–69.0)	59.5 (43.1–74.9)	51.1 (34.7–63.3)
High frequency (nu)	52.9 (40.7–66.2)	49.4 (36.2–63.6)	55.5 (44.9–70.7)
LF nu/HF nu ratio	1.1 (0.6–1.6)	1.2 (0.7–1.9)	0.9 (0.5–1.5)

Grouping based on BMI ranges ≥ 18.5 and < 25 as normal BMI group and BMI < 18.5 as low BMI group.
Data are mean ± standard deviation/median (interquartile range), $^*P < 0.05$.

data. Correlation of log transformed HRV indices with BMI, WC, and ABSI was examined using Pearson's correlation coefficient. Further, correlation of HRV indices with BMI, WC, and ABSI was examined in low BMI and normal BMI study groups. The correlation of HRV with age, ABSI, and WC was examined using linear regression models in the two BMI groups. Statistical significance was considered at $P < 0.05$. All statistical analyses were performed using SPSS v18, SPSS, Chicago, IL.

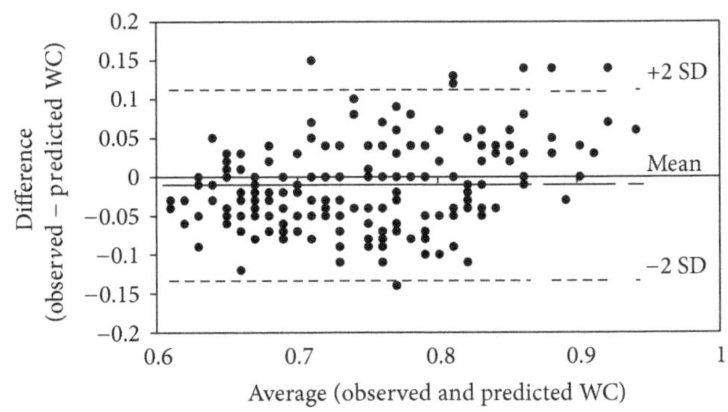

Figure 1: Bland Altman plot of WC differences between observed and predicted values. The solid line is at the bias and the dashed lines are at ±95% limits of agreement.

3. Results

Table 1 represents the anthropometric and HRV indices in absolute units. Out of the 178 subjects, based on their BMI, 73 subjects were categorized as low BMI group and 105 subjects as normal BMI group. Predicted WC was within limits of agreement when compared with measured WC. Mean of difference between observed WC from the study population and predicted WC was −0.01 and the limits of agreement were 0.10 and −0.12 (Figure 1). BMI was positively and significantly correlated with WC ($r = 0.77$, $P < 0.001$) and not with ABSI ($r = 0.028$, $P = 0.71$). HRV indices were significantly and negatively correlated with WC (log LF, $r = -0.25$, $P < 0.01$; log HF, $r = -0.30$, $P < 0.01$; log TP, $r = -0.22$, $P < 0.01$) and ABSI (log LF, $r = -0.54$, $P < 0.01$; log HF, $r = -0.59$, $P < 0.01$; log TP, $r = -0.55$, $P < 0.01$) but not with BMI (log LF, $r = 0.06$, $P = 0.4$; log HF, $r = 0.03$, $P = 0.7$; log TP, $r = 0.1$, $P = 0.2$). The correlation of ABSI with HRV remained significant even after controlling for age but not that of WC with HRV (Figure 2). When this correlation was examined separately within low BMI and normal BMI groups, ABSI was significantly correlated with all HRV indices in the normal group after controlling for age, whereas Log HF alone was correlated with ABSI in the low BMI group (Table 2). None of the HRV indices were correlated with WC in either of the groups after controlling for age. There was no significant correlation between the normalized values (nu of LF and HF) and ABSI, BMI or WC.

4. Discussion

Current study demonstrated that ABSI was significantly and negatively correlated with LF, HF, and TP indices of HRV. When grouped based on BMI, HF (an indicator of cardiac PNS activity) was negatively and significantly correlated with ABSI in both low BMI and normal BMI study group. However, LF (by far an indicator of cardiac SNS activity) and TP (total variability) were negatively correlated with ABSI only in normal BMI study group.

Nutritional transition is one of the major concerns in India [22]. Increased migration of individuals from rural to urban setup and increased urbanization has caused easy access to high caloric foods [23]. This has caused individuals with low BMI to shift towards either becoming normal BMI or even obese [24]. With reduced physical activity and

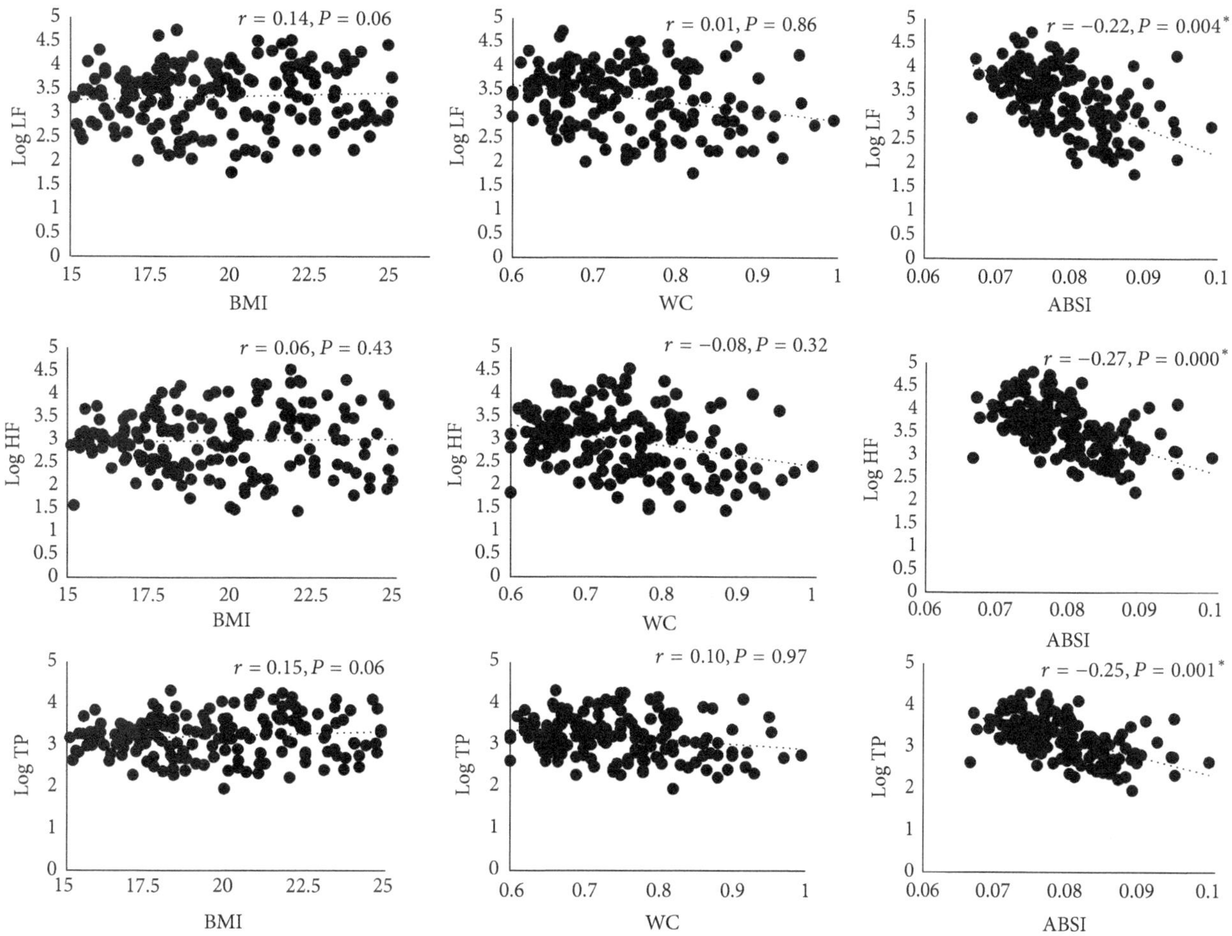

FIGURE 2: Scatter plots representing the partial correlation between BMI, WC, ABSI, and HRV after adjusting for age. Variables are represented after adjusting for age. $^*P < 0.05$ is considered statistically significant. Correlations without adjusting for age are represented in the results.

TABLE 2: Linear regression models describing the associations between HRV indices versus ABSI and WC in normal BMI and low BMI study groups.

Linear regression			Log LF β (Seβ)		Log HF β (Seβ)		Log TP β (Seβ)	
			Low BMI	Normal BMI	Low BMI	Normal BMI	Low BMI	Normal BMI
ABSI	Model 1	ABSI	−43.15 (8.5)*	−46.47 (7.1)*	−47.51 (7.7)*	−63.93 (8.6)*	−37.33 (7.3)*	−48.61 (6.9)*
	Model 2	ABSI	−13.73 (10.1)	−21.73 (8.9)*	−24.20 (9.4)*	−23.41 (10.1)*	−15.60 (8.9)	−20.33 (8.4)*
		AGE	−0.01 (0.0)*	−0.01 (0.0)*	−0.01 (0.0)*	0.02 (0.0)*	−0.01 (0.0)*	−0.01 (0.0)*
WC	Model 1	WC	−0.36 (0.0)*	−0.02 (0.0)*	−0.04 (0.0)*	−0.03 (0.0)*	−0.03 (0.0)*	−0.02 (0.0)*
	Model 2	WC	−0.01 (0.0)	−0.01 (0.0)	−0.02 (0.0)	−0.01 (0.0)	−0.01 (0.0)	−0.01 (0.0)
		AGE	−0.01 (0.0)*	−0.01 (0.0)*	−0.01 (0.0)*	−0.02 (0.0)*	−0.01 (0.0)*	−0.01 (0.0)*

Model 1 represents either ABSI or WC. Model 2 represents either ABSI or WC after controlling for age; values represent regression coefficient β (standard error of β) $^*P < 0.05$.

increased caloric consumption with preexisting low muscle mass, susceptibility to develop chronic disease is increasing in Indians [22–24]. One of the contributing factors is increased body fat accumulation, particularly fat within tissues, which is commonly termed as ectopic/visceral fat. The visceral deposition of fat is considered to be critical in defining the risk for metabolic disease [25] and is relatively more pronounced in Asian Indians compared to other ethnic groups [26]. Visceral fat carries greater risk of insulin resistance, CVS disease, metabolic syndrome [27, 28].

Measures of visceral fat are difficult, tedious and require high-end instruments [29]. Although these techniques might help us give accurate measures, their feasibility and application at a population level are questionable. Simple indices like WC have long been used as a surrogate of central adiposity [30]. Using WC, a simple index called ABSI has emerged of late, which could by far act as a surrogate of visceral fat [18]. In this study there was a good agreement between measured and predicted WC. This allowed us to utilize WC in deriving ABSI for Indian population.

Increasing central adiposity (measured using waist circumference) has been associated with a reduced autonomic functionality [17]. This has been demonstrated in individuals who are obese [31] or who are on a weight loss program [32]. There is lack of understanding of association between indices of central or visceral adiposity and autonomic nervous activity particularly in individuals with low-normal BMI.

Reduced autonomic function predominantly parasympathetic nervous activity has been linked to visceral fat [33]. Christou et al. reported higher abdominal-to-peripheral body fat distributions measured by DXA were strongly correlated with lower sympathetic and parasympathetic function in young and old healthy men [34]. Other studies have also shown that body fat distribution is a major factor, particularly visceral fat carrying the greatest risk for cardiovascular morbidity and mortality [26, 27, 35]. In keeping with this, data from the current study demonstrated that cardiac autonomic activity particularly HF (cardiac parasympathetic activity) was negatively correlated with ABSI across BMI range. This suggests that one of the possible explanations for reduced cardiac parasympathetic activity could well be due to increased ABSI which is by far an indicator of visceral fat.

Greater degree of autonomic involvement, that is, both cardiac sympathetic and parasympathetic nervous activity, is suggested to be associated with higher visceral adiposity [10, 34]. The fact that normal BMI group were having significantly greater WC and marginally higher ABSI though not significant suggests the possibility of greater visceral adiposity in this group. In addition, data from the current study demonstrated a negative association with cardiac sympathetic activity and ABSI in the normal study group. One of the mechanisms which has been linked to greater autonomic involvement (both components of ANS) and visceral fat includes reduced arterial distensibility [36]. Carotid artery distensibility is an important physiological determinant of cardiovagal baroreflex gain [37]. Therefore, it is possible that lower levels of carotid artery distensibility with elevated visceral fat would reduce mechanical transduction of arterial blood pressure into barosensory stretch [33]. Another mechanism could be greater circulating neurohumoural factors that act centrally modifying the cardiovagal baroreflex [38, 39]. Finally, impaired muscarinic receptor function has been demonstrated in an animal model of diet-induced obesity [40]. Thus it is possible that the reduction in cardiovagal baroreflex gain observed with elevated total body and abdominal visceral fat in the present study was due, in part, to reduced cardiac muscarinic receptor number and/or sensitivity.

Thin-fat phenotype, a term coined for Indian population with low-normal BMI, might well have greater visceral adiposity. This calls for urgent need to explore the visceral adiposity status in Indians across all BMI ranges. Possible causes for greater deposition of fat at tissue level may well be due to states of positive energy balance, environmental exposures such as inflammation, stress, tobacco, alcohol, and pollution interacting, and amplifying the effect of the deposition of fat on a thin frame [24]. Studies are further required to explore the role of above factors in contributing towards the deposition of visceral adiposity. Further, studies are also required to explore implication of ABSI in larger population based follow-up studies including females to understand the role of ABSI in predicting the metabolic risk in low BMI group.

In conclusion, greater ABSI by far an indicator of visceral adiposity was associated with reduced cardiac parasympathetic and sympathetic activity even in individuals with low BMI. Thus, even with slight increase in BMI among low BMI individuals, there could be a greater risk of cardiovascular morbidity and mortality.

Conflict of Interests

The authors declare that there is no conflict of interests regarding the publication of this paper.

Acknowledgment

Part of the data presented in the current study was funded by Nestle Foundation, Switzerland.

References

[1] D. B. Allison, K. R. Fontaine, J. E. Manson, J. Stevens, and T. B. VanItallie, "Annual deaths attributable to obesity in the United States," *The Journal of the American Medical Association*, vol. 282, no. 16, pp. 1530–1538, 1999.

[2] M. Duncan, M. Griffith, H. Rutter, and M. J. Goldacre, "Certification of obesity as a cause of death in England 1979–2006," *European Journal of Public Health*, vol. 20, no. 6, pp. 671–675, 2010.

[3] T. Pischon, H. Boeing, K. Hoffmann et al., "General and abdominal adiposity and risk of death in Europe," *The New England Journal of Medicine*, vol. 359, no. 20, pp. 2105–2120, 2008.

[4] A. B. Gonzalez, P. Hartge, J. R. Cerhan et al., "Body-Mass Index and mortality among 1.46 million white adults," *New England Journal of Medicine*, vol. 363, pp. 2211–2219, 2010.

[5] J. Bigaard, A. Tjønneland, B. L. Thomsen, K. Overvad, B. L. Heitmann, and T. I. A. Sørensen, "Waist circumference, BMI, smoking, and mortality in middle-aged men and women," *Obesity Research*, vol. 11, no. 7, pp. 895–903, 2003.

[6] B. M. Curtis and J. H. O'Keefe, "Autonomic tone as a cardiovascular risk factor: the dangers of chronic fight or flight," *Mayo Clinic Proceedings*, vol. 77, no. 1, pp. 45–54, 2002.

[7] D. Ziegler, "Diabetic cardiovascular autonomic neuropathy: prognosis, diagnosis and treatment," *Diabetes/Metabolism Reviews*, vol. 10, no. 4, pp. 339–383, 1994.

[8] A. I. Vinik, R. E. Maser, B. D. Mitchell, and R. Freeman, "Diabetic autonomic neuropathy," *Diabetes Care*, vol. 26, no. 5, pp. 1553–1579, 2003.

[9] B. Zahorska-Markiewicz, L. E. Kuagowska, C. Kucio, and M. Klin, "Heart rate variability in obesity," *International Journal of Obesity*, vol. 17, no. 1, pp. 21–23, 1993.

[10] K. Laederach-Hofmann, L. Mussgay, and H. Rúddel, "Autonomic cardiovascular regulation in obesity," *Journal of Endocrinology*, vol. 164, no. 1, pp. 59–66, 2000.

[11] H. R. Peterson, M. Rothschild, C. R. Weinberg, R. D. Fell, K. R. McLeish, and M. A. Pfeifer, "Body fat and the activity of the autonomic nervous sytem," *The New England Journal of Medicine*, vol. 318, no. 17, pp. 1077–1083, 1988.

[12] A. R. Shenoy, V. Doreswamy, J. P. Shenoy, and V. S. Prakash, "Impact of obesity on cardiac autonomic functions in middle aged males," *National Journal of Physiology Pharmacy and Pharmacology*, vol. 4, no. 3, 2014.

[13] R. Rajalakshmi, Y. VijayaVageesh, S. M. Nataraj, MuraliDhar, and C. G. Srinath, "Heart rate variability in Indian obese young adults," *Pakistan Journal of Physiology*, vol. 8, no. 1, pp. 39–44, 2012.

[14] Indian Institute of Population Sciences, Mumbai, India, National Family Health Survey—3, vol. 1, pp. 303–313, 2006.

[15] M. Vaz, A. V. Bharathi, S. Sucharita, and D. Nazareth, "Heart rate variability and baroreflex sensitivity are reduced in chronically undernourished, but otherwise healthy, human subjects," *Clinical Science*, vol. 104, no. 3, pp. 295–302, 2003.

[16] A. Misra and L. Khurana, "Obesity-related non-communicable diseases: South Asians vs White Caucasians," *International Journal of Obesity*, vol. 35, no. 2, pp. 167–187, 2011.

[17] B. G. Windham, S. Fumagalli, A. Ble et al., "The relationship between heart rate variability and adiposity differs for central and overall adiposity," *Journal of Obesity*, vol. 2012, Article ID 149516, 8 pages, 2012.

[18] N. Y. Krakauer and J. C. Krakauer, "A new body shape index predicts mortality hazard independently of body mass index," *PLoS ONE*, vol. 7, no. 7, Article ID e39504, 2012.

[19] S. Sucharita, A. V. Bharathi, A. V. Kurpad, and M. Vaz, "A comparative study of tests of cardiac parasympathetic nervous activity in healthy human subjects," *Physiological Measurement*, vol. 23, no. 2, pp. 347–354, 2002.

[20] "Heart rate variability: standards of measurement, physiological interpretation and clinical use. Task Force of the European Society of Cardiology and the North American Society of Pacing and Electrophysiology," *Circulation*, vol. 93, no. 5, pp. 1043–1065, 1996.

[21] J. M. Bland and D. G. Altman, "Statistical methods for assessing agreement between two methods of clinical measurement," *Lancet*, vol. 1, no. 8476, pp. 307–310, 1986.

[22] P. S. Shetty, "Nutrition transition in India," *Public Health Nutrition*, vol. 5, no. 1, pp. 175–182, 2002.

[23] L. Bowen, S. Ebrahim, B. de Stavola et al., "Dietary intake and rural-urban migration in India: a cross-sectional study," *PLoS ONE*, vol. 6, no. 6, Article ID e14822, 2011.

[24] A. V. Kurpad, K. S. Varadharajan, and I. Aeberli, "The thin-fat phenotype and global metabolic disease risk," *Current Opinion in Clinical Nutrition and Metabolic Care*, vol. 14, no. 6, pp. 542–547, 2011.

[25] K. P. McMillan, J. L. Kuk, T. S. Church, S. N. Blair, and R. Ross, "Independent associations between liver fat, visceral adipose tissue, and metabolic risk factors in men," *Applied Physiology, Nutrition and Metabolism*, vol. 32, no. 2, pp. 265–272, 2007.

[26] S. A. Lear, K. H. Humphries, S. Kohli, A. Chockalingam, J. J. Frohlich, and C. L. Birmingham, "Visceral adipose tissue accumulation differs according to ethnic background: results of the Multicultural Community Health Assessment Trial (M-CHAT)," *The American Journal of Clinical Nutrition*, vol. 86, no. 2, pp. 353–359, 2007.

[27] M. A. Banerji, N. Faridi, R. Atluri, R. L. Chaiken, and H. E. Lebovitz, "Body composition, visceral fat, leptin, and insulin resistance in Asian Indian men," *The Journal of Clinical Endocrinology and Metabolism*, vol. 84, no. 1, pp. 137–144, 1999.

[28] G. A. Rosito, J. M. Massaro, U. Hoffmann et al., "Pericardial fat, visceral abdominal fat, cardiovascular disease risk factors, and vascular calcification in a community-based sample the framingham heart study," *Circulation*, vol. 117, no. 5, pp. 605–613, 2008.

[29] K. van der Kooy and J. C. Seidell, "Techniques for the measurement of visceral fat: a practical guide," *International Journal of Obesity*, vol. 17, no. 4, pp. 187–196, 1993.

[30] M.-C. Pouliot, J.-P. Després, S. Lemieux et al., "Waist circumference and abdominal sagittal diameter: best simple anthropometric indexes of abdominal visceral adipose tissue accumulation and related cardiovascular risk in men and women," *The American Journal of Cardiology*, vol. 73, no. 7, pp. 460–468, 1994.

[31] A. Damodaran and B. Kabali, "Autonomic dysfunction in central obesity," *World Journal of Medical Sciences*, vol. 8, no. 2, pp. 118–122, 2013.

[32] G. E. Alvarez, B. M. Davy, T. P. Ballard, S. D. Beske, and K. P. Davy, "Weight loss increases cardiovagal baroreflex function in obese young and older men," *American Journal of Physiology—Endocrinology and Metabolism*, vol. 289, no. 4, pp. E665–E669, 2005.

[33] S. D. Beske, G. E. Alvarez, T. P. Ballard, and K. P. Davy, "Reduced cardiovagal baroreflex gain in visceral obesity: implications for the metabolic syndrome," *The American Journal of Physiology—Heart and Circulatory Physiology*, vol. 282, no. 2, pp. H630–H635, 2002.

[34] D. D. Christou, P. P. Jones, A. E. Pimentel, and D. R. Seals, "Increased abdominal-to-peripheral fat distribution contributes to altered autonomic-circulatory control with human aging," *The American Journal of Physiology—Heart and Circulatory Physiology*, vol. 287, no. 4, pp. H1530–H1537, 2004.

[35] B. Larsson, "Regional obesity as a health hazard in men—prospective studies," *Acta Medica Scandinavica*, vol. 223, no. 723, pp. 45–51, 1988.

[36] L. M. Resnick, D. Militianu, A. J. Cunnings, J. G. Pipe, J. L. Evelhoch, and R. L. Soulen, "Direct magnetic resonance determination of aortic distensibility in essential hypertension: relation to age, abdominal visceral fat, and in situ intracellular free magnesium," *Hypertension*, vol. 30, no. 3, pp. 654–659, 1997.

[37] I. Bonyhay, G. Jokkel, and M. Kollai, "Relation between baroreflex sensitivity and carotid artery elasticity in healthy humans," *The American Journal of Physiology—Heart and Circulatory Physiology*, vol. 271, no. 3, pp. H1139–H1144, 1996.

[38] S. Engeli and A. M. Sharma, "Role of adipose tissue for cardiovascular-renal regulation in health and disease," *Hormone and Metabolic Research*, vol. 32, no. 11-12, pp. 485–499, 2000.

[39] V. van Harmelen, M. Elizalde, P. Ariapart et al., "The association of human adipose angiotensinogen gene expression with abdominal fat distribution in obesity," *International Journal of Obesity*, vol. 24, no. 6, pp. 673–678, 2000.

[40] M. Pelat, P. Verwaerde, C. Merial et al., "Impaired atrial M2-cholinoceptor function in obesity-related hypertension," *Hypertension*, vol. 34, no. 5, pp. 1066–1072, 1999.

A Modern Cohort of Duodenal Obstruction Patients: Predictors of Delayed Transition to Full Enteral Nutrition

Sigrid Bairdain,[1] David C. Yu,[2] Chueh Lien,[1] Faraz Ali Khan,[1] Bhavana Pathak,[3] Matthew J. Grabowski,[1] David Zurakowski,[1,4] and Bradley C. Linden[5]

[1] *Department of Pediatric Surgery, Boston Children's Hospital, Harvard Medical School, 300 Longwood Avenue, Fegan Building, 3rd Floor, Boston, MA 02215, USA*
[2] *Department of Pediatric Surgery, Children's Hospital of Alabama, Birmingham, AL 35233, USA*
[3] *Department of Medicine, Johns Hopkins University, Baltimore, MD 21218, USA*
[4] *Department of Anesthesia, Boston Children's Hospital, Harvard Medical School, Boston, MA 02215, USA*
[5] *Pediatric Surgical Associates, Children's Hospitals and Clinics of Minnesota, MN 55404, USA*

Correspondence should be addressed to Sigrid Bairdain; sigrid.bairdain@childrens.harvard.edu

Academic Editor: Johannes B. van Goudoever

Background. A common site for neonatal intestinal obstruction is the duodenum. Delayed establishment of enteral nutritional autonomy continues to challenge surgeons and, since early institution of nutritional support is critical in postoperative newborns, identification of patients likely to require alternative nutritional support may improve their outcomes. Therefore, we aimed to investigate risk factors leading to delayed establishment of full enteral nutrition in these patients. *Methods.* 87 patients who were surgically treated for intrinsic duodenal obstructions from 1998 to 2012 were reviewed. Variables were tested as potential risk factors. Median time to full enteral nutrition was estimated using the Kaplan-Meier method. Independent risk factors of delayed transition were identified using the multivariate Cox proportional hazards regression model. *Results.* Median time to transition to full enteral nutrition was 12 days (interquartile range: 9–17 days). Multivariate Cox analysis identified three significant risk factors for delayed enteral nutrition: gestational age (GA) ≤ 35 weeks ($P < .001$), congenital heart disease (CHD) ($P = .02$), and malrotation ($P = .03$). *Conclusions.* CHD and Prematurity are most commonly associated with delayed transition to full enteral nutrition. Thus, in these patients, supportive nutrition should strongly be considered pending enteral nutritional autonomy.

1. Introduction

The duodenum is a common location for intestinal obstruction in the newborn requiring surgical intervention [1]. Intrinsic obstruction ranges from complete atresia with a gap to membranous web and to stenosis. In contrast to small intestinal atresia, ontogeny is thought to be secondary to failure of luminal recanalization [2]. Treatment often consists of operative duodenoplasty or bypass with a duodenoduodenostomy and has traditionally been performed through a laparotomy; however, laparoscopic approach has previously been shown to be a viable option [3]. Overall mortality for this congenital anomaly has improved; moreover, associated medical conditions are now responsible for most deaths [4].

Prolonged postoperative enteral feeding intolerance continues to challenge surgeons caring for this cohort of patients. Since early institution of nutritional support is critical in postoperative newborns, identification of patients likely to require alternative nutritional support is likely of benefit to this subset of patients. Therefore, our aim was to describe the short term nutritional outcomes and investigate risk factors leading to delayed establishment of full enteral nutrition in a modern cohort of patients with duodenal atresia at a single-institution.

2. Materials and Methods

Following approval from our Institutional Review Board (IRB number M08-08-0381), we performed a retrospective review

Table 1: Patients' characteristics with duodenal obstruction (N = 87).

Characteristic	Median	IQR	Range
Age at surgery, d	5	2–12	0–730
Gestational age, weeks	37	35–39	25–42
Birth weight, grams	2680	2145–3200	737–7000
Apgar score 1-minute	8	7-8	0–9
Apgar score 5-minutes	8	8-9	5–10
	Number	Percentage	
Gender			
Female	45	52%	
Duodenal anatomy			
Web	21	24%	
Stenosis	11	13%	
Atresia	55	63%	
Proximal bowel dilation	16	19%	
Technique of primary surgical Repair			
Duodenoduodenostomies	58	68%	
Duodenojejunostomies	6	7%	
Gastroduodenostomy	1	1%	
Duodenoplasty	15	17%	
Web excision	6	7%	
Feeding tube	13	15%	
Intestinal malrotation	27	31%	
Annular pancreas	17	20%	
Congenital heart disease	48	55%	
Down's syndrome	33	38%	
Imperforate anus	3	3%	
Hirschsprung's disease	1	1%	
EA/TEF	5	6%	

EA/TEF = esophageal atresia/tracheoesophageal fistula. IQR = interquartile range.

of records from 1998 to 2012 utilizing ICD-9 code for small intestine atresia (751.1) at Boston Children's Hospital (BCH). Patients were included if they had a primary surgical repair or surgical treatment for a complication of a previous repair of a newborn intrinsic duodenal obstruction during this 14-year period. We reviewed the following variables: gender, gestational age (GA), birth weight, Apgar 1-minute score, Apgar 5-minute score, and radiographic findings such as proximal dilatation, duodenal anatomy, technique of primary surgical repair, malrotation, annular pancreas, congenital heart disease (CHD), Down's syndrome, and esophageal atresia with or without tracheoesophageal fistula (EA/TEF).

Kaplan-Meier product-limit analysis with Greenwood estimators was used to calculate median time to full enteral nutrition, with the log-rank test to compare area under the survival curves [5]. Covariates were evaluated by univariate and multivariate analysis using the Cox proportional-hazards model to identify independent predictors with the likelihood ratio test to assess significance [6] and log-minus log plot to verify the assumption of proportional hazards over time [7]. Hazard ratios (HR) and 95% confidence intervals (CI) were determined for significant multivariate predictors of time to full enteral nutrition [8]. Statistical analysis was performed using IBM SPSS statistics (version 21.0, IBM, Armonk, NY). Two-tailed $P < .05$ was considered statistically significant. Power analysis indicated that the sample sizes of patients with potential risk factors provided 80% power for detecting a significant hazard ratio of 0.75 in delayed transition to full enteral nutrition (version 7.0, nQuery Advisor, Statistical Solutions, Saugus, MA).

3. Results

3.1. Demographics. A total of 87 patients were found to have an intrinsic duodenal obstruction and met inclusion criteria. Median age was 37 weeks (interquartile range [IQR]: 35–39 weeks). Median birth weight was 2680 grams (IQR: 2145–3200 grams). Median Apgar scores were 8 at 1 and 5 minutes (IQR: 7-8 and 8-9, resp.). There was a slight female predominance in the cohort (52% versus 48%). Proximal bowel dilatation was documented in 19% (16/86) of the patients (Table 1). A majority of the duodenal anatomy encountered was atresia (63%), followed by webs (24%) and stenosis (13%). Concurrent congenital anomalies included CHD (55%) and Down syndrome (38%) (Table 1). CHD was further subdivided into the following types encountered:

9 (19%) PDA; 21 (44%) ASDs and/or VSDs; 7 (15%) tetralogy of Fallot (TOF); 2 (4%) Coarctation of the aorta; 6 (12%) hypoplastic hearts or pulmonary arteries; 2 (4%) complete AV canal (CAVC); and 1 (2%) bicuspid aortic valves.

3.2. Primary Operative Characteristics. Choice of operation, as well as whether it was performed open or laparoscopically, was dictated by the choice of the operating surgeon. Five (6% or 5/86) of the operations were performed laparoscopically, while the remaining 81 were performed through a laparotomy. The techniques described for the primary duodenal repair included 58 duodenoduodenostomies, 6 duodenojejunostomies, 1 gastroduodenostomy, and 15 duodenoplasty repairs (with or without web excision), and 6 patients had web excision without duodenoplasty. One patient was not repaired given that she died two days into her hospital course from complex cardiac complications.

3.3. Intraoperative Findings: Single versus Multiple Intestinal Obstructions. Six (7% or 6/86) patients were found to have a second, distal, concurrent, and intestinal obstruction at the time of their proximal duodenal obstruction repair. Fifty-six percent of the operating surgeons performed a maneuver to detect a concurrent obstruction, while no maneuver was noted in the remaining 44% of operative notes. These maneuvers included inspection alone (30%), passage of small transanastomotic catheter (22%), and passage of transanastomotic catheter with small fluid bolus (3%). As a result of these maneuvers, the following anomalies were identified; two patients had a distal duodenal web, two had a second duodenal atresia, one had a jejunal atresia, and one had multiple distal atresias in both the duodenum and the jejunum.

3.4. Nutritional Perioperative Characteristics. Eighty patients received parenteral nutrition (PN). Seven patients did not receive PN; six patients never received PN and one patient never had the opportunity to receive PN. The cohort that received PN required a median of 10 days on PN (IQR: 6–12 days). The median time to 100% EN was 12 days (IQR: 9–17 days), excluding the one patient who died 2 days into her hospital course. The median time to 100% EN in those patients who never received PN was 7 days (range 5–26 days). 71 out of the 84 patients were transitioned to 100% oral feeds prior to discharge at a median of 13 days (IQR: 10–21 days (Table 2).

Thirteen patients received a transanastomotic feeding tube during their operation. Their median time on PN was 9 days (IQR: 5–16 days), with a full range 0–65 days, versus the larger cohort ($P = .29$, Mann-Whitney U-test). Regarding time to 100% enteral nutrition, the median time for those with feeding tubes was 9 days (IQR: 7–22 d), with full range 2–63 days, versus the larger cohort ($P = .28$, Mann-Whitney U-test). There were no anastomotic leaks reported in this smaller cohort of patients.

3.5. Univariate Analysis. Variables were tested as potential risk factors for delayed enteral feeding. Those included gender, GA ≤ 35 weeks, birth weight, Apgar 1 scores, duodenal anatomy, proximal bowel dilatation, technique of primary surgical repair, presence of feeding tube, malrotation, annular pancreas, CHD, Down's syndrome, and EA/TEF (Table 3). Of these, GA ≤ 35 weeks, birth weight, atresia anatomy, surgical repair, malrotation, and CHD were found to be significantly associated with delayed progression to full enteral feeding on univariate analysis.

Table 2: Nutritional characteristics of the cohort.

Characteristic	Median time	IQR	Range
Parenteral nutrition, d	10	6–12	0–258
Time to 100% EN, d	12	9–17	2–211
Time to 100% PO*, d	13	10–21	3–69
Discharge time, d	17	12–31	0–1852

*Based on the 71 who transitioned to 100% by mouth (PO). EN = enteral nutrition; IQR = interquartile range.

3.6. Multivariate Analysis. Multivariate Cox regression analysis revealed three independent predictors of delayed progression to full EN. Those variables were gestational age 35 weeks or less ($P < .001$), CHD ($P = .02$), and malrotation ($P = .03$). Variables such as birth weight, atresia anatomy, and surgical repair were not found to be independent predictors of delayed progression to full enteral feeding. Patients were then further classified based on the presence or absence of CHD and then stratified by GA.

Among patients without CHD who had GA > 35 weeks ($n = 30$) the median time to 100% EN progression was 10 days (IQR: 8–12 days), whereas if GA ≤ 35 weeks ($n = 9$) the median time to progression was 16 days (IQR: 11–20 days) ($P = .003$, log-rank test = 9.12) (Figure 1(a)). Among patients with CHD who had GA > 35 weeks ($n = 27$), the median time to progression was 11 days (IQR: 8–14 days), whereas for those with GA ≤ 35 weeks ($n = 20$) the median time to progression was 23 days (IQR: 15–30 days) ($P < .001$, log-rank test = 16.21) (Figure 1(b)).

3.7. Short-Term Mortality and Outcomes. Three patients died, all of whom had significant CHD. The first patient died two days into her hospital course. The second patient, born with a hypoplastic left heart and aortic coarctation, died of uncontrollable pulmonary hypertension shortly following an open duodenal atresia repair. The third patient, born with tetralogy of Fallot (TOF) and a chromosome 12 deletion, died almost one year following surgery while being in the hospital for a cardiac procedure. Among the 84 survivors, the median followup was 20 months (IQR: 5–48 months).

There were 13 (15%) major surgical complications following initial operative experience (Table 4). The 13 surgical complications included six bowel obstructions, four anastomotic leaks, two late-strictures, and one incisional hernia. Twenty-four percent (10/42 patients) occurred between 1998 and 2005 versus 3% (3/44) between 2006 and 2012 ($P = .04$). The earlier cohort excluded one patient from analysis as she was an early mortality and was never repaired. None of the patients with anastomotic leaks had a feeding tube present.

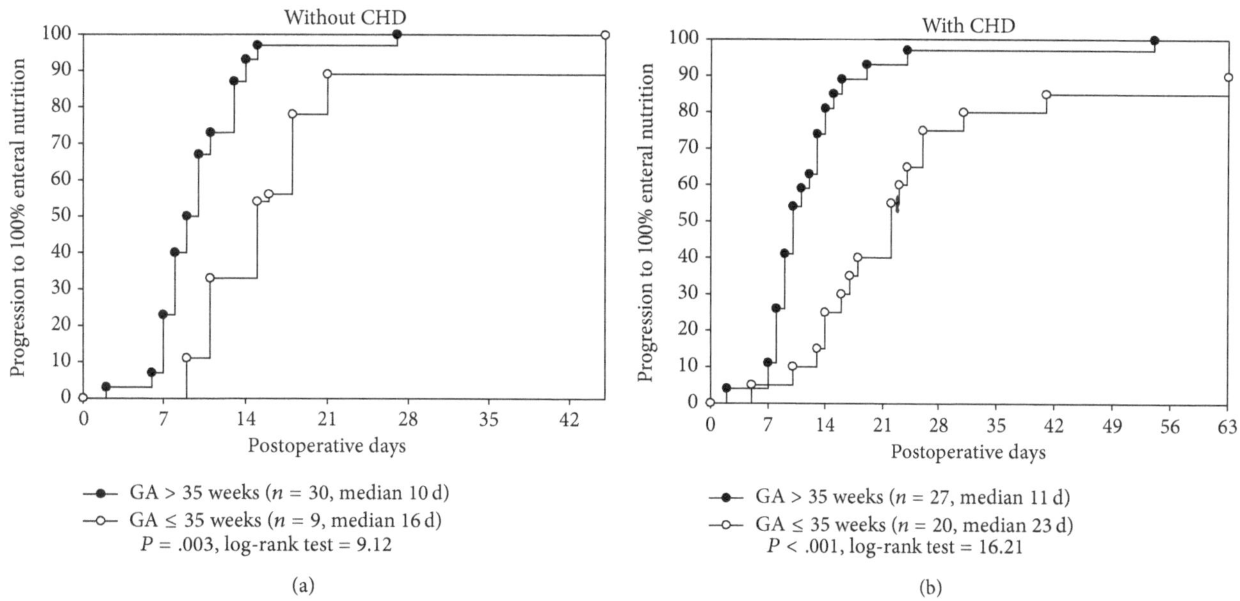

Figure 1: (a) Kaplan-Meier curves illustrating the progression to 100% enteral nutrition (100% EN) in patients without congenital heart disease (CHD). The importance of low gestational age (GA) is that if a patient was GA > 35 weeks ($n = 30$) then the median time to progression was 10 days (IQR: 8–12 days). However, if the GA ≤ 35 weeks ($n = 9$) then the median time to progression was 16 days (IQR: 11–20 days) ($P = .003$, log-rank test = 9.12). (b) Kaplan-Meier curves illustrating the progression to 100% enteral nutrition (EN) in patients with congenital heart disease (CHD) and illustrates the synergistic effect of CHD on prematurity. For those patients with GA > 35 weeks ($n = 27$) and CHD, the median time to progression was 11 days (IQR: 8–14 days). For those patients with GA ≤ 35 weeks ($n = 20$) and CHD, their median time to progression was 23 days (IQR: 15–30 days) ($P < .001$, log-rank test = 16.21). Both prematurity and the presence of CHD translated into a much longer delay in progression to 100% EN.

Medical complications were also reviewed in this cohort. The medical complications included 2 episodes of line sepsis and one episode of necrotizing enterocolitis (NEC). Treatment of the medical complications was as follows: (1) the two episodes of line sepsis were treated with antibiotics; and (2) the one episode of NEC was treated with a 14-day course of antibiotics and bowel rest.

4. Discussion

This study describes a cohort of 87 patients who were treated for intrinsic duodenal obstruction in the newborn period at a single institution. This may be one of the largest series of repairs as compared to other previously published series [1]. Six patients (7%) did have a concurrent second intestinal obstruction, while 13 patients had intraoperative nasojejunal transanastomotic feeding tubes placed without anastomotic complications. Variables that had previously been associated with delayed transition to full enteral feedings were not important in our study, including proximal bowel dilatation, duodenal anatomy, or technique of primary surgical repair. We concluded that the independent predictors for longer postoperative progression to full enteral nutrition were prematurity, CHD, and malrotation in our cohort.

In this cohort of patients, the mortality rate across 14 years was 3%, which is comparable to the approximately 4% early operative mortality reported in the available literature [1, 2]. Presence of complex cardiac conditions, Down's syndrome, and other intestinal atresia and trachea-esophageal fistula has been reported to place these patients at higher risk of late morbidity and mortality [9]. 55% of patients in this cohort had CHD, while 38% patients had Down's syndrome. Contrary to the findings of worse outcomes with associated Down's syndrome, no such association was elucidated in this cohort [10].

Recent literature has suggested that the repair of duodenal atresia can be effectively and safely performed laparoscopically [11–15]. Our study only had 5 laparoscopic operations performed; however, many laparoscopic-based studies are suggesting an early return of bowel function congruent with laparoscopic procedures in general. For example, Spilde et al. showed that the length of postoperative hospitalization (20.1 versus 12.9 days; $P = .01$), time to initial feeding (11.3 versus 5.4 days; $P = .002$), and time to full oral intake (16.9 versus 9 days; $P = .007$) were shorter in the group undergoing laparoscopic repair versus open repair [11]. Further studies are needed to elucidate this point in particular.

As reported in this study, slightly prolonged transition time to full enteral feeding may be in part due to a greater proportion of patients undergoing open operative approach versus an inherent component of the study cohort, including the initiation of PN in 80 (92%) patients within the cohort. Similar to a recently published study, the small subset of patients ($n = 6$) in our study who never received PN had a faster transition to full enteral nutrition, 7 days versus 12 days for the entire cohort [16]. The two variables though,

Table 3: Predictors of delayed transition to full enteral nutrition.

Variable	Univariate analysis	Multivariate Cox regression analysis		
	P value	Hazard ratio	95% CI	*P* value
Gender	.15			.18
GA ≤ 35 weeks	<.001	0.33	0.19–0.54	<.001*
Birth weight	<.001			.79
Apgar, 1-min	.13			.52
Proximal dilatation**	.60			.59
Duodenal anatomy	<.001			.32
Technique of primary surgical repair	<.01			.16
Feeding tube	.89			.76
Malrotation	.02	0.58	0.36–0.94	.03*
Annular pancreas	.51			.47
Congenital heart disease	.003	0.59	0.37–0.93	<.02*
Down's syndrome	.74			.71
EA/TEF	.14			.40

GA = gestational age, EA/TEF = esophageal atresia/tracheoesophageal fistula.
CI = confidence interval. *Statistically significant independent predictor of delayed transition. **Proximal bowel dilation noted on radiographic or intraoperative findings.

Table 4: Perioperative complications*.

Complication	*n* (%)	Time period 1998–2005 (*N* = 42)	Time period 2006–2012 (*N* = 44)
Small bowel obstructions	6 (7%)	3	3
Anastomotic leaks	4 (5%)	4	0
Anastomotic stricture	2 (2%)	2	0
Incisional hernia	1 (1%)	1	0
Total	13 (15%)	10 (24%)	3 (7%)†

*Based on a total of 86 patients (one patient was excluded due to early death at 2 days). †Statistically significant lower complication rate since 2006 (P = .04, Fisher's exact test).

that had the greatest influence overall, were prematurity and presence of CHD; both were noted to be associated with increased odds of delayed transition to full enteral nutrition and outweighed all other variables. Even with this being a retrospective study and there being an absence of a standardized approach to feeding advancement, we do believe that there is the potential of significant benefit of early institution of appropriate nutritional support and using these risk factors to tailor the appropriate nutritional regimen.

From a surgical perspective, several important findings also arose from this study. First, intraoperatively, six patients (7%) were noted to have a second, distal, concurrent, and intestinal obstruction, representing a higher rate of concurrent intestinal obstruction as compared to (1–3%) the rate reported in the available literature [2, 13]. This does suggest that the evaluation for presence of concurrent obstruction may be worthwhile to avoid the morbidity associated with a missed lesion. Other major complications did include line sepsis in two patients; however, there was not an increased rate in the highest risk groups [10, 16]. Secondly, there was a small degree of a time-epoch phenomena. For example, it appeared that between the two epochs 1998–2005 and 2006–2012, there was a significant decrease in the number of associated in-hospital complications (P = .04) in the time period between 2006 and 2012. This may be due in part to both a learning curve and a higher volume of cases performed, which is congruent with more recent studies but further studies are needed [17].

Lastly, none of the patients in our cohort who had received a transanastomotic tube developed a leak or developed a stricture and had a slightly decreased transition to full enteral feeds. More recent studies have shown that use of a transanastomotic feeding tube decreases not only the postoperative time to initiation of enteral nutrition, but also the time to full enteral nutrition delivered proximal to the anastomosis [18–21]. There was a trend in decreased number of days within our subset of patients who received transanastomotic feeding tubes but did not remain on multivariate analysis. We do believe that with a larger number of patients and a standardized placement of these transanastomotic feeding tubes, we would anticipate a significant difference in transition to full enteral nutrition. Further studies on a larger scale are needed.

This study has several limitations. These include its respective design, despite medical records being reviewed in full detail. Another limitation was that there was no overt standardization for repair guidelines (open versus laparoscopic or type of repair employed), or whether a patient received a transanastomotic feeding tube versus an access line for nutrition. However, these patients were followed postoperatively with minimal loss to follow-up and no significant long-term complications.

5. Conclusion

Based on our findings, prematurity, CHD, and malrotation predict a longer time to reach full enteral nutrition; therefore, supportive nutrition is suggested. Furthermore, we did not see an associated increase in leak rate in the small subset of patients who received transanastomotic feeding tubes. Given the aforementioned findings and limitations, further studies are needed to elucidate a standardized methodology for nutritional support during clinical management of these patients in the setting of single or multiple intestinal obstructions. In the absence of prospectively collected data, these risk factors can be used to potentially identify patients with a greater risk of delayed transition to full enteral nutrition and therefore tailor their nutritional support regimen accordingly.

Disclosure

IRB approval M08-08-0381 was obtained.

Conflict of Interests

The authors declare that there is no conflict of interests regarding the publication of this paper.

References

[1] L. K. Dalla Vecchia, J. L. Grosfeld, K. W. West, F. J. Rescorla, L. R. Scherer, and S. A. Engum, "Intestinal atresia and stenosis: a 25-year experience with 277 cases," *Archives of Surgery*, vol. 133, no. 5, pp. 490–497, 1998.

[2] J. L. Grosfeld and F. J. Rescorla, "Duodenal atresia and stenosis: reassessment of treatment and outcome based on antenatal diagnosis, pathologic variance, and long-term follow-up," *World Journal of Surgery*, vol. 17, no. 3, pp. 301–309, 1993.

[3] S. S. Rothenberg, "Laparoscopic duodenoduodenostomy for duodenal obstruction in infants and children," *Journal of Pediatric Surgery*, vol. 37, no. 7, pp. 1088–1089, 2002.

[4] R. Murshed, G. Nicholls, and L. Spitz, "Intrinsic duodenal obstruction: trends in management and outcome over 45 years (1951–1995) with relevance to prenatal counselling," *The British Journal of Obstetrics and Gynaecology*, vol. 106, no. 11, pp. 1197–1199, 1999.

[5] E. L. Kaplan and P. Meier, "Nonparametric estimation from incomplete observations," *Journal of the American Statistical Association*, vol. 53, pp. 457–481, 1958.

[6] D. R. Cox, "Regression models and life-tables," *Journal of the Royal Statistical Society B: Methodological*, vol. 34, pp. 187–220, 1972.

[7] M. H. Katz and W. W. Hauck, "Proportional hazards (Cox) regression," *Journal of General Internal Medicine*, vol. 8, no. 12, pp. 702–711, 1993.

[8] M. H. Katz, *Multivariable Analysis: A Practical Guide for Clinicians and Public Health Researchers*, Cambridge University Press, New York, NY, USA, 3rd edition, 2011.

[9] M. A. Escobar, A. P. Ladd, J. L. Grosfeld et al., "Duodenal atresia and stenosis: long-term follow-up over 30 years," *Journal of Pediatric Surgery*, vol. 39, no. 6, pp. 867–871, 2004.

[10] R. Niramis, M. Anuntkosol, A. Tongsin, and V. Mahatharadol, "Influence of Down's syndrome on management and outcome of patients with congenital intrinsic duodenal obstruction," *Journal of Pediatric Surgery*, vol. 45, no. 7, pp. 1467–1472, 2010.

[11] T. L. Spilde, S. D. St. Peter, S. J. Keckler, G. W. Holcomb III, C. L. Snyder, and D. J. Ostlie, "Open vs laparoscopic repair of congenital duodenal obstructions: a concurrent series," *Journal of Pediatric Surgery*, vol. 43, no. 6, pp. 1002–1005, 2008.

[12] P. A. Valusek, T. L. Spilde, K. Tsao, S. D. S. Peter, G. W. Holcomb III, and D. J. Ostlie, "Laparoscopic duodenal atresia repair using surgical U-clips: a novel technique," *Surgical Endoscopy*, vol. 21, no. 6, pp. 1023–1024, 2007.

[13] S. D. St. Peter, D. C. Little, K. A. Barsness et al., "Should we be concerned about jejunoileal atresia during repair of duodenal atresia?" *Journal of Laparoendoscopic and Advanced Surgical Techniques*, vol. 20, no. 9, pp. 773–775, 2010.

[14] N. M. Bax, B. M. Ure, D. C. van der Zee, and I. van Tuijl, "Laparoscopic duodenoduodenostomy for duodenal atresia," *Surgical Endoscopy*, vol. 15, no. 2, p. 217, 2001.

[15] J. L. Grosfeld, *Pediatric Surgery*, Mosby/Elsevier, Philadelphia, Pa, USA, 2006.

[16] M. Bishay, B. Lakshminarayanan, A. Arnaud et al., "The role of parenteral nutrition following surgery for duodenal atresia or stenosis," *Pediatric Surgery International*, vol. 29, no. 2, pp. 191–195, 2013.

[17] D. C. van der Zee, "Laparoscopic repair of duodenal atresia: revisited," *World Journal of Surgery*, vol. 35, no. 8, pp. 1781–1784, 2011.

[18] E. Arnbjörnsson, M. Larsson, Y. Finkel, and B. Karpe, "Transanastomotic feeding tube after an operation for duodenal atresia," *European Journal of Pediatric Surgery*, vol. 12, no. 3, pp. 159–162, 2002.

[19] K. P. Moriarty, N. N. Jacir, B. H. Harris, L. A. Latchaw, F. M. Robertson, and T. M. Crombleholme, "Transanastomotic feeding tubes in repair of esophageal atresia," *Journal of Pediatric Surgery*, vol. 31, no. 1, pp. 53–55, 1996.

[20] S. Mishra, R. Agarwal, M. Jeevasankar, A. K. Deorari, and V. K. Paul, "Minimal enteral nutrition," *Indian Journal of Pediatrics*, vol. 75, no. 3, pp. 267–269, 2008.

[21] G. Q. Duan, M. Zhang, X. H. Guan, and Z. Q. Yin, "Intraoperative placement of transnasal small intestinal feeding tube during the surgery in 5 cases with high position intestinal obstruction and postoperative feeding," *Chinese Journal of Pediatrics*, vol. 50, no. 9, pp. 705–707, 2012.

27

Reliability of a Fully Automated Interpretation of γ-H2AX Foci in Lymphocytes of Moderately Trained Subjects under Resting Conditions

Juliane Heydenreich,[1,2] Christoph Otto,[1] Frank Mayer,[1] and Anja Carlsohn[1,3]

[1] *Sports Medicine and Sports Orthopaedics, University Outpatient Clinic, University of Potsdam, 14469 Potsdam, Germany*
[2] *ZIEL-Research Centre for Nutrition and Food Sciences, Munich Technical University, 85354 Freising-Weihenstephan, Germany*
[3] *Institute of Health Science, Schwäbisch Gmünd University of Education, 73529 Schwäbisch Gmünd, Germany*

Correspondence should be addressed to Juliane Heydenreich; juliane.heydenreich@googlemail.com

Academic Editor: Stan Kubow

Background. Analysis of γ-H2AX foci is a promising approach to evaluate exercise-induced DNA damage. However, baseline levels and day-to-day variability of γ-H2AX foci have not been investigated in healthy subjects at rest. *Methods.* Blood was taken from eight moderately trained healthy males (29 ± 3 yrs, 1.84 ± 0.03 m, and 85 ± 6 kg) at two separate days (M1/M2) after 24-hour exercise cessation. Number of γ-H2AX foci per 100 lymphocytes (N), number of foci per affected lymphocyte (NAL), percentage of affected lymphocytes (PAL), and diameter (D) of γ-H2AX foci were analyzed (mean ± SD). Differences between M1 and M2 were analyzed using paired t-tests ($\alpha = 0.05$). Day-to-day variability was evaluated by calculating the coefficients of variation (CV%), bias, and limits of agreement (LoA). *Results.* There were no statistically significant differences between M1 (N: 7.6 ± 4.4, NAL: 1.2 ± 0.2, PAL: 5.9 ± 2.6%, and D: 0.63 ± 0.07) and M2 (N: 8.4 ± 4.6, NAL: 1.3 ± 0.1, PAL: 6.9 ± 4.2%, and D: 0.66 ± 0.06). CV was calculated to be 98.5% (N), 88.9% (PAL), 11.3% (NAL), and 8.0% (D). Bias (LoA) was 0.75 (−15.2/13.7), −0.02 (−0.36/0.33), −1.0 (−11.9/9.9), and −0.04 (−0.16/0.09), respectively. *Conclusions.* Background level in healthy subjects is approximately 0.07 to 0.09 γ-H2AX foci/cell. NAL and D are reliable measures.

1. Introduction

Exercise has been shown to induce an increased formation of reactive oxygen species (ROS) such as superoxide, hydroxyl radicals, or hydrogen peroxide [1, 2]. Sources of exercise-induced ROS production are the mitochondrial electron transport chain [3–5], the xanthine oxidase pathway [6], activated neutrophiles [6], and membrane-bound NAD(P)H oxidases of the skeletal muscle [7, 8]. Excess ROS generation may damage cell components such as lipids, proteins, and nucleic acids [9]. Exercise-induced DNA damage has been observed following acute bouts of exhaustive, prolonged, or eccentric exercises in humans [10–12]. Here, DNA double-strand breaks are the most harmful lesions and may occur due to different treatments [13].

Although acute bouts of unaccustomed exercises may induce DNA damage, chronic exercises as well as adequate antioxidant intake may alleviate oxidative damage [14–16]. For example, Miyazaki et al. have shown that twelve weeks of high-intensity endurance training in previously untrained men alleviates oxidative stress induced by an acute session of exercise [15].

Regarding antioxidant intake, Harms-Ringdahl et al. have shown that exercise-induced increase of 8-oxodG, as a marker of oxidative stress, is reduced after five weeks of daily tomato juice consumption containing approximately 15 mg lycopene per serving [17]. A daily intake of 1200 mg vitamin E for 14 days was capable of reducing exercise-induced DNA strand breakage in white blood cells of moderately trained subjects [18]. Following an acute ultraendurance event, a significant inverse correlation between oxygen radical absorbance capacity (ORAC) of the blood and indices of oxidative DNA damage was found. In the same study, a significant increase in blood antioxidants immediately after the race was observed [19].

Thus, methods that reliably measure DNA damage, despite the noise of day-to-day variability in antioxidant intake and exercise intensity, kind, and duration, are required to assess the effects of acute bouts of exercises in trained subjects.

The analysis of γ-H2AX foci is a promising marker of DNA double-strand breaks *in vitro* and *in vivo* [20]. The histone protein H2AX serves as key factor in the repair process of DNA damage, which targets damaged DNA and recruits other proteins of the DNA repair machinery [21]. As an exception among H2A proteins H2AX is phosphorylated on the serine 139 residue in the presence of damaged DNA. The phosphorylated H2AX (γ-H2AX) may be visualized using fluorescence labeled antibodies. This allows the counting of cells with damaged DNA, to calculate the number of foci per 100 cells, and to measure the diameter of γ-H2AX foci. Analysis of γ-H2AX foci has been shown to accurately measure DNA double-strand breaks [22]. Visual interpretation of fluorescent foci patterns and counting γ-H2AX foci with the naked eye is time consuming and has some limitations, such as investigator-related artefacts [23]. Recently, a fully automated device has been developed and has been shown to highly correlate with visually analyzed foci counts [23].

As the analysis of γ-H2AX foci is a sensitive marker of DNA damage and has been proven to detect DNA double-strand breaks at low levels of DNA damage [24], it promises to be a valuable tool to evaluate exercise-induced DNA damage. For well-trained athletes, however, regular exercise training seems to evoke adaptative responses capable of reducing oxidative damage [3, 25–27]. Therefore, the day-to-day variability of γ-H2AX foci in trained subjects under resting conditions needs to be investigated before assessing the effects of a given exercise bout in athletes.

Therefore, the purpose of this study was to analyze the reliability of γ-H2AX foci in lymphocytes of moderately trained subjects under resting conditions.

2. Methods

2.1. Participants. Twelve moderately trained healthy males were recruited to participate in the study. Inclusion criteria comprised an age between 25 and 40 years and a body mass index between 18 and 25 kg/m^2. Participants were moderately active with one to three exercise sessions per week. Females, smokers, and volunteers with acute or chronic diseases were excluded from the study. Changes of dietary behaviour or of exercise loads, alcohol intake, supplement use, or any medication during the study period served as exclusion criteria. Subjects who performed exercise training within 24 hours before the measurements were excluded from the analysis ($N = 4$).

Finally, eight subjects were included into the study and the data analysis (29.3 ± 3.2 yrs, 184 ± 3 cm, and 84.7 ± 6.1 kg). All subjects gave written informed consent before participating in the study. The study was approved by the institutional scientific examination board.

2.2. Study Design. The study was conducted in a test-retest design, repeating the same measurements of day 1 (M1) after seven days (M2) under standardized conditions. Before the first measurement, participants underwent a clinical examination, where inclusion and exclusion criteria were verified by means of a checklist. After refraining from exercise training for 24 hours, venous blood samples were taken. During the whole duration of the study, participants were asked to avoid strenuous physical activity. A physical activity record that included duration, kind, and ratings of perceived exertion of exercise was handed out to verify that subjects followed the instructions on physical activity.

2.3. Blood Analysis. Venous blood samples at M1 and M2 were taken from the brachial vein between 7 and 9 a.m. to minimize the effect of physical activity before the blood sample collection. Commercial tubes coated with EDTA to prevent blood clotting were used to collect 2 × 2.7 mL of whole blood. Immediately after blood draw, samples were stored at 4°C. One blood sample was used to analyze blood parameters (hematocrit [%], haemoglobin [g/dL], amount of lymphocytes [%], and number of red blood cells [10^6/mm^3]) and concentration of creatine kinase [U/L].

The other blood sample was used to assess double-strand breaks (DSB) in lymphocytes. In the first step, isolated cells were fixed in a neutral buffered 4% formaldehyde solution (Merck, Darmstadt, Germany) for 15 minutes at room temperature. Afterwards, the samples were washed three times with PBS (Merck, Darmstadt, Germany). Cell membranes of the fixed cells were incubated for five minutes with Triton X100 (Merck, Darmstadt, Germany; 0.1% in PBS v/v), followed by other three washings with PBS. Afterwards, unspecific binding was blocked using a BSA/PBS buffer. In the third step, a primary antibody targeting γ-H2AX was added which was then bound to a secondary antibody conjugated with Alexa Fluor 488 (Invitrogen, Karlsruhe, Germany). To detect antibody-bound γ-H2AX, a fluorescent mounting medium (Dako Deutschland, Hamburg, Germany) was added. 4,6′-Diamidino-2-phenylindole (DAPI, Sigma-Aldrich) was used to dye all cells' nuclei.

To analyse γ-H2AX foci in lymphocytes and to count the cells an automated indirect immune fluorescence device was adopted (AKLIDES Nuk Human Lymphocyte Complete, Medipan, Dahlewitz, Germany). A pattern recognition algorithm of the automated fluorescence interpretation system AKLIDES (Medipan, Dahlewitz, Germany) was used for evaluation of γ-H2AX foci as described elsewhere [28, 29]. For determination of γ-H2AX foci a 60x magnification of the microscope objective was used.

The following parameters were assessed: (1) number of γ-H2AX foci per 100 cells, (2) number of cells with γ-H2AX foci per 100 cells, (3) mean number of γ-H2AX foci per cell with γ-H2AX foci, and (4) the mean diameter of γ-H2AX foci.

2.4. Standardization of Diet and Physical Activity. To alleviate the acute effects of diet and physical activity, both exercise training and antioxidant intake were standardized before blood sample collection. A standardized dietary protocol was used to record the nutritional intake 24 hours prior to both measurements. Written and oral instructions on how to record the diet were given. Descriptions of portion sizes and standard household measures were provided. Dietary records from the first measurement were handed out before the second measurement, and participants were asked to follow the same diet in the 24 hours before M2. Nutrient and antioxidant intake were analyzed using nutrition analysis software based on the German food database (Prodi NutriScience 5.7 expert, Hausach, Germany). All nutritional analyses were performed by the same experienced examiner (JH).

To avoid artefacts due to exercise, subjects were requested to refrain from any exercises and strenuous physical activity for at least 24 hours before the measurements. This request was given as oral and written instructions to the participants. To verify the subjects' compliance to physical activity instructions, participants were requested to record their exercise training as well as physical activity for transportation (i.e., bicycling to work) during the whole study period. Borg's scale was used to evaluate subjective ratings of perceived exertion (RPE) [30]. To confirm resting conditions and the absence of muscle cell damage, blood samples were analysed for creatine kinase (CK) concentration, which serves as a marker of muscle cell damage [31].

2.5. Statistical Analysis. All data are given in mean ± standard deviation (mean ± SD) and 95% confidence intervals where appropriate. Shapiro-Wilk test was used to test for normal data distribution. Paired t-tests were performed to assess differences in nutrient intake, physical activity, and blood parameters between M1 and M2 ($\alpha = 0.05$). The Bland-Altman procedure was performed to assess reliability of the measurement procedure by calculating the bias (mean difference between M1 and M2) and the 95% limits of agreement [32, 33]. In addition, the coefficient of variation (CV%) was calculated with log-transformed data for the parameters of foci analysis. Statistical calculations were performed using JMP, Version 9 (SAS Institute Inc., Cary, NC).

3. Results

3.1. Physical Activity. Throughout the entire study period, subjects reported 24.2 ± 9.9 min of vigorous physical activity or exercise training per day with a subjective exertion of 13.4 ± 1.7 RPE. Subjects were compliant with the instruction to refrain from exercise the day before the measurements. There was no difference between M1 and M2 in exercise duration (M1: 1.3 ± 3.5 min/d; M2: 0 ± 0 min/d; $P = 0.35$) or perceived exertion (M1: 1.4 ± 3.9 RPE; M2: 0 ± 0 RPE; $P = 0.33$) during the 24 hours before the measurements. Blood concentration of creatine kinase did not differ between M1 (160.4 ± 85.7 U/L) and M2 (133.3 ± 73.5 U/L; $P = 0.32$).

3.2. Diet and Antioxidant Intake. There was no difference in energy intake 24 h before M1 (2407 ± 625 kcal/d) and M2 (2376 ± 517 kcal/d; $P = 0.89$). Recommendations for the daily intake of vitamins C and E and retinol equivalents were reached by all participants with no differences between M1 and M2. Mean intake of vitamin C was 137 ± 66 mg/d before M1 and 147 ± 54 mg/d before M2 ($P = 0.68$) and intake of vitamin E 16 ± 5 mg/d before M1 and 15 ± 7 mg/d before M2 ($P = 0.57$). For retinol equivalents, a mean intake of 3.2 ± 2.7 mg/d before M1 and 1.7 ± 1.0 mg/d before M2 ($P = 0.78$) was observed.

3.3. Analysis of γ-H2AX Foci. The total number of γ-H2AX foci detected per 100 lymphocytes was 7.6 ± 4.4 in M1 and 8.4 ± 4.6 in M2 ($P = 0.79$). The number of cells from 100 lymphocytes in which γ-H2AX foci were observed did not differ between M1 (5.9 ± 2.7) and M2 (6.9 ± 4.2; $P = 0.58$). In affected cells, a mean number of 1.2 ± 0.2 (M1) and 1.3 ± 0.1 (M2) γ-H2AX foci were found ($P = 0.80$). The mean diameter of γ-H2AX foci did not differ between the two measurements (M1: 0.63 ± 0.07 μg, M2: 0.66 ± 0.06 μg; $P = 0.31$). In Table 1 a detailed list of individual outcome measures for γ-H2AX foci is shown.

3.4. Reliability of the Outcome Measures. No measures of the γ-H2AX foci were significantly different between M1 and M2 ($P > 0.05$). The coefficient of variation (CV%), the bias, and the limits of agreement (LoA) for the four outcome variables are given in Table 2. CV of the particular outcome measures varied between 8.0% and 98.5%. The bias was generally close to zero except for the number of affected cells, where bias was −1.0. The 95% limits of agreement were narrow for the γ-H2AX foci diameter and the number of γ-H2AX foci per affected cell. However, wide limits of agreement were observed for the number of γ-H2AX foci per 100 lymphocytes and the percentage of affected cells.

4. Discussion

The analysis of γ-H2AX foci has been shown to accurately measure DNA double-strand breaks and to be sensitive even at low levels of DNA damage [22, 24]. As strenuous exercise is capable of inducing oxidative stress and DNA damage [12], the fully automated detection of γ-H2AX foci in human lymphocytes is a promising approach to evaluate exercise-induced DNA damage in athletes. However, until now, the day-to-day variation of this approach has not been investigated in healthy moderately active subjects.

The data of the current study suggest that the reliability of the analysis of γ-H2AX foci varies between the outcome measures. In general, there were no significant differences in the test-retest values of the four outcome parameters. However, the coefficient of variation was high for the total number of foci per 100 cells and the percentage of affected cells. For these two parameters, the bias was in an acceptable range and between zero and one, which is about 15% or less of the grand mean of the two repeatedly measured parameters. It remains unknown whether the variation between the test-retest situations is a methodological issue or reflects a real biological variation. For the COMET assay—a well-established method to quantify DNA damage in leucocytes—a high variation of values of DNA damage in healthy humans is known [34]. Additionally, Ismail et al. reported a high

Table 1: Individual values for the four parameters of the γ-H2AX foci analysis in human lymphocytes at both time points.

Subject	M1				M2			
	Number of γ-H2AX foci per 100 lymphocytes (N)	Number of γ-H2AX foci per affected cell (N)	Percentage of affected cells (%)	Diameter of γ-H2AX foci (μm)	Number of γ-H2AX foci per 100 lymphocytes (N)	Number of γ-H2AX foci per affected cell (N)	Percentage of affected cells (%)	Diameter of γ-H2AX foci (μm)
Subject 1	12	1.5	8	0.58	5	1.3	4	0.67
Subject 2	12	1.3	9	0.70	5	1.3	4	0.74
Subject 3	5	1.0	5	0.53	15	1.1	14	0.63
Subject 4	4	1.3	3	0.70	15	1.3	12	0.68
Subject 5	7	1.2	6	0.70	4	1.3	3	0.60
Subject 6	5	1.0	5	0.57	11	1.2	9	0.58
Subject 7	14	1.6	9	0.67	7	1.4	5	0.76
Subject 8	2	1.0	2	0.57	5	1.3	4	0.63

Table 2: Data of the four parameters of the γ-H2AX foci analysis in human lymphocytes and their test-retest reliability (bias: mean difference between the two measurements; LoA: 95% limits of agreement; CV: coefficient of variation).

γ-H2AX foci parameter	M1 (mean ± SD)	M2 (mean ± SD)	Bias	LoA (lower/upper)	CV (%)	P value
Number of γ-H2AX foci per 100 lymphocytes (N)	7.6 ± 4.4	8.4 ± 4.6	0.75	−15.2/13.7	98.5	0.792
Number of γ-H2AX foci per affected cell (N)	1.2 ± 0.2	1.3 ± 0.1	−0.02	−0.36/0.33	11.3	0.800
Percentage of affected cells (%)	5.9 ± 2.6	6.9 ± 4.2	−1.0	−11.9/9.9	88.9	0.579
Diameter of γ-H2AX foci (μm)	0.63 ± 0.07	0.66 ± 0.06	−0.04	−0.16/0.09	8.0	0.334

interindividual variability of γ-H2AX foci in human white blood cells in response to ionizing radiation, indicating that the number γ-H2AX produced during the early repair process of DNA double-strand breaks is different between subjects [22]. However, others have shown that the total number of γ-H2AX foci significantly correlates with the number of double-strand breaks [35].

In the present study large inter- and intraindividual differences in outcome values have been observed. For total number of foci per 100 cells individual values ranged from 2 to 15, whereas the range for percentage of affected cells was 2–14%, for number of foci per affected cell 1.0–1.6, and for diameter 0.53–0.76 μm, respectively.

Little data are available about the total number of γ-H2AX foci in untreated human cells of healthy subjects. *In vitro*, Rothkamm and Löbrich found a mean of 0.05 foci per cell in a cell culture of human fibroblasts that was not treated with irradiation [36]. Approximately five γ-H2AX foci per 100 cells were regarded as the background level of γ-H2AX foci in untreated cells by the authors [36]. *In vivo*, under normal conditions, γ-H2AX foci values less than one focus per cell are reported [37–39]. Unfortunately, most of the studies assessing γ-H2AX foci in lymphocytes are dealing with patients and/or older subjects of a mixed population with unequal gender. It remains unclear whether the observed background levels of γ-H2AX foci (0.07 to 0.08 γ-H2AX foci per cell) in moderately trained males are physiological. Rothkamm and Löbrich determined the variation in the levels of spontaneous γ-H2AX focus formation and found a mean of 0.04 and 0.06 foci per cell in different *in vitro* experiments [36]. Thus, the day-to-day variability of the number of γ-H2AX foci found by the authors in human cell lines is slightly lower but similar to the total number of foci found in the present study of human lymphocytes in healthy subjects under resting conditions.

Regarding reliability, a low systematic error with a bias very close to zero (−0.02 and −0.04) was found for both the number of γ-H2AX foci per affected cell and the mean diameter of the foci. Here, the coefficients of variation were within a typical range observed for day-to-day variability of biological outcome measures and below 12 CV%. However, it remains unknown whether the number of foci per affected cell or the diameter of the foci changes due to strenuous exercise.

Limitations of the study are the low number of subjects in the test-retest design and the missed opportunity to distinguish between biological variation and technical measurement error. Here, at each measurement, a second blood sample should have been collected immediately after the first blood sample to analyse the technical error.

The background level of 0.07 to 0.08 γ-H2AX foci per cell in the present study is comparable to values reported elsewhere and varies inter- and intraindividually under resting conditions. Future studies should be conducted, which include more subjects, to determine the resting level of γ-H2AX foci and accompanied parameters.

In conclusion, the analysis of the γ-H2AX focus diameter and the number of γ-H2AX foci per affected cell are reliable measures in healthy moderately active subjects. To analyse the effect of exercise on DNA damage the high day-to-day variability of total number of γ-H2AX foci needs to be considered. Whether exercise affects the prevalence of γ-H2AX in lymphocytes of healthy trained subjects needs to be investigated.

Conflict of Interests

The authors declare that there is no conflict of interests regarding the publication of this paper.

Acknowledgments

The authors thank all volunteers who participated in this study. They also thank Professor Dirk Roggenbuck (PhD), Peter Bauer (PhD), and Caroline Eberle for their skilled technical support in analysing the γ-H2AX foci.

References

[1] D. M. Bailey, I. S. Young, J. McEneny et al., "Regulation of free radical outflow from an isolated muscle bed in exercising humans," *American Journal of Physiology: Heart and Circulatory Physiology*, vol. 287, no. 4, pp. H1689–H1699, 2004.

[2] C. K. Sen, "Oxidants and antioxidants in exercise," *Journal of Applied Physiology*, vol. 79, no. 3, pp. 675–686, 1995.

[3] A. K. Banerjee, A. Mandal, D. Chanda, and S. Chakraborti, "Oxidant, antioxidant and physical exercise," *Molecular and Cellular Biochemistry*, vol. 253, no. 1-2, pp. 307–312, 2003.

[4] T. L. Clanton, L. Zuo, and P. Klawitter, "Oxidants and skeletal muscle function: physiologic and pathophysiologic implications," *Proceedings of the Society for Experimental Biology and Medicine*, vol. 222, no. 3, pp. 253–262, 1999.

[5] P. M. Clarkson and H. S. Thompson, "Antioxidants: what role do they play in physical activity and health?" *The American Journal of Clinical Nutrition*, vol. 72, no. 2, pp. 637S–646S, 2000.

[6] Y. Hellsten, U. Frandsen, N. Ørthenblad, B. Sjødin, and E. A. Richter, "Xanthine oxidase in human skeletal muscle following eccentric exercise: a role in inflammation," *Journal of Physiology*, vol. 498, no. 1, pp. 239–248, 1997.

[7] D. Javesghani, S. A. Magder, E. Barreiro, M. T. Quinn, and S. N. A. Hussain, "Molecular characterization of a superoxide-generating NAD(P)H oxidase in the ventilatory muscles," *American Journal of Respiratory and Critical Care Medicine*, vol. 165, no. 3, pp. 412–418, 2002.

[8] U. Kerkweg, F. Petrat, H. G. Korth, and H. de Groot, "Disruption of skeletal myocytes initiates superoxide release: contribution of NAD(P)H oxidase," *Shock*, vol. 27, no. 5, pp. 552–558, 2007.

[9] H. M. Alessio, "Exercise-induced oxidative stress," *Medicine and Science in Sports and Exercise*, vol. 25, no. 2, pp. 218–224, 1993.

[10] K. Briviba, B. Watzl, K. Nickel et al., "A half-marathon and a marathon run induce oxidative DNA damage, reduce antioxidant capacity to protect DNA against damage and modify immune function in hobby runners," *Redox Report*, vol. 10, no. 6, pp. 325–331, 2005.

[11] A. Mastaloudis, T. Yu, R. P. O'Donnell, B. Frei, R. H. Dashwood, and M. G. Traber, "Endurance exercise results in DNA damage as detected by the comet assay," *Free Radical Biology and Medicine*, vol. 36, no. 8, pp. 966–975, 2004.

[12] A. M. Niess, M. Baumann, K. Roecker, T. Horstmann, F. Mayer, and H.-H. Dickhuth, "Effects of intensive endurance exercise on DNA damage in leucocytes," *Journal of Sports Medicine and Physical Fitness*, vol. 38, no. 2, pp. 111–115, 1998.

[13] D. V. Firsanov, L. V. Solovjeva, and M. P. Svetlova, "H2AX phosphorylation at the sites of DNA double-strand breaks in cultivated mammalian cells and tissues," *Clinical Epigenetics*, vol. 2, no. 2, pp. 283–297, 2011.

[14] C. Di Giacomo, R. Acquaviva, V. Sorrenti et al., "Oxidative and antioxidant status in plasma of runners: effect of oral supplementation with natural antioxidants," *Journal of Medicinal Food*, vol. 12, no. 1, pp. 145–150, 2009.

[15] H. Miyazaki, S. Oh-ishi, T. Ookawara et al., "Strenuous endurance training in humans reduces oxidative stress following exhausting exercise," *European Journal of Applied Physiology*, vol. 84, no. 1-2, pp. 1–6, 2001.

[16] S. Sachdev and K. J. A. Davies, "Production, detection, and adaptive responses to free radicals in exercise," *Free Radical Biology and Medicine*, vol. 44, no. 2, pp. 215–223, 2008.

[17] M. Harms-Ringdahl, D. Jenssen, and S. Haghdoost, "Tomato juice intake suppressed serum concentration of 8-oxodG after extensive physical activity," *Nutrition Journal*, vol. 11, no. 1, article 29, 2012.

[18] A. Hartmann, A. M. Niess, M. Grunert-Fuchs, B. Poch, and G. Speit, "Vitamin E prevents exercise-induced DNA damage," *Mutation Research*, vol. 346, no. 4, pp. 195–202, 1995.

[19] O. Neubauer, S. Reichhold, L. Nics et al., "Antioxidant responses to an acute ultra-endurance exercise: impact on DNA stability and indications for an increased need for nutritive antioxidants in the early recovery phase," *The British Journal of Nutrition*, vol. 104, no. 8, pp. 1129–1138, 2010.

[20] E. P. Rogakou, W. Nieves-Neira, C. Boon, Y. Pommier, and W. M. Bonner, "Initiation of DNA fragmentation during apoptosis induces phosphorylation of H2AX histone at serine 139," *Journal of Biological Chemistry*, vol. 275, no. 13, pp. 9390–9395, 2000.

[21] J. Kobayashi, "Molecular mechanism of the recruitment of NBS1/hMRE11/hRAD50 complex to DNA double-strand breaks: NBS1 binds to γ-H2AX through FHA/BRCT domain," *Journal of Radiation Research*, vol. 45, no. 4, pp. 473–478, 2004.

[22] I. H. Ismail, T. I. Wadhra, and O. Hammarsten, "An optimized method for detecting gamma-H2AX in blood cells reveals a significant interindividual variation in the gamma-H2AX response among humans," *Nucleic Acids Research*, vol. 35, no. 5, article e36, 2007.

[23] R. Runge, R. Hiemann, M. Wendisch et al., "Fully automated interpretation of ionizing radiation-induced γH2AX foci by the novel pattern recognition system AKLIDES," *International Journal of Radiation Biology*, vol. 88, no. 5, pp. 439–447, 2012.

[24] J. P. Banáth, S. H. MacPhail, and P. L. Olive, "Radiation sensitivity, H2AX phosphorylation, and kinetics of repair of DNA strand breaks in irradiated cervical cancer cell lines," *Cancer Research*, vol. 64, no. 19, pp. 7144–7149, 2004.

[25] A. Carlsohn, S. Rohn, F. Bittmann, J. Raila, F. Mayer, and F. J. Schweigert, "Exercise increases the plasma antioxidant capacity of adolescent athletes," *Annals of Nutrition and Metabolism*, vol. 53, no. 2, pp. 96–103, 2008.

[26] A. Carlsohn, S. Rohn, F. Mayer, and F. J. Schweigert, "Physical activity, antioxidant status, and protein modification in adolescent athletes," *Medicine and Science in Sports and Exercise*, vol. 42, no. 6, pp. 1131–1139, 2010.

[27] Z. Radák, H. Naito, T. Kaneko et al., "Exercise training decreases DNA damage and increases DNA repair and resistance against oxidative stress of proteins in aged rat skeletal muscle," *Pflugers Archiv European Journal of Physiology*, vol. 445, no. 2, pp. 273–278, 2002.

[28] R. Hiemann, T. Büttner, T. Krieger, D. Roggenbuck, U. Sack, and K. Conrad, "Challenges of automated screening and differentiation of non-organ specific autoantibodies on HEp-2 cells," *Autoimmunity Reviews*, vol. 9, no. 1, pp. 17–22, 2009.

[29] R. Hiemann, N. Hilger, J. Michel et al., "Automatic analysis of immunofluorescence patterns of HEp-2 cells," *Annals of the New York Academy of Sciences*, vol. 1109, pp. 358–371, 2007.

[30] G. Borg, "Psychophysical scaling with applications in physical work and the perception of exertion," *Scandinavian Journal of Work, Environment and Health*, vol. 16, supplement 1, pp. 55–58, 1990.

[31] P. Brancaccio, N. Maffulli, and F. M. Limongelli, "Creatine kinase monitoring in sport medicine," *The British Medical Bulletin*, vol. 81-82, no. 1, pp. 209–230, 2007.

[32] J. M. Bland and D. G. Altman, "Statistical methods for assessing agreement between two methods of clinical measurement," *The Lancet*, vol. 1, no. 8476, pp. 307–310, 1986.

[33] J. M. Bland and D. G. Altman, "Measuring agreement in method comparison studies," *Statistical Methods in Medical Research*, vol. 8, no. 2, pp. 135–160, 1999.

[34] P. Møller, L. Möller, R. W. L. Godschalk, and G. D. D. Jones, "Assessment and reduction of comet assay variation in relation to DNA damage: studies from the European Comet Assay Validation Group," *Mutagenesis*, vol. 25, no. 2, pp. 109–111, 2010.

[35] E. P. Rogakou, D. R. Pilch, A. H. Orr, V. S. Ivanova, and W. M. Bonner, "DNA double-stranded breaks induce histone H2AX phosphorylation on serine 139," *The Journal of Biological Chemistry*, vol. 273, no. 10, pp. 5858–5868, 1998.

[36] K. Rothkamm and M. Löbrich, "Evidence for a lack of DNA double-strand break repair in human cells exposed to very low X-ray doses," *Proceedings of the National Academy of Sciences of the United States of America*, vol. 100, no. 9, pp. 5057–5062, 2003.

[37] F. Zwicker, B. Swartman, F. Sterzing et al., "Biological in-vivo measurement of dose distribution in patients' lymphocytes by

gamma-H2AX immunofluorescence staining: 3D conformal- vs. step-and-shoot IMRT of the prostate gland," *Radiation Oncology*, vol. 6, no. 1, article 62, 2011.

[38] M. A. Kuefner, M. Brand, J. Ehrlich, L. Braga, M. Uder, and R. C. Semelka, "Effect of antioxidants on X-ray-induced gamma-H2AX foci in human blood lymphocytes: preliminary observations," *Radiology*, vol. 264, no. 1, pp. 59–67, 2012.

[39] D. Geisel, J. T. Heverhagen, M. Kalinowski, and H. Wagner, "DNA double-strand breaks after percutaneous transluminal angioplasty," *Radiology*, vol. 248, no. 3, pp. 852–859, 2008.

HO-1 Upregulation Attenuates Adipocyte Dysfunction, Obesity, and Isoprostane Levels in Mice Fed High Fructose Diets

Zeid Khitan,[1] Mohit Harsh,[1] Komal Sodhi,[1] Joseph I. Shapiro,[1] and Nader G. Abraham[1,2]

[1] *Department of Internal Medicine, Marshall University Joan C. Edwards School of Medicine, Huntington, WV 25701, USA*
[2] *Departments of Medicine and Pharmacology, New York Medical College, Valhalla, NY 10595, USA*

Correspondence should be addressed to Nader G. Abraham; nader_abraham@nymc.edu

Academic Editor: Stan Kubow

Background. Fructose metabolism is an unregulated metabolic pathway and excessive fructose consumption is known to activate ROS. HO-1 is a potent antioxidant gene that plays a key role in decreasing ROS and isoprostanes. We examined whether the fructose-mediated increase in adipocyte dysfunction involves an increase in isoprostanes and that pharmacological induction of HO-1 would decrease both isoprostane levels and adipogenesis. *Methods and Results.* We examined the effect of fructose, on adipogenesis in human MSCs in the presence and absence of CoPP, an inducer of HO-1. Fructose increased adipogenesis and the number of large lipid droplets while decreasing the number of small lipid droplets ($P < 0.05$). Levels of heme and isoprostane in fructose treated MSC-derived adipocytes were increased. CoPP reversed these effects and markedly increased HO-1 and the Wnt signaling pathway. The high fructose diet increased heme levels in adipose tissue and increased circulating isoprostane levels ($P < 0.05$ versus control). Fructose diets decreased HO-1 and adiponectin levels in adipose tissue. Induction of HO-1 by CoPP decreased isoprostane synthesis ($P < 0.05$ versus fructose). *Conclusion.* Fructose treatment resulted in increased isoprostane production and adipocyte dysfunction, which was reversed by the increased expression of HO-1.

1. Introduction

Over the past several decades the use of fructose, as a sweetener, has emerged in processed foods, soft drinks, and an assortment of desserts [1–3]. Fructose has been linked to the development and exacerbation of centripetal obesity and metabolic syndrome, in both animals and humans [3]. In addition, excessive fructose consumption (>50 g/d) is involved in the development of metabolic imbalance and diabetes, independent of caloric restriction and weight gain [2, 4]. Sucrose, commonly found in table sugar, is a disaccharide comprising equimolar amounts of fructose and glucose and is the worldwide primary source for dietary fructose [1]. In light of the correlation between fructose and obesity, it is imperative to examine the pathogenesis by which fructose may lead to metabolic syndrome and adiposity.

Fructose is absorbed by the small intestine via passage through GLUT 5 transporters and is metabolized through the glycolytic pathway [1, 5]. Excessive fructose intake is associated with elevated levels of inflammatory mediators and reactive oxygen species (ROS), thereby contributing to the onset of metabolic syndrome and obesity [6]. Obesity is associated with oxidative stress, increased levels of ROS, and an increase in free heme, a known prooxidant [7, 8].

During oxidative stress, heme is released from cells and tissue and contributes to the growing "free heme pool." Free heme is a potent prooxidant and increases ROS levels [8–11]. Elevated levels of heme are toxic to the body and catalyze oxidative damage to cells, tissues, and organs. Obesity and oxidative stress increase cellular heme and the heme-mediated increased generation of isoprostanes [8, 10, 11]. Isoprostanes are prostaglandin-like compounds formed *in vivo* from the free radical-catalyzed peroxidation of the essential fatty acid, arachidonic acid [12]. These nonclassical eicosanoids have strong biological activity as inflammatory mediators [13] and are recognized as markers of lipid peroxidation in both animal and human models of oxidative stress [11, 14, 15]. In obese patients, urine isoprostane levels increase with increasing BMI in both male and female subjects [16]. Plasma concentration of isoprostane increases during acute fructose loading [17]. Fructose-induced oxidative stress and insulin resistance in humans are prevented by grape polyphenols [18].

Excessive fructose consumption leads to the elevation of ROS [19] and may lead to increased lipid peroxidation, altered metabolic homeostasis, and hypertrophic adipogenesis. Adipogenesis begins with the commitment of MSCs to the adipocyte lineage, followed by terminal differentiation of preadipocytes to mature adipocytes [20, 21]. Fat tissue derived adipocytes express several regulatory proteins such as Wnts and β-catenin, which repress adipogenic differentiation in preadipocytes [22, 23].

Obesity increases oxidative stress and concurrently decreases expression and activity of key cytoprotective enzyme systems, including heme oxygenase-1 (HO-1) [24, 25]. Heme oxygenase-1 is a stress-response protein and decreased expression results in an increase in cellular heme content, resulting in oxidative stress and adipocyte dysfunction [7, 26]. Induction of HO-1 decreases endogenous ROS and the protective effect of increased HO-1 expression is ascribed to several factors that include a reduction in cellular heme, the induction of the iron binding protein, ferritin, and the increased formation of bilirubin, which efficiently scavenges toxic oxygen species [7]. Induction of HO-1 expression *in vivo* and *in vitro* is associated with an increase in preadipocytes, a reduction in the number of enlarged adipocytes, elevated adiponectin secretion, and an increase in the number of small adipocytes, which are regarded as "healthy" adipocytes [27–31]. More importantly, heterozygous HO-1 knockout mice treated with an inducer of HO-1 failed to increase the production of nitric oxide [32]. Improvement in nitric oxide bioavailability and vascular function are critical to arterial induction of HO-1 [32].

The goal of this study was to determine the role of HO-1 gene expression in the fructose-mediated increase in adiposity, inflammation, and metabolic imbalance, in *in vitro* and *in vivo* models of excessive fructose intake. This is the first study to demonstrate that increased expression of HO-1 plays a significant role in modulating the phenotype of adipocytes via a decrease in heme and isoprostane levels in MSC-derived adipocytes treated with fructose and mice fed a high fructose diet.

2. Materials and Methods

2.1. Experimental Design for In Vitro Experiment

2.1.1. Human Bone Marrow-Derived MSC Culture. Frozen bone marrow mononuclear cells were purchased from Allcells (Emeryville, CA). After thawing, mononuclear cells (MSCs) were resuspended in α-minimal essential medium (α-MEM, Invitrogen, Carlsbad, CA) that was supplemented with 20% FBS and 1% A/A. The cells were transferred into a 75 cm^2 flask, cultured in α-MEM with supplements, and maintained at 37°C in a 5% CO_2 incubator. Medium was changed after 24 hours and every 2 days thereafter. Once 80% confluent, the MSCs were recovered by the addition of 0.25% trypsin/EDTA, split, and partitioned into additional 75 cm^2 flasks.

2.1.2. Human Bone Marrow-Derived MSC Adipocyte Differentiation. MSCs (passage 2-3) were plated in 75 cm^2 flasks at a density of 1-2 $\times$ 10^4 cells per cm^2 and were again cultured in their respective maintenance media until cells were 80% confluent. The maintenance medium was then replaced with adipogenic medium and the cells were cultured for an additional 14 days. The adipogenic medium consisted of complete culture medium supplemented with DMEM (Invitrogen, Carlsbad, CA), 10% FBS, 1% A/A, 10 μg/mL insulin, 1 μM dexamethasone (Sigma-Aldrich, St. Louis, MO), and 0.1 mM indomethacin (Sigma-Aldrich, St. Louis, MO). Human MSCs were cultured in this adipogenic differentiation medium along with fructose in the presence and absence of the HO-1 inducer, cobalt protoporphyrin (CoPP) (5 μM), and the HO activity inhibitor, tin (Sn^{4+})-mesoporphyrin (SnMP) (5 μM). Medium and treatments were renewed daily. Concurrently, human bone marrow-derived MSCs were plated in 24-well plates at a density of 1 $\times$ 10^4 cells per cm^2 and maintained for 24 hours in their respective maintenance media. Following this initial 24 hours, cells were treated with the aforementioned adipogenic differentiation medium in conjunction with fructose in the presence and absence of CoPP (5 μM) and SnMP (5 μM) for 14 days. MTT assay was done to determine the influence of all treatments on cell viability and no significant change was seen between the groups (data not shown).

2.2. Experimental Design for In Vivo Experiment.

All animal studies were approved by the Marshall University Animal Care and Use Committee in accordance with the National Institutes of Health Guidelines for Care and Use of Laboratory Animals. Forty eight-week-old C57Bl6 male mice were used in the studies. Mice were divided into 4 groups: 6 mice/group, control animals, mice fed a high fructose diet, and mice fed a high fructose diet and treated with CoPP, in the absence or presence of SnMP. Mice were fed either a normal chow diet containing 11% fat, 62% carbohydrate, and 27.0% protein with total calories of 12.6 KJ/g or a high fructose diet containing 60% fructose, 13.6% starch, 7% total fat, and 19.4% protein. The mice had free access to water. After four months of high fructose diet, the average weight of the fructose group showed a 1.4-fold increase in body weight compared to control animals. Administration of CoPP reversed the effect of fructose to the levels of animals fed regular chow (2 mg/100 g body weight) (4 mg/100 g bodyweight for the last month). Concurrent administration of SnMP resulted in an increase in body weight compared to mice fed fructose diets alone. We monitored food intake in mice receiving a normal chow diet or a high fructose diet and food intake did not change in any treatment group including those administered CoPP-SnMP. Blood samples were collected in K_3EDTA tubes at sacrifice and the plasma was separated. Samples were flash frozen in liquid nitrogen and maintained at −80°C until assayed.

2.3. Effect of Varying Fructose Concentrations on Adipogenesis. To determine the optimal effect of fructose on MSC-derived adipocyte differentiation, cells were treated with fructose at concentrations of 5, 50, and 500 μM. After 14 days, cells were stained with Oil Red O solution. Optimum concentration for fructose was found to be 500 μM and this concentration was utilized during treatment of cells throughout the experimentation.

2.4. Oil Red O Staining. 0.21% Oil Red O in 100% isopropanol (Sigma Aldrich, St. Louis, MO) was used for staining. Briefly, adipocytes were fixed in 4% formaldehyde, stained with Oil Red O for 10 min, and rinsed with pH 7.2 PBS (1x) (Invitrogen, Carlsbad, CA). Pictures were captured using an Olympus IX81 motorized inverted microscope. Oil Red O was eluted by adding 100% isopropanol for 10 min and OD was measured at 490 nm 24 hours later.

2.5. Measurement of Lipid Droplet Size. Cell size was measured using *ImagePro* Analyzer (Media Cybernetics, Inc., MD). The classification of the size of lipid droplets was based on size by area (pixels).

2.6. Detection of Isoprostane, Heme, and Cytokine Levels for In Vitro and In Vivo Experiments. Isoprostane levels were determined in conditioned media and in mouse plasma using an ELISA assay (Cayman Chemical, Ann Arbor, MI). Heme content in MSCs and mouse adipose tissue was determined by the pyridine hemochromogen method as described previously [33, 34]. The absorbance difference between λ 557 and λ 530 nm was used to calculate hepatic heme using an extinction coefficient of 20.7 $mM^{-1} cm^{-2}$. TNFα, IL-1β, and IL-6 levels were determined in conditioned media and mouse serum using an ELISA assay (Pierce Biotechnology, Woburn, MA).

2.7. Preparation of Adipocyte Matrix from Adipose Tissues Using Collagenase. Subcutaneous and visceral fat visible in the abdomen mesenteric fat, fat around the liver, kidney, and spleen was dissected free at the time of sacrifice and used to extract fat cells for RNA. Fat tissue was minced and incubated in 20 mL of Hank's balanced salt solution (HBSS) 1x (without calcium or magnesium), buffer containing 1% BSA and 0.1% collagenase at 1-2 gm of fat/10 mL. The fat tissues were placed in 15 mL conical tubes containing RNA inhibitors for RNA extraction and incubated in a warm water bath for 2 minutes. Wash fat in 3 serial dilutions of Betadine solution in PBS and then mince fat with sterile scissors for 2-3 minutes followed by RNA.

2.8. RNA Extraction and Real-Time PCR for In Vitro and In Vivo Experiments. Total RNA was extracted from mice adipose tissue using RNeasy Protect Mini kit (QIAGEN, Maryland, USA) according to manufacturer's instructions. Total RNA (1 μg) was transcribed into cDNA using GeneAmp kit (Applied Biosystems, Branchburg, NJ, USA) reverse transcription reagents. Total RNA was analyzed by a quantitative real-time polymerase chain reaction (qRT-PCR). Real-time PCR was performed using SYBR Green PCR Master Mix (Applied Biosystems) on a 7500 HT Fast Real-Time PCR System (Applied Biosystems). Specific primers for human MSCs were the following, with actin being used as an endogenous control described in Table 1. Specific primers for mouse tissue were as in Table 1, with actin being used as an endogenous control.

2.9. Statistical Analyses. Statistical significance was determined using one-way analysis of variance followed by Tukey-Kramer post hoc test. $P < 0.05$ was considered to be significant. Data are expressed as means ± SEM.

3. Results

3.1. Fructose Increases Adipogenesis and Isoprostane Release in a Dose-Dependent Manner in MSCs. In order to determine the optimal concentration for adipocyte differentiation in cultured human MSCs, cells were treated with fructose, 5, 50, and 500 μM. This resulted in an increase in large lipid droplets in MSCs and a decrease in small lipid droplets in a dose-dependent manner (Figures 1(a) and 1(b)). 500 μM of fructose was determined to be the optimal concentration for stimulating adipogenesis ($P < 0.05$ versus control) (Figures 1(a) and 1(b)). Isoprostane levels increased in a dose-dependent manner in conditional media of cultured MSCs, with 50 and 500 μM, the concentration at which there was the significant increase in isoprostane levels compared to control as determined by ELISA (Figure 1(c), $P < 0.05$ versus control).

3.2. Isoprostane Release and Heme Content Increase during Adipogenesis. We examined if a relationship with adipogenesis and isoprostane and cellular heme levels existed in MSCs. We observed a significant increase in isoprostane levels from day 0 to day 14 (Figure 2(a)) along with an increase in cellular heme content (Figure 2(b)). These results show, for the first time, an increase in isoprostane production and heme content with fructose-mediated adipogenesis.

3.3. HO-1 Overexpression Attenuates Fructose-Mediated Adipogenesis and Lipid Accumulation. In order to examine the effect of overexpression of HO-1 on adipogenesis, we treated cells with CoPP with or without SnMP in the presence of 500 μM fructose (Figure 3). After 14 days, the number of lipid droplets stained with Oil Red O was lower after CoPP when compared to control (Figure 3(a), $P < 0.05$). In contrast, SnMP resulted in higher levels of Oil Red O staining when compared to control ($P < 0.05$). CoPP decreased the number of large lipid droplets and increased the number of small lipid droplets when compared to control (Figures 3(b) and 3(c), $P < 0.05$). SnMP reversed these effects when compared to both CoPP and control ($P < 0.05$).

3.4. HO-1 Overexpression Reduces Isoprostane Production and Adipogenic Marker Expression and Concurrently Increases Canonical Wnt Signaling in Human MSCs. To verify that fructose-mediated adipocyte dysfunction can be reversed via

Table 1: The sequence of primers used for real-time PCR.

Primer	Forward	Reverse
Actin	5′-AGGCACCAGGGCGTGAT-3′	5′-GCCCACATAGGAATCCTTCTGAC-3′
HO-1	5′-TGAAGGAGGCCACCAAGGAG-3′	5′-CCCCTGAGAGGTCACCCAGG-3′
Mest	5′-CCAGCACATCCCGGTGCTT-3′	5′-TTCCATGAGTGCAGAGCAGG-3′
β-Catenin	5′-GGCAGGATACACGGTGCCGC-3′	5′-CTTCAGGTACCCTCAGGCCCGC-3′
Wnt10b	5′-GGTCTCGGCCTCCGCCCTTA-3′	5′-ACGGAAACCGCGCTTGAGGA-3′
Adiponectin	5′-GACGACACCAAAAGGGCTCAGG-3′	5′-CATCACGGCCTGGTGTGCCA-3′
aP2	5′-TGATGCCTTTGTGGGAACCT-3′	5′-GCAAAGCCCACTCCCACTT-3′
FAS	5′-GGACATGGTCACAGACGATGAC-3′	5′-GTCGAACTTGGACAGATCCTTCA-3′

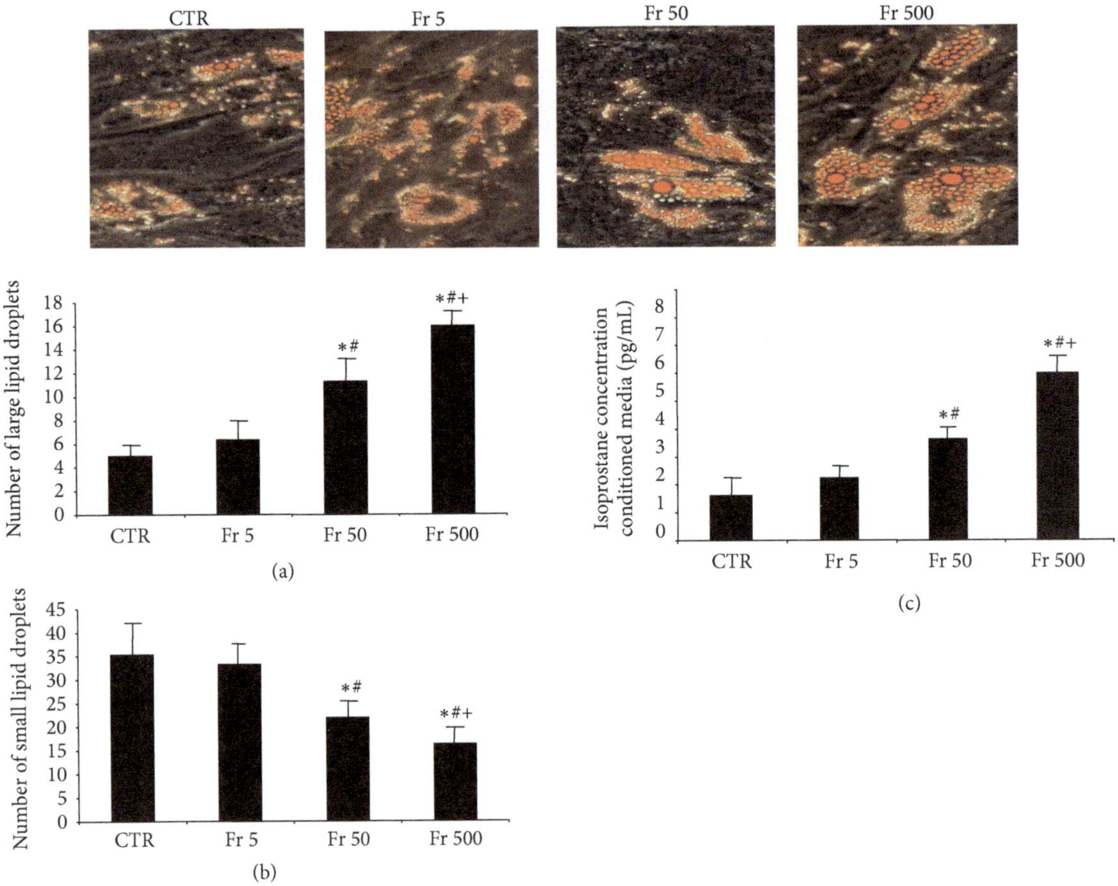

Figure 1: (a) Number of large lipid droplets from Oil Red O stained MSCs with 0, 5, 50, or 500 μM of fructose, magnifications: 40x (n = 3), and (b) number of small lipid droplets from Oil Red O stained MSCs with 0, 5, 50, or 500 μM of fructose, magnifications: 40x (n = 3). A representative section for each group is shown. (c) Measurement of isoprostane levels in conditioned media of differentiating MSCs. Values represent means ± SEM of five independent treatments. $^*P < 0.05$ versus control or $^\#P < 0.05$ versus Fr 5 μM or $^+P < 0.05$ versus Fr 50 μM.

overexpression of HO-1, we measured the expression of HO-1 and adiponectin in pelleted cells, along with isoprostane and IL-6 levels in collected conditioned media after 14 days. Fructose treatment decreased the levels of both HO-1 and adiponectin compared to control (Figures 4(a) and 4(b), $P < 0.05$). Parallel to this finding, isoprostane and IL-6 levels in cells treated with fructose increased compared to control (Figures 4(c) and 4(d), $P < 0.05$). CoPP produced an increase in HO-1 and adiponectin expression compared to fructose alone ($P < 0.05$). There was a concomitant decrease in isoprostane and IL-6 levels compared to the fructose group ($P < 0.05$). SnMP reversed the effect of CoPP, suggesting a regulatory role of HO activity in attenuating fructose-mediated adipocyte damage. We then measured mRNA levels of the canonical Wnt signaling molecules β-catenin and Wnt10b, along with aP2 and Peg-1/Mest. Fructose decreased the expression of β-catenin and Wnt10b in pelleted cells compared to control (Figures 5(a) and 5(b), resp., $P < 0.05$) and increased the expression of aP2 and Peg-1/Mest compared to control (Figures 5(c) and 5(d), $P < 0.05$). CoPP

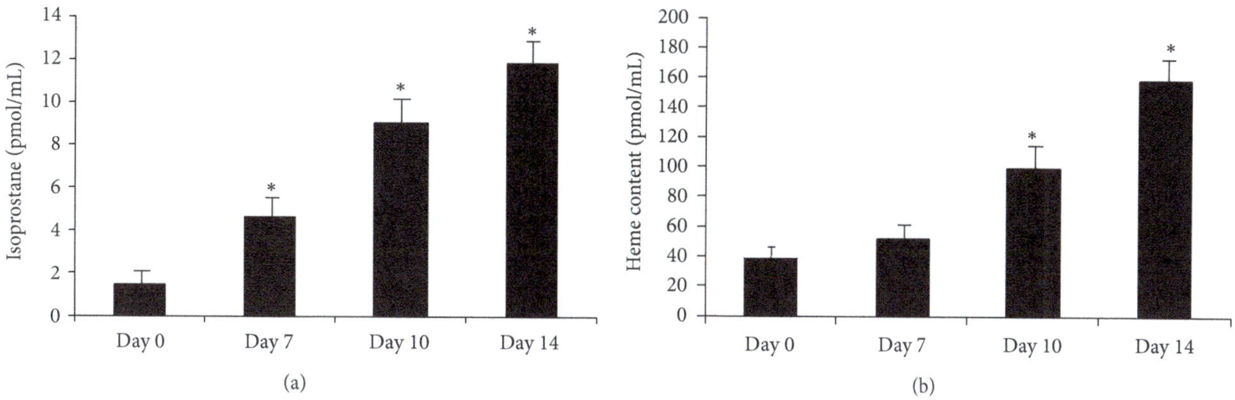

FIGURE 2: Measurement of 8-isoprostane in conditioned media and heme content in pelleted cells at days 0, 7, 10, and 14 of adipogenesis in human MSCs treated with 500 μM fructose (Figures 2(a) and 2(b), resp.). Values represent means ± SEM of five independent treatments. $^*P < 0.05$ versus control.

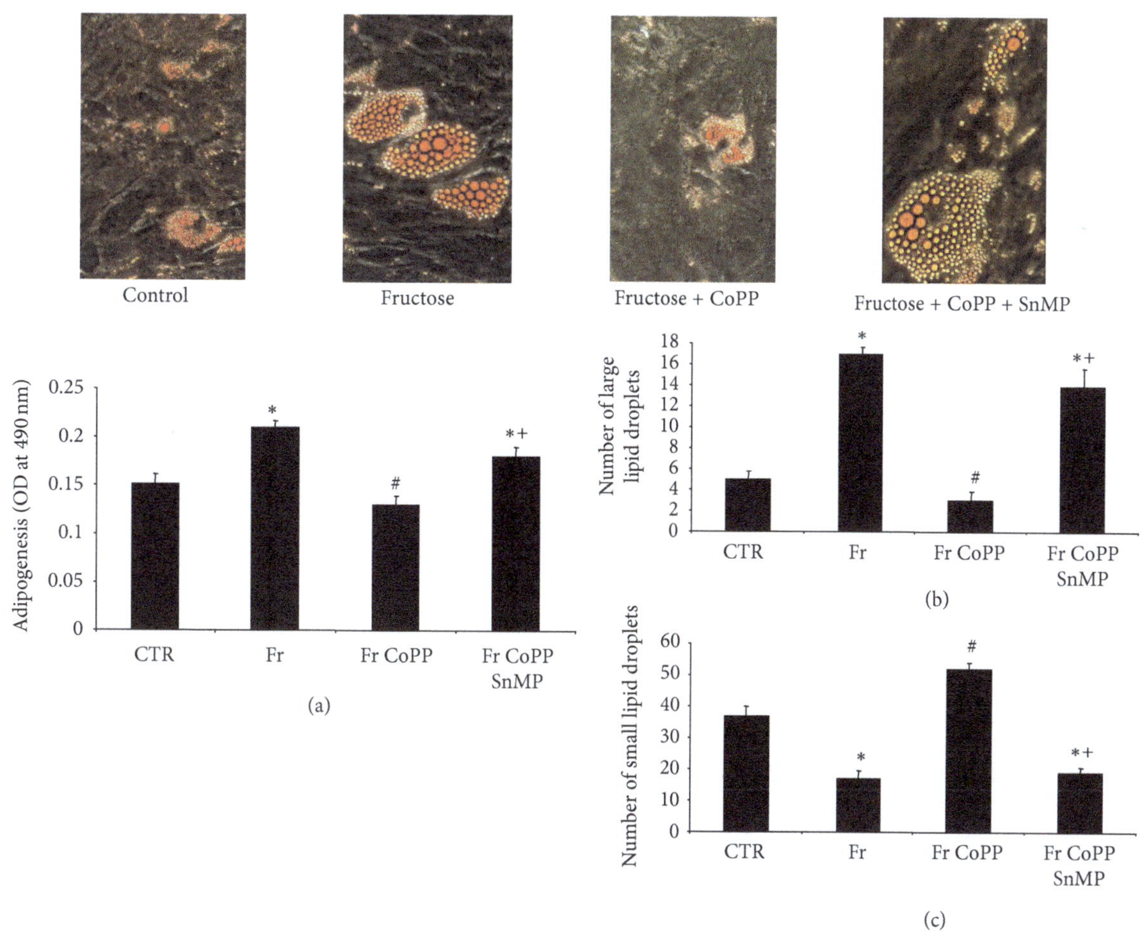

FIGURE 3: (a) Adipogenesis as shown by Oil Red O staining in MSCs treated with 500 μM fructose with or without CoPP, in the presence or absence of SnMP, magnifications and lipid droplet size from Oil Red O stained MSCs: 40x ($n = 3$). A representative section for each group is shown. (b) Number of large lipid droplets from Oil Red O stained MSCs treated with 500 μM fructose with or without CoPP, in the presence or absence of SnMP. (c) Number of small lipid droplets from Oil Red O treated with 500 μM fructose with or without CoPP, in the presence or absence of SnMP. Values represent means ± SEM of five independent treatments. $^*P < 0.05$ versus control or $^\#P < 0.05$ versus fructose or $^+P < 0.05$ versus fructose + CoPP.

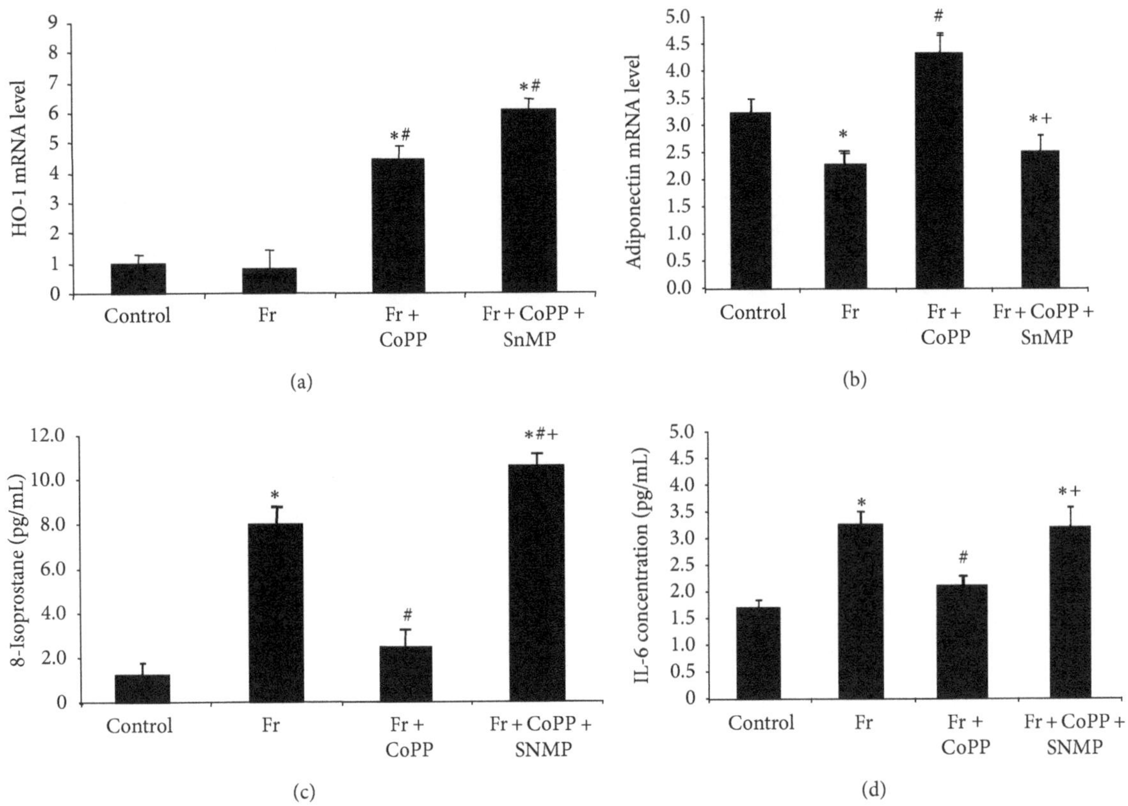

FIGURE 4: (a) Real-time PCR of HO-1 expression; (b) adiponectin expression; (c) concentration of 8-isoprostane in MSCs separated into control, fructose, fructose treated with CoPP, and fructose treated with CoPP and SnMP; and (d) concentration of IL-6 in conditioned media. Values represent means ± SEM of five independent treatments. $^*P < 0.05$ versus control or $^\#P < 0.05$ versus fructose or $^+P < 0.05$ versus fructose with CoPP.

increased expression of β-catenin and Wntl0b compared to fructose treatment ($P < 0.05$) while simultaneously decreasing the expression of aP2 and Peg-1/Mest compared to the fructose group ($P < 0.05$). The beneficial effect of CoPP was reversed by concurrent treatment with SnMP ($P < 0.05$).

3.5. High Fructose Diet Increases Adipose Heme Content and Serum Isoprostane Levels In Vivo. Adipose tissue heme content increased compared to control in animals fed a high fructose diet (Figure 6, $P < 0.05$). CoPP ameliorated this increase ($P < 0.05$) and the concurrent administration of SnMP reversed the effects of CoPP ($P < 0.05$). Isoprostane levels in plasma were increased in animals fed a high fructose diet (Figure 7(a), $P < 0.05$ versus control); however, CoPP decreased isoprostane levels ($P < 0.05$); SnMP attenuated the beneficial effect of CoPP ($P < 0.05$). TNFα, IL-1, and IL-6 serum levels increased compared to control in mice fed a high fructose diet (Figures 7(b)–7(d), $P < 0.05$). CoPP ameliorated this increase in inflammatory cytokines ($P < 0.05$) to levels below control and the concurrent administration of SnMP reversed the effects of CoPP ($P < 0.05$). These findings support HO-1 as a key regulatory factor in reducing fructose-mediated increases in heme and isoprostane production *in vivo*.

3.6. HO-1 Overexpression Ameliorates Fructose-Mediated Adiposity and Adipocyte Dysfunction In Vivo. HO-1, adiponectin, and the adipogenic markers fatty acid synthase (FAS) and aP2 were measured in adipose tissue. A fructose diet decreased the levels of HO-1 and adiponectin compared to animals fed a normal chow diet (Figures 8(a) and 8(b), $P < 0.05$). Mice fed a high fructose diet exhibited a concurrent increase in FAS and aP2 gene expression in adipose tissue compared to mice fed a normal chow diet (Figures 8(c) and 8(d), $P < 0.05$). CoPP increased HO-1 and adiponectin expression compared to both the control and fructose groups ($P < 0.05$) and decreased the levels of FAS and aP2 compared to the fructose group ($P < 0.05$). In contrast, SnMP reversed the beneficial effects of CoPP ($P < 0.05$).

4. Discussion

This is the first study to show that upregulation of HO-1 gene expression decreased fructose-mediated isoprostane and heme levels in MSC-derived adipocytes and in mice fed a high fructose diet, thereby decreasing adipocyte dysfunction. These findings are associated with decreased inflammatory cytokine levels, increased adiponectin, and the elevated expression of genes of the canonical Wnt signaling cascade. Thus, the upregulation of HO-1 may provide a potential target

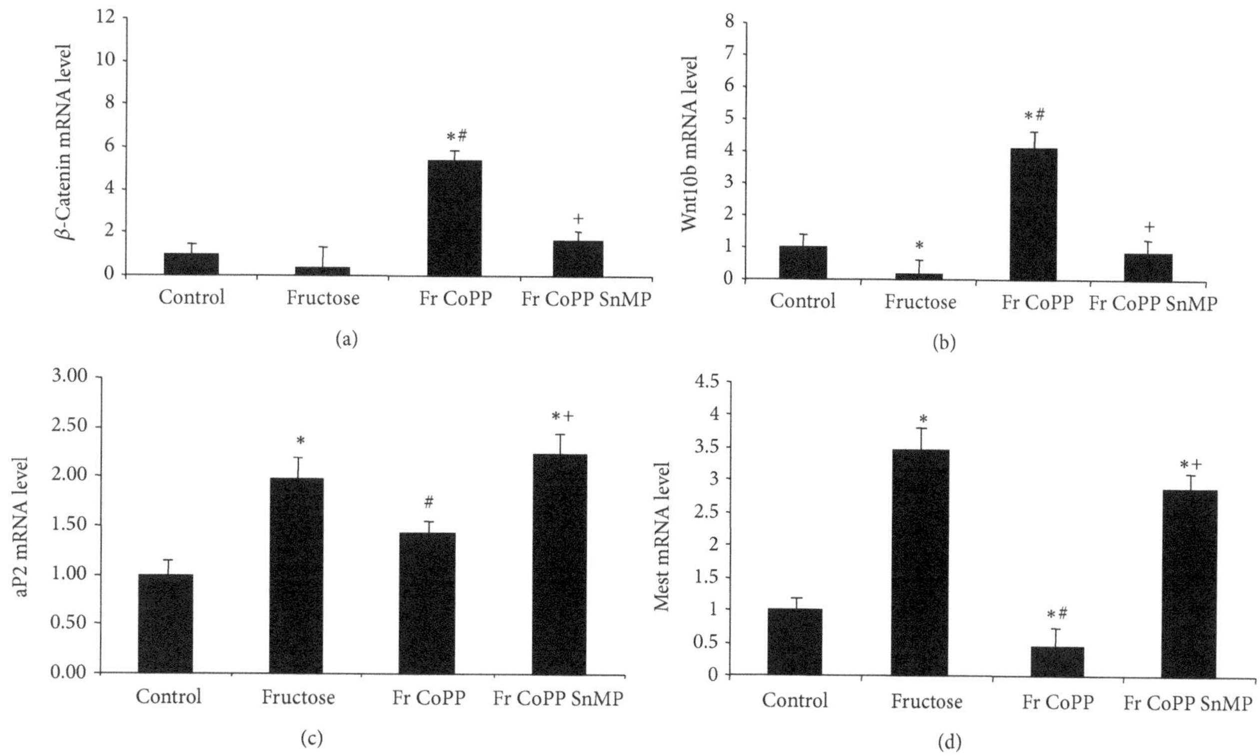

FIGURE 5: (a) Real-time PCR of β-catenin; (b) Wntl0b; (c) aP2; and (d) Peg-1/Mest expression in MSCs separated into control, fructose, fructose treated with CoPP, and fructose treated with CoPP and SnMP. Values represent means ± SEM of five independent treatments. $^{*}P < 0.05$ versus control or $^{\#}P < 0.05$ versus fructose or $^{+}P < 0.05$ versus fructose with CoPP.

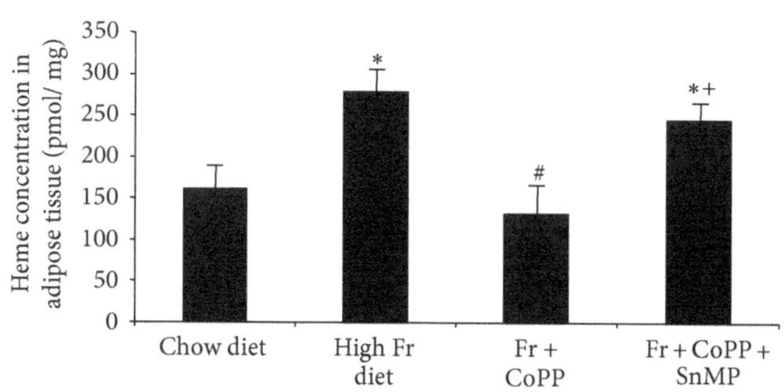

FIGURE 6: Cellular heme content from adipose tissue in mice fed chow diet, high fructose diet, and high fructose diet treated with CoPP, in the presence or absence of SnMP. Values represent means ± SEM of four independent mice per group. $^{*}P < 0.05$ versus chow diet or $^{\#}P < 0.05$ versus high fructose diet or $^{+}P < 0.05$ versus high fructose with CoPP.

for pharmacological intervention in the treatment of patients with diet-induced obesity and metabolic syndrome without side effect. Mechanisms that govern metabolic adaptation to food intake during CoPP-mediated increase in HO-1 induced weight loss are not well understood. The loss of fat suggests that CoPP treated animals preferably used fatty acid as energy sources during weight loss. Weight loss with a normal appetite is a common specific response that is well documented [35–38].

The increased expression of HO-1 resulted in decreased isoprostane levels in fructose-mediated adiposity both *in vitro* and *in vivo*, whereas inhibition of HO activity, by SnMP, increased adipocyte differentiation and isoprostane production. The perturbations in isoprostane levels and adipogenesis as a result of modulation of HO-1 levels are in part related to adipose heme content. Heme, a known prooxidant, contributes to the generation of ROS [9] and increased amounts of heme cause stem cell differentiation and the release of inflammatory cytokines [26, 39–41]. In concordance with previous reports, our results show that the increase in cellular heme appears to constitute a major factor in the fructose-mediated isoprostane release, the increase in oxidative stress, and adipocyte dysfunction. Our results suggest that one of the mechanisms by which HO-1 modulation affects adiposity and decreases isoprostane levels is via the control of the cellular heme pool. We have previously shown that CoPP increased HO-1 protein expression in the adipocyte, in adipose tissue of human MSC-derived adipocytes, and in mice fed high fat diets [25, 42]. CoPP, through the induction of HO-1, attenuated the "molecular switch" of isoprostane and heme to genetically reprogram the adipocyte phenotype and prevent fructose-mediated adiposity and adipocyte dysfunction.

Fructose bypasses the two highly regulated steps of glycolysis, catalyzed by glucokinase/hexokinase and phosphofructokinase. Instead, fructose enters the pathway at a level that is not regulated and excessive fructose metabolism leads to the direct activation of ROS through NADPH oxidase

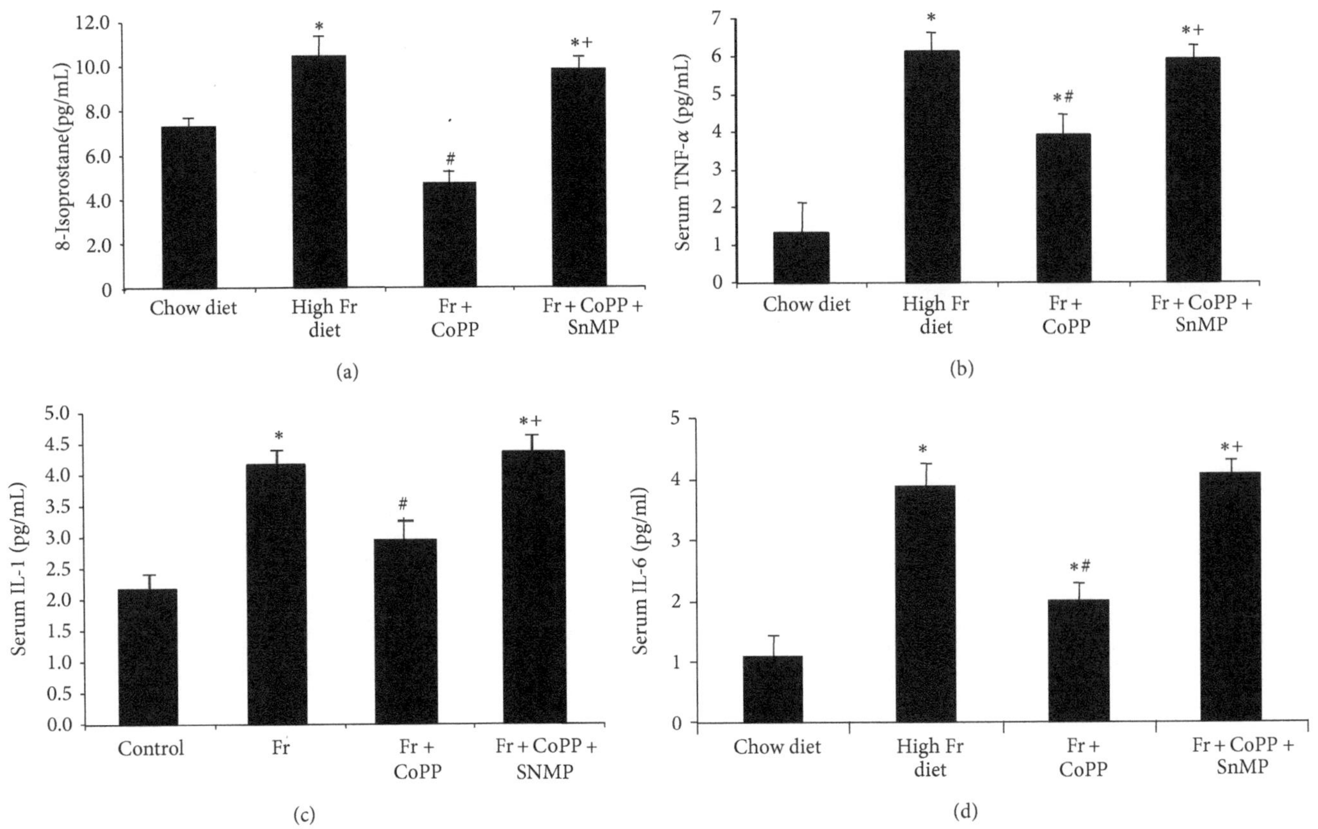

Figure 7: Assay of plasma concentrations of (a) isoprostane, (b) TNFα, (c) IL-1, and (d) IL-6 in mice fed chow diet or high fructose diet and high fructose diet treated with CoPP, in the presence or absence of SnMP. Values represent means ± SEM of four independent mice per group. $^{*}P < 0.05$ versus chow diet or $^{\#}P < 0.05$ versus high fructose diet or $^{+}P < 0.05$ versus high fructose with CoPP.

[43]. Uncontrolled fructose metabolism leads to postprandial hypertriglyceridemia, which increases ROS. Previous findings have shown that obesity and ROS decreased HO-1 expression [24, 25]. Elevated glucose levels decrease HO-1 expression [44, 45] analogous to what is observed in humans and rodents with obesity. In concordance with these findings, our results show that MSC-derived adipocytes treated with fructose and mice fed a high fructose diet resulted in diminished HO-1 expression.

Further, our results show that isoprostane levels increased in a dose-dependent manner with fructose treatment in MSCs-derived adipocytes and in mice fed a high fructose diet. The increase in isoprostanes was associated with a decrease in HO-1 expression and an increase in heme levels. Isoprostanes and heme are regarded as valid markers of oxidative stress [8, 10, 46, 47]. Isoprostane production increases in experimental models of diabetes and metabolic syndrome and is ameliorated by the induction of HO-1 [8, 46, 48]. Polyphenol derived diets prevent vascular dysfunction and increase nitric oxide associated with arterial induction of HO-1 [32]. Here, we show, for the first time, that isoprostane production is increased in MSCs treated with fructose in a dose-dependent manner and in mice fed a high fructose diet and that CoPP reverses this effect thus suggesting a potential mechanism by which fructose may lead to adipocyte dysfunction in obesity.

Several key findings in this report offer potential insights into how increased HO-1 expression resulted in decreased levels of fructose-mediated adipogenesis and obesity. Firstly, our *in vitro* and *in vivo* studies show that an increase in HO-1 gene expression resulted in an increase in adiponectin levels in the presence of excessive fructose intake. In addition, upregulation of HO-1 decreased serum isoprostane and heme content, while simultaneously diminishing the levels of inflammatory cytokines. It is noteworthy that heme levels are increased during adipogenesis [26, 49] while HO-1 protein levels and HO activity are decreased in models of obesity and metabolic syndrome [24, 27, 28]. Another mechanism by which upregulation of HO-1 may decrease adiposity is related to an increase in the level of CO and bilirubin, the latter being a potent antioxidant. Bilirubin has been shown to scavenge ROS [50, 51] and play a role in the inhibition of NADPH oxidase [52] and protein kinase C (PKC) [53], which are key modulators of oxidant-induced adipocyte dysfunction. Inhibition of biliverdin reductase, the enzyme that converts biliverdin to bilirubin, and the consequent decrease in bilirubin production increase ROS 3-fold [54]. Therefore, increased levels of bilirubin may play a significant role in decreasing oxidants in adiposity.

Another potential mechanism via which upregulation of HO-1 decreases adipocyte hypertrophy may be the activation of the canonical Wnt pathway. HO-1 acts upstream of

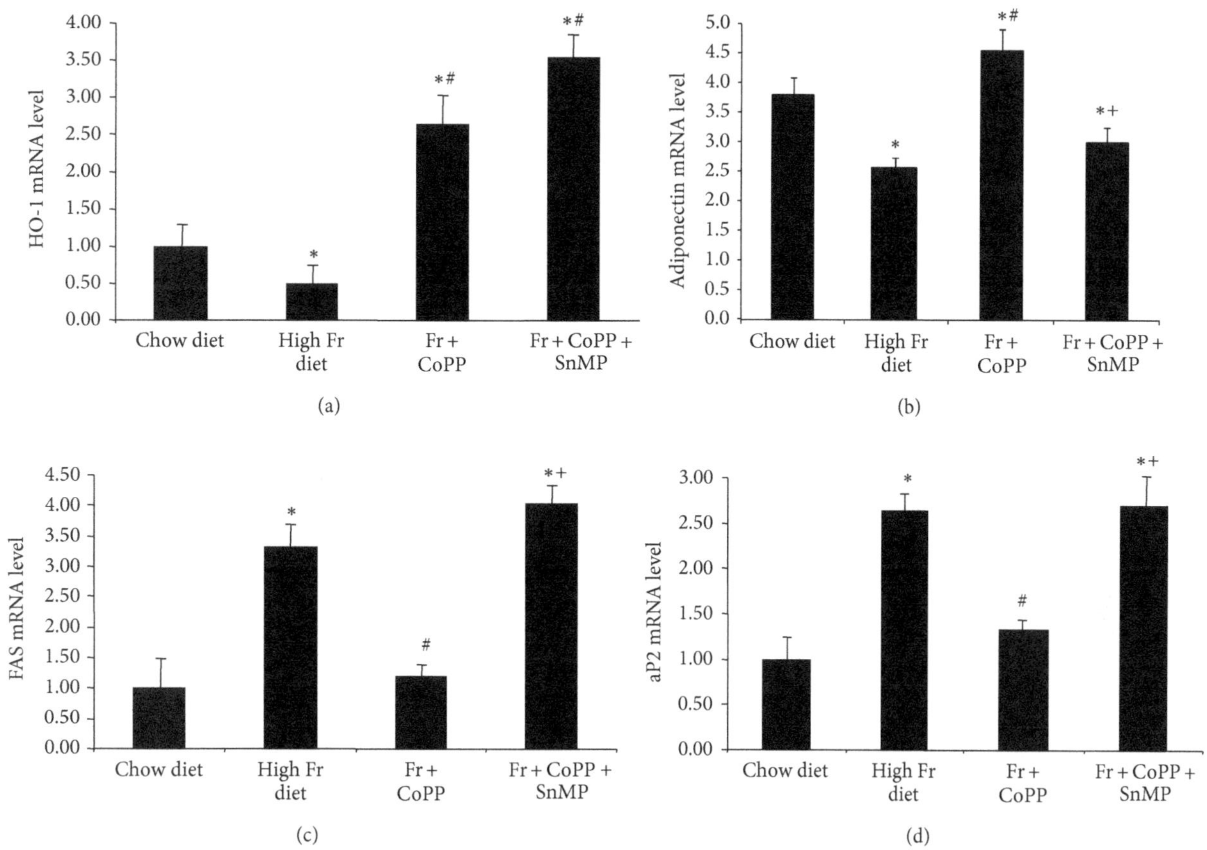

FIGURE 8: (a) Real-time PCR of HO-1, (b) adiponectin, (c) FAS, and (d) aP2 expression in chow diet, high fructose diet, and fructose diet treated with CoPP, in the presence or absence of SnMP. Values represent means ± SEM of five independent treatments. $^{*}P < 0.05$ versus chow diet or $^{\#}P < 0.05$ versus high fructose diet or $^{+}P < 0.05$ versus high fructose with CoPP.

the canonical Wnt cascade and decreases lipogenesis and adipocyte differentiation in MSCs [55]. Wnts regulate gene expression through either the canonical (β-catenin dependent) or the noncanonical pathway (β-catenin independent) [22, 23]. An increase in Wnt/β-catenin signaling inhibits the adipogenic transcription factors C/EBPα and PPARγ [56, 57]. C/EBPα and PPARα are central to growth arrest, that is, a prerequisite for adipocyte differentiation. Additionally, Peg-1/Mest upregulation results in the enlargement of adipocytes during adipose tissue expansion. On accumulation of triglycerides, Peg-1/Mest levels are increased with a parallel signal to preadipocytes to enlarge in order to accommodate more triglycerides [58]. Adipocyte enlargement is associated with an increase in the levels of inflammatory cytokines and oxidative stress [27, 30, 59].

Our results show that the increase in HO-1 gene expression resulted in an increase in the canonical Wnt signaling cascade and a decrease in Peg-1/Mest levels. A decrease in Peg-1/Mest is beneficial in the control of obesity, since upregulation of Peg-1/Mest occurs in obese adipose tissue in several models of obesity [60, 61]. Our data demonstrate that the overexpression of HO-1 was effective in suppressing adipocyte differentiation, as evidenced by an increase in the canonical Wnt cascade and a decrease in Peg-1/Mest. These effects were reversed by SnMP, further demonstrating that the HO-1-mediated increase in negative regulators of adipogenesis and decrease in Peg-1/Mest resulted in the attenuation of adipogenesis and lipid accumulation. Taken together, these observations provide compelling evidence that the HO-1-mediated increase in Wnt signaling pathway modulates fructose-mediated adipogenesis.

A final mechanism by which CoPP-mediated upregulation of HO-1 and HO activity could decrease lipid accumulation is through elevated levels of adiponectin. A decrease in oxidative stress results in adipocytes secreting more adiponectin [27]. Thus, adiponectin may be considered a therapeutic target for HO-1 in reducing fructose-mediated adipocyte dysfunction and lipid accumulation, as previous findings have shown that induction of HO-1 reduces isoprostane and inflammatory cytokine levels through an increase in adiponectin [29]. In contrast, inhibition of HO activity decreased the expression of adiponectin confirming that adipocyte hypertrophy results in a diminished secretion of adiponectin [55]. Furthermore, increased heme levels, as a result of reduced HO-1 expression and HO activity, decrease adiponectin secretion from adipose tissue and increase adipocyte dysfunction and inflammation [26]. Similarly, our results show that upregulation of HO-1 in *in vitro* and *in* vivo models of high fructose consumption resulted in an increase in adiponectin levels along with a decrease in isoprostane and

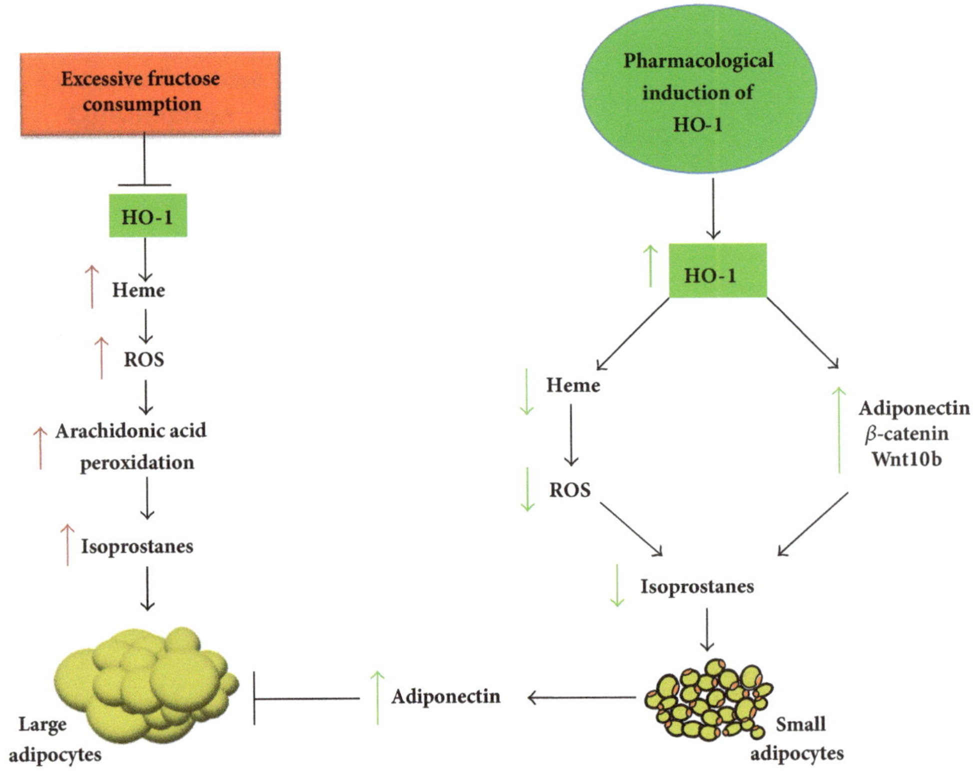

FIGURE 9: Schematic diagram of fructose-mediated adipocyte expansion and potential mechanisms underlying HO-1-mediated improvement of fructose-induced adipocyte hypertrophy. Excessive fructose consumption is accompanied by decreases in HO-1 gene expression and a concurrent increase in cellular heme levels. Increase in cellular heme results in lipid peroxidation of the essential fatty acid, arachidonic acid, via activation of ROS. The metabolite of this process is isoprostane and, therefore, heme catalyzes isoprostane synthesis. Isoprostane release induces adipocyte hypertrophy and dysfunction. In contrast, the increase in HO-1 expression and HO activity by pharmacological agents, for example, CoPP, leads to a decrease in the cellular heme pool and an increase in the canonical Wnt pathway and of adiponectin. Adiponectin enhances the antioxidant mechanisms, thereby decreasing lipid droplet size along with attenuation of isoprostane production. An increase in HO activity lowers lipid peroxidation, thereby decreasing isoprostane production and diminishing expression of the adipogenic markers Peg-1/Mest, FAS, and aP2, resulting in the lowering of adipocyte hypertrophy. Thus, activation of the HO-1-adiponectin axis along with an increase in the canonical Wnt cascade results in decreased heme (prooxidant) levels and isoprostane production in adipocytes and a decrease in fructose-mediated adipocyte dysfunction and lipid accumulation.

heme levels. Thus, activation of the HO-1-adiponectin axis may decrease fructose-mediated adipocyte dysfunction and metabolic imbalance.

In summary, this study demonstrates that the pharmacological induction of HO-1 results in a decrease in heme and isoprostane levels and a reduction in fructose-mediated adipocyte hypertrophy (as represented in Figure 9). We demonstrate that the stimulatory effect of exogenous fructose on isoprostane release exacerbates lipid accumulation and adipocyte dysfunction in MSC-derived adipocytes and mice fed a high fructose diet. Our results confirm the proadipogenic effects of fructose and account for the positive relationship between a high fructose diet and the presence of obesity. This may prove crucial in the control of obesity and the onset of metabolic syndrome as a result of a western high fructose diet.

What is already known about this subject is as follows.

(i) Fructose consumption causes adiposity and increases reactive oxygen species (ROS) levels.

(ii) Induction of heme oxygenase-1 (HO-1) decreases adiposity.

(iii) Isoprostanes are a marker of oxidative stress.

What this study adds is as follows.

(i) Fructose-mediated increase in adipogenesis is associated with an increase in isoprostane production in stem cell-derived adipocyte (MSCs).

(ii) Fructose increases isoprostane production in a dose-dependent manner.

(iii) Fructose increases the level of cellular heme, a prooxidant, in adipose tissue and the release of isoprostane into plasma.

(iv) An increase in HO-1 expression reduces obesity and F2-isoprostane plasma levels in mice fed high fructose diets.

Abbreviations

aP2: Adipocyte protein 2
β-Catenin: Beta-cadherin associated protein
C/EBPα: CCAAT/enhancer binding protein alpha
CoPP: Cobalt protoporphyrin IX dichloride
FAS: Fatty acid synthase
GLUT5: Glucose transporter member 5
HO: Heme oxygenase
HO-1: Heme oxygenase isozyme 1, inducible form
HO-2: Heme oxygenase isozyme 2, constitutive form
MSC: Mesenchymal stem cell
NADPH: Nicotinamide adenine dinucleotide phosphate (reduced form)
Peg-1/Mest: Paternally expressed gene 1/mesoderm specific transcript
PPARγ: Peroxisome proliferator activated receptor gamma
ROS: Reactive oxygen species
SnMP: Tin mesoporphyrin IX dichloride
TNFα: Tumor necrosis factor-alpha
Wnt10b: Wingless-type MMTV integration site family, member 10b.

Disclosure

None of the authors have any financial, commercial, publication, or personal benefit.

Conflict of Interests

The authors declare that there is no conflict of interests regarding the publication of this paper.

Acknowledgments

All authors had full access to the data and take responsibility for its integrity. All authors have read and agree with the paper as written. The authors thank Mrs. Jennifer Brown for secretarial support. This work was supported by National Institutes of Health Grants HL55601 and HL34300 (NGA) and HL109015 (JIS).

References

[1] J. J. Rumessen and E. Gudmand-Hoyer, "Absorption capacity of fructose in healthy adults. Comparison with sucrose and its constituent monosaccharides," *Gut*, vol. 27, no. 10, pp. 1161–1168, 1986.

[2] R. J. Johnson, S. E. Perez-Pozo, Y. Y. Sautin et al., "Hypothesis: could excessive fructose intake and uric acid cause type 2 diabetes?" *Endocrine Reviews*, vol. 30, no. 1, pp. 96–116, 2009.

[3] R. J. Johnson, M. S. Segal, Y. Sautin et al., "Potential role of sugar (fructose) in the epidemic of hypertension, obesity and the metabolic syndrome, diabetes, kidney disease, and cardiovascular disease," *The American Journal of Clinical Nutrition*, vol. 86, no. 4, pp. 899–906, 2007.

[4] T. Ishimoto, M. A. Lanaspa, M. T. Le et al., "Opposing effects of fructokinase C and A isoforms on fructose-induced metabolic syndrome in mice," *Proceedings of the National Academy of Sciences of the United States of America*, vol. 109, no. 11, pp. 4320–4325, 2012.

[5] F. Q. Zhao and A. F. Keating, "Functional properties and genomics of glucose transporters," *Current Genomics*, vol. 8, no. 2, pp. 113–128, 2007.

[6] K. L. Stanhope, J. M. Schwarz, N. L. Keim et al., "Consuming fructose-sweetened, not glucose-sweetened, beverages increases visceral adiposity and lipids and decreases insulin sensitivity in overweight/obese humans," *The Journal of Clinical Investigation*, vol. 119, no. 5, pp. 1322–1334, 2009.

[7] N. G. Abraham and A. Kappas, "Pharmacological and clinical aspects of heme oxygenase," *Pharmacological Reviews*, vol. 60, no. 1, pp. 79–127, 2008.

[8] D. Sacerdoti, R. Olszanecki, G. Li Volti, C. Colombrita, G. Scapagnini, and N. G. Abraham, "Heme oxygenase overexpression attenuates glucose-mediated oxidative stress in quiescent cell phase: linking heme to hyperglycemia complications," *Current Neurovascular Research*, vol. 2, no. 2, pp. 103–111, 2005.

[9] J. Balla, H. S. Jacob, G. Balla, K. Nath, J. W. Eaton, and G. M. Vercellotti, "Endothelial-cell heme uptake from heme proteins: Induction of sensitization and desensitization to oxidant damage," *Proceedings of the National Academy of Sciences of the United States of America*, vol. 90, no. 20, pp. 9285–9289, 1993.

[10] D. Sacerdoti, C. Colombrita, M. H. Ghattas et al., "Heme oxygenase-1 transduction in endothelial cells causes downregulation of monocyte chemoattractant protein-1 and of genes involved in inflammation and growth," *Cellular and Molecular Biology*, vol. 51, no. 4, pp. 363–370, 2005.

[11] N. G. Abraham, T. Kushida, J. McClung et al., "Heme oxygenase-1 attenuates glucose-mediated cell growth arrest and apoptosis in human microvessel endothelial cells," *Circulation Research*, vol. 93, pp. 504–514, 2003.

[12] T. Yura, M. Fukunaga, R. Khan, G. N. Nassar, K. F. Badr, and A. Montero, "Free-radical-generated F2-isoprostane stimulates cell proliferation and endothelin-1 expression on endothelial cells," *Kidney International*, vol. 56, no. 2, pp. 471–478, 1999.

[13] J. D. Morrow and L. J. Roberts, "The isoprostanes: unique bioactive products of lipid peroxidation," *Progress in Lipid Research*, vol. 36, no. 1, pp. 1–21, 1997.

[14] J. A. Awad, J.-L. Horn, L. J. Roberts II, and J. J. Franks, "Demonstration of halothane-induced hepatic lipid peroxidation in rats by quantification of F 2-isoprostanes," *Anesthesiology*, vol. 84, no. 4, pp. 910–916, 1996.

[15] I. J. Tsai, K. D. Croft, T. A. Mori et al., "20-HETE and F2-isoprostanes in the metabolic syndrome: the effect of weight reduction," *Free Radical Biology and Medicine*, vol. 46, no. 2, pp. 263–270, 2009.

[16] J. F. Keaney Jr., M. G. Larson, R. S. Vasan et al., "Obesity and systemic oxidative stress: clinical correlates of oxidative stress in the Framingham study," *Arteriosclerosis, Thrombosis, and Vascular Biology*, vol. 23, no. 3, pp. 434–439, 2003.

[17] P. Faure, C. Polge, D. Monneret, A. Favier, and S. Halimi, "Plasma 15-F2t isoprostane concentrations are increased during acute fructose loading in type 2 diabetes," *Diabetes & Metabolism*, vol. 34, no. 2, pp. 148–154, 2008.

[18] M. Hokayem, E. Blond, H. Vidal et al., "Grape polyphenols prevent fructose-induced oxidative stress and insulin resistance in first-degree relatives of type 2 diabetic patients," *Diabetes Care*, vol. 36, no. 6, pp. 1454–1461, 2013.

[19] Z. Khitan and D. H. Kim, "Fructose: a key factor in the development of metabolic syndrome and hypertension," *Journal of Nutrition and Metabolism*, vol. 2013, Article ID 682673, 12 pages, 2013.

[20] P. Vestergaard, "Bone metabolism in type 2 diabetes and role of thiazolidinediones," *Current Opinion in Endocrinology, Diabetes and Obesity*, vol. 16, no. 2, pp. 125–131, 2009.

[21] S. Gesta, Y.-H. Tseng, and C. R. Kahn, "Developmental origin of fat: tracking obesity to its source," *Cell*, vol. 131, no. 2, pp. 242–256, 2007.

[22] H. Clevers, "Wnt/β-catenin signaling in development and disease," *Cell*, vol. 127, no. 3, pp. 469–480, 2006.

[23] K. M. Cadigan and Y. I. Liu, "Wnt signaling: complexity at the surface," *Journal of Cell Science*, vol. 119, no. 3, pp. 395–402, 2006.

[24] J. Cao, S. J. Peterson, K. Sodhi et al., "Heme oxygenase gene targeting to adipocytes attenuates adiposity and vascular dysfunction in mice fed a high-fat diet," *Hypertension*, vol. 60, no. 2, pp. 467–475, 2012.

[25] L. Vanella, D. H. Kim, K. Sodhi et al., "Crosstalk between EET and HO-1 downregulates Bach1 and adipogenic marker expression in mesenchymal stem cell derived adipocytes," *Prostaglandins and Other Lipid Mediators*, vol. 96, no. 1–4, pp. 54–62, 2011.

[26] N. Puri, K. Sodhi, M. Haarstad et al., "Heme induced oxidative stress attenuates sirtuin1 and enhances adipogenesis in mesenchymal stem cells and mouse pre-adipocytes," *Journal of Cellular Biochemistry*, vol. 113, no. 6, pp. 1926–1935, 2012.

[27] M. Li, D. H. Kim, P. L. Tsenovoy et al., "Treatment of obese diabetic mice with a heme oxygenase inducer reduces visceral and subcutaneous adiposity, increases adiponectin levels, and improves insulin sensitivity and glucose tolerance," *Diabetes*, vol. 57, no. 6, pp. 1526–1535, 2008.

[28] S. J. Peterson, D. H. K. Kim, M. Li et al., "The L-4F mimetic peptide prevents insulin resistance through increased levels of HO-1, pAMPK, and pAKT in obese mice," *Journal of Lipid Research*, vol. 50, no. 7, pp. 1293–1304, 2009.

[29] A. Nicolai, M. Li, D. H. Kim et al., "Heme oxygenase-1 induction remodels adipose tissue and improves insulin sensitivity in obesity-induced diabetic rats," *Hypertension*, vol. 53, no. 3, pp. 508–515, 2009.

[30] S. J. Peterson, G. Drummond, D. H. Kim et al., "L-4F treatment reduces adiposity, increases adiponectin levels, and improves insulin sensitivity in obese mice," *Journal of Lipid Research*, vol. 49, no. 8, pp. 1658–1669, 2008.

[31] A. L. Kruger, S. J. Peterson, M. L. Schwartzman et al., "Up-regulation of heme oxygenase provides vascular protection in an animal model of diabetes through its antioxidant and anti-apoptotic effects," *Journal of Pharmacology and Experimental Therapeutics*, vol. 319, no. 3, pp. 1144–1152, 2006.

[32] Y. Shen, N. C. Ward, J. M. Hodgson et al., "Dietary quercetin attenuates oxidant-induced endothelial dysfunction and atherosclerosis in apolipoprotein E knockout mice fed a high-fat diet: a critical role for heme oxygenase-1," *Free Radical Biology & Medicine*, vol. 65, pp. 908–915, 2013.

[33] J.-L. da Silva, M. Tiefenthaler, E. Park et al., "Tin-mediated heme oxygenase gene activation and cytochrome P450 arachidonate hydroxylase inhibition in spontaneously hypertensive rats," *American Journal of the Medical Sciences*, vol. 307, no. 3, pp. 173–181, 1994.

[34] N. G. Abraham, J.-M. Camadro, S. T. Hoffstein, and R. D. Levere, "Effects of iron deficiency and chronic iron overloading on mitochondrial heme biosynthetic enzymes in rat liver," *Biochimica et Biophysica Acta—Protein Structure and Molecular Enzymology*, vol. 870, no. 2, pp. 339–349, 1986.

[35] B. M. Grossman, M. L. Devore, E. W. Kelso, and R. J. Martin, "Effect of glucose and 2-deoxyglucose on hypothalamic GABA release in lactating rats," *Physiology & Behavior*, vol. 61, no. 2, pp. 169–173, 1997.

[36] E. Scharrer, "Control of food intake by fatty acid oxidation and ketogenesis," *Nutrition*, vol. 15, no. 9, pp. 704–714, 1999.

[37] M. A. Rupnick, D. Panigrahy, C.-Y. Zhang et al., "Adipose tissue mass can be regulated through the vasculature," *Proceedings of the National Academy of Sciences of the United States of America*, vol. 99, no. 16, pp. 10730–10735, 2002.

[38] M. Morimura, O. Ishiko, T. Sumi, H. Yoshida, and S. Ogita, "Angiogenesis in adipose tissues and skeletal muscles with rebound weight-gain after diet-restriction in rabbits," *International Journal of Molecular Medicine*, vol. 8, no. 5, pp. 499–503, 2001.

[39] S. R. Datla, G. J. Dusting, T. A. Mori, C. J. Taylor, K. D. Croft, and F. Jiang, "Induction of heme oxygenase-1 in vivo suppresses NADPH oxidase-derived oxidative stress," *Hypertension*, vol. 50, no. 4, pp. 636–642, 2007.

[40] T. Chabrashvili, C. Kitiyakara, J. Blau et al., "Effects of ANG II type 1 and 2 receptors on oxidative stress, renal NADPH oxidase, and SOD expression," *The American Journal of Physiology—Regulatory Integrative and Comparative Physiology*, vol. 285, no. 1, pp. R117–R124, 2003.

[41] K. Sodhi, N. Puri, D. H. Kim et al., "PPARδ binding to heme oxygenase 1 promoter prevents angiotensin II-induced adipocyte dysfunction in Goldblatt hypertensive rats," *International Journal of Obesity*, vol. 38, no. 3, pp. 456–465, 2014.

[42] A. Burgess, M. Li, L. Vanella et al., "Adipocyte heme oxygenase-1 induction attenuates metabolic syndrome in both male and female obese mice," *Hypertension*, vol. 56, no. 6, pp. 1124–1130, 2010.

[43] K. L. Stanhope and P. J. Havel, "Fructose consumption: potential mechanisms for its effects to increase visceral adiposity and induce dyslipidemia and insulin resistance," *Current Opinion in Lipidology*, vol. 19, no. 1, pp. 16–24, 2008.

[44] N. G. Abraham, T. Kushida, J. McClung et al., "Heme oxygenase-1 attenuates glucose-mediated cell growth arrest and apoptosis in human microvessel endothelial cells," *Circulation Research*, vol. 93, no. 6, pp. 507–514, 2003.

[45] A. L. Kruger, S. Peterson, S. Turkseven et al., "D-4F induces heme oxygenase-1 and extracellular superoxide dismutase, decreases endothelial cell sloughing, and improves vascular reactivity in rat model of diabetes," *Circulation*, vol. 111, no. 23, pp. 3126–3134, 2005.

[46] N. G. Abraham, "Heme oxygenase attenuated angiotensin II-mediated increase in cyclooxygenase activity and decreased isoprostane F2α in endothelial cells," *Thrombosis Research*, vol. 110, no. 5-6, pp. 305–309, 2003.

[47] C. Colombrita, G. Lombardo, G. Scapagnini, and N. G. Abraham, "Heme oxygenase-1 expression levels are cell cycle dependent," *Biochemical and Biophysical Research Communications*, vol. 308, no. 4, pp. 1001–1008, 2003.

[48] N. G. Abraham, R. Rezzani, L. Rodella et al., "Overexpression of human heme oxygenase-1 attenuates endothelial cell sloughing in experimental diabetes," *The American Journal of Physiology—Heart and Circulatory Physiology*, vol. 287, no. 6, pp. H2468–H2477, 2004.

[49] N. Kumar, L. A. Solt, Y. Wang et al., "Regulation of adipogenesis by natural and synthetic REV-ERB ligands," *Endocrinology*, vol. 151, no. 7, pp. 3015–3025, 2010.

[50] J. E. Clark, R. Foresti, C. J. Green, and R. Motterlini, "Dynamics of haem oxygenase-1 expression and bilirubin production in cellular protection against oxidative stress," *Biochemical Journal*, vol. 348, no. 3, pp. 615–619, 2000.

[51] J. Neuzil and R. Stocker, "Free and albumin-bound bilirubin are efficient co-antioxidants for α-tocopherol, inhibiting plasma and low density lipoprotein lipid peroxidation," *The Journal of Biological Chemistry*, vol. 269, no. 24, pp. 16712–16719, 1994.

[52] J. Y. Kwak, K. Takeshige, B. S. Cheung, and S. Minakami, "Bilirubin inhibits the activation of superoxide-producing NADPH oxidase in a neutrophil cell-free system," *Biochimica et Biophysica Acta—Protein Structure and Molecular Enzymology*, vol. 1076, no. 3, pp. 369–373, 1991.

[53] K. Sano, H. Nakamura, and T. Matsuo, "Mode of inhibitory action of bilirubin on protein kinase C," *Pediatric Research*, vol. 19, no. 6, pp. 587–590, 1985.

[54] D. E. Barañano, M. Rao, C. D. Ferris, and S. H. Snyder, "Biliverdin reductase: a major physiologic cytoprotectant," *Proceedings of the National Academy of Sciences of the United States of America*, vol. 99, no. 25, pp. 16093–16098, 2002.

[55] L. Vanella, K. Sodhi, D. H. Kim et al., "Increased heme-oxygenase 1 expression in mesenchymal stem cell-derived adipocytes decreases differentiation and lipid accumulation via upregulation of the canonical Wnt signaling cascade," *Stem Cell Research and Therapy*, vol. 4, no. 2, article 28, 2013.

[56] M. Kawai, S. Mushiake, K. Bessho et al., "Wnt/Lrp/β-catenin signaling suppresses adipogenesis by inhibiting mutual activation of PPARγ and C/EBPα," *Biochemical and Biophysical Research Communications*, vol. 363, no. 2, pp. 276–282, 2007.

[57] M. Okamura, H. Kudo, K.-I. Wakabayashi et al., "COUP-TFII acts downstream of Wnt/β-catenin signal to silence PPARγ gene expression and repress adipogenesis," *Proceedings of the National Academy of Sciences of the United States of America*, vol. 106, no. 14, pp. 5819–5824, 2009.

[58] M. Takahashi, Y. Kamei, and O. Ezaki, "Mest/Peg1 imprinted gene enlarges adipocytes and is a marker of adipocyte size," *The American Journal of Physiology—Endocrinology and Metabolism*, vol. 288, no. 1, pp. E117–E124, 2005.

[59] L. Vanella, D. H. Kim, D. Asprinio et al., "HO-1 expression increases mesenchymal stem cell-derived osteoblasts but decreases adipocyte lineage," *Bone*, vol. 46, no. 1, pp. 236–243, 2010.

[60] Y. Kamei, T. Suganami, T. Kohda et al., "Peg1/Mest in obese adipose tissue is expressed from the paternal allele in an isoform-specific manner," *FEBS Letters*, vol. 581, no. 1, pp. 91–96, 2007.

[61] R. A. Koza, L. Nikonova, J. Hogan et al., "Changes in gene expression foreshadow diet-induced obesity in genetically identical mice," *PLoS Genetics*, vol. 2, no. 5, article e81, 2006.

29

Moringa oleifera Supplemented Diets Prevented Nickel-Induced Nephrotoxicity in Wistar Rats

O. S. Adeyemi and T. C. Elebiyo

Department of Biological Sciences, Landmark University, Omu-Aran 370102, Kwara State, Nigeria

Correspondence should be addressed to O. S. Adeyemi; yomibowa@yahoo.com

Academic Editor: Joel B. Mason

Background. The *Moringa oleifera* plant has been implicated for several therapeutic potentials. *Objective.* To evaluate whether addition of *M. oleifera* to diet has protective effect against nickel-induced nephrotoxicity in rats. *Methodology.* Male Wistar rats were assigned into six groups of five. The rats were given oral exposure to 20 mg/kg nickel sulphate ($NiSO_4$) in normal saline and sustained on either normal diet or diets supplemented with *Moringa oleifera* at different concentrations for 21 days. 24 hours after cessation of treatments, all animals were sacrificed under slight anesthesia. The blood and kidney samples were collected for biochemical and histopathology analyses, respectively. *Results.* $NiSO_4$ exposure reduced the kidney-to-body weight ratio in rats and caused significant elevation in the levels of plasma creatinine, urea, and potassium. Also, the plasma level of sodium was decreased by $NiSO_4$ exposure. However, addition of *M. oleifera* to diets averted the nickel-induced alteration to the level of creatinine and urea. The histopathology revealed damaged renal tubules and glomerular walls caused by $NiSO_4$ exposure. In contrast, the damages were ameliorated by the *M. oleifera* supplemented diets. *Conclusion.* The addition of *M. oleifera* to diet afforded significant protection against nickel-induced nephrotoxicity.

1. Introduction

The *Moringa oleifera* is a significant medicinal plant belonging to the family Moringaceae. The *M. oleifera* is recognized for its vast therapeutic properties since ancient times. It is also known as drumstick tree or horseradish tree; the leaves are very beneficial and offer important source of beta-carotene, vitamin C, protein, iron, and potassium [1]. The *M. oleifera* plant is native to the Indian subcontinent and has been used by the Indians for almost 5000 years [2]. *M. oleifera* tree can grow well in the humid tropic or hot dry land and it can survive in harsh climatic condition including destitute soil [2, 3]. The root, bark, gum, leaf, pods, flowers, seed, and seed oil are used in traditional medicine for treatment of various ailments [4]. The leaves, flowers, roots, gums, fruits, and seeds of *M. oleifera* are extensively used in the treatment of inflammation, cardiovascular dysfunction, liver disease, and hematological and renal malfunction [5–7]. The leaves of *M. oleifera* (g/g) have the calcium four times that of milk, vitamin C seven times that of oranges, and potassium three times that of bananas, three times the iron of spinach, four times the vitamin A in carrots, and two times the protein in milk [8, 9]. Studies have attributed the medicinal benefits of *M. oleifera* to its anti-inflammatory, antioxidant, and antipathogenic constituents [10, 11]. Also, the strong antioxidant and scavenging ability of *M. oleifera* has been linked to chemoprevention of diseases like cancer [12].

Research efforts have continued to unearth the medicinal and nutritional potential of *M. oleifera* in several significant ways. The *M. oleifera* plant has been used as antispasmodic, stimulant, cough expectorant, and diuretic agent [13]. The use of *M. oleifera* as diuretic agent is applicable in the treatment of prostatitis, kidney stones, bladder ache, scalding urine, water retention, and obesity [14]. In a separate study, the extract of *M. oleifera* showed antinephrotoxic effect against DMBA-induced renal cancer [15]. In a similar work by Owolabi et al., [16], *M. oleifera* extracts reduced the severity of lead-induced nephrotoxicity.

In furtherance of exploring and accentuating the medicinal potential of *M. oleifera* in protein against heavy metal toxicity, the present study evaluated the nephroprotective capacity of *M. oleifera* supplemented diets in Wistar rats orally administered with nickel sulphate ($NiSO_4$).

2. Methods

2.1. Chemicals and Reagents. All chemicals and reagents used were of analytical grade. Commercial reagent kits for the determination of plasma urea, creatinine, potassium, sodium, and protein were either as supplied by Randox diagnostic laboratory, Crumlin, UK, or the Agape laboratory, Switzerland.

2.2. Experimental Animals. Thirty male rats of Wistar strain of weight between 190 and 200 g were obtained from the Experimental Animal Farm at the Department of Biochemistry, University of Ilorin, Ilorin, Nigeria. The Wistar rats were housed in plastic animal cages in a well-ventilated experimental room. Rats were allowed to acclimatize for a period of 14 days before the commencement of treatments. During the period of acclimatization, the animals had unlimited access to rat feed pellets and clean water. Handling of animals was in accordance with relevant institutional and ethical guidelines as approved for scientific study.

2.3. Moringa oleifera. The leaves of *Moringa oleifera* were harvested at the Landmark University Farm, Omu-Aran, Nigeria. The leaves were identified and authenticated by Mr. Bolu Ajayi at the Herbarium Unit, Department of Plant Biology, University of Ilorin, Ilorin, Nigeria. Specimen voucher number is VIH 001/1011.

The *M. oleifera* leaves were air-dried and ground into powder. The powdered material was stored at room temperature in cool dry environment until being required for use.

2.4. Feed Formulation. The composition of the feed is as shown in Table 1.

2.5. Determination of Proximate Composition. Protein content, crude fat and fibre, moisture content, carbohydrate, and ash content were determined following established protocols as described by the Association of Official Analytical Chemists [17].

2.6. Experimental Design. Thirty male Wistar rats of weight between 195 and 200 g were randomly distributed into six experimental groups of five and fed as follows:

- control received $NiSO_4$ and fed normal rat chow;
- group 2 received $NiSO_4$ and fed with 5% *M. oleifera* supplemented diet;
- group 3 received $NiSO_4$ and fed with 10% *M. oleifera* supplemented diet;
- group 4 received $NiSO_4$ and fed with 15% *M. oleifera* supplemented diet;
- group 5 received normal saline and fed with 15% *M. oleifera* supplemented diet;
- group 6 received normal saline and fed with normal rat chow.

The rats were exposed to daily administration of 20 mg/kg body weight $NiSO_4$ by oral gavage. Treatments lasted for 21 days.

2.7. Necroscopy. 24 hours after the last treatment, the animals were sacrificed under anesthesia in slight diethyl ether. Blood samples were collected by cardiac puncture into clean EDTA bottles. The blood samples were spun at 5000 g for 10 minutes in a refrigerated centrifuge (Anke TDL-5000B, Shanghai, China) to yield the plasma which was used for the biochemical determinations. The kidneys from each animal were excised into iced physiological solution (pH 7.4), weighed immediately, and fixed in buffered formalin for histopathology examinations.

2.8. Biochemical Assays. The biochemical indices were determined in rat plasma using a UV/Vis spectrophotometer (Jenway, Staffordshire, UK) where applicable. The levels of rat plasma total protein (TP), creatinine, and urea were determined using Randox assay kits (Crumlin, UK). The rat plasma electrolytes, sodium, and potassium were determined using reagent assay kits by the Agape laboratory (Switzerland). The intraday and interday coefficient of variation (CV) is less than 5%.

2.9. Histopathology Examination. The rat kidney was fixed in 10% buffered neutral formalin (BNF) immediately following excision from animals. Fixed tissues were subsequently processed for histopathology examinations as previously described [18]. Capture and scoring for morphological changes were done by a pathologist blind to the treatments, the Pathology Unit, University of Ilorin Teaching Hospital, Ilorin, Nigeria.

2.10. Data analysis. Data were analysed on GraphPad Prism 3 (GraphPad Software Inc., San Diego, CA) using the one-way analysis of variance (ANOVA). Post hoc tests were conducted using the Tukey test. Data are reported as mean value ± standard error of mean (SEM). Mean values at $P < 0.05$ are significant.

3. Results

3.1. Proximate Composition. Diets were formulated consisting of different concentrations of *M. oleifera*. All diets used for the treatment were evaluated for proximate composition by AOAC protocols. The contents of crude fat, fibre, moisture, and ash in the normal rat chow and *M. oleifera* supplemented diets showed no significant difference (Figure 1). However, the carbohydrate content of supplemented formulated diets decreased with increasing concentration of *M. oleifera* relative to the normal diet. Conversely, protein content was higher in the *M. oleifera* supplemented diets.

TABLE 1: Formulation of experimental feed.

Feeds component (g)	Normal feed	5% *Moringa oleifera*	10% *Moringa oleifera*	15% *Moringa oleifera*
Sucrose	300	300	300	300
Cellulose	120	120	120	120
Soybean	750	750	750	750
Moringa oleifera	—	300	600	900
Vitamin mix	150	150	150	150
D-methionine	12	12	12	12
Corn starch	1518	1218	918	618
Soybean oil	150	150	150	150
Total	3000	3000	3000	3000

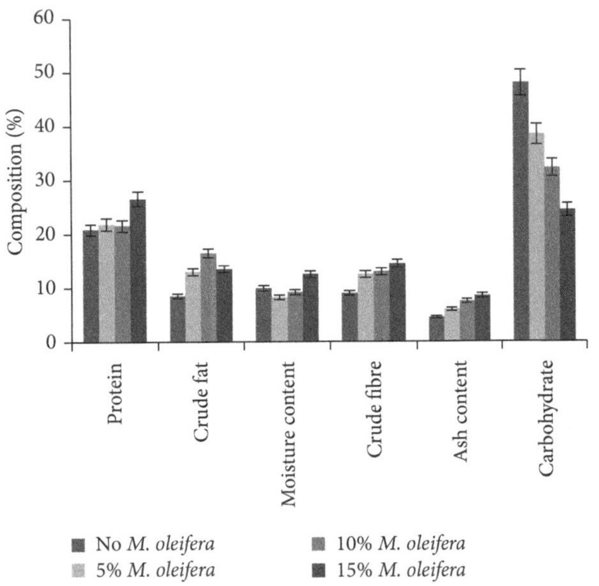

FIGURE 1: Proximate composition of experimental feeds consisting of various concentrations of *Moringa oleifera* leaves. Data are presented as mean values ± standard error of mean (SEM), $n = 3$.

3.2. Kidney and Body Weights. The various treatments did not evoke any appreciable change in average rat weights over the course of the exposure (Figure 2). However, there was 12% decline in rat kidney-to-body weight ratio for the group given $NiSO_4$ and sustained on normal rat diet. Conversely, there were improvements in the kidney weight with restoration to near normal level for groups maintained on *M. oleifera* supplemented diets.

3.3. Plasma Protein. Diets containing the 15% *M. oleifera* caused significant increments in protein level relative to the control group (Figure 2). The other treatment groups did not show any appreciable changes.

3.4. Plasma Creatinine. In order to determine whether addition of *M. oleifera* to diets could reverse or modulate the events of nickel-induced nephrotoxicity, the level of plasma creatinine was determined. For the control group which received $NiSO_4$ and normal rat diet, the plasma creatinine level was significantly elevated (Figure 3). On the contrary, groups given $NiSO_4$ and sustained on *M. oleifera* supplemented diets showed significantly lower plasma creatinine level. The restoration or modulation of the rat plasma creatinine to normal level shows correlation with increasing concentration of *M. oleifera* in the diets given to rats.

3.5. Plasma Urea. Furthermore, nickel exposure in the control group sustained on normal diet led to increased ($P < 0.05$) level of plasma urea (Figure 3). On the other hand, rats orally exposed to nickel salt and sustained on *M. oleifera* supplemented diets had significantly reduced level of plasma urea.

3.6. Plasma Electrolytes. Oral exposure to $NiSO_4$ and daily sustenance on normal diet in the control group for 21 days lowered ($P < 0.05$) the level of rat plasma sodium (Figure 4). However, for rats given $NiSO_4$ and sustained on *M. oleifera* supplemented diets, there were significant elevations in levels of plasma sodium.

Also, the control group given $NiSO_4$ and normal diet had elevated ($P < 0.05$) level of rat plasma potassium compared to groups sustained on *M. oleifera* supplemented diets in which the rat plasma potassium was appreciably maintained at near normal level.

3.7. Histopathology Examinations. The examination of rat renal sections for morphological changes revealed inimical cellular alterations in control group caused by $NiSO_4$ exposure. The cellular alterations caused included swollen renal tubules, hyaline change, thickening of glomerular wall, mild nephritis, and necrosis (Figures 5, 6, 7, 8, 9, and 10). However, all of the cellular damages caused by $NiSO_4$ exposure were nonexistent or ameliorated in the groups sustained on the *M. oleifera* diets.

4. Discussion

The *M. oleifera* has shown remarkable potential which could be harnessed for medicinal and nutritional purposes. In recent times, several studies have been able to demonstrate the medicinal significance of *M. oleifera* [19–22]. In separate

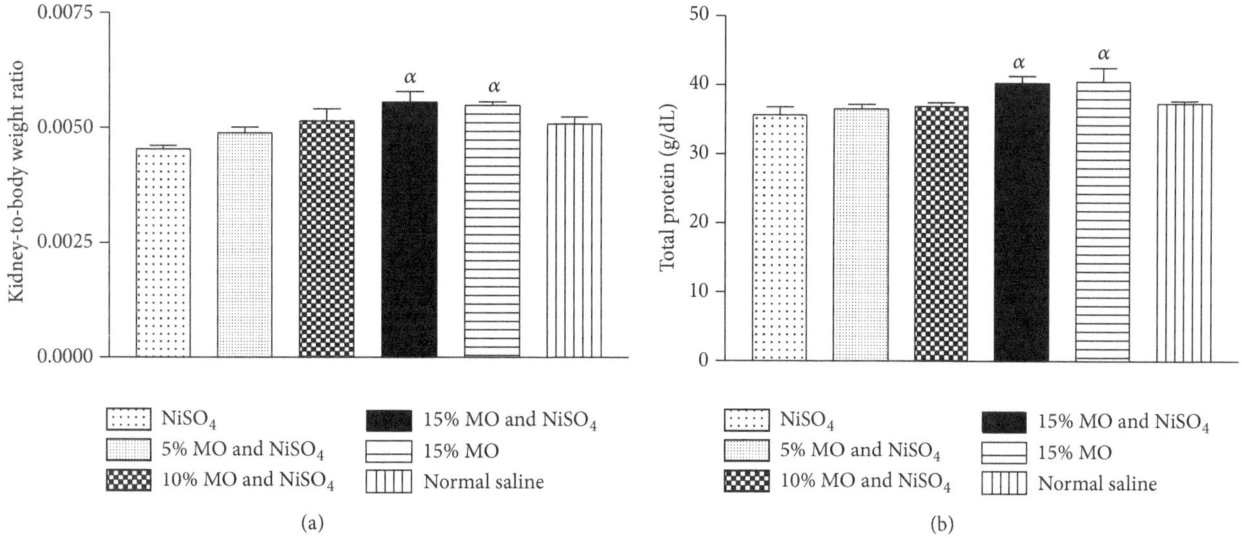

FIGURE 2: (a) Effect of *Moringa oleifera* supplemented diets on rat kidney-to-body weight ratio following exposure to nickel sulphate; (b) effect of *Moringa oleifera* supplemented diets on levels of rat plasma protein following exposure to nickel sulphate. Data are presented as mean values ± standard error of mean (SEM), $n = 6$. α is significantly ($P < 0.05$) relative to nickel sulphate.

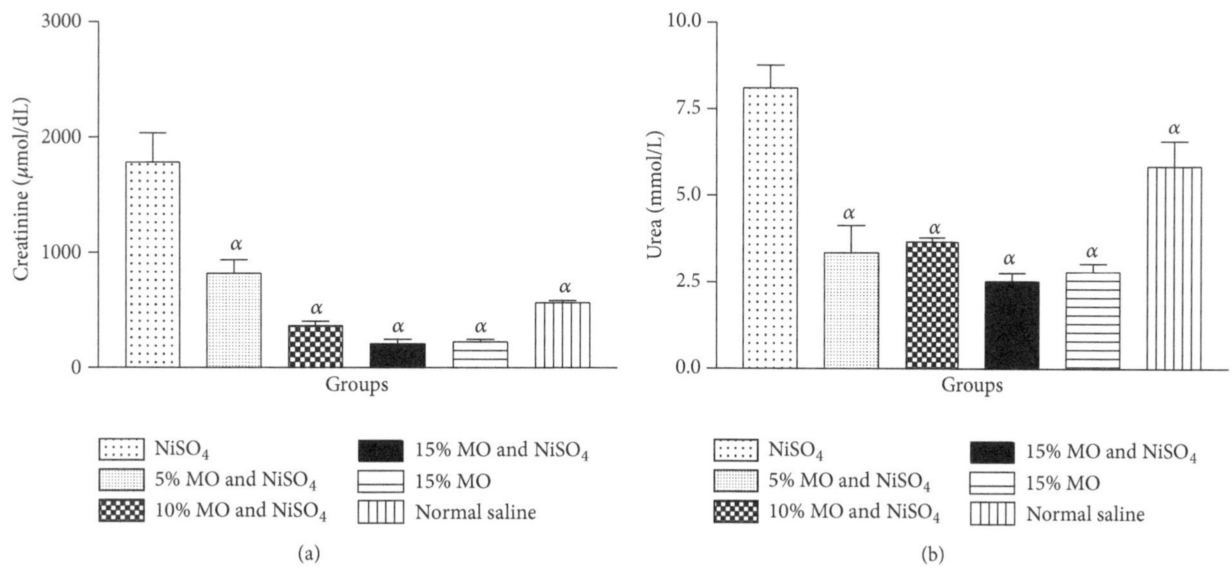

FIGURE 3: (a) Effect of *Moringa oleifera* supplemented diets on levels of rat plasma creatinine following exposure to nickel sulphate; (b) effect of *Moringa oleifera* supplemented diets on levels of rat plasma urea following exposure to nickel sulphate. Data are presented as mean values ± standard error of mean (SEM), $n = 6$. α is significantly ($P < 0.05$) relative to nickel sulphate.

studies, addition of *M. oleifera* to diet has revealed considerable promises as an adjunct to improving health in a variety of important ways [22, 23].

In furtherance of the evidence of medicinal and nutritional benefits of *M. oleifera*, diets consisting of different concentrations of *M. oleifera* were compounded and evaluated for potential of attenuating nickel-induced nephrotoxicity in rats.

Proximate analysis of the diets revealed higher protein content in the *M. oleifera* supplemented diets. The higher protein content may be due to addition of *M. oleifera* to the diets. Earlier studies have revealed *M. oleifera* as being rich in protein content [8, 9]. To further underscore the nutritional value of *M. oleifera* diets, the proximate analysis showed reduced carbohydrate content with increasing percentage of *M. oleifera*. This may imply nutritional relevance for management of diabetics or related metabolic disorder.

Measurement of body weight and/or organ weight may be used to evaluate toxic events due to exposure to a toxicant. Alteration in body weight and/or organ to body weight ratio

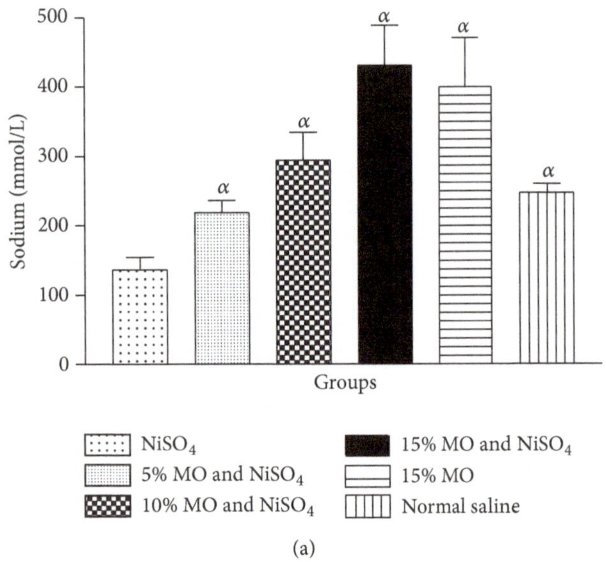

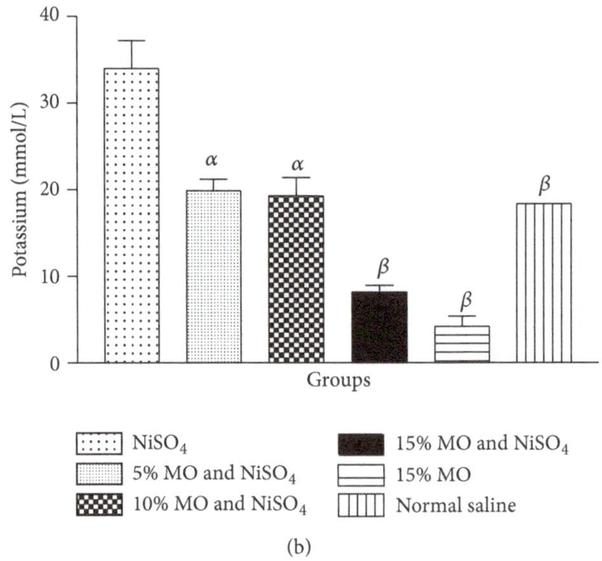

FIGURE 4: (a) Effect of *Moringa oleifera* supplemented diets on levels of rat plasma sodium following exposure to nickel sulphate; (b) effect of *Moringa oleifera* supplemented diets on levels of rat plasma potassium following exposure to nickel sulphate. Data are presented as mean values ± standard error of mean (SEM), $n = 6$. α is significantly ($P < 0.05$) relative to nickel sulphate; β is significantly ($P < 0.01$) relative to nickel sulphate.

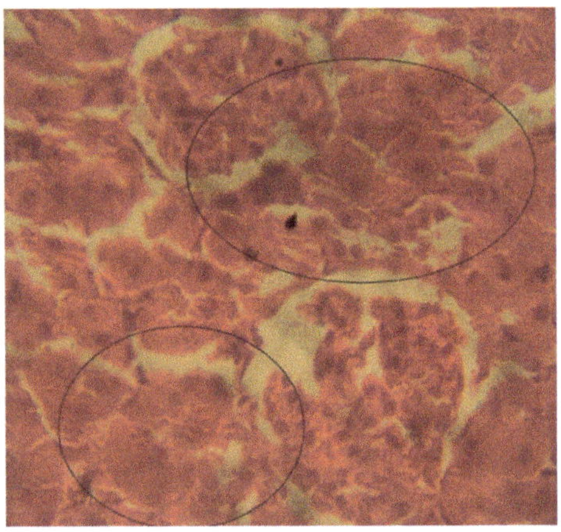

FIGURE 5: Photomicrographs of rat renal sections following oral exposure to nickel sulphate and sustenance on normal diets. H&E staining (×400).

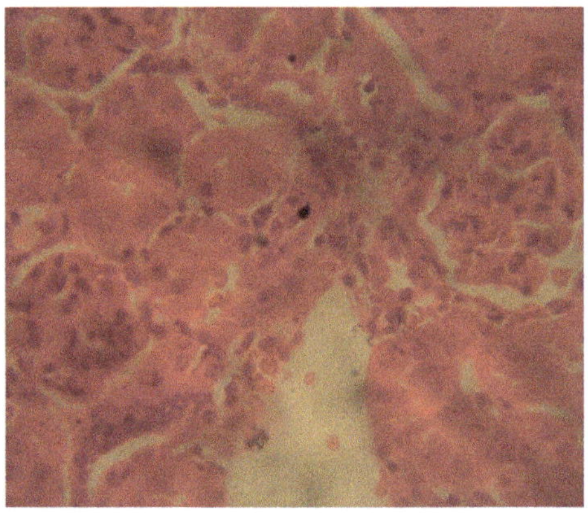

FIGURE 6: Photomicrographs of rat renal sections following oral exposure to nickel sulphate and sustenance on 5% *Moringa oleifera* supplemented diets. H&E staining (×400).

has been shown to imply toxicity arising from exposure to a toxicant [24, 25]. No appreciable weight change was recorded for all treatment groups. However, declining kidney-to-body weight ratio in the control group given $NiSO_4$ and normal diet was improved significantly in groups exposed to $NiSO_4$ and *M. oleifera* supplemented diets. This could be an indication that *M. oleifera* addition to diet was able to attenuate the adverse effect of $NiSO_4$ administration on rat renal tissues.

On the other hand, rat plasma protein levels were not appreciably affected by $NiSO_4$ exposure. However, sustaining animals on diets containing 15% *M. oleifera* increased the protein level. This may be due to the fact that *M. oleifera* has a high content of protein [8, 9]. This is even more plausible if consideration is given to the proximate composition data in which the 15% *M. oleifera* diet had the highest protein content.

Plasma levels of creatinine and urea are among the major biochemical indices commonly used to evaluate renal functions [18, 26]. Creatinine is a by-product of muscle metabolism and under normal physiological condition the amount excreted per day is constant and correlates with body mass [26]. Conversely, when creatinine level is elevated due to retention in the blood, it could be used to evaluate glomerular filtration rate [18]. Urea, on the other hand, is formed as means to rid body of nitrogenous waste from protein degradation. It is formed in the liver and excreted by the kidney in urine. Elevated plasma urea level has been linked to reduced renal function [18, 26].

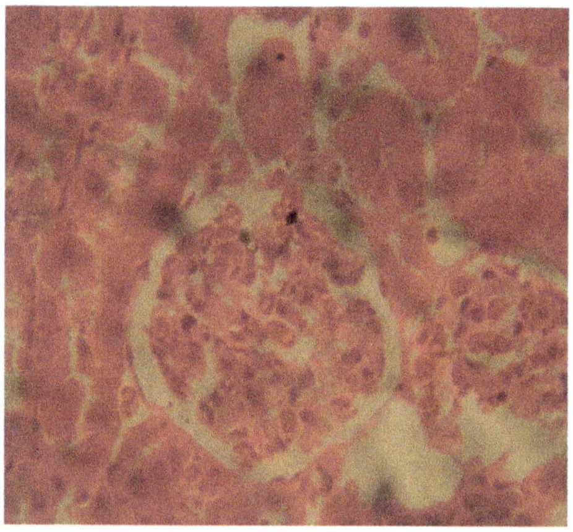

FIGURE 7: Photomicrographs of rat renal sections following oral exposure to nickel sulphate and sustenance on 10% *Moringa oleifera* diets. H&E staining (×400).

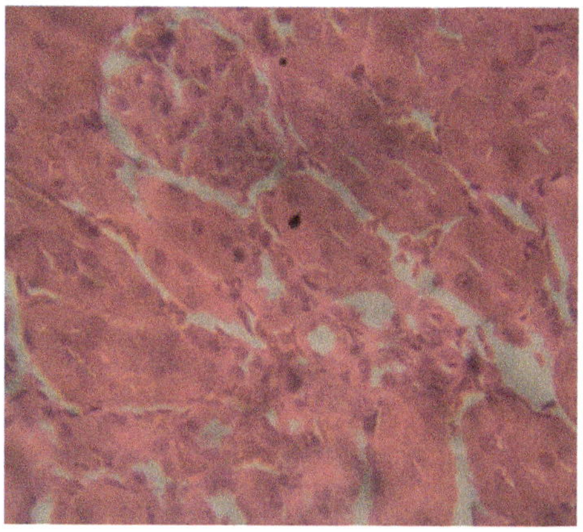

FIGURE 8: Photomicrographs of rat renal sections following oral exposure to nickel sulphate and sustenance on 15% *Moringa oleifera* diets. H&E staining (×400).

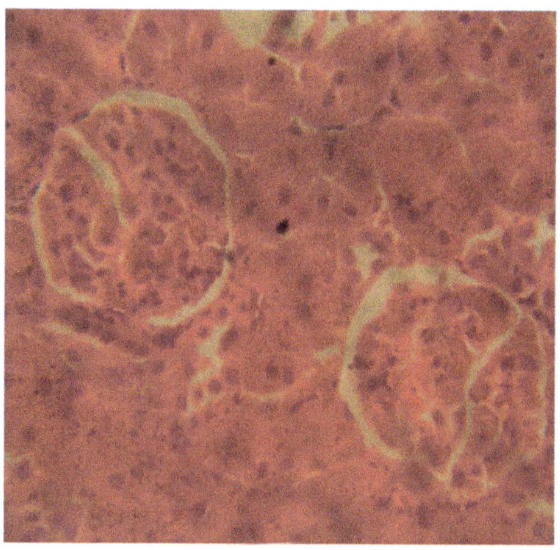

FIGURE 9: Photomicrographs of rat renal sections following oral exposure to normal saline and sustenance on 15% *Moringa oleifera* diets. H&E staining (×400).

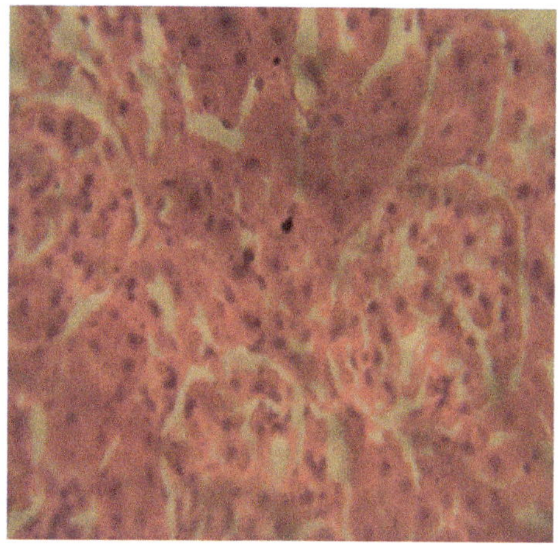

FIGURE 10: Photomicrographs of rat renal sections following oral exposure to normal saline and sustenance on normal diets. H&E staining (×400).

In the present study, levels of plasma creatinine and urea were determined. The oral exposure of $NiSO_4$ in rats caused significant elevation to the plasma levels of creatinine and urea in a manner reminiscent of compromised renal integrity. However, the feeding of diets containing different concentrations of *M. oleifera* to rats prevented the elevation of the indices for renal dysfunction. These findings are in support of previous studies which showed that *M. oleifera* offered nephro- and hepatoprotection [15, 20, 21, 27]. Also, in a recent study, the protective effect of *M. oleifera* against lead-induced nephrotoxicity has been demonstrated [16]. The evidences are demonstrations which strengthen further the belief that addition of *M. oleifera* to diets could improve health status.

To further evaluate nephroprotective effect of diets containing *M. oleifera*, we determined the level of rat plasma electrolytes. Our results revealed significant alteration to levels of rat plasma electrolytes caused by oral exposure to $NiSO_4$. On the contrary, the significant reductions in levels of plasma sodium and elevations in potassium level caused by $NiSO_4$ were averted by *M. oleifera* supplemented diets. The maintenance of normal levels of plasma electrolytes is crucial to homeostatic balance. Perturbation of plasma electrolyte balance may affect the pH, osmolality, and blood volume with adverse impact on the kidney and other related body organs. *M. oleifera* has been shown to possess diuretic effect [13, 14] and this may have contributed to protecting against homeostatic imbalance imposed by nickel exposure. Moreover, the maintenance of normal level of plasma sodium and potassium by *M. oleifera* diet is consistent with reported therapeutic potential including the improvement of thc health of renal tissues and general well-being [20, 21]. Although, at 10 and 15% *M. oleifera* supplementation, there were increases or decreases in the level of sodium and potassium, respectively, the alterations were found not to be significant when compared to the untreated group given normal saline and 0% *M. oleifera* diet.

The histopathology of renal sections revealed inimical cellular lesions caused by $NiSO_4$ exposure in rats. The hyaline change, thickening of glomerular wall, mild nephritis, and necrotic areas caused by $NiSO_4$ were conspicuously absent or attenuated in the groups fed on diets containing *M. oleifera*. The histopathology presentations support the biochemical findings and confirm further the medicinal protection afforded by *M. oleifera* against drug-induced tissue damage [15, 16, 19, 21, 27].

To our knowledge, this is the first study which revealed that the addition of *M. oleifera* to diet protected rat kidney against the toxic events caused by $NiSO_4$ exposure. The data are further scientific demonstration which underpins the medicinal and nutritional potential of *M. oleifera*. For people working in the mining industries or other related sectors, where the frequency of exposure to heavy metals is high, consumption of diet containing *M. oleifera* may help protect against occupational health risks.

Highlights

(1) Male Wistar rats were daily exposed to oral administration of $NiSO_4$ and sustained on either normal diets or *Moringa oleifera* supplemented diets.

(2) In the control group which was sustained on normal diets, $NiSO_4$ exposure altered the levels of plasma metabolites including urea, creatinine, and electrolytes in a manner that is reminiscent of impaired renal function.

(3) In the other treatment groups sustained on *Moringa oleifera* supplemented diets, there were no alterations to the levels of plasma urea, creatinine, and electrolytes revealing nephroprotective potential of the supplemented diets.

(4) Oral exposure of $NiSO_4$ caused damage to renal tubules and glomerular walls. However, *Moringa oleifera* supplemented diets protected against the renal damage.

(5) The addition of *Moringa oleifera* to diet protected against nickel-induced nephrotoxicity.

Conflict of Interests

The authors declare that there is no conflict of interests regarding the publication of this paper.

Acknowledgment

The authors thank the laboratory staff of the Department of Biological Sciences, Landmark University, Nigeria, for the technical support.

References

[1] A. R. Verma, M. Vijayakumar, C. S. Mathela, and C. V. Rao, "*In vitro* and *in vivo* antioxidant properties of different fractions of *Moringa oleifera* leaves," *Food and Chemical Toxicology*, vol. 47, no. 9, pp. 2196–2201, 2009.

[2] J. F. Morton, "The horseradish tree, *Moringa pterygosperma* (Moringaceae)—a boon to arid lands?" *Economic Botany*, vol. 45, no. 3, pp. 318–333, 1991.

[3] F. Anwar, M. Ashraf, and M. I. Bhanger, "Interprovenance variation in the composition of *Moringa oleifera* oilseeds from Pakistan," *Journal of the American Oil Chemists' Society*, vol. 82, no. 1, pp. 45–51, 2005.

[4] A. Odebiyi and E. A. Sofowora, "Pytochemical screenings of Nigerian medicinal plants part 11," *Lyodia*, vol. 44, pp. 234–246, 1999.

[5] U. K. Mazumder, M. Gupta, S. Chakrabarti, and D. Pal, "Evaluation of hematological and hepatorenal functions of methanolic extract of *Moringa oleifera* Lam. root treated mice," *Indian Journal of Experimental Biology*, vol. 37, no. 6, pp. 612–614, 1999.

[6] K. S. Rao and S. H. Mishra, "Anti-inflammatory and antihepatotoxic activities of the rats of moringa pterygosperma geaertn," *Indian Journal of Pharmaceutical Sciences*, vol. 60, no. 1, pp. 12–16, 1998.

[7] S. G. Mahajan and A. A. Mehta, "Effect of *Moringa oleifera* Lam. seed extract on ovalbumin-induced airway inflammation in guinea pigs," *Inhalation Toxicology*, vol. 20, no. 10, pp. 897–909, 2008.

[8] W. J. Asante, I. L. Nasare, D. Tom-Dery, K. Ochire-Boadu, and K. B. Kentil, "Nutrient composition of Moringa oleifera leaves from two agro ecological zones in Ghana," *African Journal of Plant Science*, vol. 8, no. 1, pp. 65–71, 2014.

[9] E. I. Bamishaiye, F. F. Olayemi, E. F. Awagu, and O. M. Bamshaiye, "Proximate and phytochemical composition of Moringa oleifera leaves at three stages of maturation," *Advance Journal of Food Science and Technology*, vol. 3, no. 4, pp. 233–237, 2011.

[10] N. A. Kumar and L. Pari, "Antioxidant action of *Moringa oleifera* Lam. (drumstick) against antitubercular drugs induced lipid peroxidation in rats," *Journal of Medicinal Food*, vol. 6, no. 3, pp. 255–259, 2003.

[11] R. Bharali, J. Tabassum, and M. R. H. Azad, "Chemomodulatory effect of *Moringa oleifera*, Lam, on hepatic carcinogen metabolising enzymes, antioxidant parameters and skin papillomagenesis in mice," *Asian Pacific Journal of Cancer Prevention*, vol. 4, no. 2, pp. 131–139, 2003.

[12] S. E. Atawodi, J. C. Atawodi, G. A. Idakwo et al., "Evaluation of the polyphenol content and antioxidant properties of methanol extracts of the leaves, stem, and root barks of *Moringa oleifera* Lam," *Journal of Medicinal Food*, vol. 13, no. 3, pp. 710–716, 2010.

[13] M. Mbikay, "Therapeutic potential of *Moringa oleifera* leaves in chronic hyperglycemia and dyslipidemia: a review," *Frontiers in Pharmacology*, vol. 3, article 24, 2012.

[14] P. S. Kumar, D. Mishra, G. Ghosh, and G. S. Panda, "Medicinal uses and pharmacological properties of *Moringa oleifera*," *International Journal of Phytomedicine*, vol. 2, no. 3, pp. 210–216, 2010.

[15] V. Sharma and R. Paliwal, "Chemo protective role of *Moringa oleifera* and its isolated saponin against DMBA induced tissue damage in male mice: a histopathological analysis," *International Journal of Drug Development & Research*, vol. 4, no. 4, pp. 215–228, 2012.

[16] J. O. Owolabi, O. K. Ghazal, F. E. Williams, and O. O. Gurusa, "Assessment of the prophylactic and rejuvenative effects of moringa oleifera phytochemicals extracts on lead-induced renal

tissue disruption in adults male wistar rats models," in *Proceedings of the Moringa at the Leading Edge: International Conference on Moringa Oleifera*, vol. 1, pp. 1–12, 2012.

[17] AOAC, *Official Methods of Analysis*, Association of Official Analytical Chemists, Washington, DC, USA, 15th edition, 1995.

[18] O. S. Adeyemi and M. A. Akanji, "Psidium guajava leaf extract: effects on rat serum homeostasis and tissue morphology," *Comparative Clinical Pathology*, vol. 21, no. 4, pp. 401–407, 2012.

[19] L. Pari and N. A. Kumar, "Hepatoprotective activity of *Moringa oleifera* on antitubercular drug-induced liver damage in rats," *Journal of Medicinal Food*, vol. 5, no. 3, pp. 171–177, 2002.

[20] F. Anwar, S. Latif, M. Ashraf, and A. H. Gilani, "*Moringa oleifera*: a food plant with multiple medicinal uses," *Phytotherapy Research*, vol. 21, no. 1, pp. 17–25, 2007.

[21] M. Ndong, M. Uehara, S.-I. Katsumata, and K. Suzuki, "Effects of oral administration of *Moringa oleifera* Lam on glucose tolerance in Goto-Kakizaki and wistar rats," *Journal of Clinical Biochemistry and Nutrition*, vol. 40, no. 3, pp. 229–233, 2007.

[22] C. Promkum, P. Kupradinun, S. Tuntipopipat, and C. Butryee, "Nutritive evaluation and effect of *Moringa oleifera* pod on clastogenic potential in the mouse," *Asian Pacific Journal of Cancer Prevention*, vol. 11, no. 3, pp. 627–632, 2010.

[23] B. C. Johnson, *Clinical Perspectives on the Health Effects of Moringa oleifera : A Promising Adjunct for Balanced Nutrition and Better Health*, KOS Health Publications, 2005.

[24] O. E. Orisakwe, D. C. Husaini, and O. J. Afonne, "Testicular effects of sub-chronic administration of *Hibiscus sabdariffa* calyx aqueous extract in rats," *Reproductive Toxicology*, vol. 18, no. 2, pp. 295–298, 2004.

[25] O. S. Adeyemi, M. Fambegbe, O. R. Daniyan, and I. Nwajei, "Yoyo Bitters, a polyherbal formulation influenced some biochemical parameters in Wistar rats," *Journal of Basic and Clinical Physiology and Pharmacology*, vol. 23, no. 4, pp. 135–138, 2012.

[26] J. L. Gross, M. J. de Azevedo, S. P. Silveiro, L. H. Canani, M. L. Caramori, and T. Zelmanovitz, "Diabetic nephropathy: diagnosis, prevention, and treatment," *Diabetes Care*, vol. 28, no. 1, pp. 164–176, 2005.

[27] O. Awodele, I. A. Oreagba, S. Odoma, J. A. da Silva, and V. O. Osunkalu, "Toxicological evaluation of the aqueous leaf extract of *Moringa oleifera* Lam. (Moringaceae)," *Journal of Ethnopharmacology*, vol. 139, no. 2, pp. 330–336, 2012.

30

Effect on Blood Pressure of Daily Lemon Ingestion and Walking

Yoji Kato,[1] Tokio Domoto,[1] Masanori Hiramitsu,[2] Takao Katagiri,[2] Kimiko Sato,[1] Yukiko Miyake,[1] Satomi Aoi,[1] Katsuhide Ishihara,[1] Hiromi Ikeda,[1] Namiko Umei,[1] Atsusi Takigawa,[1] and Toshihide Harada[1]

[1] *Faculty of Health and Welfare, Prefectural University of Hiroshima, Mihara, Hiroshima 723-0053, Japan*
[2] *Central Laboratory, Research & Development Division, Pokka Sapporo Food and Beverage Ltd., Kitanagoya, Aichi 150-0013, Japan*

Correspondence should be addressed to Yoji Kato; ykato@pu-hiroshima.ac.jp

Academic Editor: H. K. Biesalski

Background. Recent studies suggest that the daily intake of lemon (*Citrus limon*) has a good effect on health, but this has not been confirmed in humans. In our previous studies, it was observed that people who are conscious of their health performed more lemon intake and exercise. An analysis that took this into account was required. *Methodology.* For 101 middle-aged women in an island area in Hiroshima, Japan, a record of lemon ingestion efforts and the number of steps walked was carried out for five months. The change rates (Δ%) of the physical measurements, blood test, blood pressure, and pulse wave measured value during the observation period were calculated, and correlations with lemon intake and the number of steps walked were considered. As a result, it was suggested that daily lemon intake and walking are effective for high blood pressure because both showed significant negative correlation to systolic blood pressure Δ%. *Conclusions.* As a result of multiple linear regression analysis, it was possible that lemon ingestion is involved more greatly with the blood citric acid concentration Δ% and the number of steps with blood pressure Δ%, and it was surmised that the number of steps and lemon ingestion are related to blood pressure improvement by different action mechanisms.

1. Introduction

It has been reported that food materials such as tea catechins, cacao lignin, and wheat bran hemicellulose had a suppressive effect on blood pressure in spontaneously hypertensive rats (SHR) [1–3].

Several studies highlighted lemon as an important health-promoting fruit rich in phenolic compounds as well as vitamins, minerals, dietary fibers, essential oils, and carotenoids [4].

The suppressive effect of lemon juice and its crude flavonoid on blood pressure has been reported in SHR [5]. The water extract of lemon peels also had a suppressive effect on blood pressure in SHR [6, 7]. Recently, we investigated the effect of daily lemon intake on parameters related to metabolic syndrome in healthy women and showed that the amount of lemon intake had a significant negative correlation with systolic blood pressure [8].

On the other hand, it has been well known that daily walking improves the blood pressure in humans [9].

In this study, we examined the contribution of daily lemon intake and daily walking to blood pressure and related parameters in middle-aged and older women in their daily life.

2. Materials and Methods

2.1. Subjects. A hundred eleven women aged 33 to 77 (mean ± SD: 60.9 ± 8.0) who resided on islands in the Seto Inland Sea of Hiroshima Prefecture were included in this study. This area is known for lemon production in Japan. In September 2011 and in March 2012, we conducted a survey of physical measurements and biochemical and serological parameters. Of these 111 women, 10 women who were treated for hypertension, diabetes, or hypercholesterolemia therapy were excluded in data analysis. In the period between the first survey (September 2011) and the second survey (March 2012), they were requested to wear the record-type pedometer (Lifecorder EX, Suzuken Co. LTD., Japan) and to self-record their daily lemon intake. The lemon ingestion method was not specified.

2.2. Methods. The physical characteristics were measured with a Body Composition Analyzer (BC-118D, Tanita Co., Ltd., Japan). Blood pressure and pulse wave velocity were measured with inspection apparatus (Form BP-203 RPE, Omron Co., Ltd., Japan). Blood samples were sent to a privately owned clinical laboratory for biochemical and serological testing. There were 32 physical characteristics and biochemical and serological measurements examined, as shown in Table 1.

2.3. Data Analysis and Statistics. The amount of lemon intake was presented as an average number of lemons per day. Movement was expressed by the average number of steps per day according to the pedometer (Lifecorder EX, Suzuken Co., Ltd., Japan). The value of physical characteristics and biochemical and serological examinations was shown as the change rate (Δ%) of the value measured in March against the value measured in September (100%). The means and standard errors of Δ% were calculated. The data were compared between average values recorded in September and in March using Student's t-test.

The correlations between the lemon intake number per day, daily number of steps, and the change rate of each physical characteristic and biochemical and serological data were tested using Pearson's correlation coefficient.

Multiple regression analysis was performed using the number of steps per day and lemon intake as the independent variable and the change rates of systolic blood pressure and blood concentration of citric acid as the dependent variable.

All statistical analyses were performed using statistical software (SPSS version 17.0, SPSS Inc., USA) on personal computers, and $P < 0.05$ was considered as significant.

2.4. Ethics Statement. Based on Declaration of Helsinki, we explained the meaning and the methods of research to all the subjects and obtained written consent. The plan of research was recognized by the Research Ethics Committee of the Faculty of Health and Welfare of the Prefectural University of Hiroshima (M11-0024).

3. Results

The lemon intake per day during the entire study was 0.47 ± 0.47 (mean ± standard deviation), but there were particularly many omissions and deficiencies in the record of the early days of the investigation because it was self-recorded by the subjects. In February, which is the last part of the investigation, a complete record was 0.43 ± 0.49. Since the lemon intake of February and the whole period had strong correlation (correlation coefficient 0.947, $P < 0.000$), the data in February was used for analysis afterward. The number of steps per day throughout the entire period of investigation was 6947.74 ± 2187.20.

Table 1 shows the averages and standard error of the Δ% of measurement items. Obesity-related indexes, such as the weight, blood fat, and insulin, significantly increased. In addition, the iron-related indexes, such as a red blood cell and the hemoglobin and citric acid, increased.

The number of lemon intake per day correlated negatively with the Δ% of systolic blood pressure and red blood cells and positively with the Δ% of blood citric acid concentration. Similarly, the number of steps per day correlated negatively with the Δ% of systolic blood pressure and positively with blood citric acid concentration. All these correlations were significant. On the other hand, the Δ% of blood citric acid concentration showed negative and significant correlation with the Δ% of systolic blood pressure and pulse pressure and significantly positive correlation with red blood cells (Figure 1, Table 2).

Table 3 shows the results of multiple regression analysis. Lemon intake and number of steps walked were identified as the significant factors associated negatively with systolic blood pressure and the number of steps was more effective than the lemon intake. On the other hand, lemon intake and number of steps were identified as the significant factors associated positively with Δ% of blood citric acid concentration, and the lemon intake was more effective than the number of steps (Table 3).

4. Discussion

Citrus fruits are often reported to be foods which are beneficial for health [10–12]. The lemon, a representative citrus fruit, includes bioactive components such as citric acid, polyphenol, and ascorbic acid, and it is reported that there are various health benefits [4]. Previously, we reported a systolic blood pressure-lowering effect by lemon ingestion in Japanese people [8]. In the present study, keeping in mind the effect due to the motion of the subject, we confirm the effect of lemon intake.

During the five months obesity-related indicators of the subjects increased (Table 1). According to animal and *in vitro* experiments, it was reported that the lemon has effective ingredients for lipid metabolism and obesity [13–15]. But this effect could not be confirmed in this investigation. This is thought to be due to the seasonal cycle of the subjects. It was reported in at least Japanese people that the percentage of body fat increases in the winter season [16]. Furthermore, subjects of this investigation were older and middle-aged farm women. Since the survey period, from September through March, included the agricultural off-season, it is inferred that exercise time through work was significantly reduced. Therefore, it is thought that skeletal muscle masses decreased and the fat mass increased. From the fact that brachial-ankle pulse wave velocity increased and diastolic blood pressure significantly decreased, it is expected that the angiosclerosis in subjects was advancing, probably due to aging. The suppression of the increase in systolic blood pressure in this study may be due to enhanced metabolism such as urination.

The amount of lemon intake of the subjects prior to the survey period is not clear as a number. According to the prior interview, although subjects were from lemon-production farmhouses, they did not usually intentionally eat lemons.

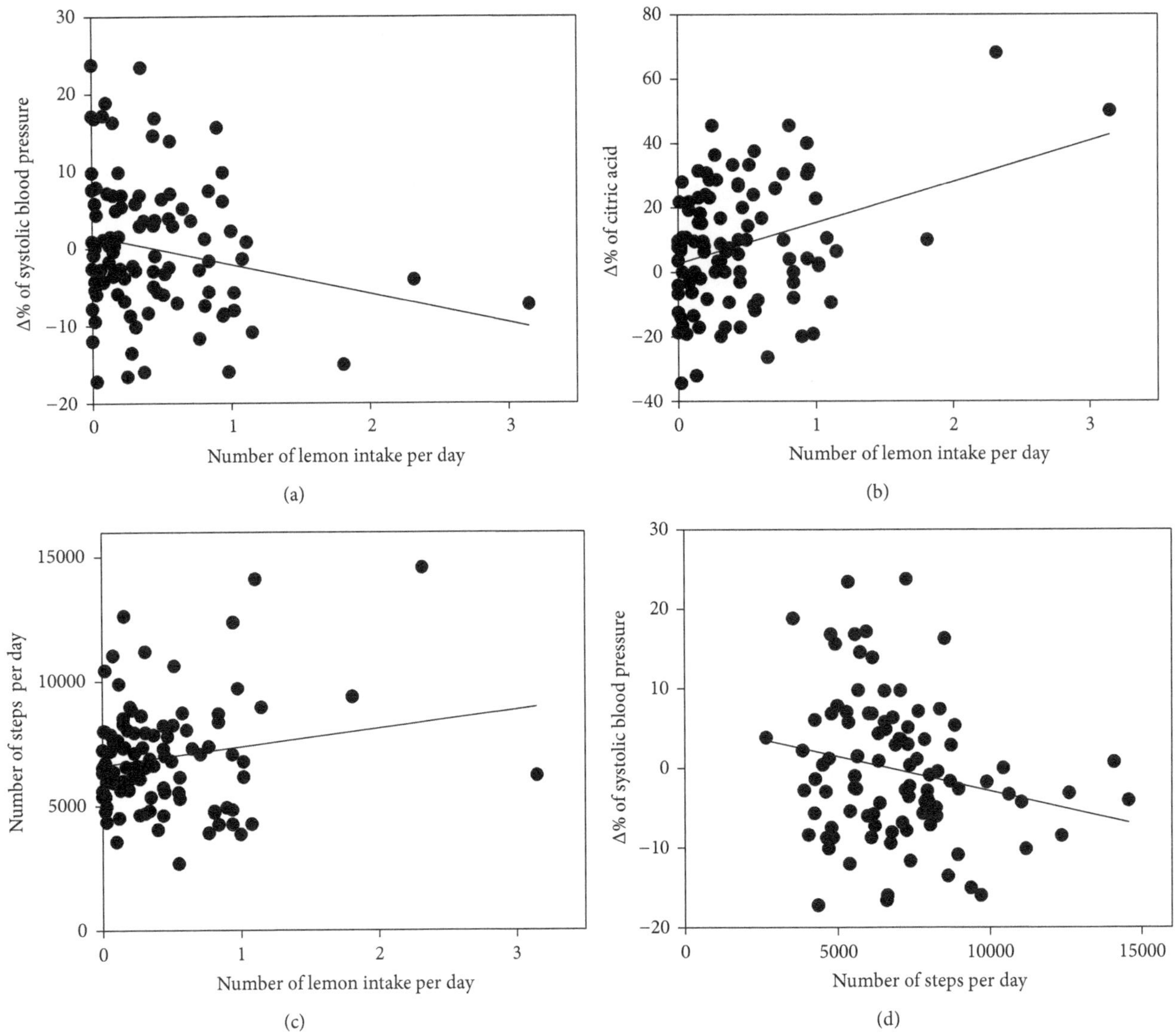

FIGURE 1: Scatter plot of lemon intake, number of steps, and Δ% of systolic blood pressure and citric acid. Linear approximations were shown in each graph as line ((a): $R^2 = 0.041$, (b): $R^2 = 0.117$, (c): $R^2 = 0.027$, and (d): $R^2 = 0.046$).

Rather, like the average Japanese, most only used them for cooking occasionally. The duty of the ingestion record of lemons and wearing of a pedometer was imposed during the survey period, and it is expected that this became the motivation to health consciousness. The increase in a lemon intake during the survey period is indirectly shown in the increase in the citrate in serum (Table 1).

The blood pressures, especially systolic pressure, showed significant negative correlation to the lemon intake and the number of steps. On the other hand, the number of steps and lemon intake tend to be in correlation (Figure 1, Table 2). It is guessed that subjects having high health consciousness ate lemons often and moved often. This means that the results shown in Table 2 are not simply dependent on the intake of lemon.

In this study, we focus on the citric acid among the components contained in the lemon. Citric acid Δ% varied strongly by the lemon intake rather than by the number of steps in a result of multiple linear regression analysis (Table 3). Therefore, it is suggested that this blood citric acid is due to the lemon.

Though the time-to-maximum blood concentration of citric acid is 0.8 hr and the half-life in blood is 2.3 hr, it is interesting that citric acid Δ% had a strong association with lemon intake and steps walked (Figure 1, Table 2) [17, 18]. That there was a cumulative effect with repeated administration is to be inferred.

Citric acid Δ% correlated with Δ% of systolic blood pressure, pulse pressure, and red blood cell count (Table 2). It is not out of the realm of analogy at all about the mechanism of action of citric acid in the present study. One of the reported properties of citric acid is that it generates calcium citrate from inorganic salts, such as calcium phosphate or calcium carbonate, to promote absorption from the intestinal

TABLE 1: Data of physical characteristics, blood vessel related values, and blood test.

	Values on September mean ± SD	Values on March mean ± SD	*P* values on between September and March	Change rate (Δ%) mean ± SE
Physical measurement				
Body weight (kg)	54.8 ± 8.3	55.4 ± 8.4	0.001	1.03 ± 0.29
Body mass index (kg/m^2)	23.3 ± 3.1	23.5 ± 3.1	0.005	1.07 ± 0.37
Body fat ratio (%)	33.1 ± 6.1	35.4 ± 6.0	<0.000	7.68 ± 1.09
Lean body mass (kg)	36.3 ± 3.4	35.4 ± 3.4	<0.000	−2.43 ± 0.24
Blood pressure, arteriosclerosis relation				
Systolic blood pressure (mmHg)	130.0 ± 16.3	130.0 ± 18.5	0.871	0.02 ± 0.89
Diastolic blood pressure (mmHg)	75.3 ± 9.2	73.4 ± 10.9	0.014	−2.37 ± 1.02
Pulse pressure (mmHg)	54.5 ± 9.8	55.9 ± 10.3	0.058	3.53 ± 1.40
Ankle Brachial Pressure Index	1.1 ± 0.1	1.1 ± 0.1	0.334	1.13 ± 0.80
Brachial-ankle pulse wave velocity (cm/sec)	1484.6 ± 286.4	1555.5 ± 322.4	0.058	5.08 ± 1.09
Blood test				
White blood cell (/μL)	5649.5 ± 1491.7	5922.8 ± 1583.5	0.013	6.49 ± 2.05
Red blood cell ($*10^4/\mu$L)	424.1 ± 29.5	437.9 ± 321.4	<0.000	3.30 ± 0.41
Hemoglobin (g/dL)	12.8 ± 1.0	13.1 ± 1.0	<0.000	3.04 ± 0.57
Hematocrit (%)	40.0 ± 2.8	41.1 ± 2.8	<0.000	3.73 ± 0.54
Platelet ($*10^4/\mu$L)	24.5 ± 5.3	25.4 ± 5.7	0.005	4.24 ± 1.31
Mean corpuscular Volume (fL)	93.7 ± 4.3	93.9 ± 4.0	0.297	0.42 ± 0.36
Mean corpuscular hemoglobin (pg)	30.1 ± 1.7	30.0 ± 1.56	0.317	−0.24 ± 0.42
Mean corpuscular hemoglobin concentration (%)	32.2 ± 0.96	32.0 ± 0.77	0.019	−0.63 ± 0.29
Hemoglobin A1c (%)	5.4 ± 0.4	5.3 ± 0.43	0.018	−0.72 ± 0.28
Serum iron (μg/dL)	86.8 ± 26.7	92.9 ± 30.0	0.032	12.99 ± 4.20
Unsaturated iron binding capacity (μg/dL)	225.1 ± 54.7	222.2 ± 47.1	0.384	0.68 ± 1.66
Reticulocyte (Ğ)	11.2 ± 3.4	10.4 ± 2.8	0.002	−3.99 ± 2.22
Total cholesterol (mg/dL)	208.0 ± 33.1	214.6 ± 33.9	0.003	3.65 ± 2.12
Neutral fat (mg/dL)	139.2 ± 79.1	139.5 ± 69.7	0.979	11.89 ± 5.05
High density lipoprotein cholesterol (mg/dL)	63.4 ± 15.4	64.5 ± 14.5	0.178	2.69 ± 1.28
Low density lipoprotein cholesterol (mg/dL)	119.5 ± 28.1	124.1 ± 30.0	0.012	4.74 ± 1.43
Uric acid (mg/dL)	4.37 ± 0.9	4.30 ± 0.9	0.245	−0.83 ± 1.31
Insulin (μU/mL)	17.1 ± 16.1	21.8 ± 19.8	0.012	95.45 ± 20.69
Cortisol (μg/dL)	50.4 ± 19.3	60.8 ± 22.7	<0.000	39.37 ± 7.70
Adiponectin (μg/mL)	9862.0 ± 5275.5	12418.0 ± 6636.6	<0.000	37.20 ± 7.42
Leptin (ng/mL)	6.4 ± 5.1	8.7 ± 5.3	<0.000	96.76 ± 21.08
Citric acid (mg/dL)	2.7 ± 0.4	2.9 ± 0.6	<0.000	8.20 ± 1.90
Renin (pg/mL)	565.0 ± 317.8	538.8 ± 318.9	0.263	3.43 ± 4.80

tract [19–21]. In addition, there is the report that the citric acid of citrus fruits promoted absorption of calcium and magnesium in food [22]. There is a possibility that calcium and magnesium which were absorbed efficiently are participating in blood pressure [23]. Also apart from the effect of citric acid in food, it is necessary to consider the effects of intracellular citric acid or the citrate in blood. The citric acid Δ% showed a positive correlation with the number of red blood cells Δ% but showed a negative correlation with lemon intake (Table 2). The mechanism is not clear, but this implies that the citric acid is involved in the blood composition. The lemon ingestion method in this research was not specified. It is believed that most subjects ingested only juice, but the ingestion of peels was not prohibited. Flavonoids of lemon

Table 2: Pearson coefficient of correlation of significantly different or tendency items.

	Lemon intake per day		Number of steps per day		Citric acid (Δ%)	
	Correlation coefficient	P value	Correlation coefficient	P value	Correlation coefficient	P value
Number of steps per day	0.167	0.099				
Citric acid (Δ%)	0.343	<0.000	0.213	0.042		
Systolic blood pressure (Δ%)	−0.204	0.040	−0.222	0.034	−0.242	0.015
Diastolic blood pressure (Δ%)	−0.171	0.088	−0.151	0.152	−0.058	0.562
Pulse pressure (Δ%)	−0.181	0.069	−0.188	0.074	−0.318	0.001
Red blood cell (Δ%)	−0.222	0.025	0.098	0.358	0.215	0.031

Table 3: Multiple regression analysis for the relationship between change rate of systolic blood pressure and blood citric acid concentration as dependent variables versus number of lemon intake and steps per day as independent variables.

	Systolic blood pressure (Δ%) (P = 0.32)		Citric acid (Δ%) (P = 0.01)	
	β coefficient	P value	β coefficient	P value
Number of steps per day	−0.190	0.060	0.154	0.113
Lemon intake per day	−0.153	0.128	0.302	0.002

are mainly contained in the peel and there are reports that the ingredient acts on the reduction in systolic blood pressure [4, 15]. It is possible that actions by ingredients other than citrate are also involved intricately.

The number of steps showed a negative correlation with systolic blood pressure Δ% and a positive correlation with citric acid Δ% (Table 2). Blood pressure improvement through walking has been reported [9]. Even in the results of heavy regression analysis of the present study, it has been shown that systolic blood pressure Δ% has been affected more strongly by the walking than by lemon ingestion (Table 3). Because of the amount of walking having been proportional to the lemon intake, but not having been related to the number of red blood cells Δ%, and the result of the multiple regression analysis, it is believed that the amount of walking is not applied to the citric acid Δ% directly.

These results suggest that the walking and lemon ingestion have the effect of lowering systolic blood pressure by, respectively, different action mechanisms. There may be an additive or synergistic effect in movement and lemon ingestion.

The habits of lemon ingestion before the investigation are not taken into consideration in this research. The method of lemon intake is also not clear, or how the peel was included. There is a need to examine blood metal ions as a clue to investigate the mechanism of action in the body of lemon components. A longer experimental period is required to study the effect on pulse wave velocity and diastolic blood pressure related to atherosclerosis [24, 25]. Noting these future studies, using a placebo in consideration of the health awareness of the subject, a long-term study of more than one year or a short-term study with less seasonal variation is desired.

Conflict of Interests

The authors declare that there is no conflict of interests regarding the publication of this paper.

Acknowledgments

The authors thank the Japan Agricultural Cooperatives (JA) Hiroshima Kajitsuren, JA Hiroshima Yutaka, JA Mihara Setoda, and the people of the region who cooperated in the research. They received generous support from Professor Timothy E. Buthod.

References

[1] Y. Hara and F. Tono-oka, "Hypotensive effect of tea catechins on blood pressure of rats," *Journal of Japanese Society of Nutrition and Food Science*, vol. 43, no. 5, pp. 345–348, 1990 (Japanese).

[2] T. Kamiwaki, K. Tsuji, and Y. Nakagawa, "Effects of cacao lignin on blood pressure and lipid metabolism in Spontaneously Hypertensive Rats," *Journal of the Agricultural Chemical Society of Japan*, vol. 68, no. 5, pp. 957–965, 1994 (Japanese).

[3] T. Kodama, K. Shiiba, and K. Tsuji, "Suppressive effect of Wheat-bran hemicellulose on blood pressure in Spontaneously Hypertensive Rats," *Journal of Japanese Society of Nutrition and Food Science*, vol. 49, no. 2, pp. 101–105, 1996 (Japanese).

[4] E. González-Molina, R. Domínguez-Perles, D. A. Moreno, and C. García-Viguera, "Natural bioactive compounds of Citrus limon for food and health," *Journal of Pharmaceutical and Biomedical Analysis*, vol. 51, no. 2, pp. 327–345, 2010.

[5] Y. Miyake, K. Kuzuya, C. Ueno et al., "Suppressive effect of components in lemon juice on blood pressure in Spontaneously Hypertensive Rats," *Food Science and Technology International*, vol. 4, no. 1, pp. 29–32, 1998.

[6] H. Kumamoto, Y. Matsubara, Y. Iizuka et al., "Structure and hypotensive effect of flavonoid glycosides in lemon peelings," *Journal of the Agricultural Chemical Society of Japan*, vol. 58, no. 2, pp. 137–143, 1984 (Japanese).

[7] H. Kumamoto, Y. Matsubara, Y. Iizuka, K. Okamoto, and K. Yokoi, "Structure and hypotensive effect of flavonoid glycosides in lemon peelings (Part II)," *Journal of the Agricultural Chemical Society of Japan*, vol. 59, no. 7, pp. 677–682, 1985 (Japanese).

[8] T. Domoto, K. Ishihara, Y. Miyake et al., "Effect of daily lemon intake on a range of parameters related to metabolic syndrome," *Health Sciences*, vol. 26, no. 4, pp. 210–218, 2010 (Japanese).

[9] L. L. Lee, M. C. Watson, C. A. Mulvaney, C. C. Tsai, and S. F. Lo, "The effect of walking intervention on blood pressure control: a systematic review," *International Journal of Nursing Studies*, vol. 47, no. 12, pp. 1545–1561, 2010.

[10] N. Martí, P. Mena, J. A. Cánovas, V. Micol, and D. Saura, "Vitamin C and the role of citrus juices as functional food," *Natural Product Communications*, vol. 4, no. 5, pp. 677–700, 2009.

[11] M. Sugiura, M. Nakamura, K. Ogawa, Y. Ikoma, and M. Yano, "High serum carotenoids associated with lower risk for bone loss and osteoporosis in post-menopausal Japanese female subjects: prospective cohort study," *PLoS ONE*, vol. 7, no. 12, Article ID e52643, 2012.

[12] J. M. Assini, E. E. Mulvihill, and M. W. Huff, "Citrus flavonoids and lipid metabolism," *Current Opinion in Lipidology*, vol. 24, no. 1, pp. 34–40, 2013.

[13] T. Tsujita, M. Sumiyoshi, L.-K. Han, T. Fujiwara, J. Tsujita, and H. Okuda, "Inhibition of lipase activities by citrus pectin," *Journal of Nutritional Science and Vitaminology*, vol. 49, no. 5, pp. 340–345, 2003.

[14] B. Morin, L. A. Nichols, K. M. Zalasky, J. W. Davis, J. A. Manthey, and L. J. Holland, "The citrus flavonoids hesperetin and nobiletin differentially regulate low density lipoprotein receptor gene transcription in HepG2 liver cells," *Journal of Nutrition*, vol. 138, no. 7, pp. 1274–1281, 2008.

[15] Y. Miyake, E. Suzuki, S. Ohya et al., "Lipid-lowering effect of eriocitrin, the main flavonoid in lemon fruit, in rats on a high-fat and high-cholesterol diet," *Journal of Food Science*, vol. 71, no. 9, pp. S633–S637, 2006.

[16] T. Oka and M. Genkai-Kato, "Variations in body weight and body composition with special attention to body fat in humans," *Kuroshio Science*, vol. 5, no. 2, pp. 161–167, 2012 (Japanese).

[17] Y. Ogawa and Y. Uji, "Impact of acute oral single administration of CG-120 to healthy humans, with special reference to stone-forming substances," *Japanese Pharmacology and Therapeutics*, vol. 14, no. 8, pp. 5251–5272, 1986 (Japanese).

[18] Y. Ogawa and Y. Uji, "Impact of oral short-term CG-120 administration to healthy humans, with special reference to stone-forming substances," *Japanese Pharmacology and Therapeutics*, vol. 14, no. 8, pp. 5273–5293, 1986 (Japanese).

[19] C. Y. C. Pak, J. A. Harvey, and M. C. Hsu, "Enhanced calcium bioavailability from a solubilized form of calcium citrate," *Journal of Clinical Endocrinology and Metabolism*, vol. 65, no. 4, pp. 801–805, 1987.

[20] J. Z. Miller, D. L. Smith, L. Flora, C. Slemenda, X. Jiang, and C. C. Johnston Jr., "Calcium absorption from calcium carbonate and a new form of calcium (CCM) in healthy male and female adolescents," *The American Journal of Clinical Nutrition*, vol. 48, no. 5, pp. 1291–1294, 1988.

[21] B. Lacour, S. Tardivel, and T. Drüeke, "Stimulation by citric acid of calcium and phosphorus bioavailability in rats fed a calcium-rich diet," *Mineral and Electrolyte Metabolism*, vol. 23, no. 2, pp. 79–87, 1997.

[22] Y. Nii, K. Fukuta, K. Sakai, and S. Yamamoto, "Japanese citrus fruit (sudachi) juice is associated with increased bioavailability of calcium from whole small fish and suppressed bone resorption in rats," *Journal of Nutritional Science and Vitaminology*, vol. 50, no. 3, pp. 177–183, 2004.

[23] D. A. McCarron, C. D. Morris, H. J. Henry, and J. L. Stanton, "Blood pressure and nutrient intake in the United States," *Science*, vol. 224, no. 4656, pp. 1392–1398, 1984.

[24] A. P. Avolio, S. G. Chen, and R. P. Wang, "Effects of aging on changing arterial compliance and left ventricular load in a northern Chinese urban community," *Circulation*, vol. 68, no. 1, pp. 50–58, 1983.

[25] G. M. London and A. P. Guerin, "Influence of arterial pulse and reflected waves on blood pressure and cardiac function," *The American Heart Journal*, vol. 138, no. 3, pp. S220–S224, 1999.

31

Trigonella Foenum-Graecum Seeds Lowers Postprandial Blood Glucose in Overweight and Obese Individuals

Sathyasurya Daniel Robert,[1] Aziz Al-Safi Ismail,[2] and Wan Ishak Wan Rosli[3]

[1] *Dietetics Program, School of Health Sciences, Health Campus, Universiti Sains Malaysia, Kubang Kerian, 16150 Kota Bharu, Kelantan, Malaysia*
[2] *Department of Community Medicine, School of Medical Sciences, Health campus, Universiti Sains Malaysia, Kubang Kerian, 16150 Kota Bharu, Kelantan, Malaysia*
[3] *Nutrition Program, School of Health Sciences, Health Campus, Universiti Sains Malaysia, Kubang Kerian, 16150 Kota Bharu, Kelantan, Malaysia*

Correspondence should be addressed to Sathyasurya Daniel Robert; dan77in@yahoo.com

Academic Editor: Pedro Moreira

This study determined the effects of fenugreek on postprandial plasma glucose (PPG) and satiety among overweight and obese individuals. Fourteen subjects were studied in the morning after overnight fasts on four separate occasions. Glycaemic responses elicited by 50 g carbohydrate portions of white bread and jam with or without 5.5 g of fenugreek and fried rice with or without 5.5 g fenugreek were determined over 2 h. The primary endpoint was the incremental area under the plasma glucose response curve (IAUC). Adding fenugreek to both foods significantly reduced the IAUC compared to the food alone: white bread and jam, 180 ± 22 versus 271 ± 23 mmol × min/L ($P = 0.001$); fried rice, 176 ± 20 versus 249 ± 25 mmol × min/L ($P = 0.001$). Fenugreek also significantly reduced the area under the satiety curve for white bread with jam (134 ± 27 versus 232 ± 33 mm × hr, $P = 0.01$) and fried rice (280 ± 37 versus 379 ± 36 mm × hr, $P = 0.01$). It is concluded that fenugreek significantly decreased the PPG response and increased satiety among overweight and obese individuals.

1. Introduction

Excessive body weight increases the risk for many chronic diseases. In 2008, more than 1.4 billion adults worldwide were overweight and of these more than 500 million were obese [1]. By 2015 it is expected that approximately 2.3 billion adults around the world will be overweight and that at least 700 million will be obese [2]. The rising trend of obesity is evident in most countries including Malaysia, where the prevalence of overweight and obesity is about 29% and 14%, respectively [3]. This obesity epidemic is mainly due to excess energy intake and decreased energy expenditure [4–6]. Hence identification of natural functional foods that could increase the satiety and control the surge in the postprandial blood glucose (PPG) may promote more effective weight management and reduce the risk of overweight or obese individuals for developing chronic metabolic diseases [7, 8]. Trigonella foenum-graecum (fenugreek) has long been used as a cooking spice and as a traditional medicine for its therapeutic properties [9, 10]. Preliminary animal and human trials suggest possible hypoglycemic properties of fenugreek when taken orally [11]. Thus, fenugreek may be an essential ingredient in cooking as well as for the development of novel highly satiating food products with a low glycaemic index (GI). Recent research findings indicate that frequent consumption of foods with low GI reduces the risk of developing chronic diseases [12–15]. Rice and bread are commonly consumed by the Malaysian population [16]. Therefore, incorporating fenugreek in these foods may provide health benefits. However, no prior Malaysian studies are done to determine the effect of fenugreek on glycaemic control or satiety among the overweight or obese individuals.

TABLE 1: Composition of control and test foods.

Food	Portion size (g)	Protein (g)	Fat (g)	Dietary fiber (g)	Available carbohydrate[a] (g)	Energy (K·cals)
White bread and jam[†]	99.3	7.2	2.2	1.3	50	257.3
White bread and jam with fenugreek[‡]	104.8	8.9	2.6	3.5	50	273.7
Fried rice[†]	62.3	5.2	5.0	3.6	50	254.9
Fried rice with fenugreek[‡]	67.8	6.9	5.4	5.8	50	271.4
Fenugreek (per 5.5 g)	—	1.7	0.4	2.2	0.4	16.4

[a]Available carbohydrates are calculated as follows: 100 – (moisture + protein + fat + dietary fiber + ash).
[†]Control foods.
[‡]Test foods.

Hence the purpose of this study was to examine the effects of untreated fenugreek seed powder on postprandial glycaemic responses and satiety in overweight and obese Malaysian individuals.

2. Materials and Methods

2.1. Subjects. Sixteen overweight or obese subjects (7 men and 9 women) aged 32–52 years were recruited through the advertisements posted in the intranet web mail of the Universiti Sains Malaysia. Subjects were excluded based on the following criteria: BMI < 25 kg/m^2, currently pregnant or lactating, intended or unintended weight loss of >5 kg in the prior 3 months, history of binge-eating disorder, smoking or alcohol abuse, diabetes, renal or hepatic diseases, use of weight loss, lipid-lowering, antihypertensive or anti-inflammatory steroid medications, and use of fiber supplements. Two subjects were eliminated during the study process because their postprandial glycaemic response was >16.9 mmol/L (304.2 mg/dL). Hence the study was completed with 14 subjects. The Universiti Sains Malaysia Ethical Committee approved the study. All subjects gave written, informed consent after reviewing the study procedures. The study was conducted at the Dietetics Laboratory, School of Health Sciences, Universiti Sains Malaysia.

2.2. Control and Test Foods. Fenugreek seeds (TESCO; Kota Bharu) were heated in a microwave oven for one minute and ground in a blender. The ground seeds were sieved to obtain a fine powder which was then stored in an air tight container. Fenugreek seed powder was added to a white bread and strawberry jam sandwich by mixing 5.5 g powder into 15.6 g jam which was spread onto 83.7 g white bread. Fenugreek seed powder was added to fried rice as follows: 62.3 g white rice was cooked in a rice cooker for 25–30 min and transferred to a hot frying pan containing 5 g of oil. Chicken flavoured seasoning (8.5 g) and 5.5 g fenugreek seed powder were added to the rice, stirred, and sautéed for 3 min. The nutrient composition of the test foods is shown in Table 1.

2.3. Feeding Protocol. Prior to the main study a pilot study was done with 3 subjects in order to determine the taste tolerance of fenugreek powder and its effect on the PPG response. No effect was noticed with 2 g of fenugreek seed powder per 50 g carbohydrate portion of white bread and jam. However, considerable reduction in the PPG was noted when the white bread and jam contained 5.5 g of fenugreek seed powder. In addition the subjects were able to tolerate the bitter taste of the fenugreek at this dose. Therefore, the main study had a crossover design in which 14 subjects participated on four separate occasions after a 10–12 hr overnight fast when they consumed, in random order, one of the following test meals: 50 g carbohydrate from white bread and jam alone or with 5.5 g fenugreek seed powder or 50 g carbohydrate from fried rice alone or with 5.5 g fenugreek seed powder. All subjects underwent 4 tests on separate mornings with at least 10 to 15 days between successive testing sessions. On each incident subjects gave a fasting blood sample and then consumed one of the 4 study meals within 10 to 13 minutes; further blood samples were collected at 15, 30, 45, 60, 90, and 120 minutes after starting to eat. Study meals were served with 250 mL of water. Capillary finger-prick blood samples (11 drops) were collected into 1.5 mL eppendorf tubes containing fluoride oxalate and centrifuged to obtain plasma, which was stored at −20°C prior to analysis. Plasma glucose was measured by using a Randox glucose GOD/PAP autoanalyzer which uses the glucose oxidase method. All subjects were asked to complete a satiety score sheet before consuming each test meal and after every blood sample thereafter for 2 h using a 6-point scale on which 0 was "extremely hungry" and 6 was "extremely full" [17]. Upon completion of the test meals, subjects rated the palatability on a Likert scale on which a score of 1 was "strongly dislike," 2 was "dislike," 3 was "neutral," 4 was "like," and 5 was "strongly like" [18].

2.4. Statistical Analysis. Data analysis was conducted using Microsoft Excel Spreadsheets and Graph Pad Prism version 5 (GraphPad Software, San Diego, CA, USA). The IAUC values and satiety scores were subjected to repeated measures ANOVA and, after demonstrating significant heterogeneity, the least significant differences between individual means were obtained by Bonferroni multiple comparisons. A value of $P < 0.05$ was considered significant. Results are expressed as means ± SEM.

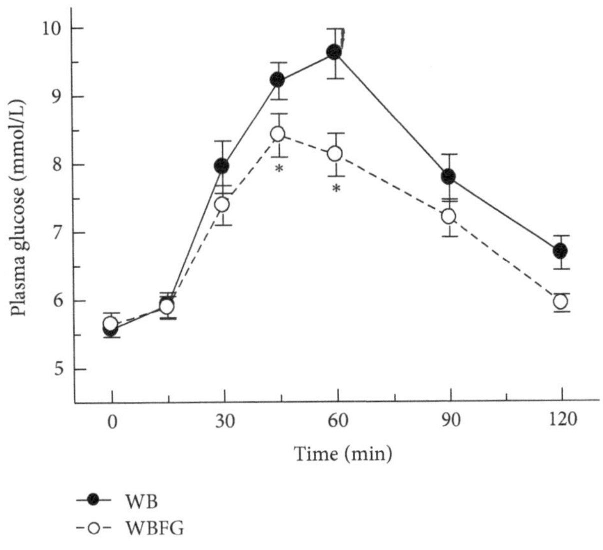

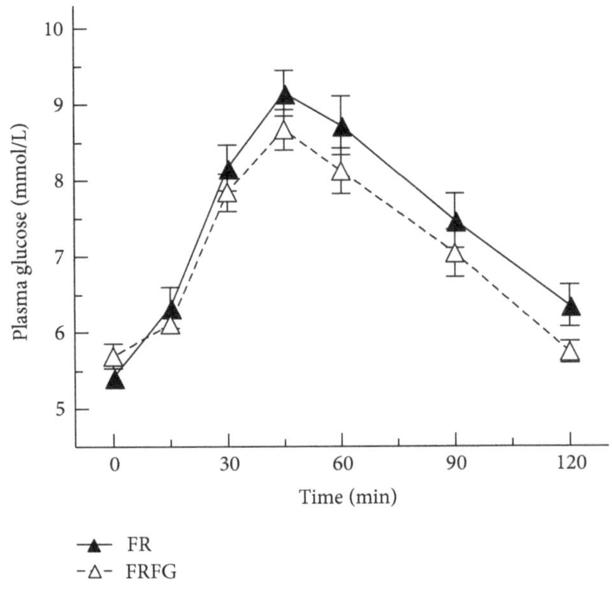

Figure 1: (a) Mean plasma glucose response of white bread and white bread with fenugreek. Values are means ± SEM ($n = 14$). Comparison of plasma glucose concentrations ($^*P < 0.05$). WB: white bread; WBFR: white bread with fenugreek. (b) Mean plasma glucose response of fried rice and fried rice with fenugreek. Values are means ± SEM ($n = 14$). Comparison of plasma glucose concentrations. FR: fried rice; FRFG: fried rice with fenugreek.

3. Results

The 14 subjects (7 women and 7 men) who completed the study were aged 38.6 ± 1.4 years and had a BMI of 29.9 ± 1.3 kg/m^2. Among the subjects 28.6% (3 men and 1 woman) were obese and the rest were overweight. Their mean systolic blood pressure was 122.6 ± 3.1 mm Hg and diastolic blood pressure was 78 ± 1.7 mm Hg.

3.1. Blood Glucose Responses. Plasma glucose after consuming white bread with fenugreek was significantly lower at 45 min and 60 min when compared to white bread alone (Figure 1(a)). Plasma glucose after fried rice with fenugreek tended to be lower at 45 min and 60 min when compared to fried rice alone but the difference was not statistically significant (Figure 1(b)). However, significant differences were seen in IAUC between the control and fenugreek-enriched foods for both bread and rice ($P = 0.001$) (Table 2).

3.2. Satiety and Taste. The satiety incremental area under the curve (SIAUC) of both test foods was significantly higher than the SIAUC of the control foods ($P = 0.01$) (Table 2). Palatability ratings of white bread with fenugreek showed that 57.1% of the subjects choose the option dislike, 28.6% choose neutral, and 14.3% choose like. Whereas the palatability ratings of fried rice with fenugreek revealed that 14.3% of the subjects choose the option dislike, 21.4% choose neutral, and 64.3% choose like. Paired t-test showed that there were no statistical differences in palatability score between the 2 test foods, $P = 1.0$.

Table 2: Incremental area under the curve and satiety area under the curve of test foods and control foods.

Foods	Area under the curve (mmol × min/L)	Satiety area under the curve (mm × hr)
White bread and jam†	271 ± 23	134 ± 27
White bread and jam with fenugreek‡	180 ± 22*	232 ± 33*
Fried rice†	249 ± 25	280 ± 37
Fried rice with fenugreek‡	176 ± 20*	379 ± 36*

Mean ± SEM.
*Mean values within a column are significantly different from the respective control foods ($P < 0.05$).
†Control foods.
‡Test foods.

4. Discussion

This crossover study demonstrated that incorporation of natural fenugreek seed powder to the test foods lowered the surge in PPG response and increased satiety in overweight and obese subjects. It was noted that there was a reduction of 14.4 mg/dL (PPG) at 45 min and a reduction of 27 mg/dL (PPG) at 60 min after the consumption of white bread with fenugreek, when compared with that of the control food. On the other hand, it was noted that after the consumption of fried rice with fenugreek the PPG decreased by 10.8 mg/dL at 60 min, when compared to the control food, but it was not significantly lower. A previous study by Mathern et al. found no reductions in PPG in response to a beverage supplemented with 8 g of isolated fenugreek fiber among obese subjects [19]. In another study, Sharma and Raghuram found that 100 g of defatted fenugreek seed powder, when given for 20 days, produced a significant fall in fasting blood glucose levels and an improvement in glucose tolerance test among subjects with type 2 diabetes [20]. Results from several additional studies also suggest that fenugreek seeds may improve glycaemic control in patients with type 2 diabetes [21–24]. However, in all those studies the patients took their regular glucose lowering medications. The hypoglycemic action of fenugreek has been related to several processes. In vitro studies demonstrated that the amino acid 4-hydroxyisoleucine in fenugreek seeds increased insulin release from human and rat pancreas. This amino acid acted only on pancreatic beta cells, but the levels of other hormones such as somatostatin and glucagon were not changed [10]. Other human studies reported that fenugreek seeds increased the number of insulin receptors, thereby decreasing the postprandial area under the curve [10]. Fenugreek seeds can also exert hypoglycemic effects by inhibiting the activities of alpha-amylase and sucrase, the two intestinal enzymes that are involved in carbohydrate digestion and absorption [10]. In clinical studies the hypoglycemic action of fenugreek has been attributed to its content of galactomannan, a water soluble fiber, which inhibits glucose absorption and slows gastric emptying [25–29]. On the other hand the satiety area under the curve after the test foods was significantly higher than that after the control foods. The increase in satiety may be due to the presence of the soluble fiber galactomannan in fenugreek. Galactomannan has gel forming characteristics which reduces gastric emptying [25]. It has been reported that 8 g of fenugreek extract, providing 7.2 g total fiber (78.9% soluble fiber), increased satiety and reduced hunger among obese subjects [19]. However, in this study we noted that addition of fenugreek seed powder to the test foods increased its total dietary fiber content by 2.18 g, which increased satiety. Several studies have stated that protein can promote satiety [30–32]. Protein induced satiety may be due to increase in thermogenesis, increase in the concentrations of the appetite suppressing hormones such as glucagon like peptide-1, cholecystokinin, and peptide tyrosine tyrosine, and decrease in appetite stimulating hormones such as ghrelin [33–38]. In a single blinded crossover design study, it was noted that satiety was increased after the subjects consumed a higher protein lunch containing 25% of the energy as dietary protein [39]. But in this study the effect of protein on satiety may be minor, because the addition of fenugreek to the test foods increased their protein content by merely 13% (white bread) and 10.2% (rice) of the energy.

In the earlier studies, defatted or isolated or extracted fenugreek was used as experimental food [20]. In contrast, untreated fenugreek seed powder was given to the subjects of this study. Processed fenugreek seed powder or any processed functional food may not be easily available for the general public. But our study showed that untreated fenugreek (functional food) can easily be obtained from the food market and by following simple cooking methods can be added in the daily diet, which is the strength of this study. The subjects of this study were able to tolerate the taste of the fenugreek well when it was added with fried rice than when it was added with white bread and jam. This shows that the bitterness of the fenugreek can be masked when it is added to spicy foods. The limitation of this study is the bitterness of the fenugreek which may limit its daily use.

5. Conclusion

In conclusion, addition of 5.5 g of untreated fenugreek seed powder to rice or bread containing 50 g available carbohydrate reduced PPG responses and increased satiety in overweight and obese individuals. As obesity is a component of metabolic syndrome which eventually leads to type 2 diabetes, incorporation of fenugreek in the daily diet may assist with weight management and hence may help prevent or delay the onset of chronic diseases.

Further research can be done to measure insulin index in order to learn more about the mechanism of action. In addition novel food products enriched with fenugreek can be developed.

Conflict of Interests

The authors have no conflict of interests. The authors alone are responsible for the content and writing of the paper.

Acknowledgments

The authors thank the study participants for their cooperation and all those who helped them in conducting this study. This study was funded by Universiti Sains Malaysia Short Term Grant (304/PPSK/61312036).

References

[1] "Obesity and overweight," 2013, http://www.who.int/mediacentre/factsheets/fs311/en/index.html.

[2] D. M. Nguyen and H. B. El-Serag, "The epidemiology of obesity," *Gastroenterology Clinics of North America*, vol. 39, no. 1, pp. 1–7, 2010.

[3] A. Z. Khambalia and L. S. Seen, "Trends in overweight and obese adults in Malaysia (1996–2009): a systematic review," *Obesity Reviews*, vol. 11, no. 6, pp. 403–412, 2010.

[4] A. K. Kant and B. I. Graubard, "Secular trends in patterns of self-reported food consumption of adult Americans: NHANES 1971–1975 to NHANES 1999–2002," *The American Journal of Clinical Nutrition*, vol. 84, no. 5, pp. 1215–1223, 2006.

[5] A. M. Prentice and S. A. Jebb, "Obesity in britain: gluttony or sloth?" *The British Medical Journal*, vol. 311, no. 7002, pp. 437–439, 1995.

[6] W. H. Dietz Jr. and S. L. Gortmaker, "Do we fatten our children at the television set? Obesity and television viewing in children and adolescents," *Pediatrics*, vol. 75, no. 5, pp. 807–812, 1985.

[7] C. M. Hasler and A. C. Brown, "Position of the American Dietetic Association: functional foods," *Journal of the American Dietetic Association*, vol. 109, no. 4, pp. 735–746, 2009.

[8] S. Henson, O. Masakure, and J. Cranfield, "The propensity for consumers to offset health risks through the use of functional foods and nutraceuticals: the case of lycopene," *Food Quality and Preference*, vol. 19, no. 4, pp. 395–406, 2008.

[9] R. Nahas and M. Moher, "Complementary and alternative medicine for the treatment of type 2 diabetes," *Canadian Family Physician*, vol. 55, no. 6, pp. 591–596, 2009.

[10] E. Basch, C. Ulbricht, G. Kuo, P. Szapary, and M. Smith, "Therapeutic applications of fenugreek," *Alternative Medicine Review*, vol. 8, no. 1, pp. 20–27, 2003.

[11] N. Z. Baquer, P. Kumar, A. Taha, R. K. Kale, S. M. Cowsik, and P. McLean, "Metabolic and molecular action of Trigonella foenum-graecum (fenugreek) and trace metals in experimental diabetic tissues," *Journal of Biosciences*, vol. 36, no. 2, pp. 383–396, 2011.

[12] L. Barclay and D. Lie, "High glycemic index foods may increase risk for type 2 diabetes in Chinese women," *Archives of Internal Medicine*, vol. 167, pp. 2310–2316, 2007.

[13] R. Villegas, S. Liu, Y. T. Gao et al., "Prospective study of dietary carbohydrates, glycemic index, glycemic load, and incidence of type 2 diabetes mellitus in middle-aged Chinese women," *Archives of Internal Medicine*, vol. 167, no. 21, pp. 2310–2316, 2007.

[14] P. A. Nancy and C. R. Rhymer, "Issues surrounding health claims for barley," *The Journal of Nutrition*, vol. 138, no. 6, pp. 1237S–1243S, 2008.

[15] A. Mente, L. De Koning, H. S. Shannon, and S. S. Anand, "A systematic review of the evidence supporting a causal link between dietary factors and coronary heart disease," *Archives of Internal Medicine*, vol. 169, no. 7, pp. 659–669, 2009.

[16] J. A. Howden, Y. H. Chong, S. F. Leung et al., "Breakfast practices in the Asian Region," *Pacific Journal of Clinical Nutrition*, vol. 2, no. 2, pp. 77–84, 1993.

[17] S. H. A. Holt, J. C. Brand Miller, P. Petocz, and E. Farmakalidis, "A satiety index of common foods," *European Journal of Clinical Nutrition*, vol. 49, no. 9, pp. 675–690, 1995.

[18] H. N. Boone Jr. and D. A. Boone, "Analyzing Likert data," *Journal of Extension*, vol. 50, no. 2, pp. 1–5, 2012.

[19] J. R. Mathern, S. K. Raatz, W. Thomas, and J. L. Slavin, "Effect of fenugreek fiber on satiety, blood glucose and insulin response and energy intake in obese subjects," *Phytotherapy Research*, vol. 23, no. 11, pp. 1543–1548, 2009.

[20] R. D. Sharma and T. C. Raghuram, "Hypoglycaemic effect of fenugreek seeds in non-insulin dependent diabetic subjects," *Nutrition Research*, vol. 10, no. 7, pp. 731–739, 1990.

[21] Z. Madar, R. Abel, S. Samish, and J. Arad, "Glucose-lowering effect of fenugreek in non-insulin dependent diabetics," *European Journal of Clinical Nutrition*, vol. 42, no. 1, pp. 51–54, 1988.

[22] A. Bordia, S. K. Verma, and K. C. Srivastava, "Effect of ginger (Zingiber officinale Rosc.) and fenugreek (*Trigonella foenum-graecum* L.) on blood lipids, blood sugar and platelet aggregation in patients with coronary artery disease," *Prostaglandins Leukotrienes and Essential Fatty Acids*, vol. 56, no. 5, pp. 379–384, 1997.

[23] R. D. Sharma, "Effect of fenugreek seeds and leaves on blood glucose and serum insulin responses in human subjects," *Nutrition Research*, vol. 6, no. 12, pp. 1353–1364, 1986.

[24] R. D. Sharma, A. Sarkar, D. K. Hazra et al., "Use of fenugreek seed powder in the management of non-insulin dependent diabetes mellitus," *Nutrition Research*, vol. 16, no. 8, pp. 1331–1339, 1996.

[25] J. M. A. Hannan, L. Ali, B. Rokeya et al., "Soluble dietary fibre fraction of Trigonella foenum-graecum (fenugreek) seed improves glucose homeostasis in animal models of type 1 and type 2 diabetes by delaying carbohydrate digestion and absorption, and enhancing insulin action," *The British Journal of Nutrition*, vol. 97, no. 3, pp. 514–521, 2007.

[26] S. J. French and N. W. Read, "Effect of guar gum on hunger and satiety after meals of differing fat content: relationship with gastric emptying," *The American Journal of Clinical Nutrition*, vol. 59, no. 1, pp. 87–91, 1994.

[27] P.-H. Groop, A. Aro, S. Stenman, and L. Groop, "Long-term effects of guar gum in subjects with non-insulin-dependent diabetes mellitus," *The American Journal of Clinical Nutrition*, vol. 58, no. 4, pp. 513–518, 1993.

[28] P. Würsch and F. X. Pi-Sunyer, "The role of viscous soluble fiber in the metabolic control of diabetes: a review with special emphasis on cereals rich in β-glucan," *Diabetes Care*, vol. 20, no. 11, pp. 1774–1780, 1997.

[29] M. Chandalia, A. Garg, D. Lutjohann, K. von Bergmann, S. M. Grundy, and L. J. Brinkley, "Beneficial effects of high dietary fiber intake in patients with type 2 diabetes mellitus," *The New England Journal of Medicine*, vol. 342, no. 19, pp. 1392–1398, 2000.

[30] A. Astrup, "The satiating power of protein—a key to obesity prevention?" *The American Journal of Clinical Nutrition*, vol. 82, no. 1, pp. 1–2, 2005.

[31] E. C. Westman, W. S. Yancy, J. S. Edman, K. F. Tomlin, and C. E. Perkins, "Effect of 6-month adherence to a very low carbohydrate diet program," *The American Journal of Medicine*, vol. 113, no. 1, pp. 30–36, 2002.

[32] W. A. M. Blom, A. Lluch, A. Stafleu et al., "Effect of a high-protein breakfast on the postprandial ghrelin response," *The American Journal of Clinical Nutrition*, vol. 83, no. 2, pp. 211–220, 2006.

[33] A. Raben, L. Agerholm-Larsen, A. Flint, J. J. Holst, and A. Astrup, "Meals with similar energy densities but rich in protein, fat, carbohydrate, or alcohol have different effects on energy expenditure and substrate metabolism but not on appetite and energy intake," *The American Journal of Clinical Nutrition*, vol. 77, no. 1, pp. 91–100, 2003.

[34] L. Tappy, "Thermic effect of food and sympathetic nervous system activity in humans," *Reproduction Nutrition Development*, vol. 36, pp. 391–397, 1996.

[35] B. Parker, M. Noakes, N. Luscombe, and P. Clifton, "Effect of a high-protein, high-monounsaturated fat weight loss diet on glycemic control and lipid levels in type 2 diabetes," *Diabetes Care*, vol. 25, no. 3, pp. 425–430, 2002.

[36] D. R. Gustafson, D. J. McMahon, J. Morrey, and R. Nan, "Appetite is not influenced by a unique milk peptide: caseinomacropeptide (CMP)," *Appetite*, vol. 36, no. 2, pp. 157–163, 2001.

[37] A. M. Wren, L. J. Seal, M. A. Cohen et al., "Ghrelin enhances appetite and increases food intake in humans," *Journal of Clinical Endocrinology and Metabolism*, vol. 86, no. 12, pp. 5992–5995, 2001.

[38] W. L. Hall, D. J. Millward, S. J. Long, and L. M. Morgan, "Casein and whey exert different effects on plasma amino acid profiles, gastrointestinal hormone secretion and appetite," *British Journal of Nutrition*, vol. 89, no. 2, pp. 239–248, 2003.

[39] A. J. Smeets, S. Soenen, N. D. Luscombe-Marsh, Ø. Ueland, and M. S. Westerterp-Plantenga, "Energy expenditure, satiety, and plasma ghrelin, glucagon-like peptide 1, and peptide tyrosine-tyrosine concentrations following a single high-protein lunch," *The Journal of Nutrition*, vol. 138, no. 4, pp. 698–702, 2008.

Permissions

All chapters in this book were first published in JNME, by Hindawi Publishing Corporation; hereby published with permission under the Creative Commons Attribution License or equivalent. Every chapter published in this book has been scrutinized by our experts. Their significance has been extensively debated. The topics covered herein carry significant findings which will fuel the growth of the discipline. They may even be implemented as practical applications or may be referred to as a beginning point for another development.

The contributors of this book come from diverse backgrounds, making this book a truly international effort. This book will bring forth new frontiers with its revolutionizing research information and detailed analysis of the nascent developments around the world.

We would like to thank all the contributing authors for lending their expertise to make the book truly unique. They have played a crucial role in the development of this book. Without their invaluable contributions this book wouldn't have been possible. They have made vital efforts to compile up to date information on the varied aspects of this subject to make this book a valuable addition to the collection of many professionals and students.

This book was conceptualized with the vision of imparting up-to-date information and advanced data in this field. To ensure the same, a matchless editorial board was set up. Every individual on the board went through rigorous rounds of assessment to prove their worth. After which they invested a large part of their time researching and compiling the most relevant data for our readers.

The editorial board has been involved in producing this book since its inception. They have spent rigorous hours researching and exploring the diverse topics which have resulted in the successful publishing of this book. They have passed on their knowledge of decades through this book. To expedite this challenging task, the publisher supported the team at every step. A small team of assistant editors was also appointed to further simplify the editing procedure and attain best results for the readers.

Apart from the editorial board, the designing team has also invested a significant amount of their time in understanding the subject and creating the most relevant covers. They scrutinized every image to scout for the most suitable representation of the subject and create an appropriate cover for the book.

The publishing team has been an ardent support to the editorial, designing and production team. Their endless efforts to recruit the best for this project, has resulted in the accomplishment of this book. They are a veteran in the field of academics and their pool of knowledge is as vast as their experience in printing. Their expertise and guidance has proved useful at every step. Their uncompromising quality standards have made this book an exceptional effort. Their encouragement from time to time has been an inspiration for everyone.

The publisher and the editorial board hope that this book will prove to be a valuable piece of knowledge for researchers, students, practitioners and scholars across the globe.

List of Contributors

Nathania Wonoputri, Julistio T. B. Djais and Ina Rosalina
Department of Child Health, Universitas Padjadjaran, Dr. Hasan Sadikin General Hospital, Jalan Pasteur No. 38, Bandung 40161, Indonesia

Geórgia Craveiro Holanda Malveira Maia, Juan Eduardo Lucas Castillo, Murilo Soares Faleiros, Rejane Souza de Aquino Sales, Denise Moraes Lopes Galeno and Rosany Piccolotto Carvalho
Programa Multi-Institucional de Pós-graduação em Biotecnologia, Universidade Federal do Amazonas, Avenue General Rodrigo Octóvio Jordão Ramos, 3000 Coroado, Manaus, AM, Brazil

Mozer da Silva Campos
Departamento de Ciências Fisiológicas, Instituto de Ciências Biológicas, Universidade Federal do Amazonas, Avenue General Rodrigo Octóvio Jordão Ramos, 3000 Coroado, Manaus, AM, Brazil

Janice Barros-Monteiro
Biochemistry Department, Ponce School of Medicine and Health Sciences, P.O. Box 7004, Ponce, PR 00732-7004, USA

Edson Lira
Fundação de Hematologia e Hemoterapia do Estado do Amazonas (FHEMOAM), Nucleo de Estatística, Avenue Constantino Nery 3240, 69050-001 Manaus, Brazil

Francisca das Chagas do Amaral Souza
Instituto Nacional de Pesquisas da Amazônia, Coordenação Sociedade Saúde Ambiente-Laboratório de Alimentos, Avenue André Araújo 2.936, Petrópolis, 69067-375 Manaus, AM, Brazil

Carmen Ortiz
Physiology, Pharmacology, and Toxicology Department, Ponce School of Medicine and Health Sciences, P.O. Box 7004, Ponce, PR 00732-7004, USA

Luisa Morales
Public Health Department, Ponce School of Medicine and Health Sciences, P.O. Box 7004, Ponce, PR 00732-7004, USA

Lovisa Heyman, Ulrika Axling, Cecilia Holm and Karin Berger
Department of Experimental Medical Science, Lund University, Lund, Sweden

Narda Blanco and Olov Sterner
Centre for Analysis and Synthesis Organic Chemistry, Department of Chemistry, Lund University, P.O. Box 124, 22100 Lund, Sweden

Rebecca Seguin and Leah Connor
Division of Nutritional Sciences, Cornell University, Ithaca, NY 14853, USA

Miriam Nelson
Friedman School of Nutrition Science and Policy, Tufts University, Boston, MA 02111, USA

Andrea La Croix
University of California, San Diego, CA 92093, USA

Galen Eldridge
Montana State University, Bozeman, MT 59717, USA

Majid Mohammadshahi
Hyperlipidemia Research Center, Ahvaz Jundishapur University of Medical Sciences, Ahvaz 61357-15794, Iran

Fatemeh Haidari and Majid Karandish
Nutrition and Metabolic Diseases Research Center, Ahvaz Jundishapur University of Medical Sciences, Ahvaz 61357-15794, Iran

Sara Ebrahimi
Department of Nutritional Science, Arvand International Division of Ahvaz Jundishapur University of Medical Sciences, Ahvaz 61357-15794, Iran

Mohammad-Hosein Haghighizadeh
School of Health, Ahvaz Jundishapur University of Medical Sciences, Ahvaz 61357-15794, Iran

Marina de Figueiredo Ferreira, Filipe Detrano, Maria Elisa Barros, Emilson Souza Portella, Haydée Serrão Lanzillotti and Eliane de Abreu Soares
Nutrition and Health of Nutrition Institute, State University of Rio de Janeiro, St. São Francisco Xavier, No. 524, BlockD, 12* Floor, Maracanã, 20559-900 Rio de Janeiro, RJ, Brazil

Gabriela Morgado de Oliveira Coelho
Nutrition and Health of Nutrition Institute, State University of Rio de Janeiro, St. São Francisco Xavier, No. 524, BlockD, 12* Floor, Maracanã, 20559-900 Rio de Janeiro, RJ, Brazil
Nutrition Institute, Federal University of Rio de Janeiro, Aloízio Avenue, No. 50, Granja dos Cavaleiros, 27930-560 Macaé, RJ, Brazil
Nutrition Department, Estácio de Sá University, St. Bispo, No. 83, 20261-063 Rio de Janeiro, RJ, Brazil
Nutrition Department, Arthur Sá Earp Neto College, Barão do Rio Branco Avenue, No. 1003, 25680-120 Petrópolis, RJ, Brazil

Regina Serrão Lanzillotti
Mathematics and Statistics Institute, State University of Rio de Janeiro, St. São Francisco Xavier, No. 524, Block B, 6* Floor, Maracanã, 20559-900 Rio de Janeiro, Brazil

José Firmino Nogueira Neto
Lipid Laboratory, Faculty of Medical Science, State University of Rio de Janeiro, Avenida Marechal Rondon, No. 381, 20950-003 Rio de Janeiro, RJ, Brazil

Alexandra Bédard, Benoît Lamarche and Simone Lemieux
Institute of Nutrition and Functional Foods (INAF), Laval University, 2440 Hochelaga Boulevard, Québec, QC, Canada G1V 0A6
Department of Food Science and Nutrition, Pavillon Paul-Comtois, Laval University, 2425 rue de l'Agriculture, Québec, QC, Canada G1V 0A6

Louise Corneau
Institute of Nutrition and Functional Foods (INAF), Laval University, 2440 Hochelaga Boulevard, Québec, QC, Canada G1V 0A6

Sylvie Dodin
Department of Obstetrics and Gynaecology, Pavillon Ferdinand-Vandry, Laval University, 1050 Medicine Avenue, Québec, QC, Canada G1V 0A6

Rachel A. Millstein
SDSU/UCSD Joint Doctoral Program in Clinical Psychology, 6363 Alvarado Court, San Diego, CA 92120, USA

Hasanain Faisal Ghazi, Zaleha Md. Isa, Shamsul Azhar Shah, Mohammed A. Abdal Qader, Tiba Nezar Hasan and AL-abed Ali AL-abed
Department of Community Health, Universiti Kebangsaan Malaysia Medical Centre, Bandar Tun Razak, Cheras, 56000 Kuala Lumpur, Malaysia

Shinichi Katsumata, HiroshiMatsuzaki, Rie Katsumata-Tsuboi and Kazuharu Suzuki
Department of Nutritional Science, Faculty of Applied Bioscience, Tokyo University of Agriculture, 1-1-1 Sakuragaoka, Setagaya-ku, Tokyo 156-8502, Japan

Mariko Uehara
Department of Nutritional Science and Food Safety, Faculty of Applied Bioscience, Tokyo University of Agriculture, 1-1-1 Sakuragaoka, Setagaya-ku, Tokyo 156-8502, Japan

Kellie Bilinski
Westmead Breast Cancer Institute, Westmead Hospital and The University of Sydney, P.O. Box 143,Westmead, NSW2145, Australia

Peter Talbot
Department of Dietetics & Nutrition, Westmead Hospital, Westmead, NSW2145, Australia

Yeri Kombe, Lydia U. Kaduka, Zipporah N. Bukania, MosesMwangi and Richard Mutisya
Centre for Public Health Research, Kenya Medical Research Institute, P.O. Box 20752, Nairobi 00200, Kenya

Robert M. Karanja
Centre for Biotechnology Research and Development, Kenya Medical Research Institute, P.O. Box 54840, Nairobi 00200, Kenya

Timothy Johns
School of Dietetics and Human Nutrition, McGill University, Sainte Anne de Bellevue, QC, Canada H9X 3V9

Anja Carlsohn
University of Education Schwäbisch Gmünd, Institute of Health Sciences, Oberbettringer Straße, 200 73525 Schwäbisch Gmünd, Germany

Wolfram Müller
Medical University of Graz, Institute of Biophysics, Harrachgasse 21, 8010 Graz, Austria

Haji Kedir
College of Health and Medical Sciences, Haramaya University, P.O. Box 235, Harar, Ethiopia

Yemane Berhane
Addis Continental Institute of Public Health, Addis Ababa, Ethiopia

Alemayehu Worku
School of Public Health, Addis Ababa University, P.O. Box 26751/1000, Addis Ababa, Ethiopia

Eric P. Plaisance and Gordon Fisher
Department of Human Studies, Exercise and Nutritional Physiology Laboratory, University of Alabama at Birmingham, Birmingham, AL 35294, USA

Katiya Ivanovitch, Jeeranun Klaewkla and Rewadee Chongsuwat
Department of Nutrition, Faculty of Public Health, Mahidol University, 420/1 Ratchawithi Road, Ratchathewi, Bangkok 10400, Thailand

Chukiat Viwatwongkasem
Department of Biostatistics, Faculty of Public Health, Mahidol University, 420/1 Ratchawithi Road, Ratchathewi, Bangkok 10400, Thailand

Wanicha Kitvorapat
Regional Health Promotion Center 8, Nakhon Sawan, 157 Moo 1, Phahonyothin Road, Muang District, Nakhon Sawan 6000, Thailand

N. M. R. Sales
Universidade Anhembi Morumbi/Gerência de Nutrição Enteral e Parenteral, Rua Pedro Ivo 22, 01323-070 São Paulo, SP, Brazil

P. B. Pelegrini
BioLife Brasil Ltda, Distrito Agroindustrial de Formosa, Quadra 01, Lote 15, 73801-970 Formosa, GO, Brazil

M. C. Goersch
Gerência de Nutrição Enteral e Parenteral-GANEP, Rua Pedro Ivo 22, 01323-070 São Paulo, SP, Brazil

Patricia L. Brown and Chunzi Liang
Department of Nutrition, University of Tennessee, Knoxville, 1215W. Cumberland Avenue, 229 Jessie Harris Building, Knoxville, TN 37996-1920, USA

Benjamin J. Curry
Ension, Inc., 11020 Solway School Road, Suite 108, Knoxville, TN 37931, USA

Michael B. Zemel
NuSirt Biopharma, 11020 Solway School Road, Suite 109, Knoxville, TN 37931, USA

Gurudath Gundurao Sreekantamurthy, N Biplab Singh, Th Bhimo Singh, Th Suraj Singh and Karam Romeo Singh
Department of General Medicine, Regional Institute of Medical Sciences, Imphal 795004, India

Chou, Anna Hara, Dong Dong Du, Namiko Shimizu, Hiroe Sakuyama and Yoshio Uehara
Division of Clinical Nutrition, Faculty of Home Economics, Kyoritsu Women's University, 2-2-1 Hitotsubashi, Chiyoda, Tokyo 101-8437, Japan

Mavil May C. Cervo and Leonora N. Panlasigui
School of Nutrition, Philippine Women's University, 1004 Manila, Philippines

Luisito O. Llido
Clinical Nutrition Services, St. Luke's Medical Center, 1102 Quezon City, Philippines

Erniel B. Barrios
School of Statistics, University of the Philippines-Diliman, 1101 Quezon City, Philippines

Sarah A. Elliott
Children's Nutrition Research Centre, Queensland Children's Medical Research Institute, Royal Children's Hospital, The University of Queensland, Herston, QLD 4029, Australia

Peter S.W. Davies and Kimberley A. Baxter
Children's Nutrition Research Centre, Queensland Children's Medical Research Institute, Royal Children's Hospital, The University of Queensland, Herston, QLD 4029, Australia

Helen Truby
Department of Nutrition and Dietetics, Faculty of Medicine, Nursing and Health Sciences, Monash University, Notting Hill, VIC 3168, Australia

Bashair Al Riyami, Sultan Al Maskari and Jumana Saleh
Department of Biochemistry, College of Medicine, Sultan Qaboos University, Al-Khod, 123 Muscat, Oman

Marah El-Tahir
Department of Microbiology and Immunology, College of Medicine, Sultan Qaboos University, Al-Khod, 123 Muscat, Oman

Eugene H. Johnson
Department of Animal and Veterinary Sciences, College of Agricultural and Marine Sciences, Sultan Qaboos University, Al-Khod, 123 Muscat, Oman

Marquitta C. Webb and Safiya E. Beckford
Department of Agricultural Economics and Extension, Faculty of Food and Agriculture, The University of the West Indies, St. Augustine, Trinidad and Tobago

Sharma Sowmya
Department of Physiology, St. John's Medical College, Bangalore 560034, India

Tinku Thomas
Division of Epidemiology and Statistics, St. John's Research Institute, Bangalore 560034, India

Ankalmadagu Venkatsubbareddy Bharathi
Clinical Physiology Unit, Department of Physiology, St. John's Medical College, Bangalore 560034, India

Sambashivaiah Sucharita
Department of Physiology, St. John's Medical College, Bangalore 560034, India
Clinical Physiology Unit, Department of Physiology, St. John's Medical College, Bangalore 560034, India

Sigrid Bairdain, Chueh Lien, Faraz Ali Khan and Matthew J. Grabowski
Department of Pediatric Surgery, Boston Children's Hospital, Harvard Medical School, 300 Longwood Avenue, Fegan Building, 3rd Floor, Boston, MA 02215, USA

David C. Yu
Department of Pediatric Surgery, Children's Hospital of Alabama, Birmingham, AL 35233, USA

Bhavana Pathak
Department of Medicine, Johns Hopkins University, Baltimore, MD 21218, USA

David Zurakowski
Department of Pediatric Surgery, Boston Children's Hospital, Harvard Medical School, 300 Longwood Avenue, Fegan Building, 3rd Floor, Boston, MA 02215, USA
Department of Anesthesia, Boston Children's Hospital, Harvard Medical School, Boston, MA 02215, USA

Bradley C. Linden
Pediatric Surgical Associates, Children's Hospitals and Clinics of Minnesota, MN 55404, USA

Juliane Heydenreich
Sports Medicine and Sports Orthopaedics, University Outpatient Clinic, University of Potsdam, 14469 Potsdam, Germany
ZIEL-Research Centre for Nutrition and Food Sciences, Munich Technical University, 85354 Freising-Weihenstephan, Germany

Christoph Otto and Frank Mayer
Sports Medicine and Sports Orthopaedics, University Outpatient Clinic, University of Potsdam, 14469 Potsdam, Germany

Anja Carlsohn
Sports Medicine and Sports Orthopaedics, University Outpatient Clinic, University of Potsdam, 14469 Potsdam, Germany
Institute of Health Science, Schwäbisch Gmünd University of Education, 73529 Schwäbisch Gmünd, Germany

Zeid Khitan, Mohit Harsh, Komal Sodhi and Joseph I. Shapiro
Department of Internal Medicine, Marshall University Joan C. Edwards School of Medicine, Huntington, WV 25701, USA

Nader G. Abraham
Department of Internal Medicine, Marshall University Joan C. Edwards School of Medicine, Huntington, WV 25701, USA
Departments of Medicine and Pharmacology, New York Medical College, Valhalla, NY 10595, USA

O. S. Adeyemi and T. C. Elebiyo
Department of Biological Sciences, Landmark University, Omu-Aran 370102, Kwara State, Nigeria

Sathyasurya Daniel Robert
Dietetics Program, School of Health Sciences, Health Campus, Universiti Sains Malaysia, Kubang Kerian, 16150 Kota Bharu, Kelantan, Malaysia

Aziz Al-Safi Ismail
Department of Community Medicine, School of Medical Sciences, Health campus, Universiti Sains Malaysia, Kubang Kerian, 16150 Kota Bharu, Kelantan, Malaysia

Wan Ishak Wan Rosli
Nutrition Program, School of Health Sciences, Health Campus, Universiti Sains Malaysia, Kubang Kerian, 16150 Kota Bharu, Kelantan, Malaysia